VW Automotive Repair Manual

by J H Haynes
Member of the Guild of Motoring Writers
Bruce Gilmour
and A K Legg T Eng MIMI

Models covered
VW Rabbit, Pick-up and Jetta diesel,
90 and 97 cu in (1471 and 1588 cc)
1977 thru 1984
Covers all versions of above
Does not cover petrol (gasoline) engined models

ISBN 0 85696 993 1

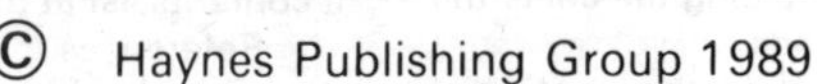

© Haynes Publishing Group 1989

All rights reserved. No part of this book may be reproduced or transmitted in any form or by any means, electronic or mechanical, including photocopying, recording or by any information storage or retrieval system, without permission in writing from the copyright holder.

Printed in England *(8R3-451)*

Haynes Publishing Group
Sparkford Nr Yeovil
Somerset BA22 7JJ England

Haynes Publications, Inc
861 Lawrence Drive
Newbury Park
California 91320 USA

Library of Congress
Catalog card number
85-80066

Acknowledgements

Thanks are due to the VW organisation for the supply of technical information and certain illustrations. Castrol Limited provided lubrication details, and Lucas CAV Limited supplied information on the testing and servicing of the CAV fuel injection pump and injectors.

Lastly thanks are due to all of those people at Sparkford who helped in the production of this manual.

About this manual

Its aims

The aim of this manual is to help you get the best value from your car. It can do so in several ways. It can help you decide what work must be done (even should you choose to get it done by a garage), provide information on routine maintenance and servicing, and give a logical course of action and diagnosis when random faults occur. However, it is hoped that you will use the manual by tackling the work yourself. On simpler jobs it may even be quicker than booking the car into a garage and going there twice to leave and collect it. Perhaps most important, a lot of money can be saved by avoiding the costs the garage must charge to cover its labour and overheads.

The Manual has drawings and descriptions to show the function of the various components so that their layout can be understood. Then the tasks are described and photographed in a step-by-step sequence so that even a novice can do the work.

Its arrangement

The manual is divided into eleven Chapters, each covering a logical sub-division of the vehicle. The Chapters are each divided into Sections, numbered with single figures, eg 5; and the Sections into paragraphs (or sub-sections), with decimal numbers following on from the Section they are in, eg 5.1, 5.2, 5.3 etc.

It is freely illustrated, especially in those parts where there is a detailed sequence of operations to be carried out. There are two forms of illustration: figures and photographs. The figures are numbered in sequence with decimal numbers, according to their position in the Chapter; eg Fig. 6.4 is the 4th drawing/illustration in Chapter 6. Photographs are numbered (either individually or in related groups) the same as the Section or sub-section of the text where the operation they show is described.

There is an alphabetical index at the back of the manual as well as a contents list at the front.

References to the 'left' or 'right' of the vehicle are in the sense of a person in the driver's seat facing forwards.

Unless otherwise stated, nuts and bolts are removed by turning anti-clockwise and tightened by turning clockwise.

Although we have tried to produce workable limits in Imperial measure the fact remains that the maker's tolerances are all in metric measure, so a set of metric feeler gauges will help to set tolerances really accurately.

Whilst every care is taken to ensure that the information in this manual is correct no liability can be accepted by the authors or publishers for loss, damage or injury caused by any errors in, or omissions from, the information given.

Introduction to the VW Golf/Rabbit/Jetta/Pick-up Diesel

The Golf (UK) or Rabbit/Jetta/Pick-up (USA) Diesel is the first small family vehicle to be released by a major manufacturer with a diesel engine which can compete with a similar sized petrol engine. Its greatest asset is the fact that it can average 50 mpg, which is an obvious advantage in times of fuel shortage. In fact, this diesel engine is most economical in town traffic where its petrol counterpart is least economical.

The hatchback body is designed with safety cell zones and the brakes and steering also incorporate safety features. The engine and transmission unit is mounted transversely at the front of the vehicle, and drive is through the front wheels.

Independent suspension all around and a well-appointed interior make the vehicle very comfortable to ride in.

Since its introduction the Golf/Rabbit/Jetta/Pick-up has undergone a number of modifications, all aimed at improving some technical aspect. With the backing of the VW organisation, this vehicle appears to be all set for a successful production run over many years.

Contents

VW Diesel Golf

General dimensions, weights and capacities

Dimensions

Overall length:	
Golf	3·705 m (145·9 in) standard, 3·725 m (146·6 in) with rubber strips
Rabbit	3·945 m (155·3 in)
Overall width	1·610 m (63·4 in)
Overall height (unladen)	1·410 m (55·5 in)
Wheelbase	2·400 m (94·5 in)
Track:	
Front	1·390 m (54·7 in)
Rear	1·350 m (53·1 in)
Ground clearance	0·122 m (4·8 in)
Turning circle	10·0 m (32·8 ft)

Weights

Kerb weight	830 kg (1830 lb)
Payload (maximum)	420 kg (925 lb)
Trailer weight (maximum) up to 12% gradient:	
Trailer without brakes	400 kg (885 lb)
Trailer with brakes	800 kg (1765 lb)
Roof rack load (maximum)	75 kg (165 lb)

Capacities

Engine oil	3·5 litres (6 Imp pints, 7·4 US pints)
Transmission oil	1·25 litres (2·2 Imp pints, 2·65 US pints)
Coolant	6·5 litres (11·0 Imp pints, 13·6 US pints)
Fuel tank	45 litres (9·9 Imp gal, 11·8 US gal)

Use of English

As this book has been written in England, it uses the appropriate English component names, phrases, and spelling. Some of these differ from those used in America. Normally, these cause no difficulty, but to make sure, a glossary is printed below. In ordering spare parts remember the parts list may use some of these words:

English	American	English	American
Accelerator	Gas pedal	Locks	Latches
Aerial	Antenna	Methylated spirit	Denatured alcohol
Anti-roll bar	Stabiliser or sway bar	Motorway	Freeway, turnpike etc
Big-end bearing	Rod bearing	Number plate	License plate
Bonnet (engine cover)	Hood	Paraffin	Kerosene
Boot (luggage compartment)	Trunk	Petrol	Gasoline (gas)
Bulkhead	Firewall	Petrol tank	Gas tank
Bush	Bushing	'Pinking'	'Pinging'
Cam follower or tappet	Valve lifter or tappet	Prise (force apart)	Pry
Carburettor	Carburetor	Propeller shaft	Driveshaft
Catch	Latch	Quarterlight	Quarter window
Choke/venturi	Barrel	Retread	Recap
Circlip	Snap-ring	Reverse	Back-up
Clearance	Lash	Rocker cover	Valve cover
Crownwheel	Ring gear (of differential)	Saloon	Sedan
Damper	Shock absorber, shock	Seized	Frozen
Disc (brake)	Rotor/disk	Sidelight	Parking light
Distance piece	Spacer	Silencer	Muffler
Drop arm	Pitman arm	Sill panel (beneath doors)	Rocker panel
Drop head coupe	Convertible	Small end, little end	Piston pin or wrist pin
Dynamo	Generator (DC)	Spanner	Wrench
Earth (electrical)	Ground	Split cotter (for valve spring cap)	Lock (for valve spring retainer)
Engineer's blue	Prussian blue	Split pin	Cotter pin
Estate car	Station wagon	Steering arm	Spindle arm
Exhaust manifold	Header	Sump	Oil pan
Fault finding/diagnosis	Troubleshooting	Swarf	Metal chips or debris
Float chamber	Float bowl	Tab washer	Tang or lock
Free-play	Lash	Tappet	Valve lifter
Freewheel	Coast	Thrust bearing	Throw-out bearing
Gearbox	Transmission	Top gear	High
Gearchange	Shift	Torch	Flashlight
Grub screw	Setscrew, Allen screw	Trackrod (of steering)	Tie-rod (or connecting rod)
Gudgeon pin	Piston pin or wrist pin	Trailing shoe (of brake)	Secondary shoe
Halfshaft	Axleshaft	Transmission	Whole drive line
Handbrake	Parking brake	Tyre	Tire
Hood	Soft top	Van	Panel wagon/van
Hot spot	Heat riser	Vice	Vise
Indicator	Turn signal	Wheel nut	Lug nut
Interior light	Dome lamp	Windscreen	Windshield
Layshaft (of gearbox)	Countershaft	Wing/mudguard	Fender
Leading shoe (of brake)	Primary shoe		

Buying spare parts and vehicle identification numbers

Buying spare parts

Spare parts are available from many sources, for example: VW garages, other garages and accessory shops, and motor factors. Our advice regarding spare part sources is as follows:

Officially appointed VW garages – This is the best source of parts which are peculiar to your vehicle and are otherwise not generally available (eg complete cylinder heads, internal gearbox components, badges, interior trim etc). It is also the only place at which you should buy parts if your car is still under warranty – non-VW components may invalidate the warranty. To be sure of obtaining the correct parts it will always be necessary to give the storeman your car's engine and chassis number, and if possible, to take the old part along for positive identification. Remember that many parts are available on a factory exchange scheme – any parts returned should always be clean! It obviously makes good sense to go straight to the specialists on your car for this type of part for they are best equipped to supply you.

Vehicle identification plate (Golf)

Chassis number (Golf)

Other garages and accessory shops – These are often very good places to buy materials and components needed for the maintenance of your car (eg oil filters, bulbs, fan belts, oils and greases, touch-up paint, filler paste etc). They also sell general accessories, usually have convenient opening hours, charge lower prices and can often be found not far from home.

Motor factors – Good factors will stock all of the more important components which wear out relatively quickly (eg clutch components, pistons, valves, exhaust system, brake cylinders/pipes/hoses/seals/shoes and pads etc). Motor factors will often provide new or reconditioned components on a part exchange basis – this can save a considerable amount of money.

Vehicle identification numbers

It is most important to identify the vehicle accurately when ordering spare parts or asking for information, since modifications are constantly being made.

The vehicle identification plate is located on the front crossmember by the bonnet lock on Golf models, and on the left-hand side doorjamb on Rabbit models.

The chassis number is located in the engine compartment on the right-hand side suspension strut mounting on Golf models, and in the passenger compartment on the left-hand side of the instrument panel on Rabbit models.

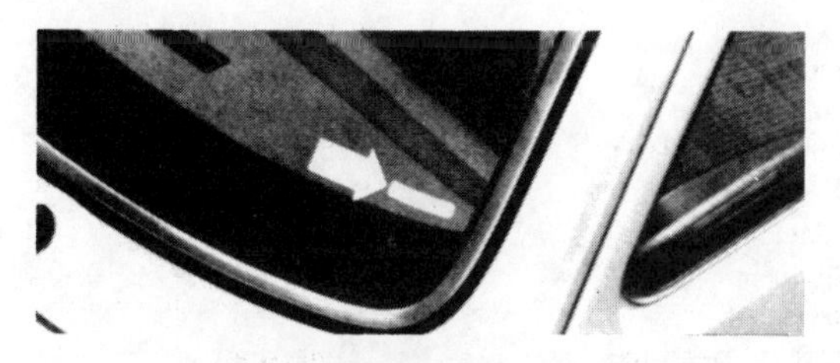

Chassis number (Rabbit)

Engine number is stamped on the side of the block

Transmission code and date of manufacture (ie 4-12-78)

The engine number is located on the front of the cylinder block, just below the cylinder head (photo).

The transmission type number is stamped on the gear housing below the left-hand driveshaft. The code letters and date of manufacture are stamped on the bottom of the bellhousing (photo).

These numbers should be identified and recorded by the owner, they are required when ordering spares, going through the customs, and regrettably, by the police, if the vehicle is stolen.

When ordering spares remember that VW output is such that inevitably spares vary, are duplicated, and are held on a usage basis. If the storeman does not have the correct identification, he cannot produce the correct item. It is a good idea to take the old part if possible to compare it with a new one. The storeman has many customers to satisfy, so be accurate and patient. In some cases, particularly in the brake system, more than one manufacturer supplies an assembly, eg both Teves and Girling supply front calipers. The assemblies are interchangeable but the integral parts are not, although both use the same brake pads. This is only one of the pitfalls in the buying of spares, so be careful to make an ally of the storeman. When fitting accessories it is best to fit VW recommended ones. They are designed specifically for the vehicle.

Tools and working facilities

Introduction

A selection of good tools is a fundamental requirement for anyone contemplating the maintenance and repair of a motor vehicle. For the owner who does not possess any, their purchase will prove a considerable expense, offsetting some of the savings made by doing-it-yourself. However, provided that the tools purchased are of good quality, they will last for many years and prove an extremely worthwhile investment.

To help the average owner to decide which tools are needed to carry out the various tasks detailed in this manual, we have compiled three lists of tools under the following headings: *Maintenance and minor repair, Repair and overhaul,* a *Special.* The newcomer to practical mechanics should start off with the *Maintenance and minor repair* tool kit and confine himself to the simpler jobs around the vehicle. Then, as his confidence and experience grows, he can undertake more difficult tasks, buying extra tools as, and when, they are needed. In this way, a *Maintenance and minor repair* tool kit can be built-up into a *Repair and overhaul* tool kit over a considerable period of time without any major cash outlays. The experienced do-it-yourselfer will have a tool kit good enough for most repair and overhaul procedures and will add tools from the *Special* category when he feels the expense is justified by the amount of use these tools will be put to.

It is obviously not possible to cover the subject of tools fully here. For those who wish to learn more about tools and their use there is a book entitled *How to Choose and Use Car Tools* available from the publishers of this manual.

Maintenance and minor repair tool kit

The tools given in this list should be considered as a minimum requirement if routine maintenance, servicing and minor repair operations are to be undertaken. We recommend the purchase of combination spanners (ring one end, open-ended the other); although more expensive than open-ended ones, they do give the advantages of both types of spanner.

Combination spanners - 6 to 19 mm
Adjustable spanner - 9 inch
Transmission drain plug key
Set of metric feeler gauges
Brake bleed nipple spanner
Screwdriver - 4 in long x $\frac{1}{4}$ in dia (flat blade)
Screwdriver - 4 in long x $\frac{1}{4}$ in dia (cross blade)
Combination pliers - 6 inch
Hacksaw, junior
Tyre pump
Tyre pressure gauge
Oil can
Fine emery cloth (1 sheet)
Wire brush (small)
Funnel (medium size)

Repair and overhaul tool kit

These tools are virtually essential for anyone undertaking any major repairs to a motor vehicle, and are additional to those given in the *Maintenance and minor repair* list. Included in this list is a comprehensive set of sockets. Although these are expensive they will be found invaluable as they are so versatile - particularly if various drives are included in the set. We recommend the $\frac{1}{2}$ in square-drive type, as this can be used with most proprietary torque wrenches. If you cannot afford a socket set, even bought piecemeal, then inexpensive tubular box spanners are a useful alternative.

The tools in this list will occasionally need to be supplemented by tools from the *Special* list.

27mm ring spanner (for fuel injectors)
Sockets (or box spanners) to cover range in previous list
Reversible ratchet drive (for use with sockets)
Extension piece, 10 inch (for use with sockets)
Universal joint (for use with sockets)
Torque wrench (for use with sockets)
'Mole' wrench - 8 inch
Ball pein hammer
Soft-faced hammer, plastic or rubber
Screwdriver - 6 in long x $\frac{5}{16}$ in dia (flat blade)
Screwdriver - 2 in long x $\frac{5}{16}$ in square (flat blade)
Screwdriver - $1\frac{1}{2}$ in long x $\frac{1}{4}$ in dia (cross blade)
Screwdriver - 3 in long x $\frac{1}{8}$ in dia (electricians)
Pliers - electricians side cutters
Pliers - needle nosed
Pliers - circlip (internal and external)
Cold chisel - $\frac{1}{2}$ inch
Scriber (this can be made by grinding the end of a broken hacksaw blade)
Scraper (this can be made by flattening and sharpening one end of a piece of copper pipe)
Centre punch
Pin punch
Hacksaw
Valve grinding tool
Steel rule/straight edge
Allen keys
Selection of files
Wire brush (large)
Axle-stands
Jack (strong scissor or hydraulic type)

Special tools

The tools in this list are those which are not used regularly, are expensive to buy, or which need to be used in accordance with their manufacturers' instructions. Unless relatively difficult mechanical jobs are undertaken frequently, it will not be economic to buy many of these tools. Where this is the case, you could consider clubbing together with friends (or a motorists' club) to make a joint purchase, or borrowing the tools against a deposit from a local garage or tool hire specialist.

The following list contains only those tools and instruments freely available to the public, and not those special tools produced by the vehicle manufacturer specifically for its dealer network. You will find occasional references to these manufacturers' special tools in the text of this manual. Generally, an alternative method of doing the job without the vehicle manufacturer's special tool is given. However, sometimes, there is no alternative to using them. Where this is the case and the relevant tool cannot be bought or borrowed you will have to entrust the work to a franchised garage.

Valve spring compressor
Piston ring compressor
Balljoint separator
Universal hub/bearing puller
Impact screwdriver
Micrometer and/or vernier gauge
Dial gauge
Universal electrical multi-meter
Cylinder compression gauge
Lifting tackle
Trolley jack
Light with extension lead

Buying tools

For practically all tools, a tool factor is the best source since he will have a very comprehensive range compared with the average garage or accessory shop. Having said that, accessory shops often offer excellent quality tools at discount prices, so it pays to shop around.

Remember, you don't have to buy the most expensive items on the shelf, but it is always advisable to steer clear of the very cheap tools. There are plenty of good tools around at reasonable prices, so ask the proprietor or manager of the shop for advice before making a purchase.

Care and maintenance of tools

Having purchased a reasonable tool kit, it is necessary to keep the tools in a clean serviceable condition. After use, always wipe off any dirt, grease and metal particles using a clean, dry cloth, before putting the tools away. Never leave them lying around after they have been used. A simple tool rack on the garage or workshop wall, for items such as screwdrivers and pliers is a good idea. Store all normal spanners and sockets in a metal box. Any measuring instruments, gauges, meters, etc, must be carefully stored where they cannot be damaged or become rusty.

Take a little care when tools are used. Hammer heads inevitably become marked and screwdrivers lose the keen edge on their blades from time to time. A little timely attention with emery cloth or a file will soon restore items like this to a good serviceable finish.

Working facilities

Not to be forgotten when discussing tools, is the workshop itself. If anything more than routine maintenance is to be carried out, some form of suitable working area becomes essential.

It is appreciated that many an owner mechanic is forced by circumstances to remove an engine or similar item, without the benefit of a garage or workshop. Having done this, any repairs should always be done under the cover of a roof.

Wherever possible, any dismantling should be done on a clean flat workbench or table at a suitable working height.

Any workbench needs a vice: one with a jaw opening of 100 mm (4 in) is suitable for most jobs. As mentioned previously, some clean dry storage space is also required for tools, as well as the lubricants, cleaning fluids, touch-up paints and so on which become necessary.

Another item which may be required, and which has a much more general usage, is an electric drill with a chuck capacity of at least (8 mm) 5/16 in. This, together with a good range of twist drills, is virtually essential for fitting accessories such as wing mirrors and reversing lights.

Last, but not least, always keep a supply of old newspapers and clean, lint-free rags available, and try to keep any working area as clean as possible.

Spanner jaw gap comparison table

Jaw gap (in)	Spanner size
0·250	1/4 in AF
0·277	7 mm
0·313	5/16 in AF
0·315	8 mm
0·344	11/32 in AF; 1/8 in Whitworth
0·354	9 mm
0·375	3/8 in AF
0·394	10 mm
0·433	11 mm
0·438	7/16 in AF
0·445	3/16 in Whitworth; 1/4 in BSF
0·472	12 mm
0·500	1/2 in AF
0·512	13 mm
0·525	1/4 in Whitworth; 5/16 in BSF
0·551	14 mm
0·563	9/16 in AF
0·591	15 mm
0·600	5/16 in Whitworth; 3/8 in BSF
0·625	5/8 in AF
0·630	16 mm
0·669	17 mm
0·686	11/16 in AF
0·709	18 mm
0·710	3/8 in Whitworth; 7/16 in BSF
0·748	19 mm
0·750	3/4 in AF
0·813	13/16 in AF
0·820	7/16 in Whitworth; 1/2 in BSF
0·866	22 mm
0.875	7/8 in AF
0·920	1/2 in Whitworth; 9/16 in BSF
0·938	15/16 in AF
0·945	24 mm
1·000	1 in AF
1·010	9/16 in Whitworth; 5/8 in BSF
1·024	26 mm
1·063	1 1/16 in AF; 27 mm
1·100	5/8 in Whitworth; 11/16 in BSF
1·125	1 1/8 in AF
1·181	30 mm
1·200	11/16 in Whitworth; 3/4 in BSF
1·250	1 1/4 in AF
1·260	32 mm
1·300	3/4 in Whitworth; 7/8 in BSF
1·313	1 5/16 in AF
1·390	13/16 in Whitworth; 15/16 in BSF
1·417	36 mm
1·438	1 7/16 in AF
1·480	7/8 in Whitworth; 1 in BSF
1·500	1 1/2 in AF
1·575	40 mm; 15/16 in Whitworth
1·614	41 mm
1·625	1 5/8 in AF
1·670	1 in Whitworth; 1 1/8 in BSF
1·688	1 11/16 in AF
1·811	46 mm
1·813	1 13/16 in AF
1·860	1 1/8 in Whitworth; 1 1/4 in BSF
1·875	1 7/8 in AF
1·969	50 mm
2·000	2 in AF
2·050	1 1/4 in Whitworth; 1 3/8 in BSF
2·165	55 mm
2·362	60 mm

Jacking and towing points

Before jacking-up the car, ensure that it is on level ground and apply the handbrake firmly. There are two jacking points on each side of the car adjacent to each wheel and marked with an embossed triangle. The jack lifting hook must engage with the ridge on the body underframe (photos).

If the car is being jacked-up for maintenance purposes, axle stands or wooden blocks must be used; never work beneath the car with only the tool kit jack supporting it.

Emergency towing points are provided at the front and rear of the car but should only be used for short distances (photos). When being towed, remember that extra effort will be required on the footbrake pedal, as the vacuum servo pump will not be working with the engine stationary.

Spare wheel location

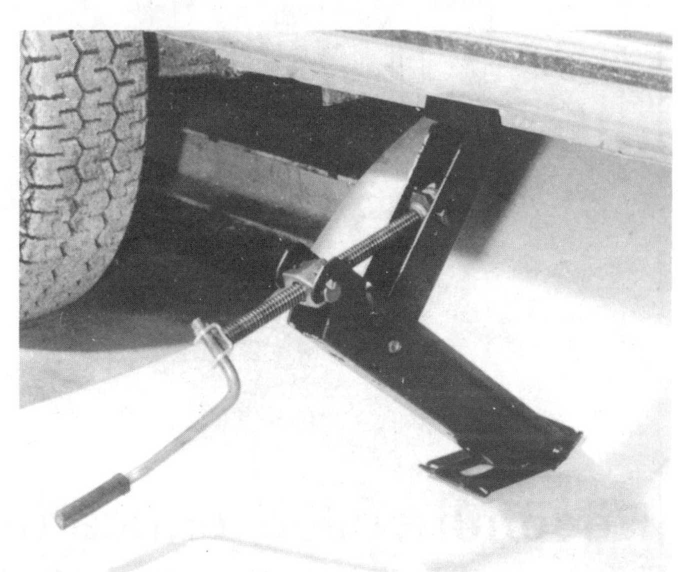

Jacking-up the car

Front towing eye

Rear towing eye

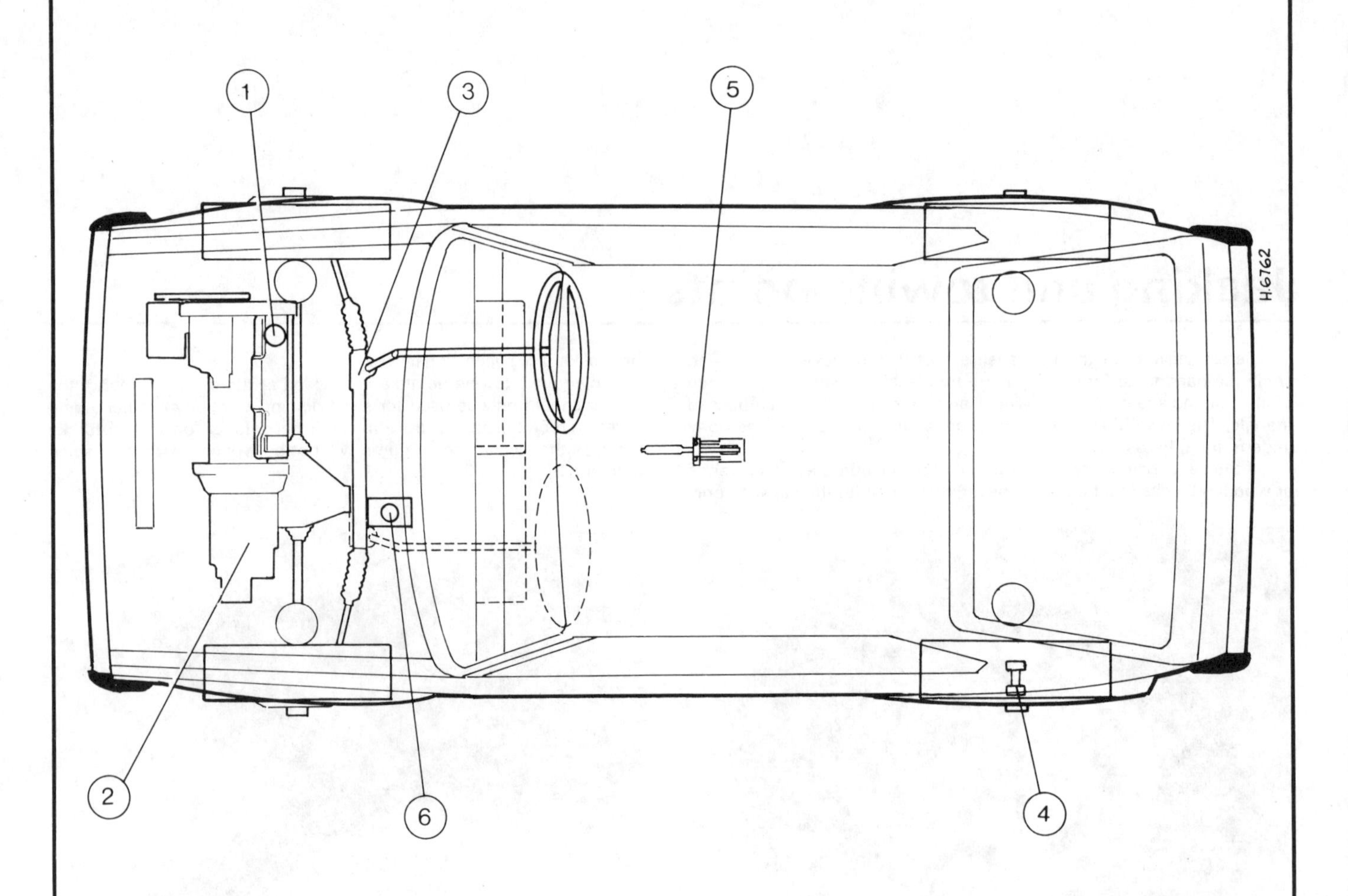

Recommended lubricants and fluids

Refer to Chapter 11 for specifications applicable to later models

Component	Castrol product
Engine (1)	Castrol GTX
Gearbox (2)	Castrol Hypoy Light (80 EP)
Rack and pinion unit (3)	Obtain special lubricant from VW agent
Rear wheel bearings (4)	Castrol LM grease
Handbrake compensator (5)	Castrol LM grease
Clutch cable	Castrol LM grease
Hinges, locks, pivots etc	Castrol GTX
Brake fluid (6)	Castrol Girling Universal Brake and Clutch Fluid

Note: The above are general recommendations only. Lubrication requirements may vary with operating conditions and from territory to territory. If in doubt consult the operator's handbook or your VW dealer.

Routine maintenance

Refer to Chapter 11 for specifications and information applicable to later models

Maintenance is essential for ensuring safety and desirable for the purpose of getting the best in terms of performance and economy from the car. Over the years the need for periodic lubrication – oiling, greasing and so on – has been drastically reduced, if not totally eliminated. This has unfortunately tended to lead some owners to think that because no such action is required the items either no longer exist or will last for ever. This is a serious delusion. It follows, therefore, that the largest initial element of maintenance is visual examination which may lead to repairs or renewals.

Every 250 miles (400 km) or weekly – whichever comes first

Engine

Check oil level and top-up if necessary (photos)
Check coolant level and top-up if necessary

Electrical system

Check operation of all lights
Check wipers and horn
Check windscreen washer fluid level (photos)
Check battery electrolyte level and top-up with distilled water if necessary

Steering

Check tyre pressures, including spare (photo)
Examine tyres for wear or damage
Check steering for smooth and accurate operation

Brakes

Check reservoir fluid level
Check brake efficiency

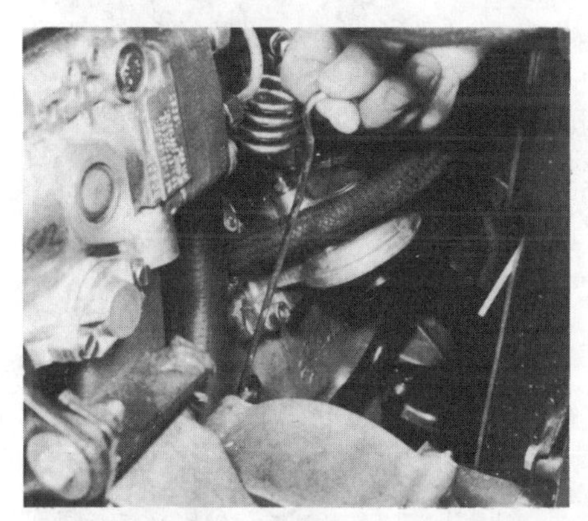

Removing the oil level dipstick

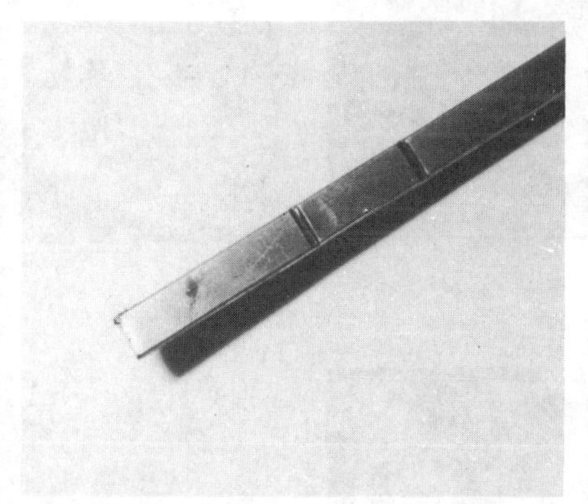

Oil level dipstick minimum and maximum marks. Difference between marks is 1 litre (1.75 Imp pints, 2 US pints)

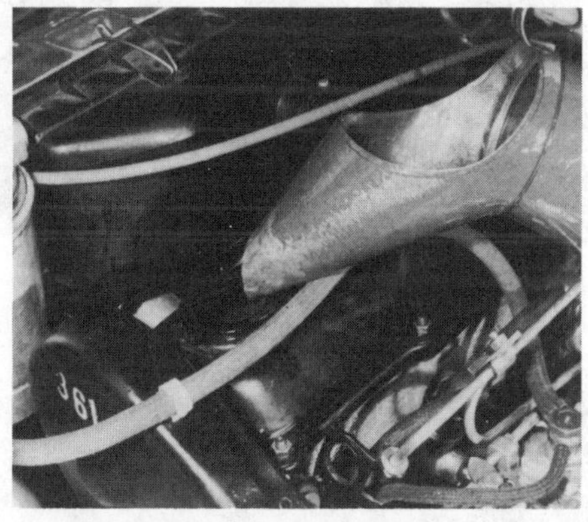

Topping-up the engine oil

Windscreen washer reservoir

Tailgate washer reservoir

Using a tyre pressure gauge

Every 7500 miles (12 000 km) or six months, whichever comes first

Engine

Change engine oil (photo)
Test strength of coolant antifreeze and top-up if necessary

Every 15 000 miles (24 000 km) or yearly, whichever comes first

Engine

Renew oil filter element
Adjust valve clearances and renew valve cover gasket
Check cylinder compression
Clean crankcase ventilation hoses
Check engine for oil leaks
Check idle speed and adjust if necessary

Transmission

Check transmission oil level and top-up if necessary (photos)
Check clutch adjustment

Fuel and exhaust system

Clean air cleaner element
Renew fuel filter element
Check fuel tank and supply lines
Check exhaust system for corrosion and leakage, check mountings for security

Electrical system

Check alternator drivebelt and adjust if necessary
Adjust headlight beam alignment

Steering

Check for wear in steering gear and balljoints, check condition of dust seals and rubber bellows
Check front wheel alignment and adjust if necessary

Brakes

Check brake hydraulic system for leaks, damaged or corroded pipes, etc
Check disc pads and brake shoes for wear
Adjust rear brakes and handbrake (if necessary)

General

Lubricate all controls and linkages, door catches and hinges

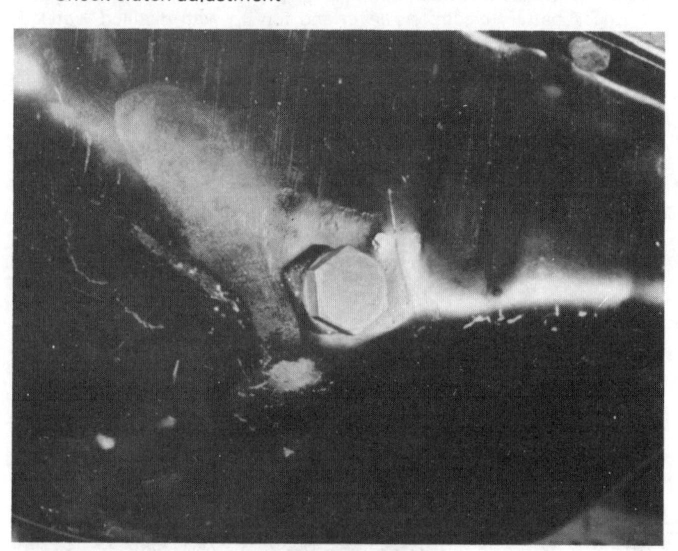

Engine sump drain plug

Transmission filler plug

Transmission drain plug

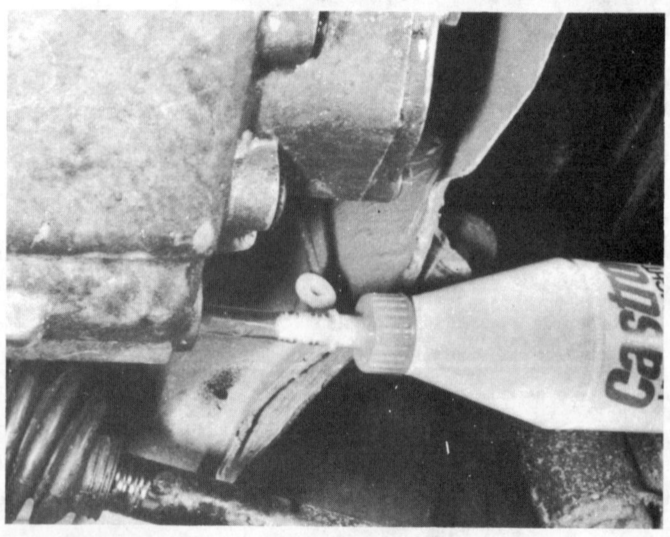

Topping-up the transmission oil

Every 30 000 miles (48 000 km) or two yearly, whichever comes first

Fuel and exhaust system
Renew air cleaner element

Electrical system
Renew alternator drivebelt

Brakes
Renew brake fluid
Check operation of brake warning light switch

Every 60 000 miles (96 000 km) or four yearly, whichever comes first

Engine
Clean glow plugs and check operation

Fuel and exhaust system
Overhaul injectors

Chapter 1 Engine

Refer to Chapter 11 for specifications and information applicable to later models

Contents

Specifications

Engine – general

Type	Four-cylinder in-line diesel, water cooled with overhead camshaft
Bore	76·5 mm (3·012 in)
Stroke	80·0 mm (3·150 in)
Capacity	1471 cc (89·8 cu in)
Compression ratio	23·5 : 1
Maximum output (DIN)	37 kW (50 bhp) @ 5000 rpm
Maximum torque (DIN)	82·0 Nm (56·5 lbf ft) @ 3000 rpm

Cylinder block and pistons

Material	Cast iron
Number of main bearings	5
Cylinder out of round – maximum	0·04 mm (0·0016 in)
Piston to cylinder clearance:	
New part	0·03 mm (0·001 in)
Wear limit	0·07 mm (0·0025 in)

Piston sizes	Piston dia (mm)	Cylinder bore (mm)
Standard – 1	76·48	76·51
Standard – 2	76·49	76·52
Standard – 3	76·50	76·53

Three oversizes are available in steps of 0·25 mm

Piston rings

Side clearance:	
Upper compression ring	0·06 – 0·09 mm (0·002 – 0·0035 in)
Lower compression ring	0·05 – 0·08 mm (0·0019 – 0·0031 in)
Oil scraper ring	0·03 – 0·06 mm (0·0011 – 0·002 in)

Wear limit:	
Compression rings	0·20 mm (0·0079 in)
Oil scraper ring	0·15 mm (0·0059 in)
Ring gap:	
Compression rings	0·3 – 0·5 mm (0·012 – 0·020 in)
Oil scraper ring	0·25 – 0·40 mm (0·010 – 0·016 in)
Wear limit	1·0 mm (0·039 in)

Crankshaft

Main bearing journals:	
Standard	54·00 mm (2·126 in)
1st undersize	53·75 mm (2·116 in)
2nd undersize	53·50 mm (2·106 in)
3rd undersize	53·25 mm (2·097 in)
Maximum out-of-round	0·03 mm (0·0012 in)
Connecting rod journals:	
Standard	46·00 mm (1·811 in)
1st undersize	45·75 mm (1·801 in)
2nd undersize	45·50 mm (1·791 in)
3rd undersize	45·25 mm (1·781 in)
Maximum out-of-round	0·03 mm (0·0012 in)

Main bearings

Bearing radial clearance	0·03 – 0·08 mm (0·0012 – 0·0031 in)
Wear limit	0·17 mm (0·0066 in)
Crankshaft endplay (measured at No 3 bearing)	0·07 – 0·17 mm (0·003 – 0·0066 in)
Wear limit	0·37 mm (0·015 in)

Connecting rod bearings

Bearing radial clearance	0·028 – 0·088 mm (0·0011 – 0·0034 in)
Wear limit	0·12 mm (0·004 in)
Maximum side clearance	0·37 mm (0·014 in)

Camshaft

Maximum endplay	0·15 mm (0·006 in)

Valves

Inlet:	
Length	104·8 mm (4·125 in)
Head diameter	34·0 mm (1·338 in)
Stem diameter	7·97 mm (0·314 in)
Seat width	2·0 mm (0·078 in)
Seat angle	45°
Exhaust:	
Length	104·6 mm (4·117 in)
Head diameter	31·00 mm (1·220 in)
Stem diameter	7·95 mm (0·313 in)
Seat width	2·4 mm (0·094 in)
Seat angle	45°
Valve clearance – cold:	
Inlet	0·15 – 0·25 mm (0·006 – 0·010 in)
Exhaust	0·35 – 0·45 mm (0·014 – 0·018 in)
Valve clearance – hot:	
Inlet	0·20 – 0·30 mm (0·008 – 0·012 in)
Exhaust	0·40 – 0·50 mm (0·016 – 0·020 in)

Lubrication system

Oil pump	Gear type
Backlash between gears	0·20 mm (0·008 in) max.
Endplay	0·15 mm (0·006 in) max.
Oil pressure	1·96 kgf/cm² (28 lbf/in²) minimum at 2000 rpm with oil temperature of 80°C (176°F)
Oil capacity	3·5 litres (6 Imp pints, 7·4 US pints)

Torque wrench settings

	kgf m	lbf ft
Camshaft bearing caps	2·0	14
Camshaft sprocket	4·5	33
Connecting rod caps	4·5	33
Crankshaft sprocket	8·0	56
Crankshaft pulley	2·0	14
Crankshaft bearing caps	6·5	47
Cylinder head bolts (engine cold):		
Stage 1	7·5	55
Stage 2	Additional ¼ turn to each bolt	
Engine to transmission	5·5	40
Engine mounting to body	4·0	29

Torque wrench settings

	kgf m	lbf ft
Driveshaft flange bolts	4·5	33
Fuel pump mounting plate	2·5	18
Fuel pump to mounting plate	2·5	18
Fuel injection pipes	2·5	18
Intermediate shaft flange	2·5	18
Intermediate shaft pulley	4·5	32
Oil drain plug	3·0	21
Oil pump to engine	2·0	14
Oil pump cover bolts	1·0	7
Oil sump bolts	2·0	14
Oil pressure switch	1·2	8·5
Oil filter adapter to block	2·0	18
Drivebelt tensioner locknut	4·5	32
Fuel injection pump sprocket	4·5	32
Fuel injectors	7·0	51
Manifolds to cylinder head	2·5	18
Exhaust pipe to manifold	2·5	18
Clutch pressure plate to crankshaft	7·5	54
Flywheel to clutch assembly	2·0	14
Vacuum pump clamp bolt	1·4	10
Valve cover bolts	0·6	4

1 General description

The four-cylinder, in-line, water-cooled engine has a cast iron cylinder block and an aluminium cylinder head. The crankshaft is supported in five main bearings.

The overhead camshaft is driven by a toothed drivebelt from a sprocket on the crankshaft. The belt also drives the intermediate shaft which in turn drives the oil pump and vacuum pump. The fuel injection pump, also driven by the toothed drivebelt, is mounted on the left-hand side of the engine.

The camshaft lobes press on the bucket type tappets which bear directly on the end of the valve stems. A steel disc, available in different thicknesses, is fitted on each tappet to provide for adjustment of the valve clearances.

The spark plugs as used on petrol engines are replaced by fuel injectors which inject atomised fuel into the pre-combustion chambers in the cylinder head when the piston is at the top of the compression stroke. The compression ratio is 23·5:1 (about three times that of petrol engines) and the heat generated by the high compression of the intake air ignites the vapourised fuel. Electrically heated glow plugs are fitted to each pre-combustion chamber to assist cold starting.

The engine is supported on flexible mountings. These are attached to a bracket at the front of the engine (drivebelt end) and one at the rear of the transmission, a side support to the frame behind the engine and torque reaction damper from the left-hand side of the engine to the frame below the radiator.

2 Major operations possible with the engine in the car

1 The following operations can be carried out with the engine in the car. Removal and refitting of the following:

*(a) Cylinder head assembly**
(b) Toothed drivebelt†
(c) Fuel injection pump
(d) Crankshaft front oil seal†
(e) Intermediate shaft oil seal†
(f) Oil sump and oil pump ‡
(g) Pistons and connecting rods‡
(h) Engine mountings

** Requires preliminary draining of coolant*
† Requires preliminary draining of coolant and removal of radiator
‡ Requires draining of coolant and oil

3 Major operations requiring engine removal

1 It is necessary to remove the engine from the car when removing and refitting the following:

(a) Crankshaft rear oil seal
(b) Crankshaft and main bearings
(c) Intermediate shaft

4 Engine removal – general

1 The engine and transmission can be removed as an assembly. Alternatively, the transmission can be removed first, as described in Chapter 5, and then the engine removed as a separate unit.
2 If the engine and transmission are being removed as a single assembly, it must be lowered out of the engine compartment and removed from under the car. If the transmission is removed first then the engine can be hoisted out of the engine compartment.
3 For removal of the engine/transmission assembly a 330 lbs (150 kg) capacity hoist will be required. The car must be raised and supported on axle stands, or other suitable supports, placed under the bodyframe. Make sure it is high enough to allow the engine/transmission assembly to be withdrawn from underneath the car.
4 When disconnecting wiring, control cables and hoses, label them so that they can be identified to their correct locations and thus avoid any confusion when the engine is being refitted.
5 Have suitable containers available to hold small items so that they will not get lost.

5 Engine/transmission assembly – removal

1 Disconnect the cables from the battery.
2 Mark the position of the bonnet hinges, then remove the four hinge to bonnet bolts and lift away the bonnet with the help of an assistant. Place the bonnet safely out of the way where it will not get scratched or damaged.
3 Unscrew the battery hold-down clamp securing bolt and remove the clamp. Lift the battery out of the engine compartment, taking care not to spill any of the electrolyte.
4 Drain the cooling system as described in Chapter 2. Remove the sump drain plug and drain the oil into a suitable container. Refit the drain plug.
5 Slacken the securing clips and remove the radiator top and bottom hoses. Disconnect the hose to the coolant expansion tank.
6 Disconnect the wiring from the fan motor at the connector and the wiring from the thermoswitch on the radiator.
7 Remove the radiator lower securing nuts. Release the radiator from the upper securing clip, then tilt the radiator towards the engine and lift it up off the bottom mounting supports. Remove it from the engine compartment complete with fan.
8 Remove the alternator as described in Chapter 7, and the air conditioning compressor (if fitted) as described in Chapter 2.
9 Disconnect the fuel supply and return lines from the fuel injection pump. Disconnect the servo vacuum hose from the vacuum pump.
10 Disconnect the accelerator and cold start cables from the levers on the fuel pump. Lift away the accelerator cable complete with support bracket.
11 Disconnect the wiring from the stop solenoid on the fuel pump and from the glow plugs on the cylinder head.
12 Disconnect the wiring from the oil pressure switch, at the rear of

Fig. 1.1 Disconnect the fuel lines and accelerator cable from the fuel injection pump (Sec 5)

6 Fuel supply line
7 Fuel return line
8 Accelerator cable
9 Cable support bracket

Fig. 1.2 Remove the exhaust pipe support and disconnect the exhaust pipe from the exhaust manifold (Sec 5)

16 Support bracket
17 Support rod
18 Flange securing nuts
19 Gearbox mounting

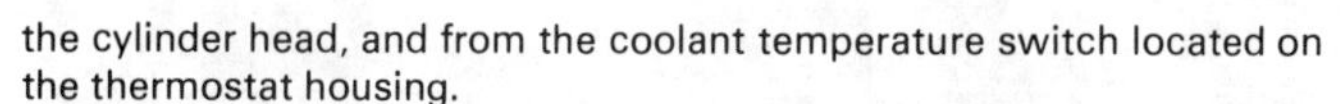

5.19 Tying the driveshafts to one side when removing the engine

Fig. 1.3 Disconnect the gear selector linkage (Sec 5)

13 Relay lever
14 Connecting rod
15 Selector rod

the cylinder head, and from the coolant temperature switch located on the thermostat housing.

13 Disconnect the wiring from the starter motor and the reversing light switch.

14 Slacken the securing clips and remove the heater hoses.

15 Remove the fuel filter mounting bolts and place the filter assembly on the wheel housing. It is not necessary to disconnect the fuel lines from the filter assembly.

16 Raise the front of the car and support it securely on axle stands.

17 Slacken the clutch cable adjuster, refer to Chapter 4, and remove the cable from the operating lever. Tie the cable out of the way.

18 Undo the clamp bolt and remove the speedometer cable drive from the transmission. Plug the hole in the casing to prevent dust or moisture entering the gearbox.

19 Disconnect the driveshafts from the transmission by removing the bolts from the driveshaft flanges with an Allen key. Support the shafts so that the rubber boots will not get damaged (photo).

20 Remove the exhaust pipe support and the exhaust pipe to exhaust manifold bolts, then lower the exhaust pipe.

21 Remove the gear shift relay lever and disconnect the shift linkage connecting rod from the transmission. Disconnect the selector rod at the rear.

22 Undo the securing bolts and remove the front transmission mounting complete by taking the bracket off the frame and the bracket from the engine/transmission. Remove the earth strap from the transmission (photos).

23 From underneath the car remove the mounting bracket which holds the unit to the body. The mounting is secured to the body by three nuts and to the transmission by two nuts (photos).

24 Position the lifting gear, fit the sling and take the weight of the engine. The engine is now supported at both ends. Undo the nuts securing the mountings at the front of the engine and at the rear of the transmission. Remove the nuts securing the bracket to the transmission. Lift slightly on the hoist, then remove the mounting bolts and pull the engine/transmission assembly away from the mounting brackets.

25 Turn the roadwheels fully to the right and position the left-hand driveshaft along the wishbone and the right-hand driveshaft pointing towards the front of the car. Make a final check that all wiring and control cables have been disconnected and tucked out of the way, then carefully lower the assembly to the ground and remove it from under the car (photos).

6 Separating the transmission from the engine

1 The engine must be supported so that the transmission can be withdrawn. Either support the engine so that the transmission overhangs the bench or remove the transmission from the engine while the

5.22A Removing the front transmission mounting bolt

5.22B Earth strap location on front transmission mounting

5.23A Rear transmission mounting location

5.23B Locating peg on rear transmission mounting

5.23C Removing the transmission rear mounting (engine out of car for clarity)

5.25A Lowering the engine to the ground with a hoist

5.25B Lifting the body over the engine with a hoist

5.25C Hook location when lifting body

6.3A TDC hole plug

assembly is suspended on the hoist.

2 Undo the securing bolts and remove the starter motor.

3 Remove the plug from the TDC marker hole in the bellhousing. Rotate the crankshaft, using a spanner on the crankshaft sprocket bolt, until the cut-out on the flywheel (76° BTDC) is aligned with the pointer on the clutch housing (photos). This is necessary to prevent the flywheel fouling the housing as the transmission is withdrawn from the engine.

4 Remove the securing bolts and lift away the cover plate (arrowed in Fig. 1.4) over the driveshaft flange.

5 Remove the transmission-to-engine securing bolts and lift away the cover plate. Withdraw the transmission from the engine. **Do not** insert wedges to separate the units or you will damage the mating faces; rock the transmission carefully to free it from the two locating dowels. Withdraw the transmission carefully, taking care not to put any stress on the transmission mainshaft or the clutch pushrod which may damage them (photo).

7 Engine ancillary components – removal

1 Before basic engine dismantling begins remove the following components;

(a) Air cleaner
(b) Vacuum pump
(c) Intake and exhaust manifolds
(d) Water pump assembly
(e) Fuel injectors
(f) Oil filter
(g) Oil pressure switch
(h) Clutch pressure plate assembly and friction disc

2 Information on the removal and refitting of these components can be found in the relevant Chapters throughout this manual.

6.3B Flywheel alignment marks when separating engine and transmission

Fig. 1.4 Removing the cover plate (arrowed) (Sec 6)

8 Engine dismantling – general

1 It is best to mount the engine on a dismantling stand but if one is not available, then place the engine on a strong bench so as to be at a comfortable working height. Failing this the engine can be stripped down on the floor, but take care to avoid grit coming in contact with the internal parts of the engine.

2 During the dismantling process the greatest care should be taken to keep the exposed parts free from dirt. As an aid to achieving this, it is a sound scheme to clean the outside of the engine thoroughly (away from the working area), removing all traces of oil and congealed dirt, before dismantling begins.

3 Use paraffin or a good grease solvent. The latter compound will make the job much easier, as, after the solvent has been applied and allowed to stand for some time, a vigorous jet of water will wash off all the solvent and the grease and dirt. If the dirt is very thick, work the solvent into it with a stiff brush.

4 As the engine is dismantled, clean each part in a bath of paraffin. Never immerse parts with oilways in paraffin, eg the crankshaft, but to clean, wipe down with a paraffin dampened rag. Oilways can be cleaned out with wire. If an air line is available all the parts can be blown dry and the oilways blown through as an added precaution.

5 Re-use of old engine gaskets and oil seals is a false economy and can give rise to oil and water leaks, if nothing worse. To avoid the possibility of trouble with leaks after the engine has been reassembled always use new gaskets and oil seals.

6 Do not throw the old gaskets away as it sometimes happens that an immediate replacement cannot be obtained and the old gasket is then very useful as a template. As the gaskets are removed hang them up on suitable hooks or nails.

7 When dismantling the engine it is best to work from the top down, with the engine supported in an upright position. Ensure that it is well supported so that it will not fall over when tight nuts and bolts are being slackened. When the stage where the oil sump has to be removed is reached, the engine can be turned on its side and all the other work carried out with it in this position.

8 Wherever possible, refit nuts, bolts and washers finger tight from where they were removed. This helps avoid possible loss, or doubt as to where they belong. If they cannot be refitted then lay them out in such a fashion that it is clear where they belong.

9 Toothed drivebelt – removal and refitting (engine in car)

1 Release the four securing clips and remove the air cleaner cover and filter. Remove the CAV type fuel filter where fitted (Chapter 3).

2 Undo the eight bolts or nuts securing the valve cover and remove them, the two strengthening strips and the valve cover (photo).

6.5 Separating the transmission from the engine

9.2 Valve cover and retaining nuts

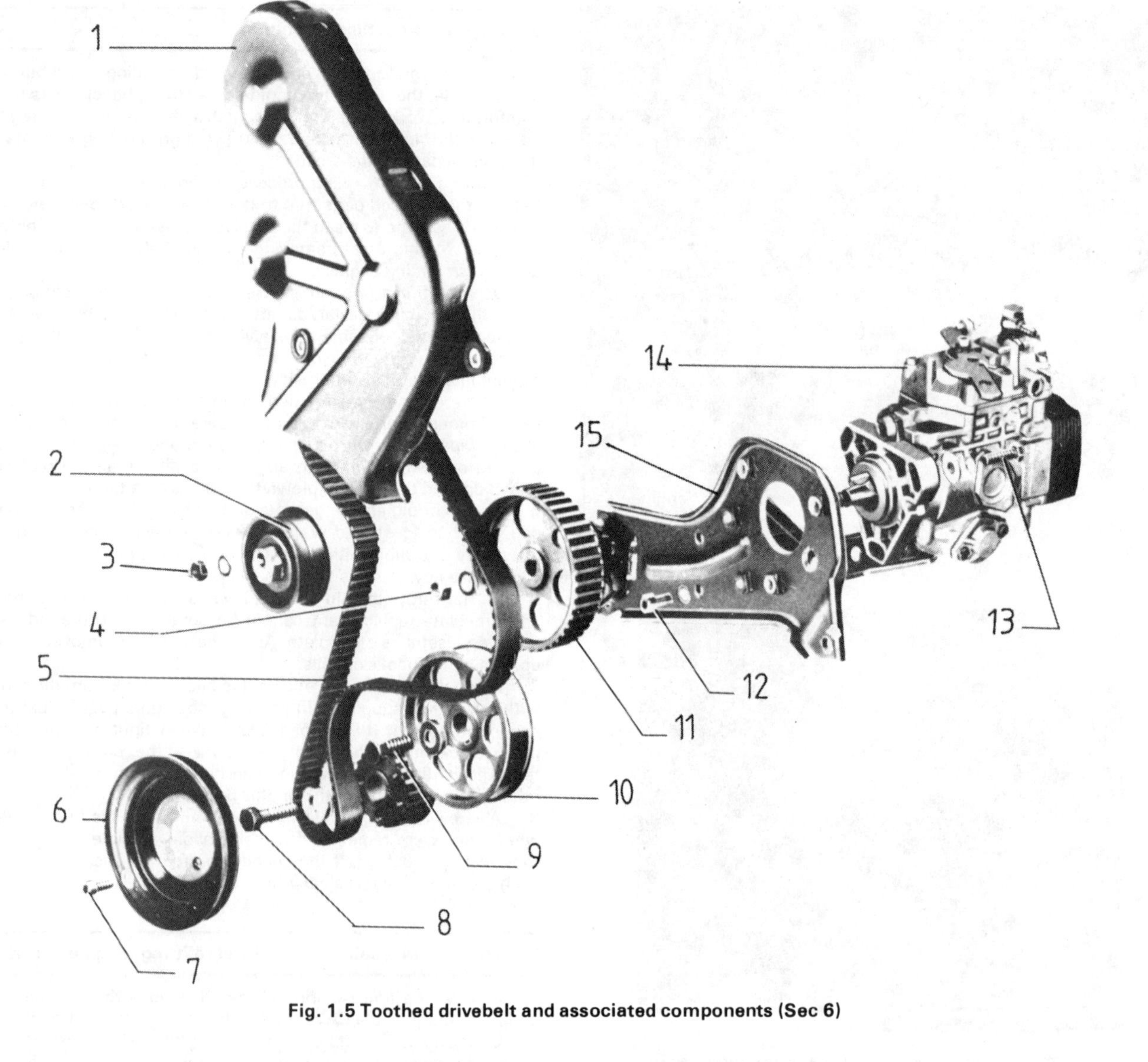

Fig. 1.5 Toothed drivebelt and associated components (Sec 6)

1 Belt cover
2 Belt tensioner
3 Tensioner locknut
4 Pump sprocket securing nut
5 Drivebelt
6 Crankshaft pulley
7 Pulley securing bolt
8 Crankshaft sprocket securing bolt
9 Intermediate shaft pulley bolt
10 Intermediate shaft pulley
11 Fuel injection pump sprocket
12 Injection pump mounting plate securing bolt
13 Injection pump securing bolt
14 Fuel injection pump
15 Mounting plate

9.4A The crankshaft pulley

9.4B Removing the crankshaft pulley

9.5 Removing the engine drivebelt cover

Remove the valve cover gasket.
3 Slacken the alternator mounting bolts, then push the alternator towards the engine and remove the alternator drivebelt.
4 Undo the crankshaft pulley securing bolts with an Allen key and remove the pulley (photos).
5 Remove the securing bolts and lift away the drivebelt cover (photo). Later models have two covers.
6 Remove the plug from the TDC marker hole on the clutch bellhousing. Rotate the crankshaft, using a spanner on the crankshaft sprocket retaining bolt, until No 1 piston is at TDC on the compression stroke, (both valves closed, cam lobes pointing upwards) and the notch on the flywheel is aligned with the pointer on the bellhousing (Fig. 1.6).
7 Lock the camshaft in position to prevent it from turning while the belt is removed. VW Tool No 2065 is specified for this operation, but a metal bar of suitable dimensions is just as good (photo). Lock the fuel injection pump sprocket in the set position with a locking pin of suitable diameter. Again a special VW tool (No. 2064) is available, but a bolt will serve just as well (photo).
8 Slacken the drivebelt tensioner locknut and unscrew the tensioner adjuster to release the tension from the belt.

9 Undo the drivebelt shield securing bolts. Remove the belt shield and the drivebelt.

10 When fitting a new belt check that the TDC mark on the flywheel is aligned with the pointer on the bellhousing, then slacken the camshaft sprocket securing bolt half a turn and free the sprocket from the camshaft by tapping it with a plastic hammer.
11 Fit the new drivebelt on the sprockets and remove the locking pin from the injection pump sprocket.
12 Turn the belt tensioner clockwise to tension the belt and check the tension between the camshaft sprocket and injection pump sprocket with tool VW210. If this tool is not available, refer to paragraph 14. Adjust the tension so that the scale on tool VW210 reads 12–13, then tighten the camshaft sprocket bolt and the tensioner locknut to the specified torque.
13 Remove the locking fixture from the camshaft.
14 Turn the engine through two revolutions in the normal direction of rotation, then strike the belt with a plastic hammer between the camshaft sprocket and the injection pump sprocket. Now re-check the belt tension and, if necessary, re-adjust. If the tool VW210 is not available a reasonably accurate tension of the belt (as a temporary measure) can be achieved by adjusting the belt tensioner until the belt will twist only 90° when held between the finger and thumb midway between the camshaft sprocket and the fuel injection pump sprocket.
15 Check the fuel injection pump timing, refer to Chapter 3.
16 The remainder of the refitting procedure is the reverse of removal. When refitting the alternator drivebelt adjust the belt tension as described in Chapter 7. Always use a new gasket when refitting the valve cover.

Fig. 1.6 TDC notch on flywheel aligned with pointer (Sec 9)

9.7A Method of locking the camshaft with a metal bar

9.7B Using a bolt to lock the fuel injection pump sprocket

Fig. 1.7 Checking the drivebelt tension with VW tool 210 (Sec 9)

10 Crankshaft pulley, drivebelt and tensioner, fuel injection pump and crankshaft sprocket – removal

1 Undo the four socket headed bolts securing the crankshaft pulley to the crankshaft sprocket and remove the pulley. Remove the belt cover securing bolts and the cover.

2 Before removing the toothed drivebelt, check the belt tension. If the belt is held between the finger and thumb, midway between the camshaft sprocket and the fuel injection pump sprocket, it should be just possible to twist it through 90°. If it is too slack, adjust it by slackening the locknut and turning the tensioner clockwise. If the belt cannot be tensioned correctly, this means that it has stretched and a new belt must be obtained.

3 Remove the two bolts securing the drivebelt shield and lift away the shield.

4 Release the drivebelt tension and remove the drivebelt.

5 Remove the drivebelt tensioner (photos).

6 Remove the crankshaft sprocket by unscrewing the securing bolt and then pulling off the sprocket, using a suitable puller. Prevent the

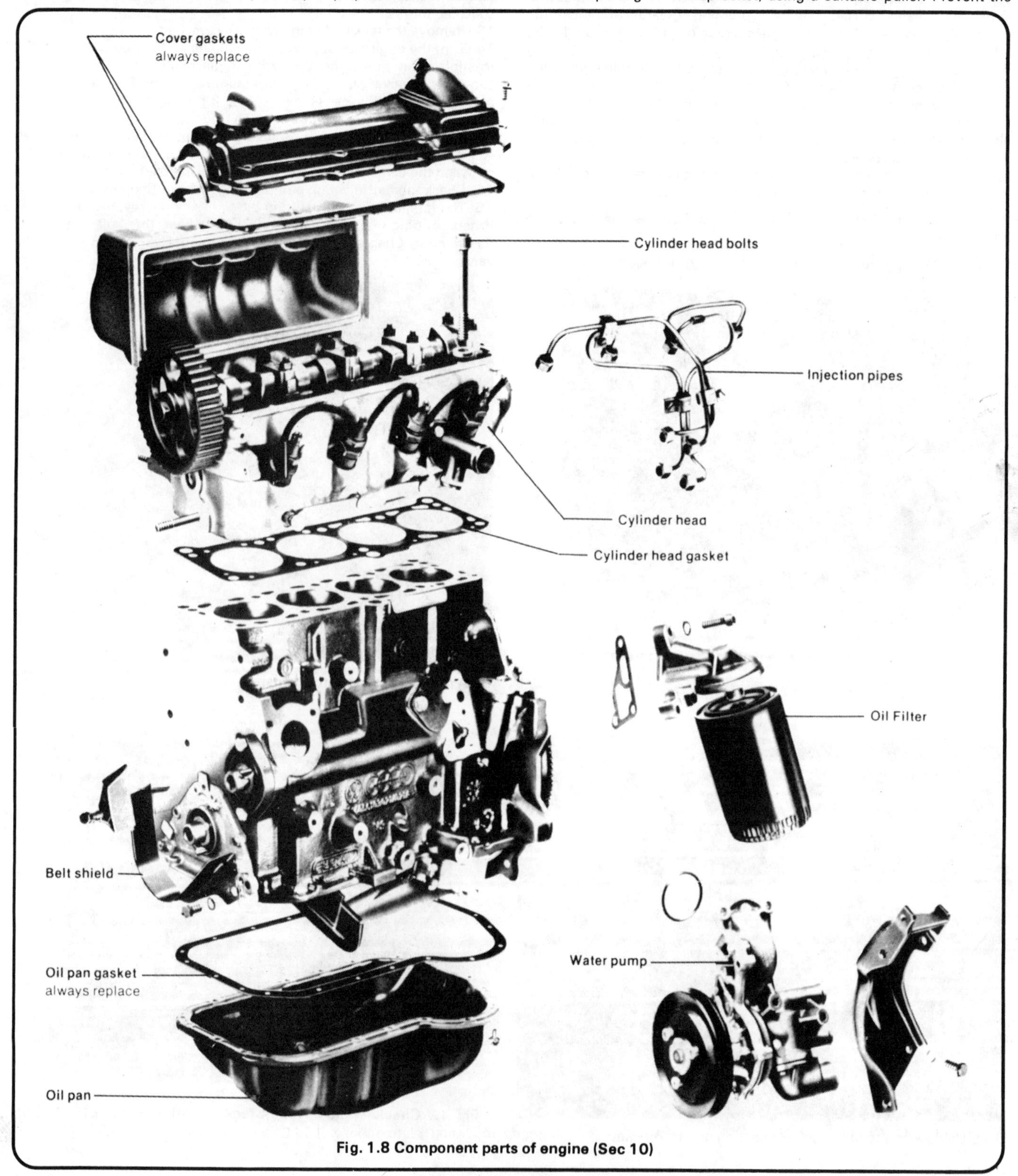

Fig. 1.8 Component parts of engine (Sec 10)

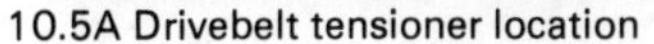

10.5A Drivebelt tensioner location

10.5B Removing the drivebelt tensioner

10.5C Removing the drivebelt tensioner pivot stud

crankshaft from rotating by jamming the flywheel starter ring gear.

7 Slacken the injection pump sprocket securing nut, but do not take it off. Using a suitable two-legged puller, with the centre screw locating on the sprocket securing nut, pull the sprocket to free it on the pump shaft, then remove the puller, nut and sprocket. Fig. 1.9 shows the sprocket being removed using tool VW203b.

8 Remove the bolts securing the injection pump to the pump mounting plate and lift away the fuel injection pump.

9 Remove the bolts attaching the alternator mounting bracket, the engine mounting and injection pump mounting plate to the engine block and lift away the bracket and plate (photos).

Fig. 1.9 Removing the injection pump sprocket with puller VW 203b (Sec 10)

11 Camshaft and cylinder head – removal

Note: *Although removal of the camshaft is described in this Section with the cylinder head fitted on the engine, the cylinder head can be removed as an assembly complete with the camshaft. Always remove the fuel injectors and glow plugs before working on the cylinder head.*

1 Undo the valve cover securing bolts, lift off the strengthening strips and remove the valve cover and valve cover gasket.

2 Remove the camshaft bearing caps. These have to be refitted in their original positions. They are numbered 1 to 5 (Fig. 1.11), the cap marked No 1 being the one at the sprocket end of the camshaft. The numbers on the caps are not always on the same side, but note that the bearing position is offset and the caps can only be fitted one way round. Undo the nuts securing bearing caps 5, 1 and 3 in that order and lift off the caps. Now undo the nuts securing 2 and 4 caps, in an alternate and diagonal sequence, and the camshaft will lift them up under pressure of the valve springs. Remove the caps and lift out the camshaft complete with sprocket and oil seal (photos).

3 The bucket-type tappets are now exposed and can be lifted out. Take out each one in turn, prise the steel disc out of the tappet by inserting s small screwdriver blade each side and lift the disc away. On the underside of the disc a number is engraved (eg 3.65), this is the thickness of the disc in millimetres. Note the number, then clean the disc and refit it on the tappet with the number side down (not visible). The eight tappets must be refitted in their original positions in the cylinder head so label them to identify their correct locations. Also record the thickness of the tappet discs for use at reassembly (photos).

4 The next job is to remove the cylinder head bolts. These are located in the well of the head and have polygon socket heads which require a special socket spanner. Undo and remove the bolts in the reverse order of the tightening sequence shown in Fig. 1.12.

5 When all ten bolts have been removed lift the cylinder head from the cylinder block. It may need tapping with a plastic hammer to loosen it, but **do not** try to prise it loose by using wedges between the head and the block as this will result in damage to the mating faces of the head and block.

10.9A Alternator mounting bracket

10.9B Injection pump mounting plate (front view)

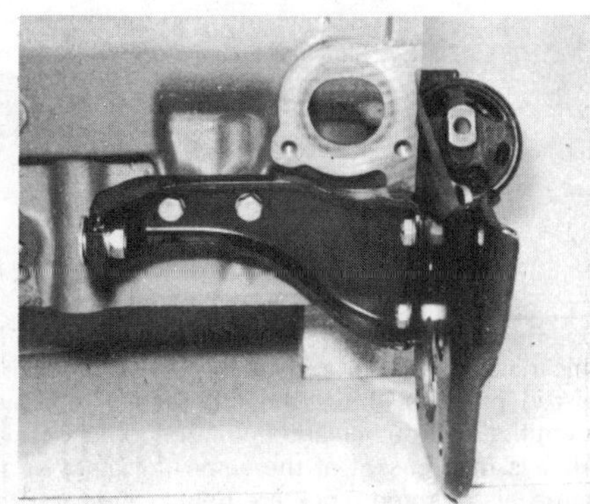

10.9C Injection pump mounting plate (side view)

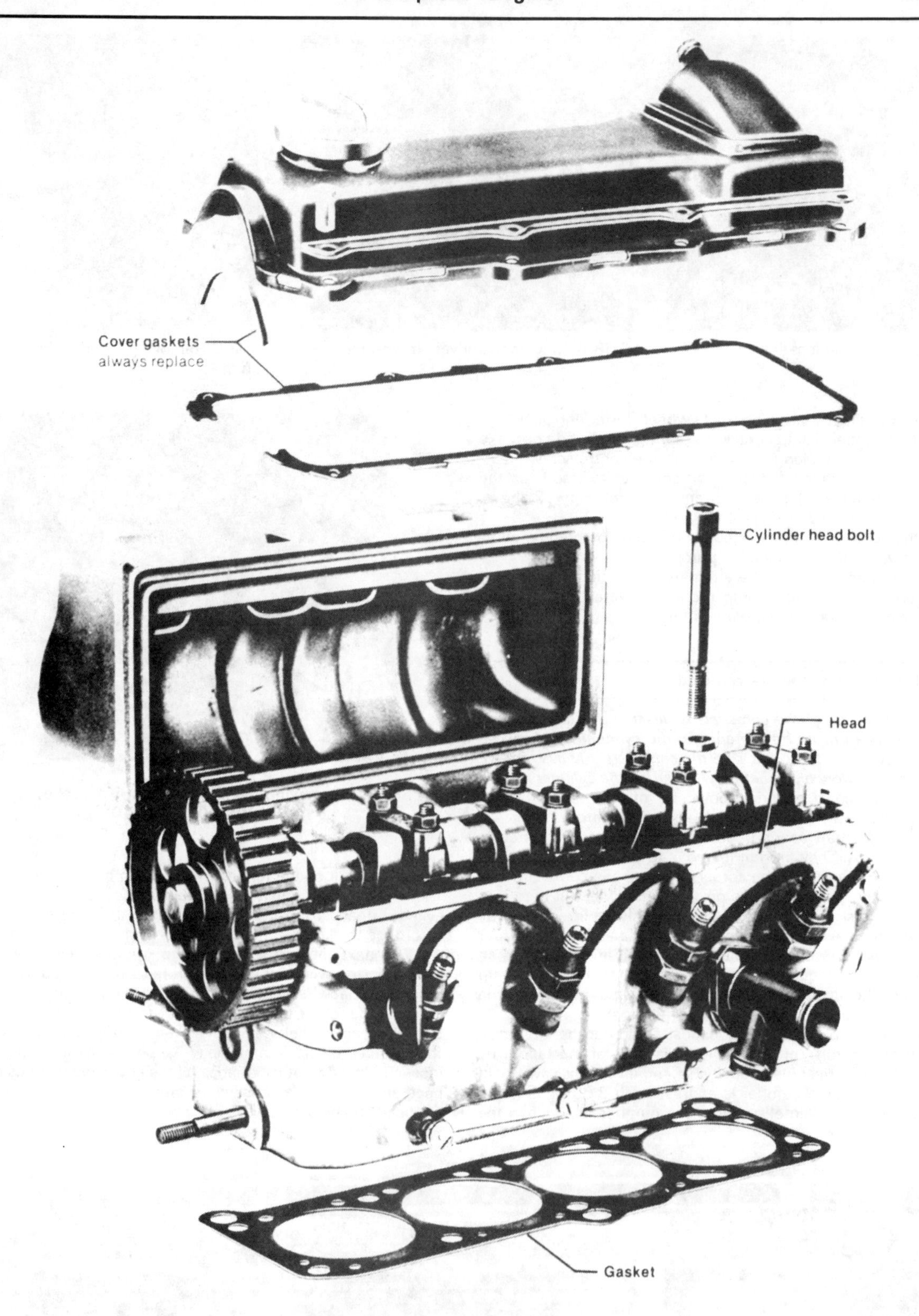

Fig. 1.10 Cylinder head assembly (Sec 11)

6 Remove the cylinder head gasket. **Note**: There are four different thicknesses of cylinder head gasket used, according to the amount the piston projects above the top face of the cylinder block. They are identified by notches and numbers on the gasket, arrowed in Fig. 1.13. Always fit a gasket of the same thickness as the one removed if the same pistons and block are being refitted. If the block or pistons are being renewed then the amount the piston projects (at TDC) above the top face of the block must be measured to determine the thickness of the cylinder head gasket required. See Section 42.

12 Oil sump and oil pump – removal

1 Turn the engine on its side. Undo the securing bolts, which may be hexagon or socket type heads, and remove the sump (photo).
2 The oil pump complete with strainer can now be removed after undoing the two socket headed bolts which secure it to the crankcase (photos).

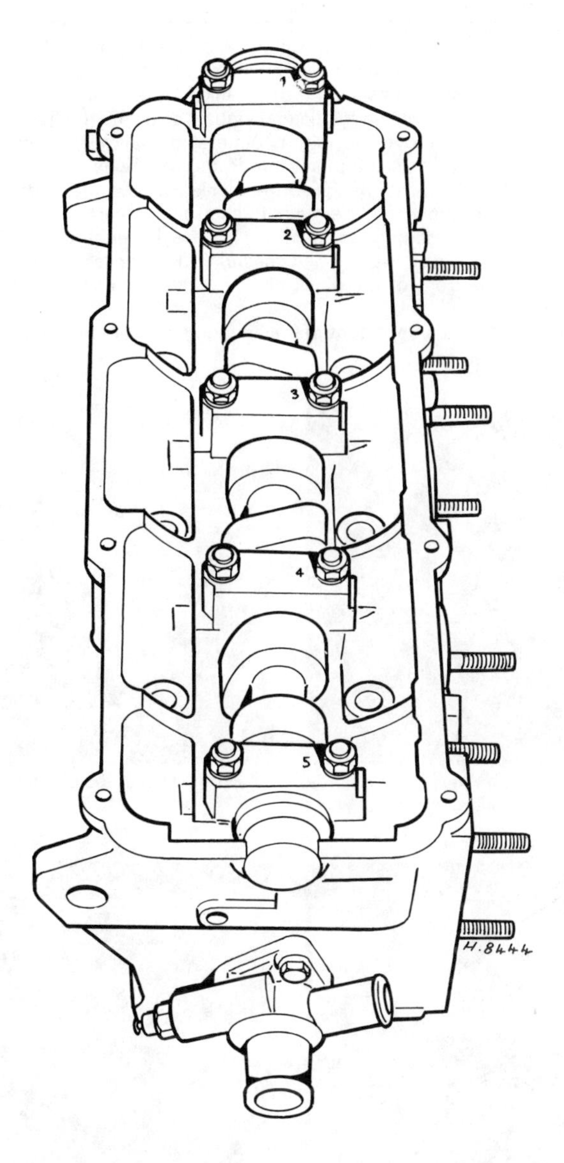

Fig. 1.11 The camshaft bearing caps are numbered 1-5 (Sec 11)

11.2A Removing a camshaft bearing cap retaining nut

11.2B Removing a camshaft bearing cap

11.3A Removing a tappet assembly

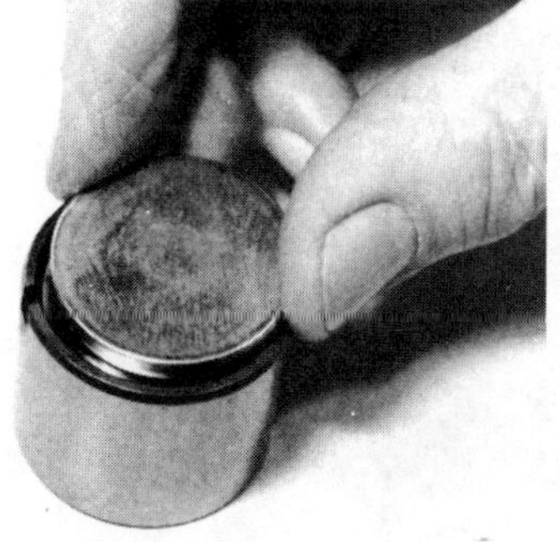

11.3B Removing a tappet disc

11.3C Tappet disc thickness identification

Fig. 1.12 Tightening sequence for cylinder head bolts (Sec 11 and 42)

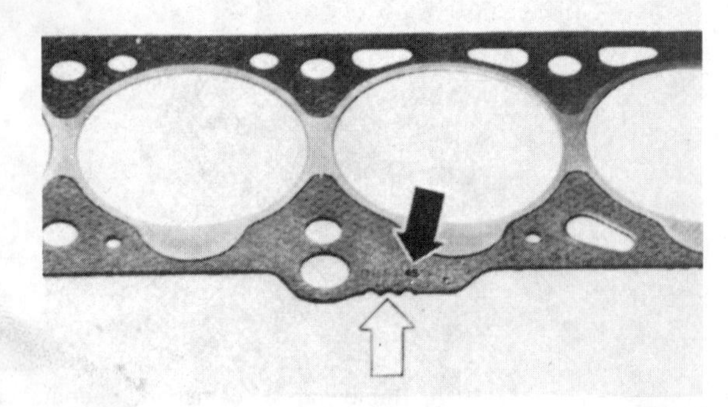

Fig. 1.13 The cylinder head gasket thickness is identified by notches (white arrow) and numbers (black arrow) (Sec 11 and 42)

13 Intermediate shaft, piston and connecting rod assemblies and crankshaft – removal

1 The intermediate plate will have been lifted off the transmission locating dowels when the clutch assembly was removed. Now remove the six bolts securing the crankshaft rear oil seal flange to the engine block. Remove the flange, with oil seal, and the gasket.
2 Remove the intermediate shaft pulley securing bolt and washer and pull off the pulley. Undo the two bolts securing the intermediate shaft oil seal flange to the block. Remove the flange and O-ring, then withdraw the intermediate shaft taking care not to lose the Woodruff key or damage the bearings in the cylinder block (photos).
3 Undo the five bolts securing the crankshaft front oil seal flange to the front of the block, then remove the flange, with oil seal, and the gasket.
4 It is important that all bearing caps are refitted in exactly the same

12.1 Removing the sump

12.2A Oil pump retaining bolts

12.2B Removing the oil pump

13.2A Removing the intermediate shaft pulley

13.2B Removing the intermediate shaft oil seal flange

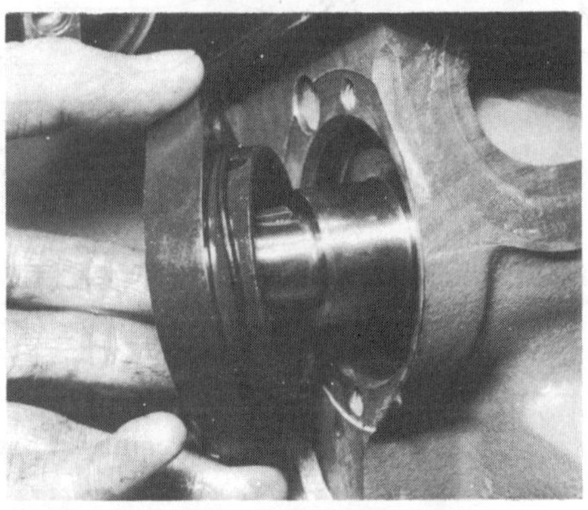

13.2C Intermediate shaft flange O-ring ...

13.2D ... and Woodruff key

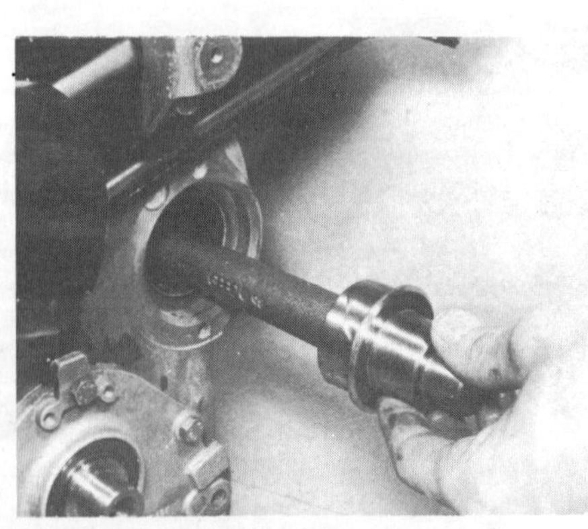

13.2E Removing the intermediate shaft

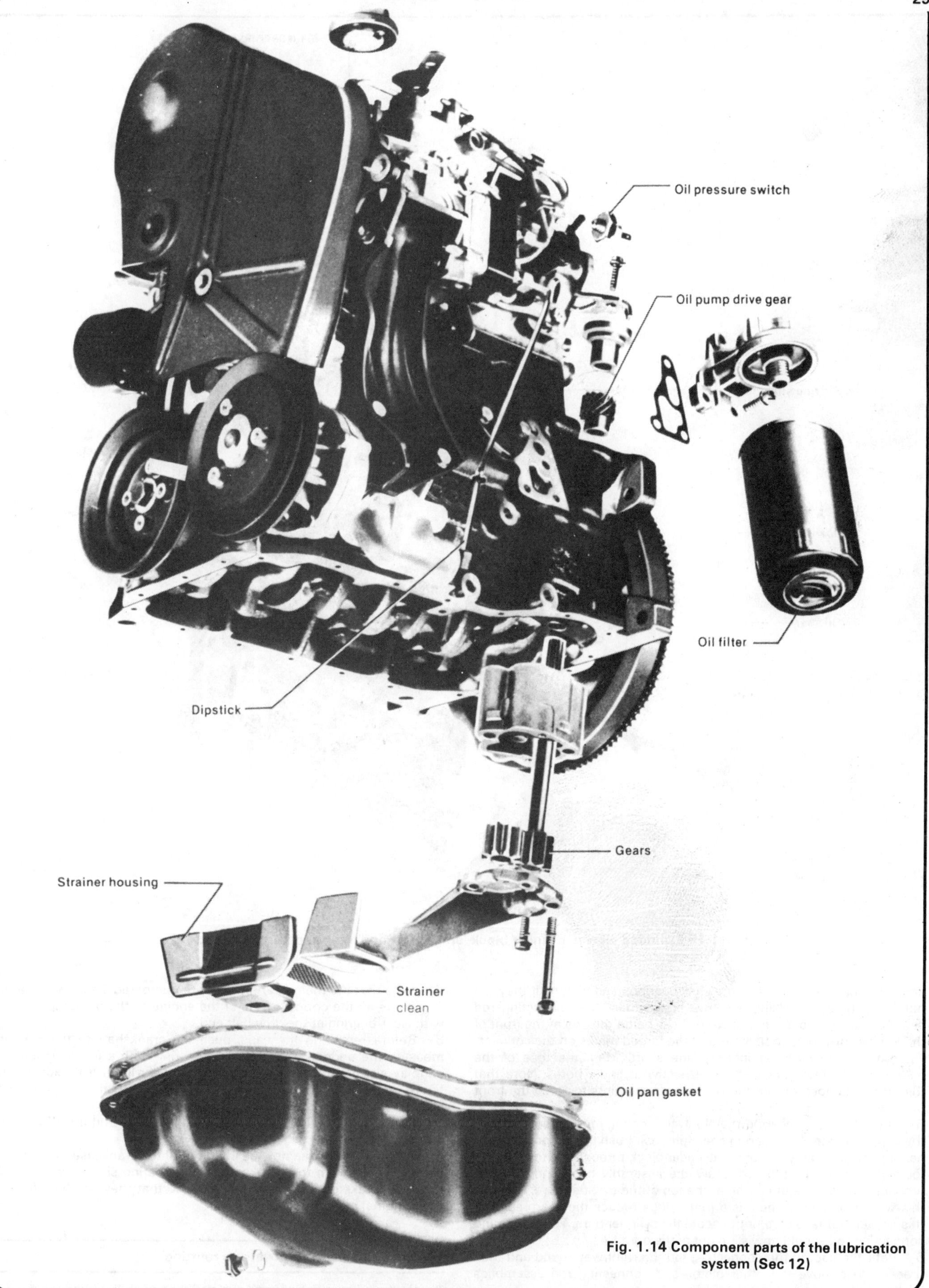

Fig. 1.14 Component parts of the lubrication system (Sec 12)

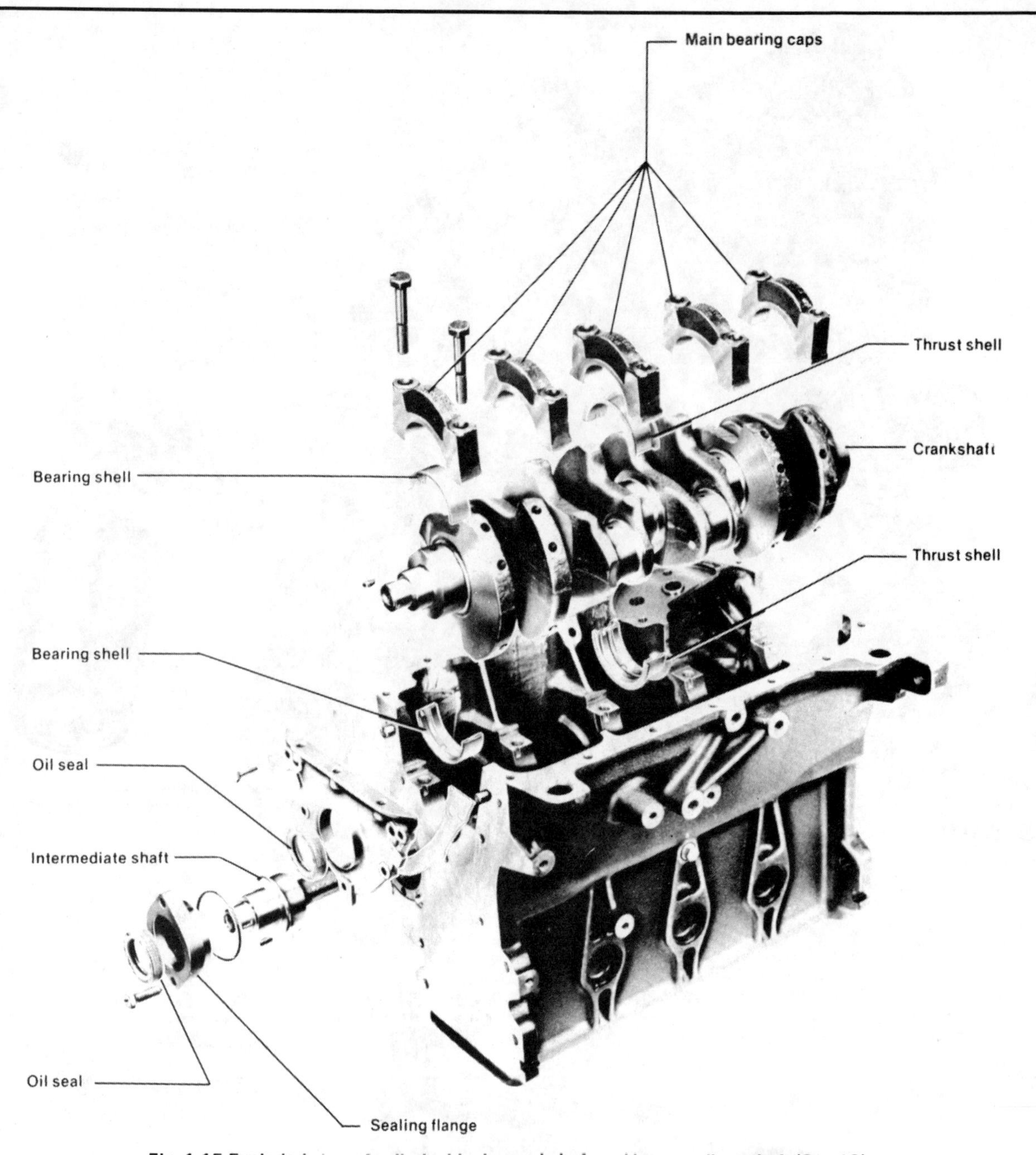

Fig. 1.15 Exploded view of cylinder block, crankshaft and intermediate shaft (Sec 13)

position. This also applies to the shell bearings and pistons if they are not being renewed. Using a centre punch mark the connecting rod bearing caps and rods No 1 to No 4, No 1 being the one at the front of the engine (drivebelt end). Note that the forged marks on the connecting rods are toward the intermediate shaft. Mark the tops of the pistons to identify them to their respective cylinder bores. Note that the arrow stamped on the top of the pistons points towards the front of the engine.

5 Undo the nuts securing the No 1 connecting rod cap and remove the cap complete with bearing shell (photos). Push the piston and connecting rod assembly out of the cylinder block through the top. Do not force it, if there is difficulty draw the assembly back and you will probably find a ridge of carbon at the top of the cylinder bore. Remove this with a scraper. If there is a metal ridge reduce the sharp edge of this as well but take care not to score the cylinder bore. The piston and connecting rod should now slide out of the bore.

6 Refit the cap on the connecting rod the right way round and then proceed to remove the other piston and connecting rod assemblies (photos).

7 Note that the main bearing caps are numbered 1 to 5 and that the number is on the opposite side of the engine to the oil pump location, with No 1 bearing at the drivebelt end.

8 Before removing the caps, push the crankshaft to the rear and measure the endplay at No 3 main bearing with a feeler gauge. The endplay should not exceed the specified wear limit. If it exceeds this figure the main bearings must be renewed irrespective of their condition.

9 Remove the main bearing cap securing bolts and lift off the caps and shell bearings.

10 The crankshaft can now be lifted out of the crankcase. Remove the bearing shells from the crankcase. If the bearing shells are not being renewed, make sure they are identified so that they can be refitted in their original positions.

14 Gudgeon pins – removal and refitting

1 Remove a circlip from one end and then push the gudgeon pin out.

Fig. 1.16 Make identifying marks on the top of the pistons (Sec 13)

13.5A Removing a connecting rod cap retaining nut ...

13.5B ... and a connecting rod cap

13.6A Connecting rod and cap identification punch marks

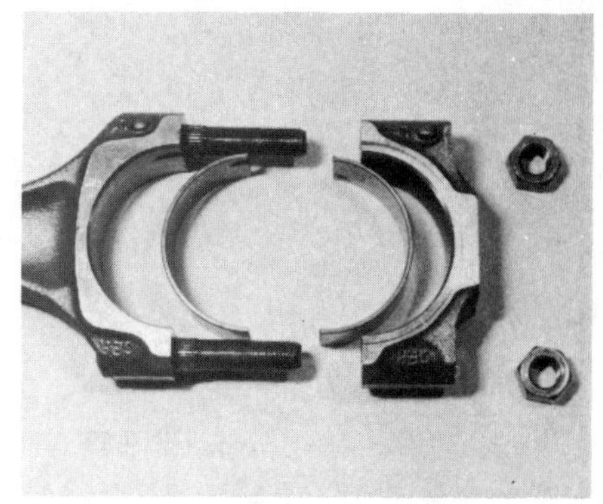

13.6B Connecting rod big-end components

14.1A Removing a gudgeon pin retaining circlip

14.1B Removing a gudgeon pin

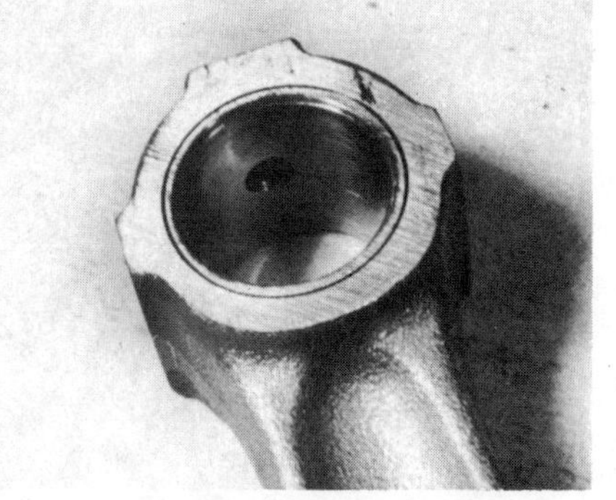

14.2 A connecting rod small-end bush

If it is tight, raise the temperature of the piston to 60° C (140° F) with hot water and the pin will slide out easily (photos).

2 Check the play in the connecting rod bush. If the pin seems loose and can be rocked at all then either the pin or the bush (or both) is worn. New bushes can be pressed into the connecting rod, if necessary, but they must then be reamed out to size to fit the gudgeon pin. This job should be left to your local VW garage as the bush must be reamed square to the connecting rod and without a suitable reaming fixture this is unlikely to be achieved (photo).

3 When assembling the piston to the connecting rod, make sure the arrow on the piston crown points to the front (drivebelt end) then push in the gudgeon pin and fit the retaining circlip.

15 Piston rings – removal

1 To remove the piston rings, slide them carefully over the top of the piston, taking care not to scratch the aluminium alloy. Do not slide them off the bottom of the piston skirt. It is very easy to break the cast iron piston rings if they are pulled off roughly, so this operation should be carried out with extreme care. It is helpful to make use of an old feeler gauge by sliding it behind the ring and removing the ring using a

twisting action.

2 Lift one end of the piston ring out of its groove and insert under it the end of the feeler gauge.

3 Run the feeler gauge slowly round the piston. As the ring comes out of its groove, apply a slight upward pressure so that it rests on the land above. It can then be eased off the piston with the feeler gauge stopping it from slipping into an empty groove if it is any but the top ring that is being removed.

16 Valves – removal

1 If the camshaft and tappets were not removed when the cylinder head was removed, refer to Section 11, paragraphs 2 and 3 and remove them now.

2 Using a valve spring compressor tool, compress the valve springs

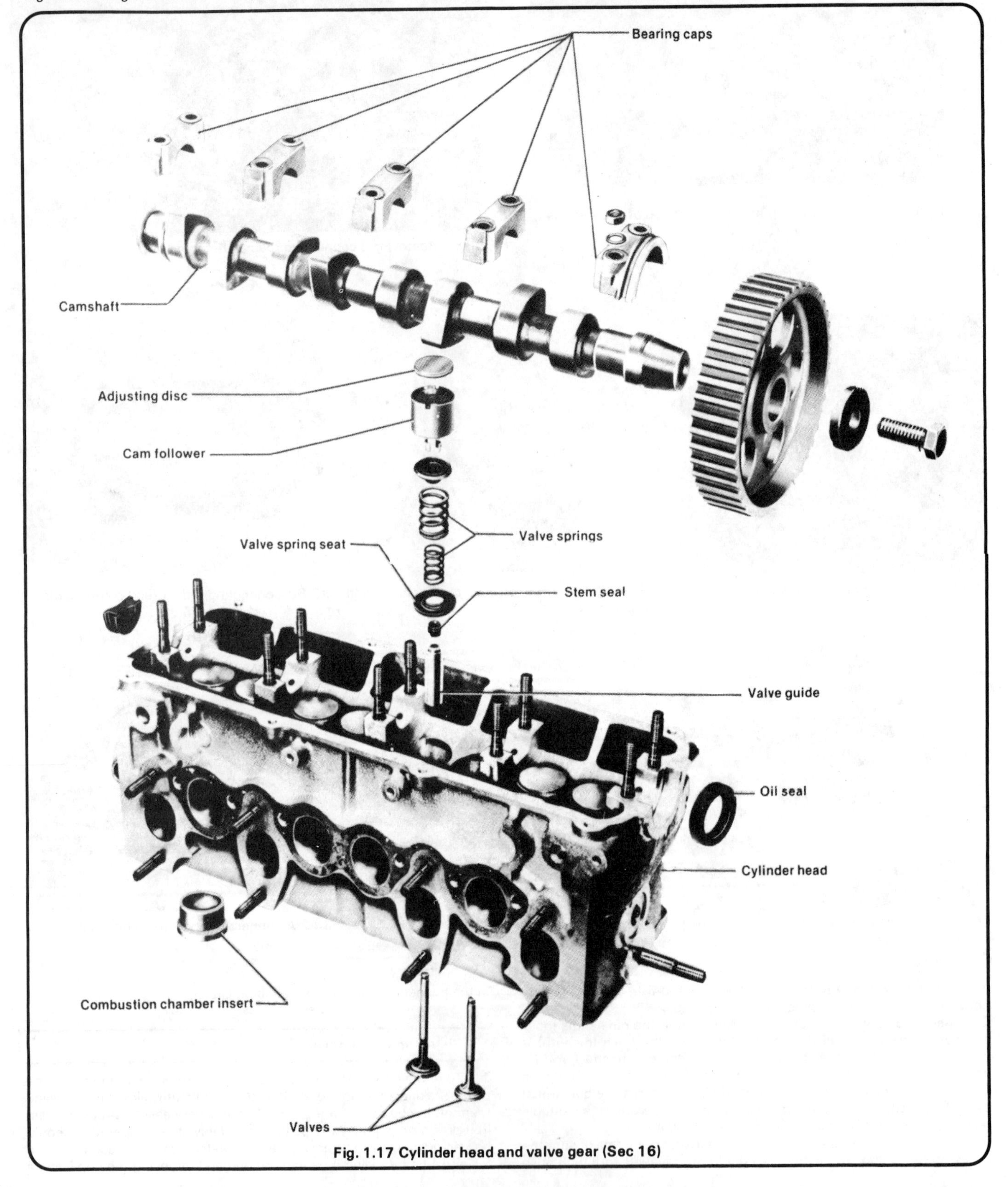

Fig. 1.17 Cylinder head and valve gear (Sec 16)

enough to free the collets, then lift out the collets (photo). When compressing the springs ensure that the valve spring retainer does not damage the valve stem.

3 Release the spring compressor and lift out the valve spring retainer, the springs and the valve spring seat (photos).

4 Remove the valve stem oil seal using needle nosed pliers and withdraw the valve from the underside of the cylinder head (photo). Repeat for the other seven valves.

5 Keep the valves, springs, retainer, valve spring seat and collets in the right order by using a piece of cardboard with holes numbered 1 to 8 (No 1 being the valve nearest the drivebelt end). Alternatively use an internally divided box.

6 Refer to Section 39 for the refitting of the valves.

16.2 Compressing a valve spring to expose the split collets

17 Lubrication system – description

1 The pressed steel oil sump, attached to the underside of the crankcase, acts as a reservoir for the engine oil. The gear type oil pump drains oil through a strainer located under the oil level in the sump, and delivers it under pressure along a short passage and into the full flow oil filter. The oil pressure is regulated by means of a pressure relief valve located within the oil pump. The valve is staked in position and cannot be serviced.

2 From the filter the oil flows to the oil pressure switch, screwed into the rear of the cylinder head, and then to the main oil gallery. From the

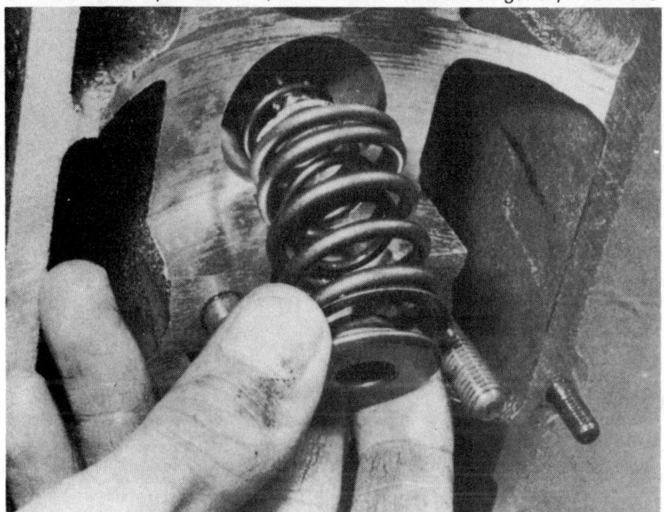
16.3A Removing a retainer and valve springs

16.3B Valve spring seat location

Fig. 1.18 Removing the valve stem seals (Sec 16)

16.4 Removing an inlet valve

main oil gallery, oil is fed to each of the five main bearings and through drillings in the engine block and cylinder head to the camshaft bearings.

3 Drillings in the crankshaft enable oil from the main bearings to be fed to the connecting rod bearings. From there, oil is splashed on to the piston pins and cyiinder walls.

18 Positive crankcase ventilation (PCV) system

1 The closed crankcase ventilation system ellminates emission of fumes and vapour from the crankcase to atmosphere by directing the fumes back through the intake system.

2 Blow-by gases pass through the hose from the valve cover to the air cleaner air intake and are then drawn into the combustion chamber.

19 Oil pump – overhaul

1 Refer to Fig. 1.14. The strainer gauze may be removed when the cap is levered off with a screwdriver. Clean the gauze, replace it and refit the cap (photos).

2 Remove the two small bolts and take the cover away from the body.

3 Remove the gears and wash the body and gears in clean paraffin. Dry them and reassemble the gears, lubricating them with clean engine oil. Measure the backlash between the gears with a feeler gauge (Fig. 1.19). This should be 0.05 to 0.20 mm (0.002 to 0.008 in).

4 Now place a straight-edge over the pump body along the line joining the centre of the two gears and measure with a feeler gauge the axial clearance between the gears and the straight-edge (Fig. 1.20). This must not be more than 0.15 mm (0.006 in).

5 If all is well, check that the shaft is not slack in its bearings, and reassemble the pump for fitting to the engine.

6 If there is any doubt about the pump it is recommended strongly that a replacement be obtained. Once wear starts in a pump it progresses rapidly. In view of the damage that may follow a loss of oil pressure, skimping the oil pump repair is a false economy.

20 Engine components – examination for wear

When the engine has been stripped down and all parts properly cleaned, decisions have to be made as to what needs renewal and the following sections tell the examiner what to look for. In any borderline case it is always best to decide in favour of a new part. Even if a part may still be serviceable its life will have been reduced by wear and the

19.1A The oil pump strainer

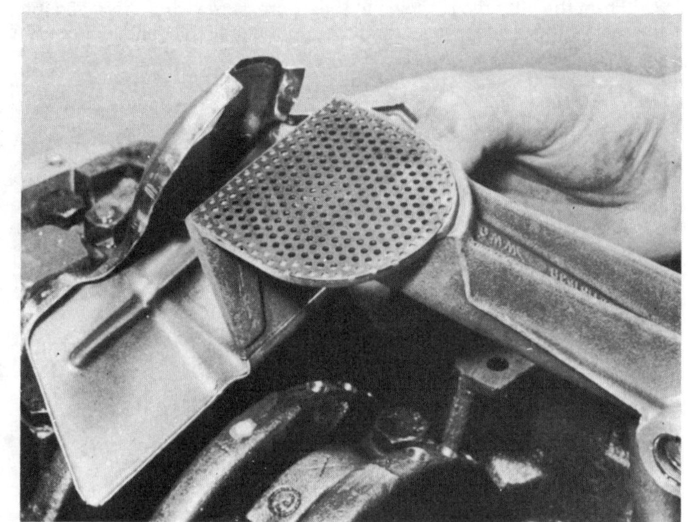

19.1B Removing the oil pump strainer cap

Fig. 1.19 Checking the backlash between the pump gears (Sec 19)

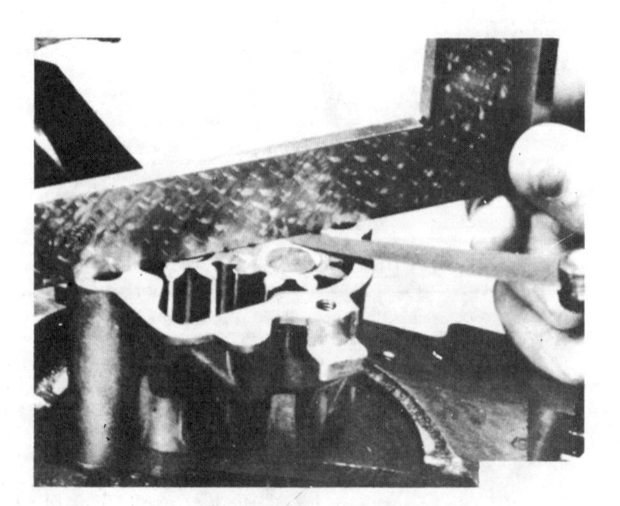

Fig. 1.20 Using a straight-edge and feeler gauge to measure the axial clearance (Sec 19)

degree of trouble needed to replace it in the future must be taken into consideration. However, these things are relative and it depends on whether a quick 'survival' job is being done or whether the car as a whole is being regarded as having many thousands of miles of useful and economical life remaining.

21 Crankshaft – examination and renovation

1 Look at the main bearing journals and the crankpins and if there are any scratches or score marks then the shaft will need regrinding. Such conditions will nearly always be accompanied by similar deterioration in the matching bearing shells.
2 Each bearing journal should also be round and can be checked with a micrometer or caliper gauge around the periphery at several points. If there is more than 0.0254 mm (0.001 in) of ovality regrinding is necessary.
3 A main VW agent or motor engineering specialist will be able to decide to what extent regrinding is necessary and also supply the special undersize shell bearing to match whatever may need grinding off.
4 Before taking the crankshaft for regrinding check also the cylinder bore and pistons as it may be advantageous to have the whole engine done at the same time.

22 Main and big-end bearings – examination and renovation

1 With careful servicing and regular oil filter changes bearings will last for a very long time but they can still fail for unforeseen reasons. With big-end bearings an indication is a regular rhythmic loud knocking from the crankcase. The frequency depends on engine speed and is particularly noticeable when the engine is under load. This symptom is accompanied by a fall in oil pressure although this is not normally noticeable unless an oil pressure gauge is fitted. Main bearing failure is usually indicated by serious vibration, particularly at higher engine revolutions, accompanied by a more significant drop in oil pressure and a 'rumbling' noise.
2 Big-end bearings can be removed with the engine still in the car. If the failure is sudden and the engine has a low mileage since new or overhaul this is possibly worth doing. Bearing shells in good condition have bearing surfaces with a smooth, even matt silver/grey colour all over. Worn bearings will show patches of a different colour when the bearing metal has worn away and exposed the underlay. Damaged bearings will be pitted or scored. It is always worthwhile fitting new shells as their cost is relatively low. If the crankshaft is in good condition, it is merely a question of obtaining another set of standard size shells. A reground crankshaft will need new bearing shells as a matter of course.

23 Cylinder bores – examination and renovation

1 A new cylinder is perfectly round and the walls parallel throughout its length. The action of the piston tends to wear the walls at right angles to the gudgeon pin due to side thrust. This wear takes place principally on that section of the cylinder swept by the piston rings.
2 It is possible to get an indication of bore wear by removing the cylinder head with the engine still in the car. With the piston down in the bore first signs of wear can be seen and felt just below the top of the bore where the top piston ring reaches and there will be a noticeable lip. If there is no lip it is fairly reasonable to expect that bore wear is not severe and any lack of compression or excessive oil consumption is due to worn or broken piston rings.
3 If it is possible to obtain a bore measuring micrometer, measure the bore diameter. Refer to Fig. 1.21 and measure at points 1, 2 and 3, first in direction A and then in direction B. If the out-of-round exceeds the maximum specified then a rebore is required.
4 Any bore which is significantly scratched or scored will need reboring. This symptom usually indicates that the piston or rings are damaged also. In the event of only one cylinder being in need of reboring it will still be necessary for all four to be bored and fitted with new oversize pistons and rings. Your VW agent or local motor engineering specialist will be able to rebore and obtain the necessary matched pistons. If the crankshaft is undergoing regrinding also, it is a good idea to let the same firm renovate and reassemble the crankshaft and pistons to the block. A reputable firm normally gives a guarantee for such work.

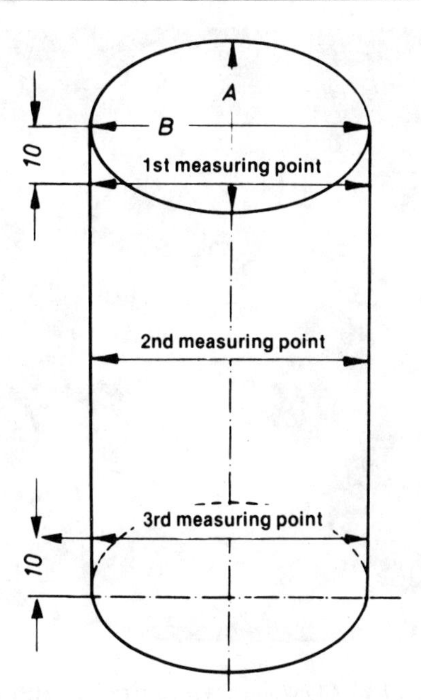

Fig. 1.21 Checking the cylinder bore for wear (Sec 23)

24 Pistons and piston rings – inspection

1 Worn pistons and rings can usually be diagnosed when the symptoms of excessive oil consumption and lower compression occur and are sometimes, though not always, associated with worn cylinder bores. Compression testers that fit into the fuel injector hole are available and these can indicate where low compression is occurring. Wear usually accelerates the more it is left so when the symptoms occur, early action can possibly save the expense of a rebore.
2 Another symptom of piston wear is piston slap – a knocking noise from the crankcase not to be confused with big-end bearing failure. It can be heard clearly at low engine speed when there is no load (idling for example) and much less audible when the engine speed increases. Piston wear usually occurs in the skirt or lower end of the piston and is indicated by vertical streaks in the worn area which is always on the thrust side. It can also be seen where the skirt thickness is different.
3 Piston ring wear can be checked by first removing the rings from the pistons as described in Section 15. Then place the rings in the cylinder bores from the top, pushing them down about 40 mm ($1\frac{1}{2}$ inches) with the head of a piston (from which the rings have been removed) so that they rest square in the cylinder. Then measure the gap at the ends of the ring with a feeler gauge (Fig. 1.22). If it exceeds that given in the Specifications at the beginning of this Chapter then new rings are required.
4 The grooves in which the rings locate in the pi ton can also become enlarged through wear. The clearance between the ring and the piston, in the groove, should not exceed that given in the Specifications (Fig. 1.23).
5 It is seldom that a piston is only worn in the ring grooves and the need to renew them for this fault alone is hardly ever encountered.

25 Connecting rods – examination

1 Connecting rods are not subject to wear but in extreme circumstances such as engine seizure they could be distorted. Such conditions may be visually apparent but where doubt exists they should be changed. The bearing caps should also be examined for indications of filing down which may have been attempted in the mistaken idea that bearing slackness could be remedied in this way. If there are such signs then the connecting rods should be renewed.
2 The connecting rods are supplied in sets of four: therefore if one rod is defective they must all be renewed.

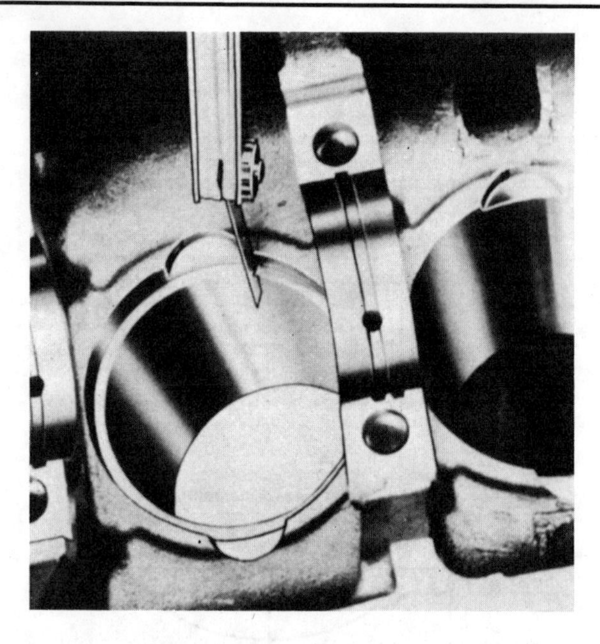

Fig. 1.22 Checking the piston ring gap (Sec 24)

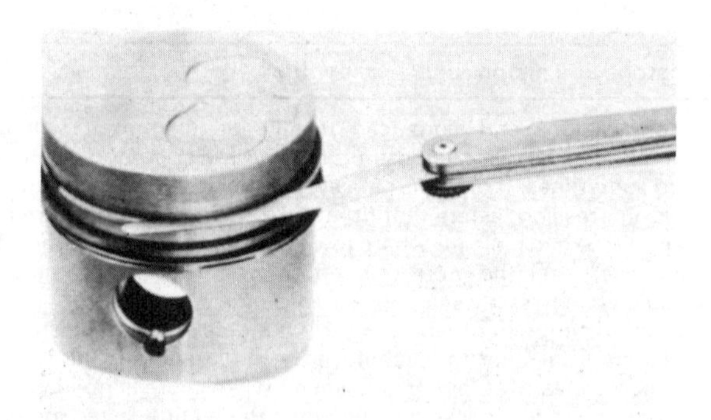

Fig. 1.23 Checking the piston ring clearance in the piston ring grooves (Sec 24)

Fig. 1.24 Checking the camshaft endplay with tool VW387 (Sec 26)

3 If the gudgeon pin bush (small-end bearing) is worn it can be renewed, refer to Section 14.

26 Camshaft and bearings – examination

1 Examine the camshaft bearing journals and cam lobes for wear or scoring. The main decision to take is what degree of wear justifies replacement, which is costly. Scoring or damage to the bearing journals cannot be removed by grinding. Renewal of the camshaft is the only solution.
2 The cam lobes may show signs of ridging or pitting on the high points. If ridging is light then it may be possible to smooth it out with a fine oilstone. The lobes, however, are surface hardened and once this is penetrated wear will take place very rapidly.
3 The camshaft bearings form part of the cylinder head and in the event of wear occurring on these, the complete cylinder head will have to be renewed.
4 Fit the camshaft in the cylinder head without the valves and check that the endplay does not exceed that given in the Specifications. Fig. 1.24 shows the camshaft endplay being checked using a dial gauge.

27 Intermediate shaft – examination and renovation

1 Check the fit of the intermediate shaft in its bearings. If there is excessive play the shaft must be compared with a new one. If the shaft is in good order and the bearings in the block are worn this job is beyond your scope, you may even need a new block so seek expert advice.
2 Check the teeth of the brake booster vacuum pump drivegear for scuffing or chipping. Check the condition of the drivebelt sprocket.
3 There is an oil seal in the intermediate shaft retaining flange. This should be renewed if there are any signs of leakage. To do this, remove the drivebelt pulley and withdraw the flange from the shaft. The oil seal can now be prised out and a new one pressed in. Note that the intermediate shaft rotates anti-clockwise, and the seal must be fitted in accordance with the arrow marked on it (photo).
4 Always fit a new O-ring on the retaining flange before assembling it to the cylinder block.

28 Flywheel – examination

Inspect the friction surface of the flywheel for cracks and scoring. Inspect the ring gear for worn or chipped teeth. If defective the complete flywheel assembly should be renewed. Reconditioning of the flywheel is not recommended due to the balancing which is carried out during manufacture. If a new flywheel is fitted, the TDC mark will be put on it during manufacture.

27.3 Intermediate shaft retaining flange. Arrow below left-hand paint spot indicates direction of shaft rotation

29 Crankshaft oil seals – removal and refitting

1 The rear (flywheel end) oil seal is located in an aluminium housing. Remove the old seal, clean the housing and press in the new seal (photo). Ease it in carefully and press it fully home using either a flat plate in a large vice or a mandrel press. If the seal housing is fitted to the crankcase, make a suitable disc and using two clutch pressure plate to crankshaft bolts, press the seal into position. Fig. 1.25 shows VW tool 2003/1 being used to fit the seal. **Do not** hammer the seal into position: if the lip of the seal is damaged, oil will seep through and find its way on to the clutch friction disc.

2 The front oil seal (sprocket end) is also located in an aluminium housing. Prise out the old seal and press in a new one (photo). Press it in to a depth of $\frac{3}{32}$ in (2 mm) below the face of the housing. This seal can be renewed with the engine in the car provided you have the special tools required: tool No 10–219 to extract the old seal and tool No 10–203 to fit the new one. Those tools are also required if the intermediate shaft oil seal is to be renewed with the engine in the car.

30 Cylinder head and piston crowns – decarbonisation

Note: *When the pistons are at TDC they project above the surface of the cylinder block. Take care when cleaning the top of the block not to damage the pistons.*

1 Always remove the injectors and glow plugs before working on the cylinder head. When decarbonising the cylinder head, insert a drift through the fuel injector holes and drive out the pre-combustion chamber inserts as shown in Fig. 1.26.

2 When the cylinder head is removed, either in the course of an overhaul or for inspection of bores or valve condition when the engine is in the car, it is normal to remove all carbon deposits from the piston crowns and heads.

3 This is best done with a cup shaped wire brush and an electric drill and is fairly straightforward when the engine is dismantled and the pistons removed. Sometimes hard spots of carbon are not easily removed except by a scraper. When cleaning the pistons with a scraper, take care not to damage the surface of the piston in any way.

4 When the engine is in the car certain precautions must be taken when decarbonising the piston crowns in order to prevent dislodged pieces of carbon falling into the interior of the engine which could cause damage to cylinder bores, piston and rings – or if allowed into the water passages – damage to the water pump. Turn the engine so that the piston being worked on is at the top of its stroke and then mask off the adjacent cylinder bores and all surrounding water jacket orifices with paper and adhesive tape. Press grease into the gap all round the piston to keep carbon particles out and then scrape all carbon away by hand carefully. Do not use a power drill and wire brush when the engine is in the car as it will be virtually impossible to keep

29.1 Engine rear oil seal and housing

Fig. 1.25 Fitting a new rear seal in its housing (Sec 29)

29.2 Engine front oil seal and housing

Fig. 1.26 Using a drift to drive out the pre-combustion chamber inserts (Sec 30)

all the carbon dust clear of the engine. When completed, carefully clear out the grease around the rim of the piston with a matchstick or something similar – bringing any carbon particles with it. Repeat the process on the other piston crown. It is not recommended that a ring of carbon is left around the edge of the piston on the theory that it will aid oil consumption. This was valid in the earlier days of long stroke low revving engines but modern engines, fuels and lubricants cause less carbon deposits anyway and any left behind tends merely to cause hot spots.

31 Valves and valve seats – examination and renovation

1 Examine the heads of the valves for pitting and burning, especially the heads of the exhaust valves. The valve seats should be examined at the same time. If the valve seats are so badly worn or burned that they cannot be refaced then the cylinder head will have to be renewed.

2 If the pitting on the valve and valve seat is very slight this can be corrected by grinding the valves and seats together with coarse, and then fine, grinding paste. Where bad pitting of the seats has occurred it will be necessary to recut the seats and fit new valves. Recutting of the valve seats should be left to your local VW garage or local engineering works. Normally it is the valves that are too badly worn for refacing, and the owner can purchase a new set of valves and match them to the seats by hand grinding. Inlet valves can be refaced on a valve grinding machine. Exhaust valves **must not** be machine ground. Grind by hand only against the valve seat.

3 Valve grinding is carried out as follows: Smear a trace of coarse carborundum paste on the seat face and apply a suction grinder tool to the valve head. With a semi-rotary motion, grind the valve head to its seat, lifting the valve occasionally to redistribute the grinding paste. When a dull matt even surface finish is produced on both the valve seat and the valve, wipe off the paste and repeat the process with fine carborundum paste, lifting and turning the valve to redistribute the paste as before. A light spring placed under the valve head will greatly ease this operation. When a smooth unbroken ring of light grey matt finish is produced, on both valve and valve seat faces, the grinding operation is completed.

4 Scrape away all carbon from the valve head and the valve stem. Carefully clean away every trace of grinding compound, taking great care to leave none in the ports or in the valve guides. Clean the valves and valve seats with a paraffin soaked rag then with a clean rag, and finally, if an air line is available, blow the valves, valve guides and valve ports clean.

Fig. 1.27 Grind in the valves with a semi-rotary motion (Sec 31)

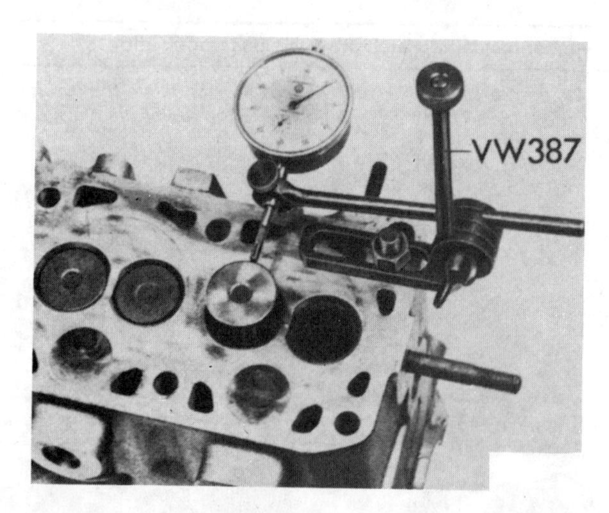

Fig. 1.28 Checking a valve guide for wear with tool VW387 (Sec 32)

32 Valve guides and springs – examination and renovation

1 Wear in the valve guides may be checked by fitting a new valve in the guide and observing the amount that the head of the valve will move sideways when the end of the valve stem is flush with the top of the valve guide. The maximum permitted movement for intake and exhaust valves is 1.30 mm (0.051 in) when a new valve is used. Fig. 1.28 shows a dial gauge being used to measure the side movement of the valve.

2 If the side movement exceeds the permitted maximum, new valve guides will be required. This is a job for your VW garage or local engineering works, as the guides have to be fitted using a press and then reamed to the specified dimension.

3 Examine the valve springs for distortion and damage. If possible compare the free length against a new spring. If it is shorter it must be renewed.

33 Cylinder head – examination

1 Examine the cylinder head for cracks or other damage. The machined face of the head should be smooth and free of pitting and burrs.

2 Check the surface of the head for distortion. Place a straight-edge along the centre of the machined face of the head. Make sure there are no ridges at the extreme ends and using a feeler gauge check for distortion. If the straight-edge is held firmly on the head and feelers in excess of 0.1 mm (0.004 in) can be inserted between the straight-edge and the head, a new cylinder head is required.

Fig. 1.29 Using a straight-edge and feeler gauge to check the cylinder head for distortion (Sec 33)

34 Engine reassembly – general

1 To ensure maximum life with minimum trouble from a rebuilt engine, not only must everything be correctly assembled, but everything must be spotlessly clean; all the oilways must be clear, locking washers and spring washers must always be fitted where

indicated and all bearing and other working surfaces must be thoroughly lubricated during assembly.

2 Before assembly begins renew any bolts or studs, the threads of which are in any way damaged, and whenever possible use new spring washers.

3 Apart from your normal tools, a supply of clean rag, an oil can filled with engine oil (an empty plastic detergent bottle thoroughly washed out and cleaned, will do just as well), a new supply of assorted spring washers, a set of new gaskets, and a torque spanner should be collected together.

4 Sit down with a pencil and paper and list all those items which you intend to renew and acquire all of them before reassembly. If you have had experience of shopping around for parts you will appreciate that they cannot be obtained quickly. Do not underestimate the cost either. Spare parts are relatively much more expensive now.

35 Crankshaft, oil seal housings and intermediate shaft – refitting

1 Place the cylinder block upside down on the bench.

2 Thoroughly clean the main bearing seatings and fit the main bearing upper shells. Note that the upper shells have lubrication grooves, whereas the shells in the bearing caps are without lubrication grooves. No 3 main bearing shell has small flanges (photo). If the old bearings are being refitted it is essential that they go back in their original locations.

3 Lubricate the bearings and crankshaft journals with engine oil and then lower the crankshaft onto the bearings (photos).

4 Fit the lower half of the bearing shells in the main bearing caps. Fit the caps in their correct locations (No 1 at the sprocket end and No 5 at the flywheel end), then insert the securing bolts and tighten them to the specified torque (photos). NOTE: The caps are fitted with the

35.2 Fitting No 3 main bearing upper shell

35.3A Lubricating the main bearing shells

35.3B Refitting the crankshaft

35.4A No 3 main bearing cap and bearing shell

35.4B Refitting No 3 main bearing cap

35.4C Main bearing caps are numbered for correct location

35.4D Tighten the retaining bolts to the specified torque

35.4E Sprocket end main bearing cap fitted position

35.5 Checking crankshaft endfloat with a feeler gauge

35.6A Fitting the rear oil seal housing gasket ...

35.6B ... and the rear oil seal and housing

35.6C Tightening the rear oil seal housing retaining bolts

35.7 Fitting the front (sprocket end) oil seal and housing

37.2 Ring compressor fitted to a piston

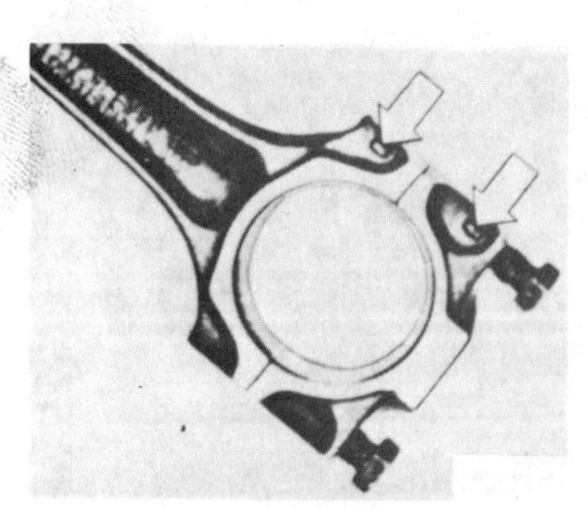

Fig. 1.30 Connecting rod casting mark locations (arrowed) (Sec 37)

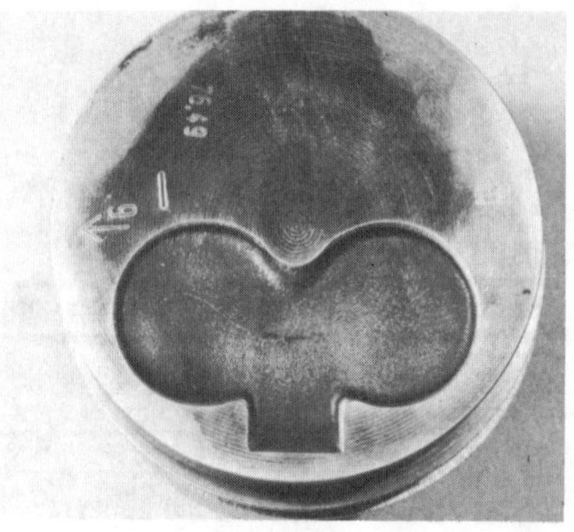

37.3 Piston crown, showing front facing arrow

37.4 Fitting a piston into a cylinder bore

37.5 Connecting rod and cap markings

37.6 Tightening a big-end bearing cap retaining nut

37.8 Checking the connecting rod side clearance with a feeler gauge

numbers on the side opposite to the oil pump location.
5 Tap the crankshaft to the rear as far as it will go and check the clearance between the flange of No 3 bearing and the crankshaft with a feeler gauge to measure the endplay (photo). The endplay should not exceed that specified; if it does then the bearings must be renewed. Check that the crankshaft is free to rotate, if it is not, then the bearings and journals must be rechecked.
6 Lubricate the rear end of the crankshaft. Using a new gasket, fit the rear oil seal housing complete with oil seal. Fit the six securing bolts (photos).
7 Lubricate the front end of the crankshaft and position a new gasket on the engine block. Fit the front oil seal housing, complete with oil seal, and securing bolts (photo).
8 Fit the intermediate shaft in the cylinder block. When inserting the shaft take care not to damage the bearings.
9 Fit the intermediate shaft retaining flange complete with oil seal. Use a new O-ring between the flange and the block. Fit the two securing bolts and tighten them to the specified torque.

36 Piston rings – refitting

1 Ensure that the piston ring grooves are clean.
2 Fit the rings on the pistons by reversing the removal procedure, refer to Section 15.
3 All three rings are marked TOP and must be fitted with the TOP mark towards the piston crown. Start with the oil scraper ring first and then the compression rings.
4 When fitted in their grooves, generously lubricate the rings and piston with engine oil and position the gaps at 120° intervals.

37 Piston and connecting rod assemblies – refitting

1 Position the engine block on its side.
2 Fit a piston ring compressor on the first piston (photo). Lubricate the crankshaft connecting rod journal with engine oil.
3 Insert the piston and connecting rod assembly in to the cylinder bore. Ensure that it is the correct assembly for that particular bore and that the arrow on the piston crown is pointing towards the drivebelt end with the casting marks on the connecting rod (see Fig. 1.30) facing towards the intermediate shaft (photo).
4 Using the shaft of a hammer, tap the piston into the cylinder bore, collecting the ring compressor as the rings go into the bore (photo).
5 Pull the connecting rod onto the crankshaft journal and fit the cap, making sure it is the right way round (photo).
6 Fit the connecting rod cap securing nuts and tighten them to the specified torque (photo). Check that the crankshaft is free to rotate and that there are no high spots causing binding.
7 Repeat this procedure for the other three piston assemblies, checking that the crankshaft is free to rotate after the fitting of each assembly.
8 After all the assemblies are fitted, check the connecting rod side clearance. Push the connecting rod against the crankshaft web and measure the gap at the other side with a feeler gauge (photo). The clearance should not exceed the specified value. If it does then the bearings must be renewed.
9 Turn the crankshaft and position No 1 piston at TDC (top dead centre).

38 Oil pump, sump and crankshaft sprocket – refitting

1 Clean the mating faces of the engine block and oil pump. Position the oil pump, complete with strainer, on the block and fit the two securing bolts. Tighten the securing bolts to the specified torque (photo).
2 Ensure that the mating faces of the engine block and the oil sump are clean and free of burrs, then position a new sump gasket on the engine block using a sealant compound to keep it it in place.
3 Apply sealant to the joint face of the sump, then assemble the sump to the block and fit the securing bolts. Tighten the bolts in a progressive and even manner (photo).
4 Fit the drivebelt sprocket on the front end of the crankshaft. Ensure that it is driven on squarely so that it does not displace or damage the Woodruff key on the crankshaft. Make sure that it is driven fully home, using a piece of suitable diameter steel tubing.

39 Valves, pre-combustion chamber inserts, fuel injectors and glow plugs – refitting

1 With the valves having been ground in (Section 31) and kept in their correct order, assemble the valves and springs in the cylinder head.
2 Fit the valve stem oil seal on the valve guide. With a packet of new oil seals is a small plastic sleeve. This is fitted over the valve stem and lubricated and then the seal should be pushed on over the plastic sleeve until it seats on the guide. This must be done with a special tool VW 10–204 which fits snugly round the outside of the seal and pushes it on squarely. If the seal is assembled without the plastic sleeve the seal will be damaged and oil consumption will become excessive. If you cannot put them on properly then ask the agent to do it for you.
3 Insert the valve in the head and fit the valve spring seat followed by the two valve springs and the valve spring retainer.
4 Using a valve spring compressor, compress the valve springs until the split collets can be slid into position. When the collets are located in the groove in the valve stem (photo), release the compressor. Using a plastic hammer tap the valve stem to check that the collets are properly seated.
5 Fit the other seven valves using the same procedure.
6 Fit the pre-combustion chamber inserts in the cylinder head.
7 Fit the fuel injectors and the glow plugs.

40 Tappets and camshaft – refitting

1 Insert the tappets in their original positions in the cylinder head. Ensure that the engraved (numbered) side of the tappet discs are facing downwards (invisible) (photo).
2 Lubricate the camshaft bearings with engine oil.

38.1 Refitting the oil pump

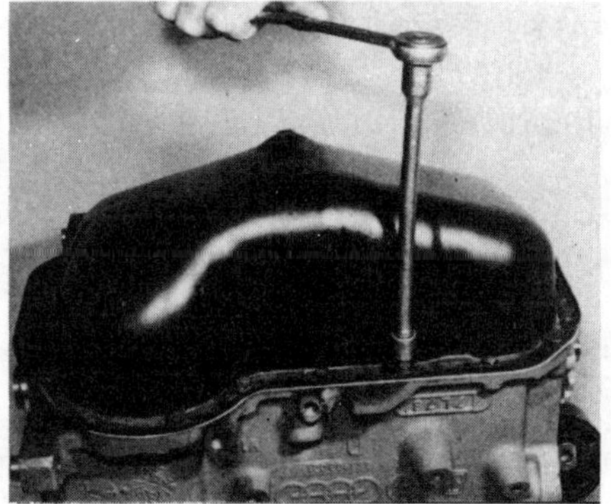
38.3 Tightening the sump retaining bolts

39.4 Fitted position of the valve split collets

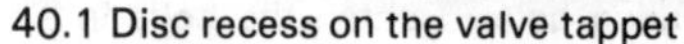

40.1 Disc recess on the valve tappet

40.3A Fitting the camshaft front oil seal

40.3B Refitting the camshaft

3 Slide a new camshaft oil seal into position on the front (drivebelt end) of the camshaft and fit the camshaft on the cylinder head (photos).
4 Fit the bearing caps and tighten them down evenly and progressively to the specified torque (photo). Before tightening the securing nuts ensure that the caps are the right way round as the bearings are off centre. Note that the numbers stamped on the caps are not always on the same side.
5 Fit the drivebelt sprocket on the camshaft and tighten the securing bolt (photos).
6 The cylinder head is now ready for fitting on the engine block, except that the valve clearances have not been adjusted. Valve clearance adjustment is described in the next Section.

41 Valve clearances – adjustment

1 Valve clearance adjustment consists of measuring the clearance between the heel of the cam and the steel disc in the top of each tappet, calculating any error and replacing the disc if necessary with one of the thickness required to obtain the specified clearance.
2 The 26 different thicknesses of tappet disc available range from 3.00 mm to 4.25 mm in steps of 0.05 mm, which means that the purchase of a set of feeler gauges graduated in millimetres would be an advantage. Otherwise a lot of calculation will be necessary and a set of conversion tables to transpose 0.05 mm steps into thousandths of an inch. Referring to the table in paragraph 7, the thinnest disc is 3.00 mm (0.1181 in). The thickness progresses at the rate of 0.05 mm (0.001969 in) and since feeler gauges measure only to the nearest thousandth, 0.001969 in may be taken for practical purposes as 0.002 in. By calculating 25 size increases at this rate the thickest disc, 4.25 mm, becomes 0.1681 in whereas its true value is 0.1673 in. The error over the total range is thus 0.0008 in which is acceptable.
3 If routine checking and adjustment of the valve clearances is to be done with the engine in the car, first remove the air cleaner and duct and the valve cover. The engine can be turned over by using a spanner on the crankshaft sprocket bolt or by engaging top gear and pushing the car as required. The procedure for checking the clearance after overhaul is the same, except that during an engine overhaul the adjustment discs will already have been removed and cleaned and the thickness etched on the back of the disc will be known.
4 If the cylinder head is on the bench the camshaft can be turned by the sprocket.
5 Using a feeler gauge measure the clearance between the tappet adjusting disc and the heel of the cam (photo). Make a note of the clearance. Counting from the timing belt end of the engine, the valves are numbered:

Inlet	*2 – 4 – 5 – 7*
Exhaust	*1 – 3 – 6 – 8*

6 Repeat this procedure for all the valves in turn, rotating the camshaft as necessary, and then compare the measurements with the clearances given in the Specifications at the beginning of this Chapter. Clearances that are set cold must be rechecked with the engine coolant temperature at approximately 35° C (95° F) and, if necessary, re-adjusted.

7 Make a table of the actual clearances and then calculate the error from the specified clearance as shown in the following example:

	Inlet valve	*Exhaust valve*
Specified clearance	0.20–0.30 mm (0.008–0.012 in)	0.40–0.50 mm (0.016–0.020 in)
Measured clearance	0.15 mm (0.006 in)	0.55 mm (0.022 in)
Insert disc that is	0.10 mm (0.004 in) thinner	0.10 mm (0.004 in) thicker

Discs are available in the following thicknesses:

Thickness (mm)	Part Number	Thickness (mm)	Part Number
3.00	*056 109 555*	*3.65*	*056 109 568*
3.05	*056 109 556*	*3.70*	*056 109 569*
3.10	*056 109 557*	*3.75*	*056 109 570*
3.15	*056 109 558*	*3.80*	*056 109 571*
3.20	*056 109 559*	*3.85*	*056 109 572*
3.25	*056 109 560*	*3.90*	*056 109 573*
3.30	*056 109 561*	*3.95*	*056 109 574*
3.35	*056 109 562*	*4.00*	*056 109 575*
3.40	*056 109 563*	*4.05*	*056 109 576*
3.45	*056 109 564*	*4.10*	*056 109 577*
3.50	*056 109 565*	*4.15*	*056 109 578*
3.55	*056 109 566*	*4.20*	*056 109 579*
3.60	*056 109 567*	*4.25*	*056 109 580*

8 As the discs are in steps of 0.05 mm, the required disc can be obtained once the thickness of the disc at present fitted is known. If you have dismantled the cylinder head you will know the thickness etched on the back of the disc (provided you remembered to record it), but if you do not then the disc will have to be removed to find out. If you have VW tools 10–209 and 10–208, the job of removing the adjusting disc from the top of the tappet is very straightforward, but we managed using the tool shown in the photograph, a small screwdriver and thin long-nosed pliers. With the camshaft positioned to give the maximum clearance (cam lobe pointing upwards) the tappet is pushed down with tool 10–209, or its counterpart, against the valve spring while the disc is removed with tool 10–208, or prised up with a small screwdriver and lifted out with the pliers. When fitting the disc always ensure that the side with the thickness marked on it is towards the tappet (photos).
9 If the adjustment is made with a cold engine it must be checked again when the engine is hot, and if the cylinder head has been overhauled it must be checked again after 1000 miles (1600 km).
10 If the adjustment was carried out with the engine in the car refit the valve cover and air cleaner.
11 It is suggested that you keep a record of the disc sizes fitted as this can save time at the next check by allowing you to calculate the thickness of disc required without having to remove the old disc.

42 Cylinder head, drivebelt tensioner, intermediate shaft pulley, fuel injection pump and drivebelt – refitting

1 Support the engine on the sump with wooden blocks. Clean the top face of the block.

40.4 Camshaft front bearing cap refitted

40.5A The camshaft drivebelt sprocket

40.5B Fitting the camshaft drivebelt sprocket

41.5 Checking the valve clearances with a feeler gauge

41.8A Tappet compressing tool

41.8B Using the tappet compressing tool

41.8C Compress the tappet ...

41.8D ... prise out the disc ...

41.8E ... and remove the disc with long nosed pliers

2 Four thicknesses of cylinder head gasket are used according to how much the pistons protrude (at TDC) above the top face of the cylinder block. They are as follows:

Piston projection in mm (in)	Gasket thickness in mm (in)	Identification (no of notches)	Part No
0.43–0.63 (0.017–0.025)	*1.3 (0.051)*	*2*	*068 103 383*
0.63–0.82 (0.025–0.032)	*1.4 (0.055)*	*3*	*068 103 383C*
0.82–0.92 (0.032–0.036)	*1.5 (0.059)*	*4*	*068 103 383G*
0.92–1.02 (0.036–0.040)	*1.6 (0.063)*	*5*	*068 103 383H*

3 If the same cylinder block and pistons are being refitted ensure that the new gasket is the same thickness as the one that was removed. The identification marks on the gasket are shown arrowed in Fig. 1.13. If either the pistons or the cylinder block is being renewed then the amount that the pistons protrude above the face of the block must be measured to determine the thickness of gasket required. VW tools 382/7 and 385/17 are specified for measuring piston projection; alternatively, use a straight-edge and feeler blades as shown (photo).

4 Position the new cylinder head gasket on the cylinder block with the side marked OBEN facing upwards (photos).

5 Lift the cylinder head on to the block and insert bolts 8 and 10 (see Fig. 1.12) first to centre the head. We used two $\frac{1}{16}$ in drills in the bolt holes to act as guides when lowering the head onto the cylinder block (photos).

6 Fit the other cylinder head bolts and tighten them in the sequence shown in Fig. 1.12 to the specified Stage 1 torque (photo). Then tighten the bolts further, in the same sequence, to the specified Stage 2 torque. **Note:** *Do not re-tighten the bolts after 1000 miles (1600 km).*

7 Fit the fuel injection pump mounting plate on the front of the engine block.

8 Fit the drivebelt tensioner on its mounting stud. Do not tighten the

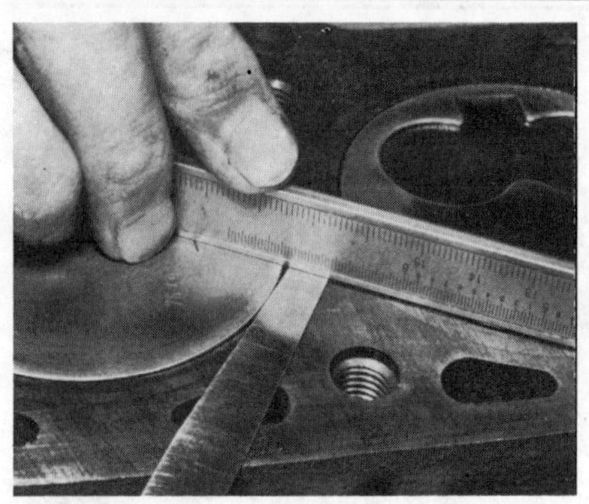
42.3 Checking the piston protrusion with a feeler gauge and straight-edge

42.4A Fitting the cylinder head gasket

42.4B Cylinder head gasket upper facing and identifying marks

42.5A Using two drills for cylinder head location

42.5B Refitting the cylinder head

42.6 Tightening the cylinder head bolts

locknut at this stage.

9 Fit the pulley on the intermediate shaft and tighten the securing bolt to the specified torque (photo).

10 Turn the camshaft so that both cams for No 1 cylinder are in the valves-closed position (both cam lobes pointing upwards) and lock the camshaft in this position, refer to Section 9.

11 Check that No 1 piston is still at TDC (see Section 37). If there is any doubt, use setting tool No 2068 after refitting the clutch (Section 45). It is not possible to judge the piston position through the swirl chambers.

12 Assemble the fuel injection pump to the mounting plate, align the marks on the pump and the mounting plate (Fig. 1.31) and fit the securing bolts.

13 Fit the sprocket on the injection pump shaft and tighten the securing nut to the specified torque.

14 Turn the pump sprocket to line up the mark on the sprocket with the mark on the mounting plate, then lock the injection pump sprockets with a suitable locking pin, refer to Section 9.

15 Slacken the camshaft sprocket securing bolt approximately half a turn and then using a plastic hammer tap the sprocket free of the taper

42.9 Tightening the intermediate shaft pulley retaining bolt

Fig. 1.31 Injection pump alignment marks (Sec 42)

on the camshaft.

16 Check that the engine is still at TDC (compression stroke) on No 1 cylinder and fit the drivebelt and the drivebelt shield (photos).

17 Remove the locking pin from the fuel injection pump sprocket and tension the drivebelt by turning the belt tension adjuster as required.

18 VW tool 210 is required to measure the belt tension. If the tool is not available, refer to Section 9, paragraph 14. Measure the tension at the point midway between the camshaft sprocket and the fuel injection pump sprocket. When the belt is correctly tensioned the scale on tool VW 210 will read 12–13. Tighten the drivebelt tensioner locknut.

19 Tighten the camshaft sprocket securing bolt to the specified torque and remove the camshaft locking tool (photo).

20 Turn the crankshaft through two complete revolutions in the normal direction of rotation, then remove the play in the drive by striking the belt once with a plastic hammer between the pump and camshaft sprockets. Recheck the belt tension and re-adjust if necessary.

21 Check the fuel injection pump timing as described in Chapter 3.

43 Water pump, crankshaft pulley, valve cover and vacuum pump – refitting

1 Fit the water pump and thermostat housing assembly, complete with alternator adjusting strap. Ensure that the mating faces of the housing and the engine block are clean and use a new O-ring seal between the pump housing and the block.

2 Bolt the crankshaft pulley to the crankshaft sprocket. Tighten the four securing bolts to the specified torque.

3 Fit the alternator mounting bracket. Don't forget the earth strap which is connected to one of the bolts.

4 Fit a new valve cover gasket on the cylinder head. Ensure that the rubber seal at the rear end and the seal on the front camshaft bearing cap are properly seated, then fit the valve cover, the strengthening strips and securing bolts.

5 Fit the drivebelt cover (photo).

6 Fit the hose between the water pump housing and the cylinder head.

7 Using a new O-ring, fit the brake servo vacuum pump. Ensure that the drive on the end of the pump shaft engages with the oil pump driveshaft and fit the clamp plate and securing bolt (photos).

8 Using a new gasket fit the heater hose connection, complete with coolant temperature switch, on the rear end of the cylinder head.

44 Oil filter adapter, fuel pipes and oil pressure switch – refitting

1 Using a new gasket, fit the oil filter adapter and tighten the securing bolts to the specified torque (photo). The oil filter cartridge screws into the adapter, and is hand tightened only, but do not fit the filter at this stage as it may get damaged when fitting the engine in the car.

2 Screw the oil pressure sender switch into the rear end of the cylinder head and tighten it to the specified torque.

3 Fit the injection pump-to-fuel injector pipes.

45 Intermediate plate and clutch and flywheel assembly – refitting

1 Fit the intermediate plate on the engine block, it locates on the two transmission locating dowels (photo).

2 The flywheel is bolted to the clutch pressure plate. For information on the fitting of the clutch assembly and the flywheel refer to Chapter 4, Section 4.

46 Inlet and exhaust manifolds and air cleaner – refitting

1 Clean the joint faces of the cylinder head and the intake and exhaust manifolds. Fit a new gasket over the exhaust manifold studs.

2 Fit the intake and exhaust manifolds and tighten the socket headed securing bolts and nuts in an even and progressive manner.

42.16A Refitting the drivebelt shield

42.16B Drivebelt location over camshaft and injection pump sprockets

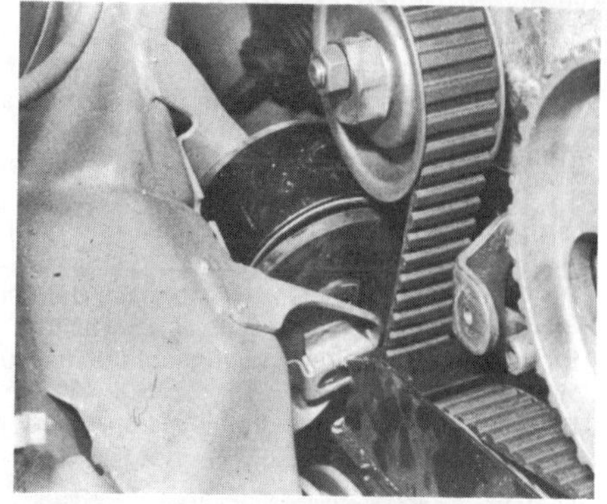

42.16C Drivebelt location over tensioner and intermediate pulley

42.19 Tightening the camshaft sprocket securing bolt

43.5 Refitting the drivebelt cover (engine in car)

43.7A Brake servo vacuum pump (exhauster) drive blade on the oil pump driveshaft (early type oil pump)

43.7B Brake servo vacuum pump drivegear, showing slot for oil pump shaft (early models)

44.1 Refitting the oil filter adaptor

45.1 Cylinder block to transmission intermediate plate retaining bolt locations

3 Fit a new filter element on the air cleaner and fit the air cleaner on the intake manifold. Secure it in position with the four spring retaining clips.

47 Engine – refitting in the car

1 Assemble the transmission to the engine in the reverse order to that described in Section 6 for separating it from the engine. If the clutch has been disturbed, centralise it as described in Chapter 4.
2 Refitting the engine/transmission assembly in the car is the reverse of the removal procedure described in Section 5.
3 When fitting the engine mounting bolts, insert all the bolts loosely and then align the power unit so that there is no twisting or undue stress on the rubber parts of the mountings, refer to Figs. 1.32, 1.33 and 1.34. Tighten the bolts to the specified torque (photo).
4 Reconnect the driveshafts and tighten the flange bolts to the specified torque, refer to Chapter 8.
5 Connect the transmission shift linkage, refer to Chapter 5 for linkage adjustment procedure.
6 Connect the clutch cable and adjust the clutch pedal free movement, refer to Chapter 4.
7 Fit the speedometer cable drive in the transmission. Connect the exhaust pipe tro the exhaust manifold flange.
8 Fit the alternator (don't forget the earth strap), refer to Chapter 7.
9 Smear a new oil filter seal with engine oil and screw the filter cartridge into the oil filter adapter. Tighten the filter by hand, do not use a strap wrench (photo).
10 Connect the accelerator cable and the cold start cable to the fuel injection pump. Adjust the cables as described in Chapter 3.
11 Refit the fuel filter and connect the fuel supply and return pipes to the fuel injection pump.
12 Refit the radiator and fan assembly, refer to Chapter 2, and the air cleaner intake duct. Fit the heater hoses.
13 Reconnect the electrical wiring to the alternator, starter motor, fan motor and fan thermoswitch, oil pressure switch, coolant temperature switch, stop solenoid on the injection pump, glow plugs and the reversing light switch.
14 Fit the vacuum pump-to-brake servo hose.
15 Refill the cooling system with coolant, refer to Chapter 2.

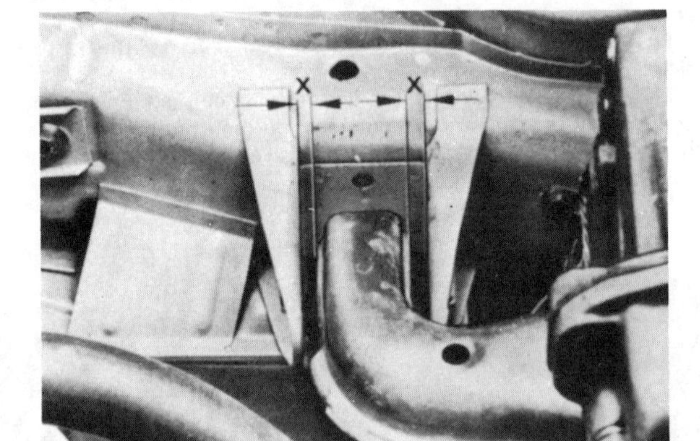

Fig. 1.33 Correct alignment of centre engine mounting (Sec 47)

Fig. 1.32 Correct alignment of rear transmission mounting (Sec 47)

Fig. 1.34 Correct alignment of front engine mounting (Sec 47)

47.3 Engine front mounting

47.9 Oil filter cartridge

16 Refill the sump with engine oil to the correct level. If the transmission was drained, don't forget to refill it with the specified grade of lubricant.
17 Fit the engine compartment bonnet and the battery. Reconnect the battery leads.
18 Finally, check that all electrical wiring and control cables have been connected, all hose clips tightened, and all tools, rags and any other loose items removed from the engine compartment.

48 Engine – initial start-up after overhaul or major repair

1 Make sure that the battery is fully charged and that all lubricants and coolants have been replenished.
2 Bleed the fuel system as described in Chapter 3, then using the glow plugs start the engine and run it to bring it up to normal operating temperature.
3 As the engine warms up there will be odd smells and some smoke from the parts getting warm, particularly the exhaust manifold, and burning off oil deposits. Check for leaks of coolant, oil and fuel which, if serious, will be very obvious. Check also the exhaust pipe and manifold connections as these do not always find their gastight position until the heat and vibration have acted on them and it is almost certain that they will need further tightening. This should be done, of course, with the engine stopped.
4 When normal operating temperature is reached, adjust the idle speed and the maximum rpm as described in Chapter 3.
5 Stop the engine and wait a few minutes, then check to see if any coolant, oil or fuel is leaking while the engine is stopped.
6 Road test the car to check that the engine is giving the necessary power and smoothness. Do not race the engine: if new bearings and/or pistons have been fitted it should be treated as a new engine and run in at a reduced speed, not more than 62 mph (100 km/h) for the first 300 miles (500 km) and not more than 75 mph (120 km/h) for the next 300 miles (500 km). It is not necessary to regulate the speed in the different gears as the engine is automatically governed to approximately 5400 rpm. After 600 miles the speed can be gradually increased to the maximum.

See overleaf for 'Fault diagnosis – engine'

49 Fault diagnosis – engine

Symptom	Reason/s
Engine will not turn over when starter switch is operated	Flat battery Bad battery connections Bad connections at solenoid switch and/or starter motor Starter motor jammed Starter motor defective
Engine turns over but fails to start	Glow plugs or associated wiring defective Injection pump solenoid defective, or mountings or connections loose No fuel reaching the injectors, air in system Injection pump timing incorrect Defective injectors Fuel injection pump defective
Engine starts but runs unevenly and misfires	Air in the fuel system Restricted fuel supply due to blocked filter or fuel lines Defective injectors Idle speed incorrectly adjusted Injection pump timing incorrect Injection pump defective
Excessive exhaust smoke – black or white	Injectors defective Injection pump timing incorrect Maximum rpm adjustment incorrect Injection pump defective
Lack of power	Maximum rpm adjustment incorrect Air in the fuel system Injectors defective Injection pump timing incorrect Injection pump defective Blocked air cleaner filter element Valve clearances incorrect Burnt out valves Worn piston rings and/or cylinder bores
Excessive oil consumption	Oil leaks from gaskets and seals Worn piston rings and/or cylinder bores resulting in oil being burnt in the engine, indicated by excessive blue exhaust smoke Worn valve guides and/or defective valve stem seals
Excessive mechanical noise from the engine	Valve clearances incorrect Worn main and/or big-end bearings Worn cylinders (piston slap)

Chapter 2 Cooling, air conditioning and heating systems

Contents

Specifications

System type	Thermo-syphon, assisted by belt-driven pump, pressurised	
Cooling fan	Electrically driven, thermostatically controlled	
Pressure cap release pressure	1·2 to 1·35 kg/cm^2 (17 to 19 lbf/in^2)	
Thermostat		
Starts to open	80°C (176°F) approx	
Fully extended	94°C (200°F) approx	
Minimum stroke	7·0 mm (0·27 in)	
System capacity (including heater)	6·5 litres (11·4 Imp pints, 13·6 US pints)	
Torque wrench settings	**kgf m**	**lbf ft**
Temperature gauge sender	0·7	5
Top hose outlet to cylinder head	2·0	14
Heater hose outlet to cylinder head	1·0	7
Water pump to block	2·0	14
Water pump housings	1·0	7
Water pump pulley to flange	2·0	14

1 General description

1 The engine coolant is circulated by the thermo-syphon method, assisted by a water pump. The radiator cap seals the cooling system which pressurises the system. This has the effect of considerably increasing the boiling point of the coolant.

2 The impeller type water pump is mounted on the side of the engine block and is driven by a V-belt from the crankshaft pulley. This belt also drives the alternator.

3 The electrically operated cooling fan is mounted on the radiator and is controlled by a thermo-switch screwed into the left-hand side of the radiator. As soon as the coolant temperature rises to 90°–95° C (194°–203° F), the fan motor comes into operation, *even if the ignition is switched off.* It is therefore important whenever working near the fan to ensure that the battery is disconnected.

4 The system functions in the following fashion. Cold coolant in the bottom of the radiator circulates up the lower radiator hose to the water pump where it is pushed round the water passages in the cylinder block, cooling the cylinder bores. The coolant then travels up into the cylinder head and circulates round the combustion chambers and valve seats, absorbing more heat, and then, when the engine is at its correct operating temperature, travels out of the cylinder head through the top radiator hose and into the radiator (photo).

5 The coolant passes through the radiator where it is rapidly cooled by the flow of cold air through the radiator core, which is created both

1.4 Cylinder head coolant hose connector.
A – Radiator top hose
B – Water pump hose

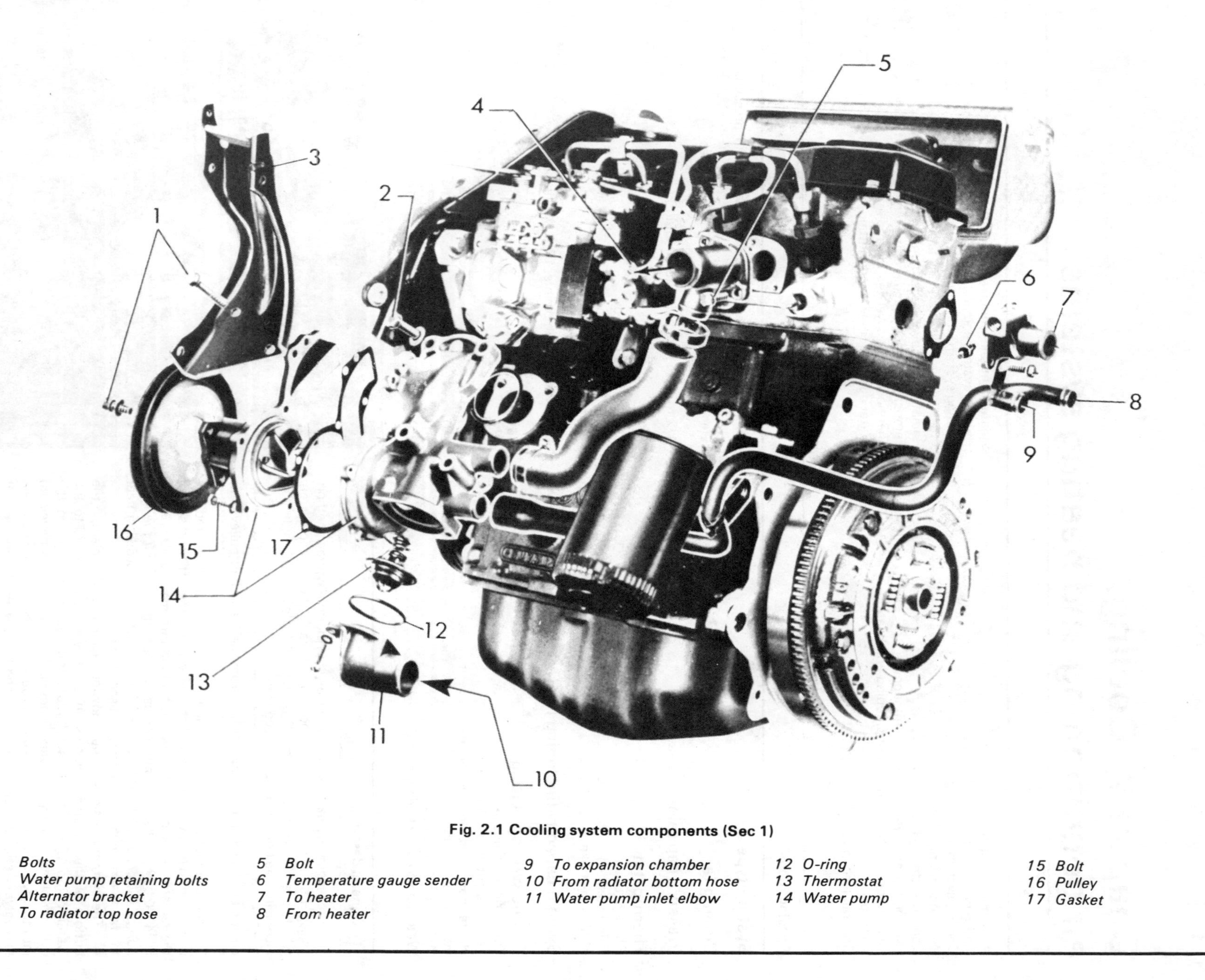

Fig. 2.1 Cooling system components (Sec 1)

1 *Bolts*
2 *Water pump retaining bolts*
3 *Alternator bracket*
4 *To radiator top hose*
5 *Bolt*
6 *Temperature gauge sender*
7 *To heater*
8 *From heater*
9 *To expansion chamber*
10 *From radiator bottom hose*
11 *Water pump inlet elbow*
12 *O-ring*
13 *Thermostat*
14 *Water pump*
15 *Bolt*
16 *Pulley*
17 *Gasket*

by the fan and by the motion of the car. The coolant, now much cooler, reaches the bottom of the radiator, when the cycle is repeated.
6 When the engine is cold the thermostat, which is a valve located in the water pump housing that opens and closes according to the temperature of the coolant, restricts the circulation of the coolant in the engine. Only when the correct minimum operating temperature has been reached, as given in the Specifications, does the thermostat begin to open, allowing the coolant to return to the radiator.
7 Engine coolant flows from an outlet connection on the rear end of the cylinder head (which also houses the temperature sender switch) through a hose to the heater radiator inside the car, and then returns through a hose to the water pump housing. The heater radiator is contained within the heater assembly which also houses a blower fan.

2 Cooling system – draining

1 It is best to drain the cooling system with the engine cold, but if this is not possible, the pressure must be released prior to draining the water. To do this, place a cloth over the expansion chamber cap and turn it slowly in an anti-clockwise direction until pressure is heard to escape, then stop turning.
2 Allow all the pressure to escape, then continue turning the cap anti-clockwise and remove it (photo). Be sure to keep your hand away from the escaping steam as otherwise you may be scalded.
3 If the coolant is to be retained for further use, place a suitable container beneath the water pump.
4 Move the heater controls inside the car to the hot position.
5 Disconnect the bottom hose from the water pump inlet elbow, and the heater hose return pipe from the water pump body (photo). Alternatively, remove the thermostat as described in Section 9.

3 Cooling system – flushing

1 After some time the radiator and engine waterways may become restricted or even blocked with scale or sediment, causing the engine to overheat. The coolant will appear rusty or dark in colour and when this occurs, the system should be flushed.
2 With the bottom hose and heater hose disconnected from the water pump and the thermostat removed, disconnect the top hose from the radiator and insert a water hose into the top of the radiator. Allow water to run through the radiator until it flows clear from the bottom hose.
3 Insert the water hose into the radiator top hose and allow water to run through the engine until it emerges clean from the water pump. Squeeze the by-pass hose occasionally to ensure that water circulates through all the engine waterways.
4 In severe cases of contamination the system should be reverse flushed. To do this, remove the radiator and insert the water hose in the bottom outlet with the radiator inverted. Continue flushing until clear water flows from the top of the radiator.

4 Cooling system – filling

1 Refit the thermostat if removed, and reconnect the radiator and heater hoses.
2 Make sure that the heater controls are set at the hot position.
3 Pour coolant of the specified mixture (see next Section) into the expansion chamber to the 'MIN' or cold water level mark (photo).
4 Fit the pressure cap to the expansion chamber and run the engine at a fast tick-over for 15 seconds.
5 Top-up the coolant level in the expansion chamber to the 'MIN' mark or slightly above and refit the pressure cap. When the engine is at normal operating temperature the coolant level will be nearer the 'MAX' level mark.
6 Never fill a hot engine with cold water, otherwise damage may occur to the cylinder head and block.

5 Antifreeze mixture

1 The coolant should be renewed every two years, not only to maintain the antifreeze properties but also to prevent corrosion in the system which would otherwise occur as the strength of the inhibitors

2.2 Removing the pressure cap

2.5 Pit view of the bottom hose and water pump inlet elbow

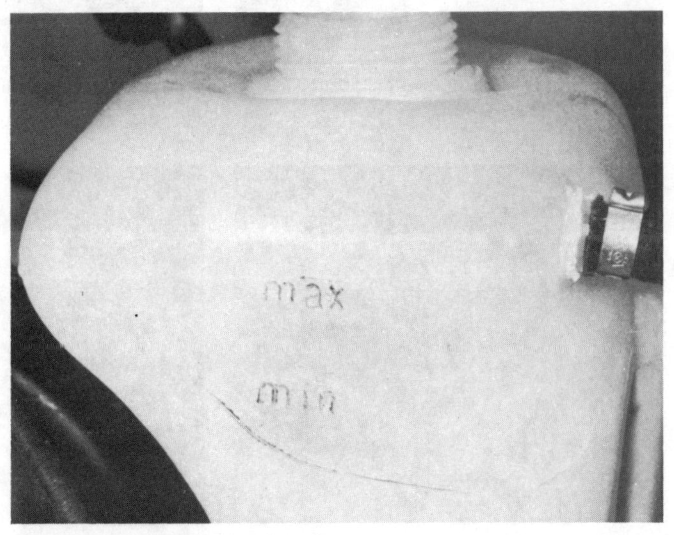

4.3 Expansion chamber level marks

in the coolant become progressively less effective.
2 Before adding antifreeze to the system, check all hoses and clips for deterioration and renew them as necessary. Flush the system as described in Section 3.
3 Mix the solution of antifreeze and water in a clean container; the total capacity of the system is given in the Specifications. The strength of the mixture can be determined from the following table:

Protection to outside temperature of	*Quantity of Antifreeze*	*Water*
-25°C (-13°F)	*40%*	*60%*
-35°C (-31°F)	*50%*	*50%*

4 When topping-up the cooling system always use a solution of the same strength as that already circulating in order to avoid dilution.

6 Radiator – removal, inspection, cleaning and refitting

1 Drain the cooling system as described in Section 2.
2 Disconnect the battery negative terminal.
3 Disconnect the thermo-switch wires from the left-hand side of the radiator (photo).
4 Disconnect the wiring plug from the radiator fan motor (photo).
5 Loosen the clips and disconnect the top and bottom hoses from the radiator (photo).
6 Disconnect the expansion chamber hose from the radiator. If the clamp type of clip is fitted, it will be necessary to obtain a screw type clip before refitting the hose.
7 Unscrew and remove the mounting nuts, slide the radiator sideways to disengage the upper mounting clip, and withdraw the radiator from the engine compartment (photos).
8 Unscrew the retaining bolts and remove the cowl complete with fan motor.
9 Radiator repair is best left to a specialist but minor leaks can be cured using a proprietary product, in many cases without removing the radiator.
10 Clear the radiator of flies and small leaves with a soft brush or by hosing, and flush it if necessary as described in Section 3.
11 Refitting is a reversal of removal.

7 Pressure cap – testing

If leakage can be heard from the pressure cap or if the cooling system requires constant topping-up, the cap should be tested. Most garages have test equipment and, if the cap does not open within the specified limits, it should be renewed.

8 Fan and thermo-switch – testing and renewal

1 To test the fan motor, disconnect the plug and connect a 12 volt supply to the terminals. If the motor does not operate the brushes may be worn or sticking. To dismantle the motor, remove the fan blades and through bolts; the end brackets can now be removed but note the location of the shims which determine the armature endfloat. Clean the commutator and renew the brushes if necessary, then reassemble the motor in the reverse order.
2 If the thermo-switch is thought to be faulty, it should be removed, after draining the cooling system, and tested in a container of heated water. With a 12 volt test lamp and lead connected to the terminals, or alternatively an ohmmeter, the switch contacts should close when the temperature of the water reaches 90° C (194° F) If the switch is proved faulty, it must be renewed.
3 Should the switch contacts fail to close at any time resulting in overheating due to the fan not operating, the two terminals can be connected together as an emergency measure in order to reach the nearest repair garage.

9 Thermostat – removal, testing and refitting

1 A faulty thermostat will prevent the engine from running at the

6.3 Radiator fan thermo-switch location

6.4 Fan motor and wiring plug

6.5 Radiator top hose connection

6.7A Radiator lower mounting

6.7B Radiator upper mounting clip

6.7C Removing the radiator assembly

most efficient operating temperature. If overheating occurs, considerable damage to the engine may result. If the thermostat is suspect it should be removed and tested.
2 Drain the cooling system as described in Section 2.
3 Unscrew the two retaining bolts and remove the inlet elbow from the bottom of the water pump (photo).
4 Prise out the rubber O-ring and withdraw the thermostat (photos).
5 To test whether the unit is serviceable, suspend it by a piece of string in a pan of water being heated. The thermostat should commence to open at the specified temperature and should extend by the specified stroke when fully open. After cooling, the thermostat should return to its fully closed position.
6 Refitting is a reversal of removal, but fit a new O-ring if necessary.

10 Water pump – removal and refitting

1 Drain the cooling system as described in Section 2.
2 Remove the alternator as described in Chapter 7.
3 Remove the air conditioner compressor where fitted (Section 15).
4 Unbolt the alternator bracket from the cylinder block and water pump and remove the earth lead.
5 Disconnect the by-pass hose from the water pump, also the heater hose if not already removed (photo).
6 Unscrew the bolts securing the water pump to the cylinder block.
7 Remove the water pump complete with pulley, then prise out the rubber O-ring.
8 Unbolt the pulley from the water pump flange.
9 Unscrew the retaining bolts and separate the two halves; if they are tight, use a wooden mallet to tap them free.
10 The removal sequence above is the best one to follow with the engine in the car. If the engine is out of the car, you may prefer to follow the sequence shown in the photographs and separate the two halves of the water pump before removing the rear half from the block (photos).
11 Remove the gasket and clean the two surfaces. Remove the thermostat as described in Section 9.
12 The impeller and housing are serviced as one unit; if the driveshaft is faulty or a leak is present, the whole assembly must be renewed.
13 Refitting is a reversal of removal, but always use a new gasket and O-rings, and tighten all bolts to the specified torque wrench settings. Adjust the drivebelt as described in Chapter 7 and refill the cooling system as described in Section 4.

11 Temperature gauge sender unit – renewal and testing

1 The unit is screwed into the coolant flange at the flywheel end of the cylinder head (photo).
2 Before removing it drain the coolant from the radiator.
3 The resistance varies as the temperature increases. To check the sender disconnect the lead and using an ohmmeter, check the resistance cold and again with the engine hot. The cold resistance should be 270 ohms ± 5% and the normal working temperature resistance 150 ohms ± 5%. If it is not within these limits it should be renewed. It will be noted that the resistance decreases with rise in temperature.

12 Heating system – description

1 The heating system consists of a heat exchanger, blower motor and fan, a control box behind the dashboard, a control valve in the engine compartment (photo) and an extensive ducting system. Refer to Figs. 2.2 and 2.3.
2 The coolant is fed to the heat exchanger from the connection on the rear end of the cylinder head. The coolant is drawn through the heat exchanger and back to the water pump housing. The temperature of the coolant is in excess of 90°C (194°F) when it leaves the cylinder head. The quantity of coolant circulating depends upon how much the valve is opened, so that the heat available to the car interior is controlled in this way.
3 The heater cover is located in the engine compartment next to the windscreen wiper motor. Under the cover, operated by a control cable

9.3 Removing the water pump inlet elbow

9.4A Removing the thermostat

9.4.B Thermostat location aperture in the water pump

10.5 Location of the by-pass hose (A) and heater hose (B) on the water pump

10.10A If the engine is out of the car, first remove the water pump pulley ...

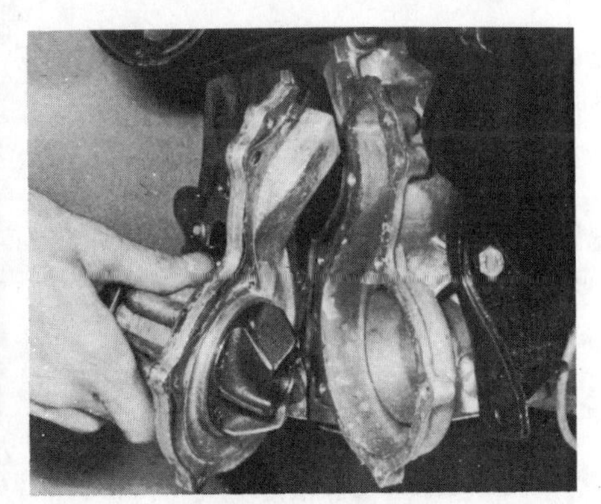
10.10B ... separate the two halves ...

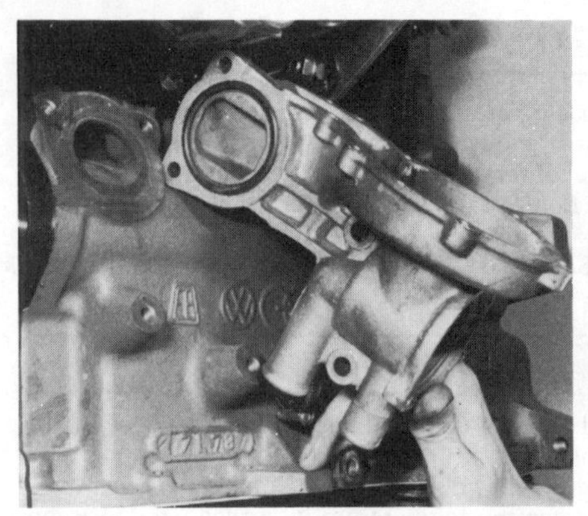
10.10C ... and remove the rear half from the block

11.1 Temperature gauge sender units
A – Engine coolant
B – Heater control

12.1 Heater control valve in the engine compartment

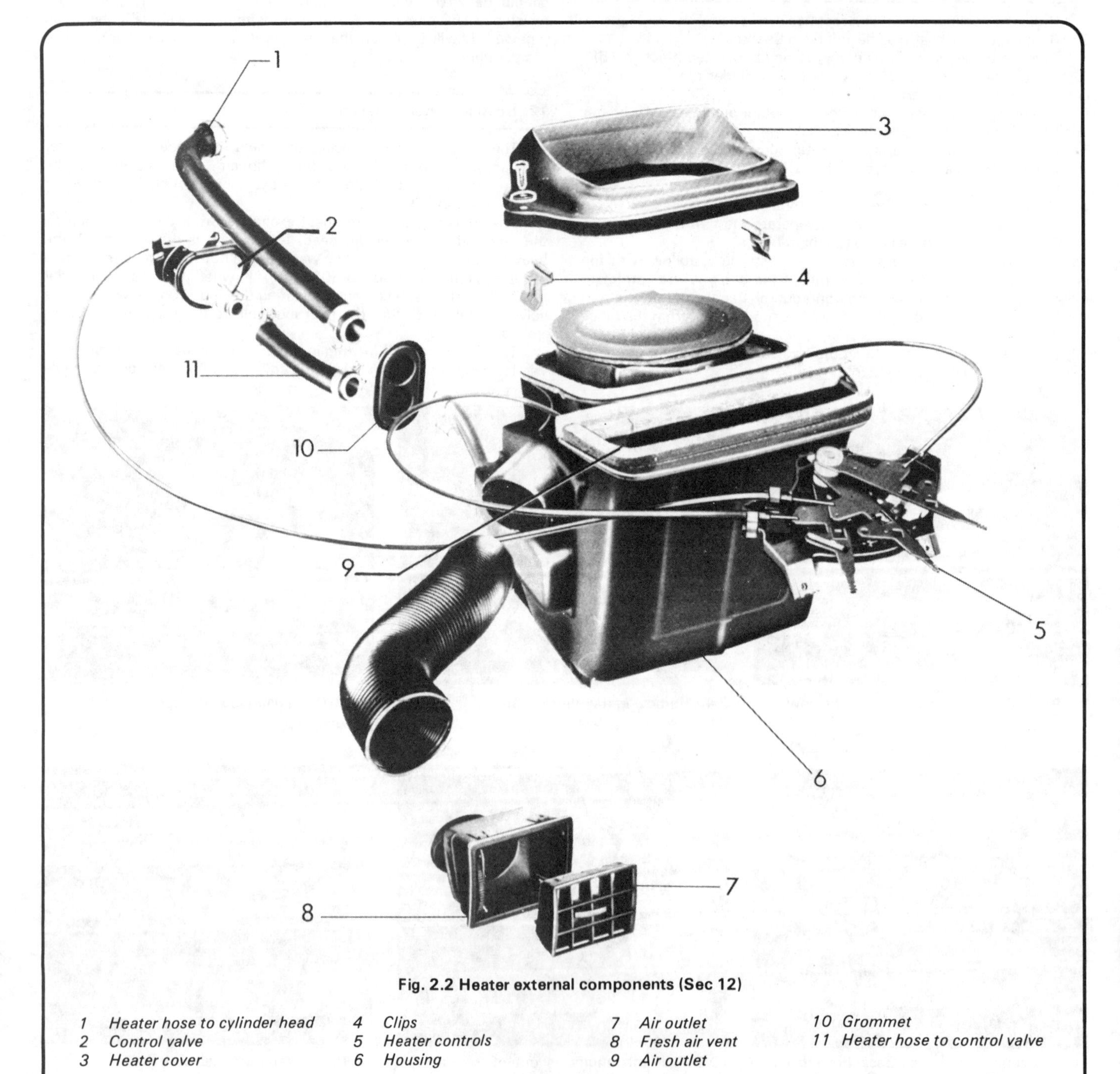

Fig. 2.2 Heater external components (Sec 12)

1 Heater hose to cylinder head
2 Control valve
3 Heater cover
4 Clips
5 Heater controls
6 Housing
7 Air outlet
8 Fresh air vent
9 Air outlet
10 Grommet
11 Heater hose to control valve

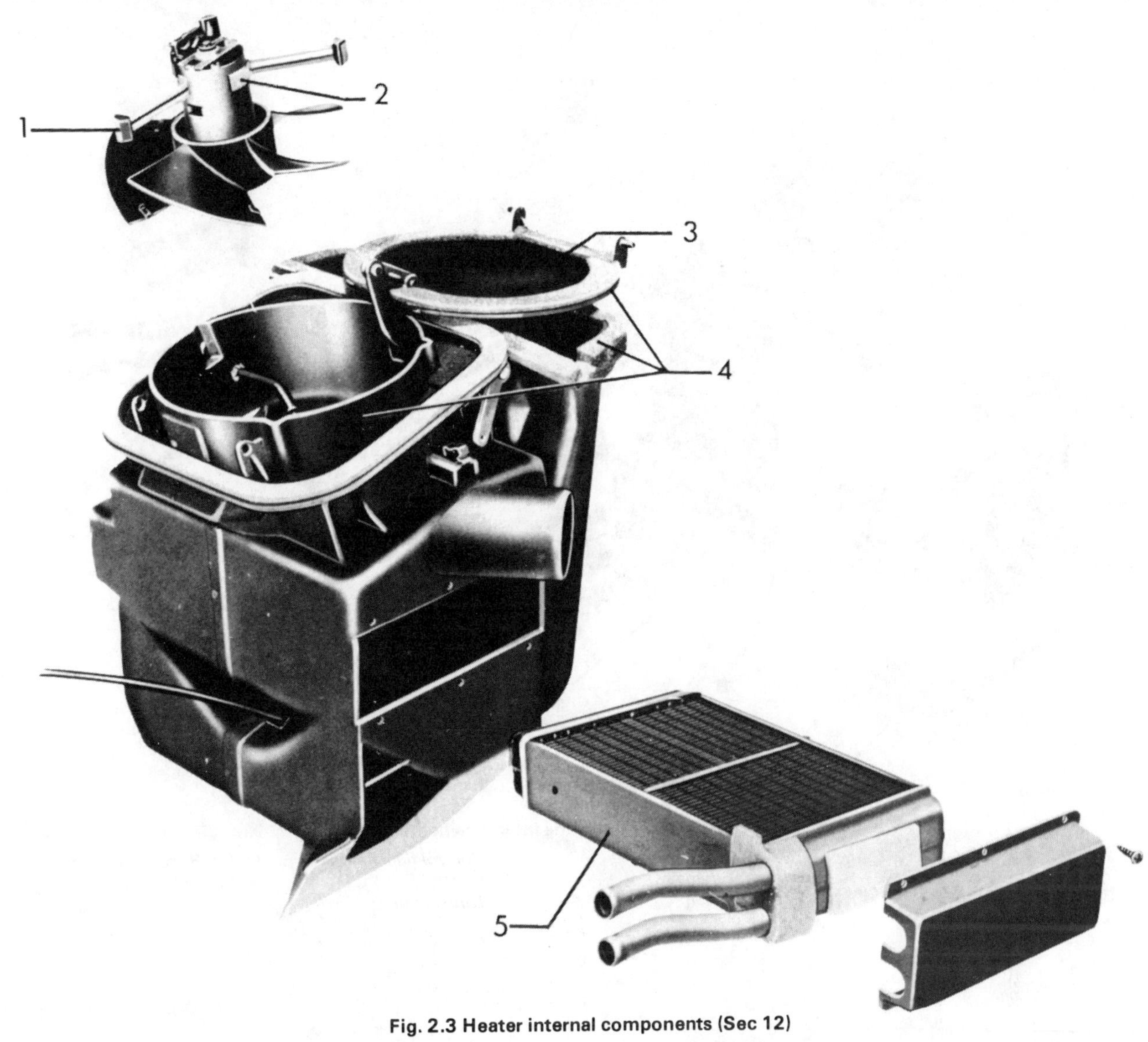

Fig. 2.3 Heater internal components (Sec 12)

1 Clip
2 Fan motor
3 Cut-off flap
4 Gaskets
5 Heat exchanger

from the control box, is a circular dished flap. This controls the amount of air admitted to the system, and can close the system completely if so wished. Directly under the flap is the fan motor and fan. The fan motor has three speeds.

4 The air is forced into the fresh air housing where it is directed to the footwell air vents by the left-hand lever A on the control panel (see Fig. 2.4) or to the mixed vents for the windscreen and side windows by the right-hand control lever A.

5 Heated air is available through all the ventilaticn vents except the ones at each end of the dashboard which are for fresh unheated air only. Fig. 2.5 shows a later type of heater which operates in a similar manner (photo).

13 Heating system components – removal and refitting

1 Refer to Figs. 2.2 and 2.3. The heater cover can be removed from the back of the engine compartment by extracting the two plastic retaining pegs. This gives access to the cut-off flap and its actuating lever. Disconnect the cable and unhook one pivot of the cut-off flap. The flap may now be removed.

2 The fan motor is located in the throat of the inlet to the fresh air housing and may be lifted out. Disconnect the fan cable by pulling off the clip.

3 The housing is held to the body of the heater by two metal clips

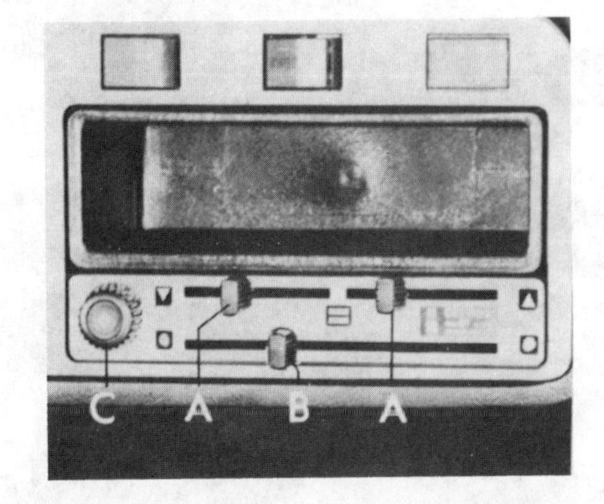

Fig. 2.4 Heater control panel (Sec 12)

A Air distribution levers
B Temperature control
C Fan speed control knob

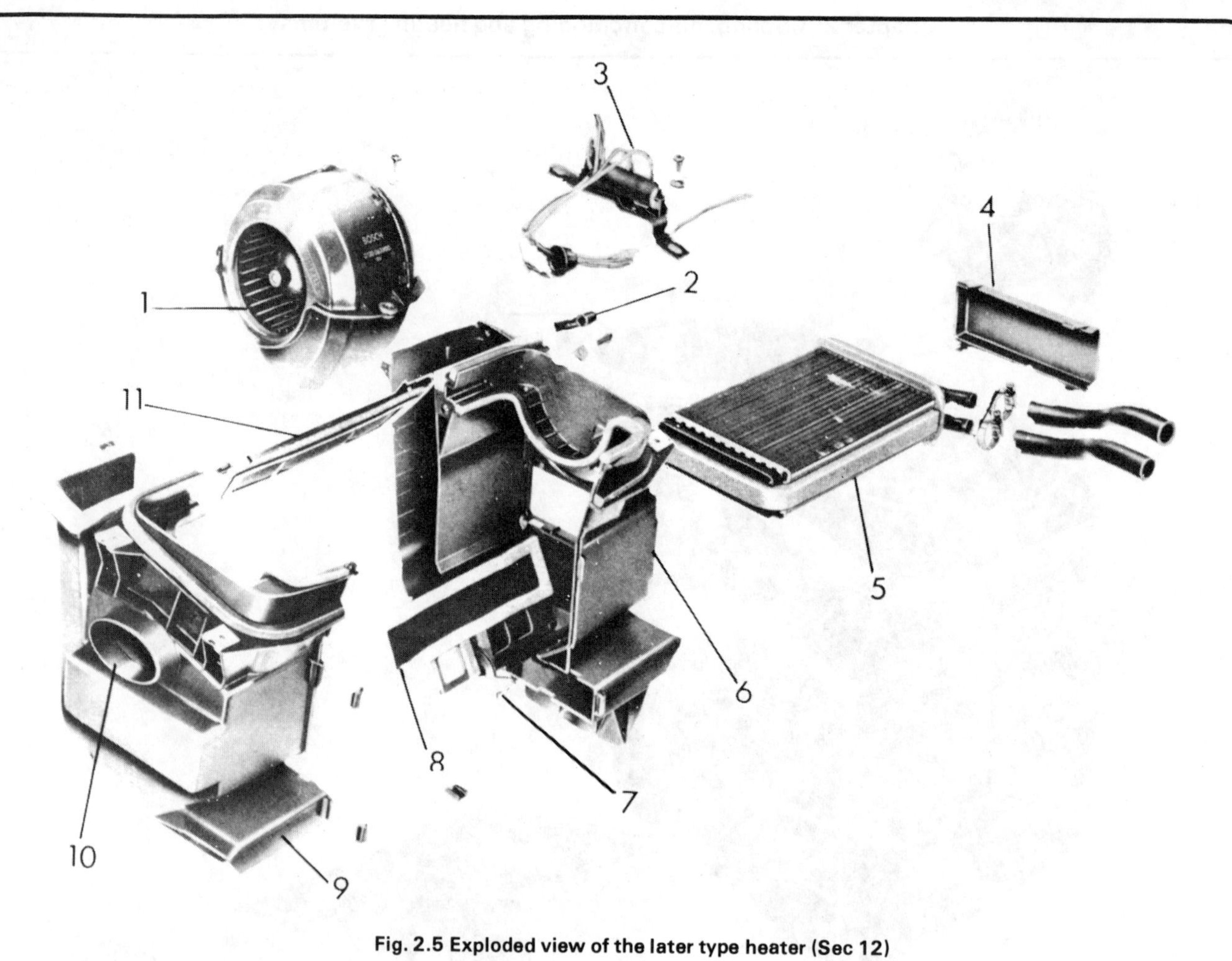

Fig. 2.5 Exploded view of the later type heater (Sec 12)

1 Rotary fan unit
2 Lever
3 Series resistance
4 Cover
5 Heat exchanger
6 Housing half
7 Control flap clamp
8 Control flap
9 Housing half
10 Hose connection
11 Cut-off flap

12.5 Later type heater motor location

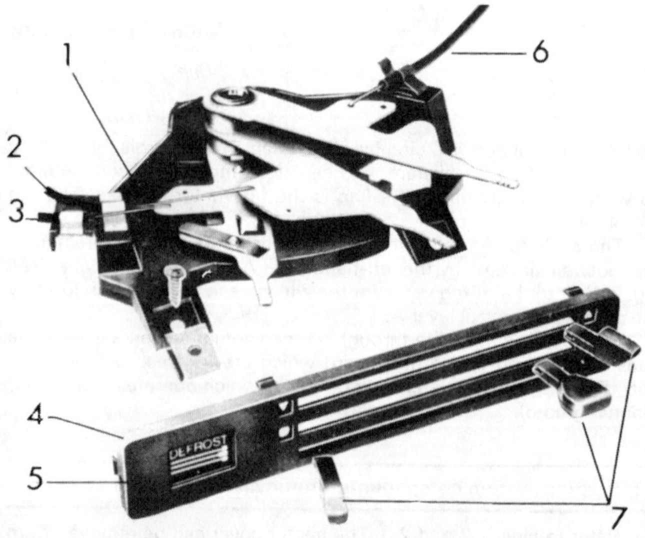

Fig. 2.6 Heater controls on early models (Sec 14)

1 Fresh air controls
2 Heater valve cable
3 Cut-off flap cable
4 Bulb holder
5 Trim plate
6 Footwell flap cable
7 Control knobs

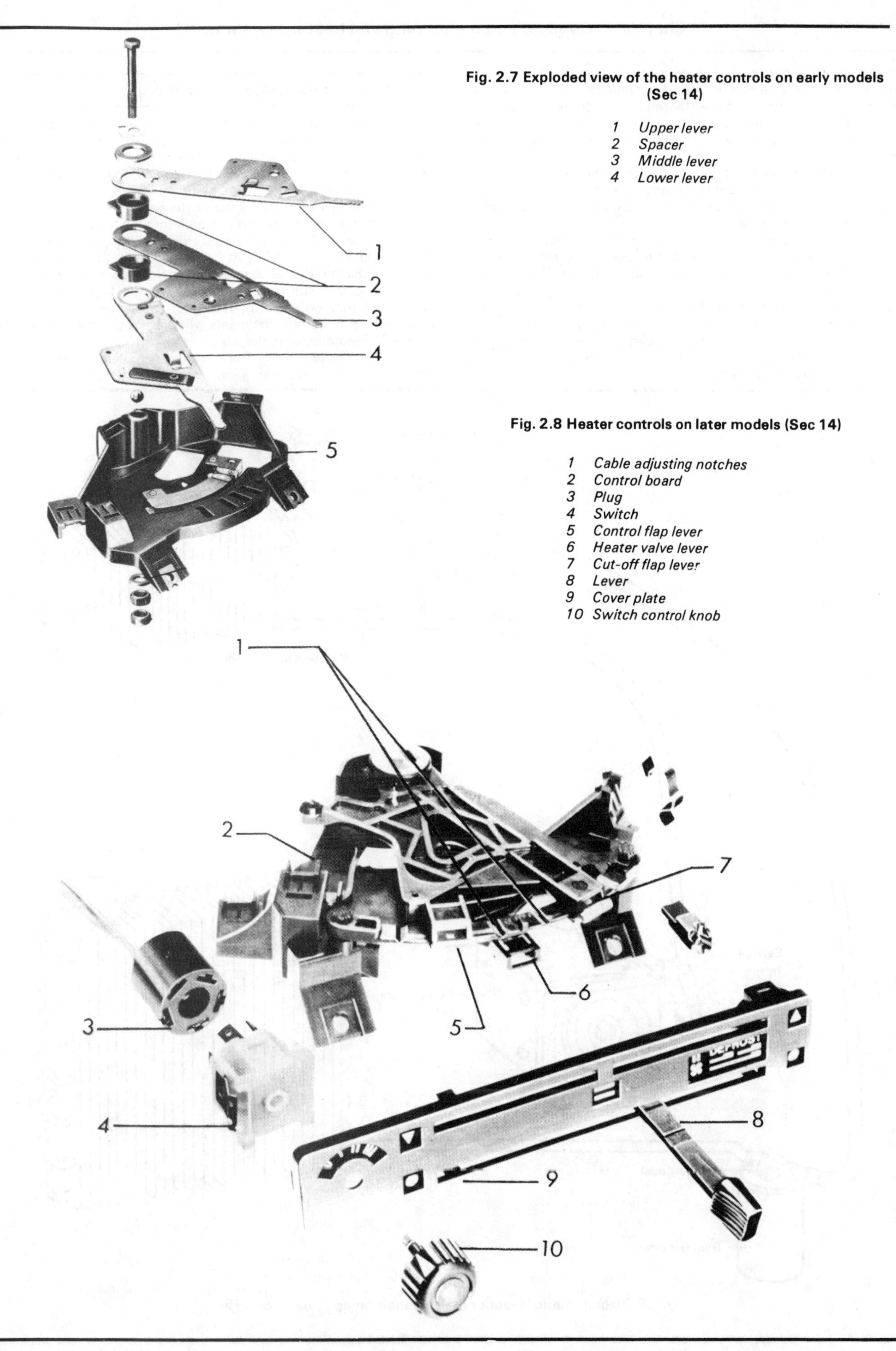

Fig. 2.7 Exploded view of the heater controls on early models (Sec 14)

1. *Upper lever*
2. *Spacer*
3. *Middle lever*
4. *Lower lever*

Fig. 2.8 Heater controls on later models (Sec 14)

1. *Cable adjusting notches*
2. *Control board*
3. *Plug*
4. *Switch*
5. *Control flap lever*
6. *Heater valve lever*
7. *Cut-off flap lever*
8. *Lever*
9. *Cover plate*
10. *Switch control knob*

which are difficult to locate and even more difficult to release. In the centre of the clip is a tongue or retaining spring, this must be pressed in and the clip will come undone. There is a clip on each side of the housing and once these are free the housing can be drawn off the heat exchanger. The large air ducts to the dashboard outlets pull out of the housing.

4 The heat exchanger must be removed after the fresh air housing. Two pipes go through the double grommet and are connected to the coolant hoses. It will be necessary to drain the cooling system before these hoses are disconnected.

5 Refitting is the reverse of the removal procedure. Fitting the clips is a bit tricky, the tab of the clip should be pushed into the retainer on the fresh air housing and the housing lifted squarely until the spring clicks into position.

6 Unless the heat exchanger leaks, the control cables break or the fan motor fails to work, it is best to leave well alone. The dismantling and reassembly is not a long job but you can get into difficulties when trying to get the housing clips off, and on, and unless the job is done carefully the housing may crack or split.

14 Heater controls – removal and refitting

1 The control unit is located in the centre of the dashboard. It is accessible after the radio has been removed, or on cars without a radio, the glovebox. Once the radio is extracted the control unit may be seen. Disconnect the battery earth strap and undo the two crosshead screws holding the control unit and it may be eased forward.

2 The middle lever controls the heater valve. It is reasonably easy to unhook the bowden cable and extract it through the bulkhead.

3 The upper and lower cables control the airflow. Although the cable for the upper lever is visible at its far end, right at the bottom of the fresh air housing it is necessary to dismantle the housing to renew it. The outer end of the lever cable operates the flap under the heater cover. This may be got at by removing the heater cover but you will probably break the plastic dowels when extracting them, so get some new ones before tackling the job.

4 It is best to renew the heater cables completely if the inner cable snaps. In this way the exact length required is obtained. We found it a

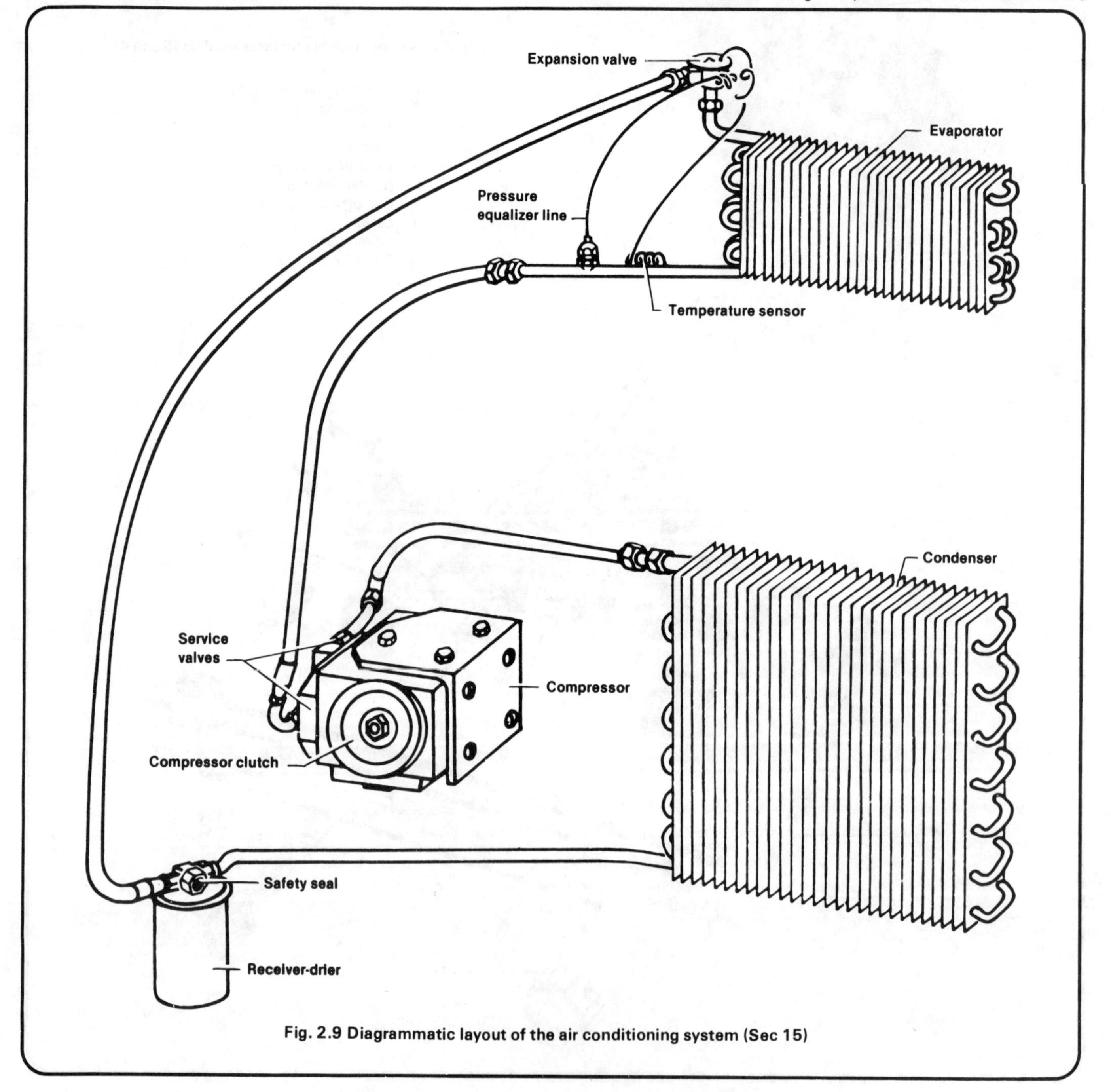

Fig. 2.9 Diagrammatic layout of the air conditioning system (Sec 15)

good thing to fit new cable clamps too, as the old ones seem to distort when removed.

5 Fig. 2.8. shows the later type of control. The removal and refitting procedure is almost identical for both types.

15 Air conditioning system – general

1 The unit works on exactly the same principle as a domestic refrigerator, having a compressor, a condenser and an evaporator. The condenser is attached to the car radiator system. The compressor, belt-driven from the crankshaft pulley, is installed on a bracket on the engine. The evaporator is installed in a housing under the dashboard which takes the place of the normal fresh air housing. This housing also contains a normal heat exchanger unit for warming the intake air. The evaporator has a blower motor to circulate cold air as required.

2 The system is controlled by a unit on the dashboard similar to the normal heater control in appearance.

3 The refrigerant used is difluorodichloromethane (CF_2CL_2) more commonly known as Frigen F12 or Freon F12. It is a dangerous substance in unskilled hands. As a liquid it is very cold and if it touches the skin there will be cold burns and frostbite. As a gas it is colourless and has no odour. Heavier than air it displaces oxygen and can cause asphyxiation if pockets of it collect in pits or similar workplaces. It does not burn, but even a lighted cigarette, causes it to breakdown into constituent gases, some of which are poisonous to the extent of being fatal. So if you have an air-conditioner and your car catches fire, you have an additional problem.

4 We strongly recommend that even trained refrigeration mechanics do not adjust the system unless they have had instruction by VW. The notes for VW personnel cover 25 pages. For the remainder we suggest that the system is left entirely alone, except for the adjustment of the compressor drivebelt, which should have 10 to 15 mm (0.4 to 0.6 inch) deflection in the centre when depressed by the thumb.

5 Removal and refitting of the air conditioner compressor is straightforward as can be seen from Fig. 2.10, but the refrigerant circuit **must**

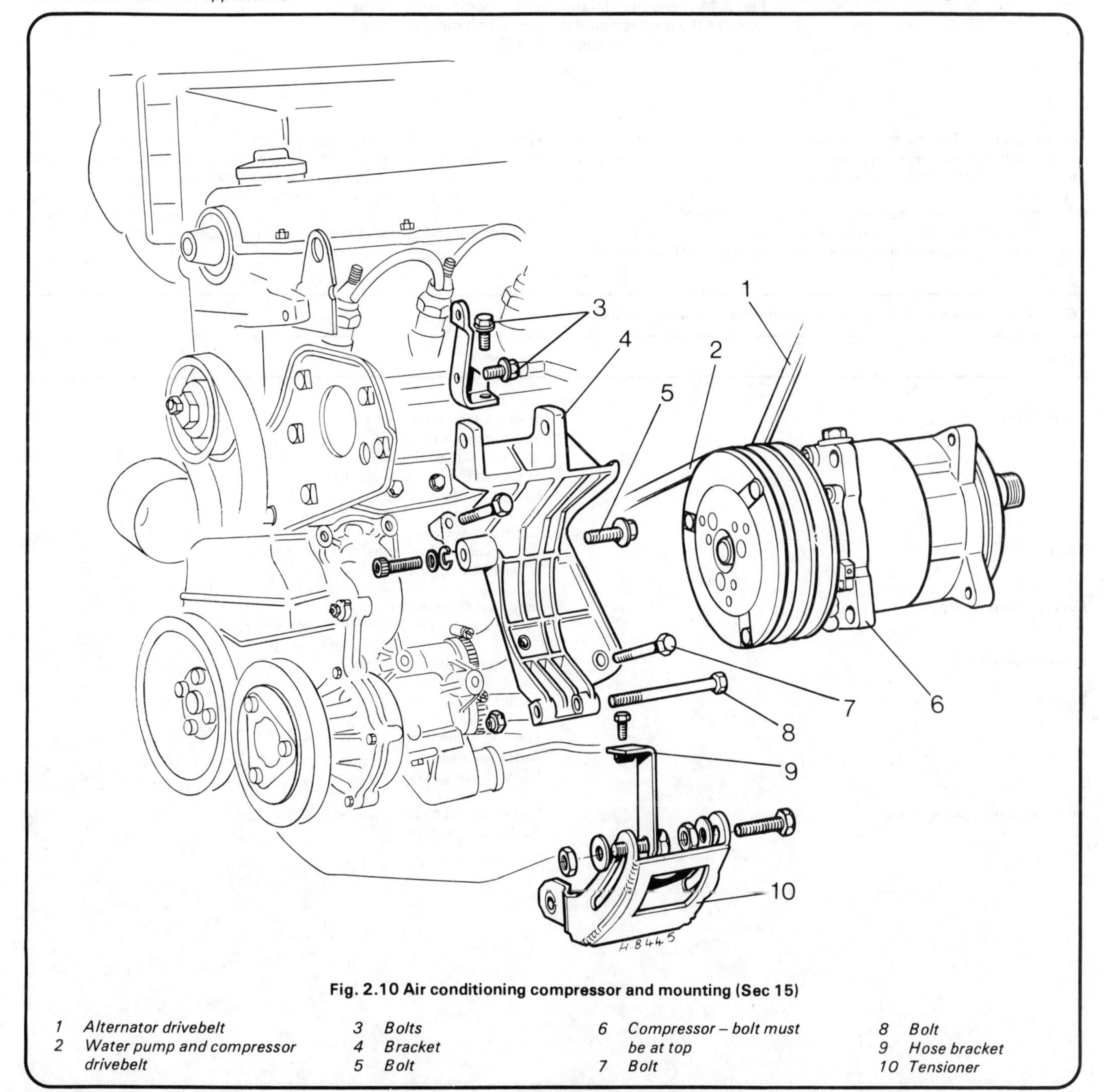

Fig. 2.10 Air conditioning compressor and mounting (Sec 15)

1 Alternator drivebelt
2 Water pump and compressor drivebelt
3 Bolts
4 Bracket
5 Bolt
6 Compressor – bolt must be at top
7 Bolt
8 Bolt
9 Hose bracket
10 Tensioner

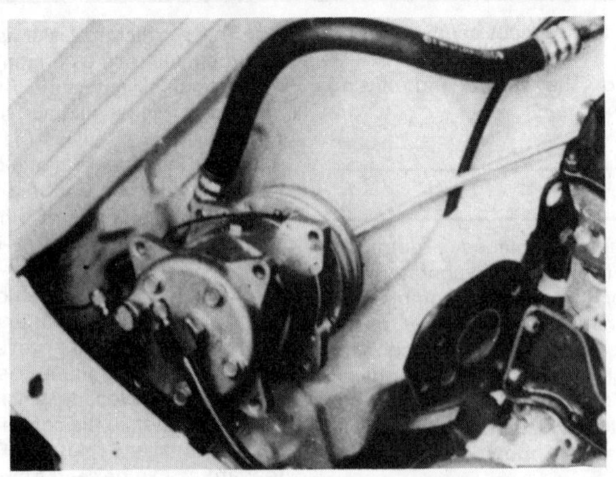

Fig. 2.11 Place the air conditioning compressor on the side of the engine compartment when removing the engine (Sec 15)

not be opened. The compressor must be placed on the side of the engine compartment when removing the engine, but only move it to the point where the flexible refrigerant hoses are in no danger of being stretched.

6 When a situation arises which calls for the removal of one of the air conditioning system components, have the system discharged by your VW agent or a qualified refrigeration engineer. Similarly have the system recharged by him on completion.

7 During the winter period operate the air conditioner for a few minutes each week to keep the system in good order.

8 Periodically, clean the condenser of dirt and insects, either by washing with a cold water hose or by air pressure. Use a soft bristly brush to assist removal of dirt jammed in the condenser fins.

16 Fault diagnosis – cooling, air conditioning and heating systems

Symptom	Reason/s
Overheating	Low coolant level Faulty pressure cap Thermostat sticking shut Drivebelt incorrectly adjusted Clogged radiator matrix Incorrect engine timing Corroded system Faulty cooling fan motor or thermo switch Blown head gasket
Cool running	Incorrect thermostat Faulty cooling fan motor or thermo switch
Slow warm-up	Thermostat sticking open
Coolant loss	Faulty pressure cap Split hose Leaking water pump Loose hose clip Blown head gasket
Air conditioning not working	Drivebelt incorrectly adjusted Faulty compressor

Chapter 3 Fuel and exhaust systems

Refer to Chapter 11 for specifications and information applicable to later models

Contents

Specifications

Fuel injection pump

Type	Distributor
Make	Bosch or CAV

Air cleaner element
Paper

Preheating
Glow plug in each cylinder

Fuel grade
CN 45 min (UK), Number 2 (USA)

Fuel tank capacity
45 litres (9·9 Imp gal, 11·8 US gal)

Torque wrench settings

	kgf m	lbf ft
Injection pump mounting bolt	2·5	18
Fuel pipes	2·5	18
Fuel pump unions (Bosch)	4·5	32
Injection pump sprocket	4·5	33
Injectors	7·0	51
Injector upper to lower section (Bosch)	7·0	51
Injector upper to lower section (CAV)	8·3	60
Fuel tank strap	2·5	18
Manifolds to cylinder head	2·5	18

1 General description

The fuel system comprises a rear mounted fuel tank located below the rear of the car, a mechanical fuel injection pump driven by the engine timing drivebelt, and injectors screwed into the cylinder head which inject fuel oil into pre-combustion or ante chambers.

A fuel filter is incorporated in the fuel line between the tank and injection pump. It is important to renew the filter at the specified intervals, since the fuel oil is used to lubricate the precision made injection pump. Both the injection pump and the injectors are connected to a fuel return pipe which allows excess fuel to return to the fuel tank.

Two types of injection pump may be fitted, one made by Bosch or one by CAV. Both are of the distributor type incorporating a distribution rotor.

The cold start system comprises preheater glow plugs located in the pre-combustion chambers, and a linked device on the injection pump for advancing the injection timing and increasing the idle speed. In addition, an internal automatic device provides extra fuel at engine cranking speed. A relay linked with the starter motor continues to operate the glow plugs during starting.

A shut-off solenoid is incorporated into the fuel supply circuit in the injection pump. The solenoid valve is shut when de-energised and open when energised, thus when the ignition key is switched on the circuit is open and when it is switched off the supply is stopped.

The servicing of fuel injection components is a complex subject and also requires the use of specialised equipment in dust-free surroundings. The home mechanic is therefore advised to attempt only the procedures described in this Chapter. Even so, the work **must** be carried out under conditions of the utmost cleanliness. Remember also that fuel oil is detrimental to rubber components such as water hoses and toothed drivebelts.

Emission control

Unlike its petrol (gasoline) engined counterpart, the design of the Golf diesel engine is such that satisfactory emission control is achieved without the addition of special equipment or devices, only a crankcase ventilation system (see Chapter 1) being fitted. In order to maintain this desirable condition however, the state of the engine and fuel system must be faultless and regular servicing and adjustment must not be neglected.

2 Air cleaner element – renewal

1 The air cleaner element should be removed and cleaned at 15 000 miles (24 000 km), and renewed at 30 000 miles (48 000 km) or every two years. Earlier renewal may be required if the car is operated in very dusty conditions.
2 To remove the element, prise the spring clips up and withdraw the housing from the inlet manifold casting (photo).
3 Remove the element and, if it is to be re-used, shake it to displace all the accumulated dust and dirt (photo).
4 Clean the interior of the housing and manifold casting with a paraffin moistened cloth and wipe dry (photo).
5 Refitting is a reversal of removal, but make sure that the retaining clips are fully engaged with the recesses on the top and bottom of the inlet manifold casting. At the same time, the crankcase ventilation hose from the casting to the valve cover can be removed and cleaned out with paraffin (photo).

3 Fuel filter – renewal and maintenance

1 The fuel filter is located on the right-hand side of the engine compartment by the suspension strut upper bearing (photos). It must be renewed every 15 000 miles (24 000 km) in order to prevent excessive wear occurring in the injection pump.
2 *On cars fitted with the Bosch system,* unscrew the filter canister with a filter wrench, or a spanner if a hexagon is provided on the bottom of the canister. Hold the canister in a cloth and do not tilt it otherwise fuel will spill from it.
3 Discard the old filter canister.
4 Smear a film of diesel fuel onto the sealing ring of the new canister, then tighten it onto the housing using hand pressure only (photo).
5 Early models were provided with a primer pump to fill the system with fuel prior to starting but this has now been discontinued. The system does not need bleeding even though a primer is provided; the engine should be turned with the starter until it starts, then accelerated several times.
6 With the engine running, check the filter canister for fuel leaks, then stop the engine.
7 A water drain tap is provided at the bottom of the filter in order to drain off any accumulated moisture from the fuel (photo). Periodically the tap should be loosened to allow the water to drain out, then tightened when clear fuel flows. Loosen the vent screw on top of the filter housing and have an assistant operate the starter without operating the glow plugs. Tighten the vent screw as soon as clear fuel flows from it.
8 *On cars fitted with the CAV system,* first loosen the filter retaining bolt located in the centre of the upper housing. Do not fully unscrew it at this stage.
9 Wipe the filter upper and lower housings clean with a paraffin moistened cloth.
10 Unscrew the filter mounting nuts and lift the filter from the mounting bracket.
11 Remove the centre bolt, separate the housings, and remove the filter element.
12 Discard the old element and clean any accumulated sediment from the lower housing.
13 When fitting the new element, make sure that the sealing rings are correctly located as shown in Fig. 3.3 and smear their contact faces with a little fuel oil. Tighten the centre bolt, then bleed the fuel system as described in Section 4.
14 The CAV filter also incorporates a water drain tap (refer to paragraph 7). To drain the water, loosen the breather screw on the upper housing, then loosen the drain plug. Tighten both screws when clear fuel emerges.

4 Bleeding the fuel system

1 *On cars fitted with the Bosch system,* bleeding is unnecessary, even if the fuel tank has run dry. However, the battery will be required to turn the engine during the time it takes to fill the system, and

2.2 Removing the air cleaner housing

2.3 The air cleaner element

2.4 Intake manifold and air cleaner base casting

2.5 Crankcase ventilation hose connection to the valve cover

3.1A Fuel filter location

3.1B Fuel filter mounting bracket

3.4 The fuel filter canister seal and housing

Fig. 3.1 Tightening the fuel filter canister on the Bosch system (Sec 3)

3.7 Fuel filter canister water drain tap

Fig. 3.2 Fuel filter element retaining bolt (1) and mounting nuts (2) on the CAV system (Sec 3)

battery discharge can be avoided where the primary facility is fitted as described in the next paragraph.

2 To fill the Bosch system, first loosen the vent screw on the filter housing. Operate the primer pump until the fuel emerging from the screw is free of air bubbles, then tighten the screw (Fig. 3.4). If the engine refuses to start after this, loosen two injector unions and turn the engine with the starter until fuel emerges, then tighten the unions.

3 *On cars fitted with the CAV system,* bleeding is recommended after removing any component or after running out of fuel. First operate the hand lift pump until fuel is visible in the inlet pipe free of air bubbles.

4 Loosen the vent screw on the injection pump next to the throttle lever 1 or 2 turns, and turn the engine with the starter for 15 seconds. If the engine starts before the end of this period, switch it off and tighten the vent screw. With the screw tightened, the engine may be started in the normal manner (Figs. 3.5 and 3.6).

5 Fuel tank – removal, servicing and refitting

1 Disconnect the battery negative terminal.

2 Unscrew the drain plug and drain the fuel into a suitable container.

3 Remove the brake fluid reservoir filler cap and tighten it down onto a piece of polythene sheeting; this will reduce the loss of brake fluid in subsequent operations.

4 Chock the front wheels. Jack-up the rear of the car and support it on axle stands.

5 Disconnect the brake line hoses on each side of the car, just forward of the rear axle mountings (Fig. 3.8). Plug the ends to prevent the entry of dirt.

6 Unscrew and remove the axle mounting nuts from each side of the car, and lower the rear axle beam as far as the handbrake cable guides will allow.

7 Remove the exhaust silencer mounting rubbers and lower the silencer.

8 Loosen the securing clip and disconnect the fuel supply hose from the tank.

9 Note the location of the engine supply and return pipes, then disconnect them from the tank.

10 Remove the retaining straps and lower the tank sufficiently to disconnect the fuel gauge sender unit wires, breather pipe and drain pipe.

11 Withdraw the fuel tank from under the car.

12 A leak in the fuel tank should be repaired by specialists or alternatively a new tank fitted. Never be tempted to solder or weld a fuel tank.

13 If the tank is contaminated with sediment or water, it can be swilled out using several changes of fuel, but if any vigorous shaking is required to dislodge accumulations of sediment, the fuel gauge sender unit should first be removed as described in Section 6.

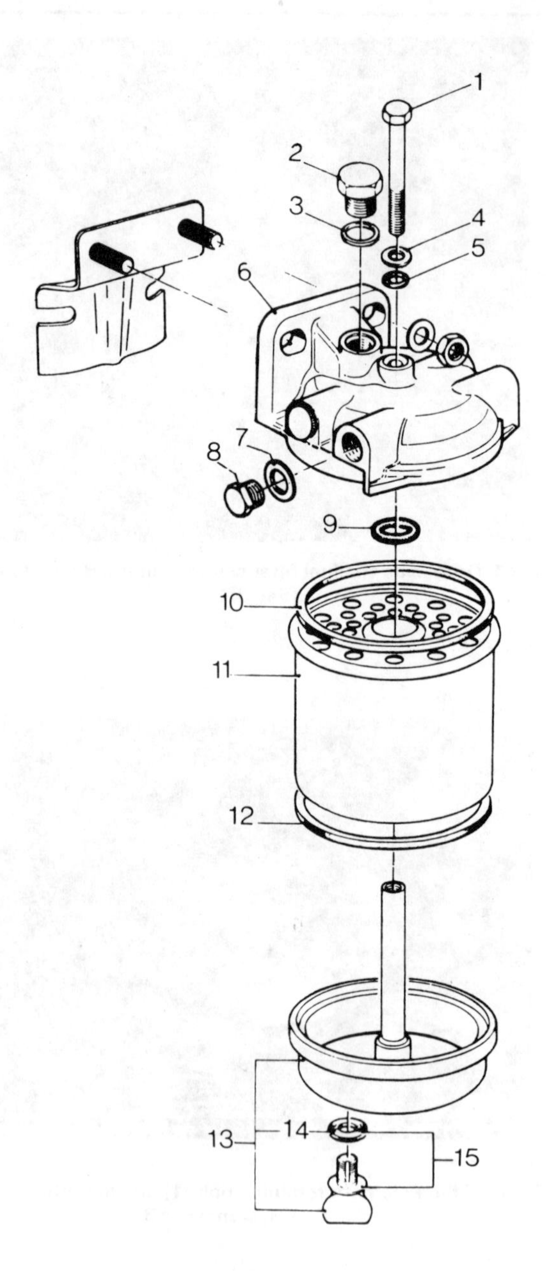

Fig. 3.3 Exploded view of the CAV fuel filter (Sec 3)

1 *Retaining bolt*
2 *Vent plug*
3 *Washer*
4 *Washer*
5 *Sealing ring*
6 *Filter head*
7 *Washer*
8 *Plug*
9 *Sealing ring*
10 *Sealing ring*
11 *Element*
12 *Sealing ring*
13 *Filter base*
14 *Sealing ring*
15 *Drain tap*

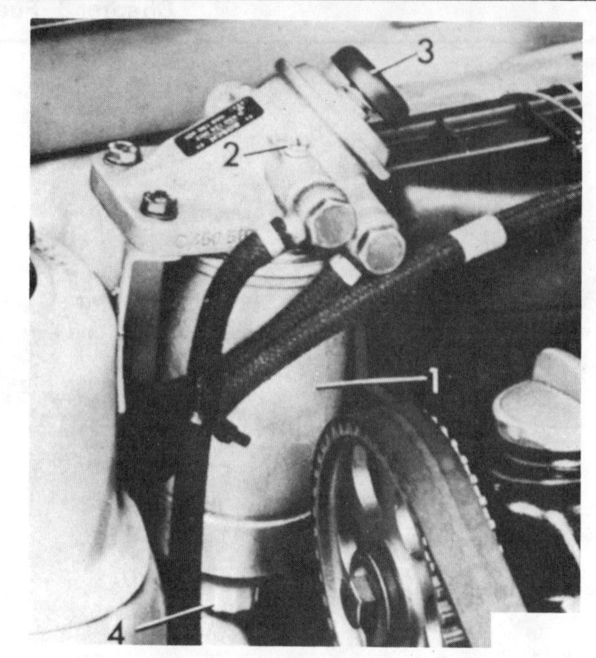

Fig. 3.4 Bosch fuel filter element (1), vent screw (2), primer (3), and drain tap (4) (Sec 4)

Fig. 3.5 CAV fuel filter primer pump (A) and outlet pipe (B) (Sec 4)

Fig. 3.6 CAV injection pump vent (1) (Sec 4)

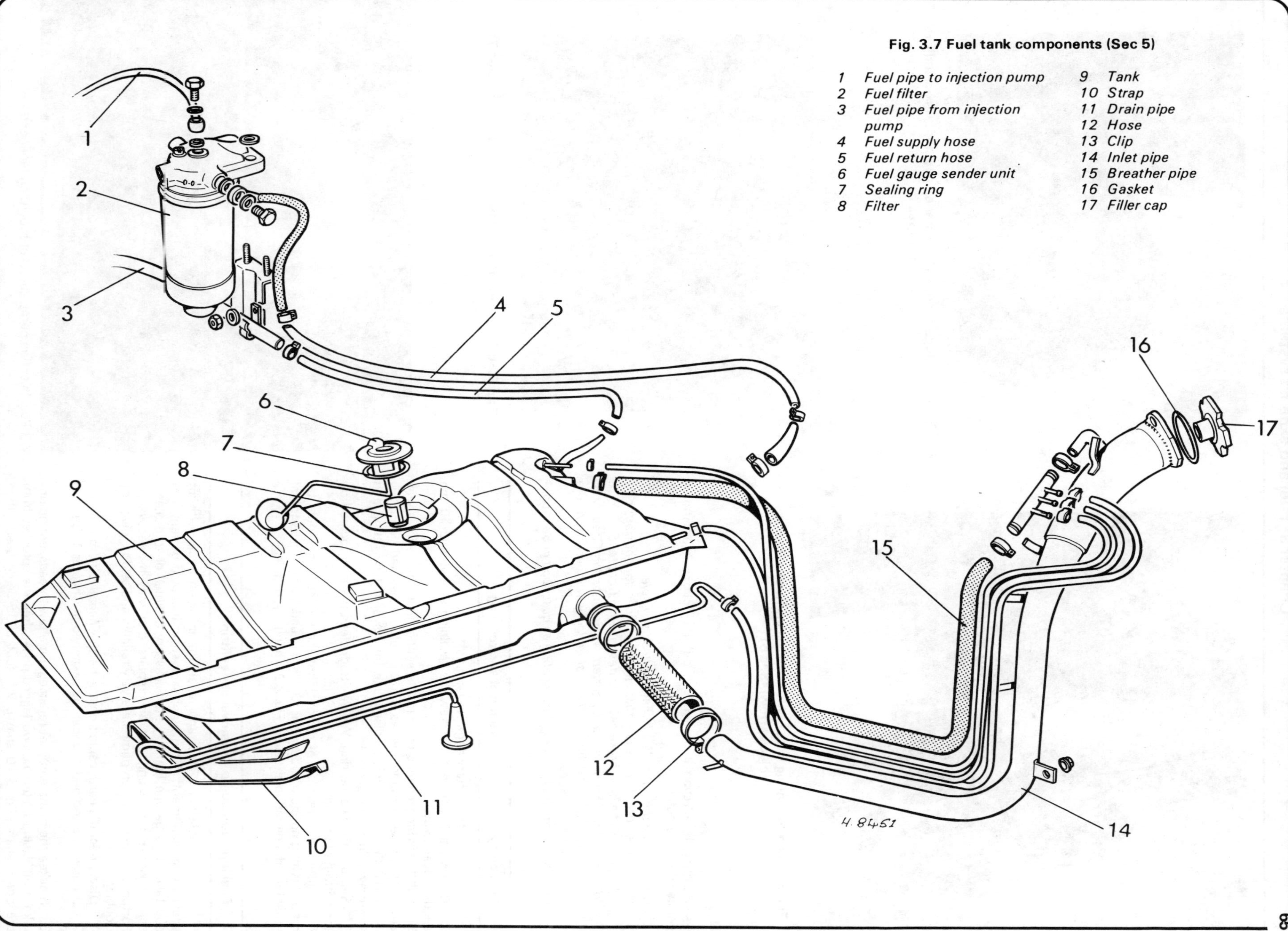

Fig. 3.7 Fuel tank components (Sec 5)

1 *Fuel pipe to injection pump*
2 *Fuel filter*
3 *Fuel pipe from injection pump*
4 *Fuel supply hose*
5 *Fuel return hose*
6 *Fuel gauge sender unit*
7 *Sealing ring*
8 *Filter*
9 *Tank*
10 *Strap*
11 *Drain pipe*
12 *Hose*
13 *Clip*
14 *Inlet pipe*
15 *Breather pipe*
16 *Gasket*
17 *Filler cap*

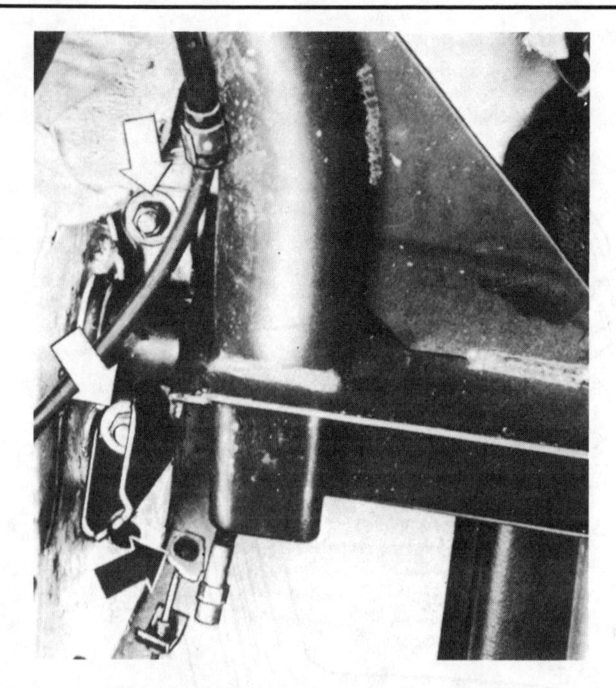

Fig. 3.8 Axle mounting nuts (white arrows) and brake line hose (black arrow) (Sec 5)

14 Refitting is a reversal of removal, but the following additional points should be noted:

(a) *Renew the hose clips if they are unserviceable*
(b) *Make sure that the breather and drain pipes are not trapped*
(c) *Adjust the handbrake cable as described in Chapter 6*
(d) *Bleed the brake hydraulic system as described in Chapter 6; make sure that the polythene sheeting is removed from the reservoir filler cap*
(e) *Bleed the fuel system (if applicable) as described in Section 4*

6 Fuel gauge sender unit – removal and refitting

1 Disconnect the battery negative terminal.
2 Remove the rear seat, unscrew the retaining screws and lift the sender cover away.
3 Pull the wiring connector from the sender unit, then wipe away any dirt from the surrounding area with a paraffin moistened cloth.
4 Twist the sender unit to disengage the bayonet type fitting and withdraw it from the fuel tank.
5 Remove the sealing ring. Where fitted, clean the filter located on the bottom of the unit.
6 Refitting is a reversal of removal, but the sealing ring should be rubbed with graphite powder prior to refitting, and the sender terminal must face the direction shown in Fig. 3.9 when fully installed.

7 Fuel injection pump – removal and refitting

It is not possible to test and repair the injection pump without special test equipment. If the pump is suspect it should be removed and taken to a garage housing the necessary facilities or to a diesel service centre for testing and any repairs that may be necessary. Fortunately fuel injection pumps are very reliable and seldom give any trouble.
1 Disconnect the earth cable from the battery negative terminal.
2 Remove the toothed drivebelt as described in Chapter 1, Section 9.
3 Slacken the pump sprocket retaining nut a few threads, but do not take it off. Using a suitable two-legged puller, free the sprocket from the pump shaft (Fig. 3.10 shows VW tool 203b being used), then remove the puller, retaining nut and sprocket (photo).

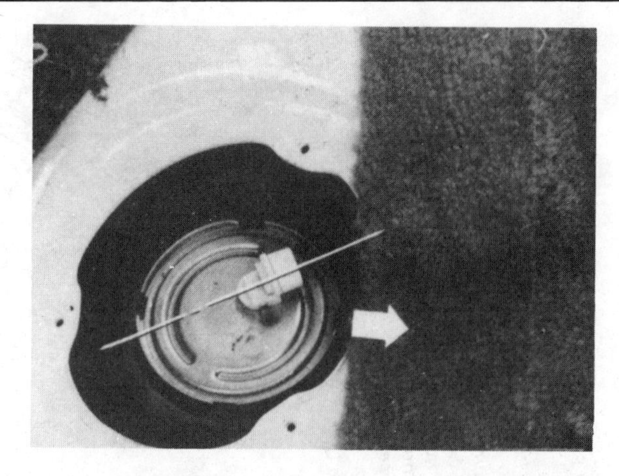

Fig. 3.9 Fuel gauge sender unit installed position. Arrow indicates front of car (Sec 6)

Fig. 3.10 Removing the injection pump sprocket (Sec 7)

7.3 Fuel injection pump sprocket mounting taper and key

7.4 Fuel distribution pipe connections on the injection pump

7.5 Fuel stop solenoid terminal (arrowed) on the Bosch injection pump

7.6 Accelerator cable connection to the throttle lever (Bosch system)

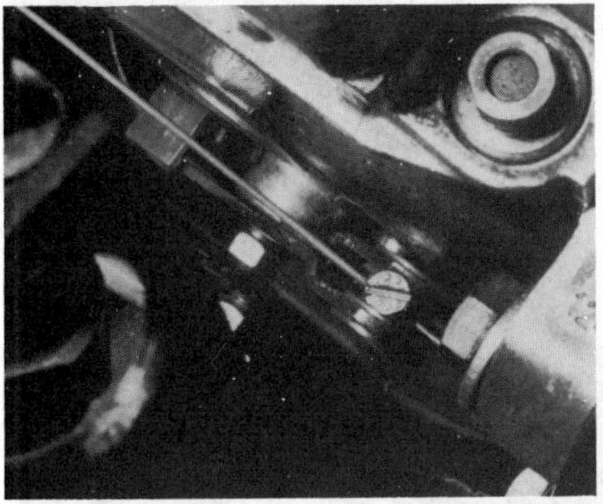

7.7 Cold start cable connection on the injection pump (Bosch system)

7.8A Injection pump front mounting plate and accelerator cable bracket (Bosch system)

7.8B Injection pump rear mounting bolt and plate (Bosch system)

4 Disconnect the fuel lines from the injection pump and cover the union with a clean cloth to prevent the entry of dirt or moisture (photo).
5 Disconnect the electrical lead from the stop solenoid on the pump (photo).
6 Disconnect the accelerator cable from the pump lever, undo the accelerator cable bracket securing nuts and lift away the cable complete with bracket (photo).
7 Disconnect the cold start cable from the pump lever (photo).
8 Remove the injection pump mounting bolts and lift away the pump (photos). Do not unscrew the bolts securing the rear mounting plate to the injection pump.
9 To refit the pump, insert the securing bolts and tighten them finger-tight. Position the injection pump so that the mark on the pump body is aligned with the mark on the mounting plate, as shown in Fig. 3.11, and tighten the bolts to the specified torque.

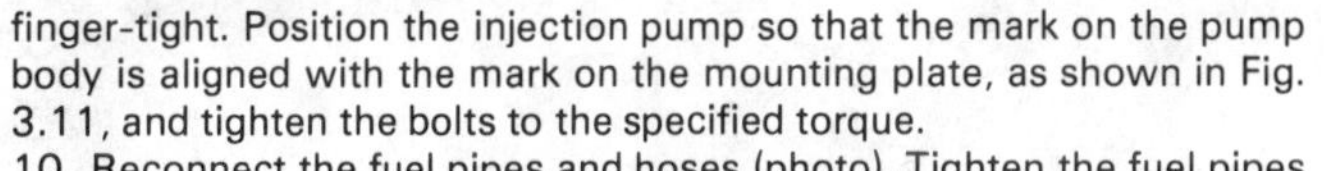

10 Reconnect the fuel pipes and hoses (photo). Tighten the fuel pipes to the specified torque. Do not overtighten as this can result in fuel leakage.
11 Fit the injection pump sprocket. Tighten the securing nut to the specified torque.
12 Check that the flywheel TDC mark is still aligned with the pointer on the bellhousing, then turn the pump sprocket to align the mark on the sprocket with the mark on the pump mounting plate, see Fig. 3.12

Fig. 3.11 Injection pump and mounting plate alignment marks

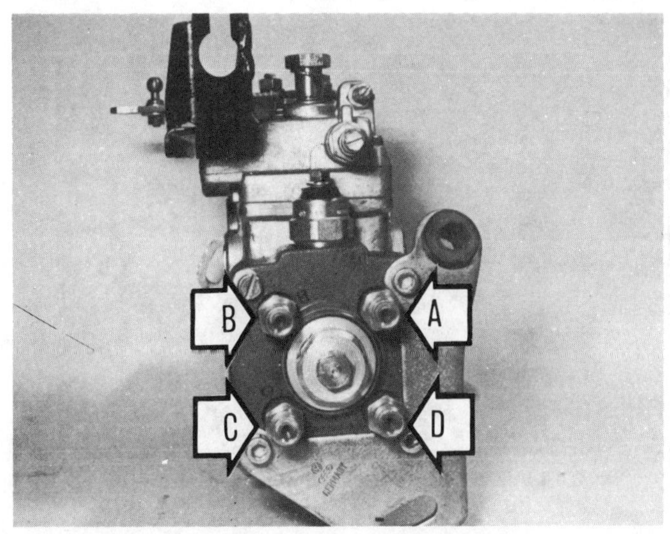

7.10 The fuel distribution pipe unions on the injection pump (Bosch system). A – No 1 injector B – No 3 injector C – No 4 injector
D – No 2 injector

Fig. 3.12 Bosch injection pump alignment marks (black arrow), and locking pin (white arrow) (Sec 7)

7.12A Flywheel TDC notch and bellhousing pointer

7.12B Injection pump timing marks (arrowed)

Fig. 3.13 TDC setting fixture for use when engine removed (Sec 8)

Fig. 3.14 Bosch injection pump timing plug (arrowed) (Sec 8)

Fig. 3.15 Checking the Bosch injection pump timing with a dial gauge (Sec 8)

(photos). Now lock the sprocket in position with a stepped pin of suitable diameter, as described in Chapter 1, Section 9.

13 Refit and tension the drivebelt as described in Chapter 1, Section 9.

14 Refer to Section 8 and adjust the injection pump timing.

15 Reconnect the electrical lead to the stop solenoid and refit the accelerator cable and bracket, and the cold start cable, to the pump. Adjust the cables, if necessary, as described in Sections 10 and 11.

16 For the remainder of the refitting procedure, refer to Chapter 1, Section 9, then bleed the fuel system if necessary as described in Section 4.

8 Injection pump timing (Bosch system) – checking and adjusting

1 Turn the engine until No 1 piston is at TDC on compression stroke (both valves closed, cam lobes pointing upwards) and the TDC notch on the flywheel is in alignment with the pointer on the bellhousing. If the transmission is removed, a setting fixture is required to determine the TDC point. Fig. 3.13 shows VW tool No 2068 being used.

2 Remove the plug from the pump cover (Fig. 3.14) and screw in an adapter and dial gauge in place of the plug as shown in Fig. 3.15. The dial gauge should have a range of 0 to 3 mm.

3 Screw the adapter in until the gauge shows a reading of 2.5 mm, then slowly turn the crankshaft anti-clockwise until the needle on the gauge stops moving. Now zero the dial gauge with a 1 mm preload.

4 Rotate the crankshaft in a clockwise direction to align the TDC notch on the flywheel with the pointer and check that the gauge reads 0.83 mm.

5 Adjust, if necessary, by slackening the pump mounting bolts and turning the pump to adjust the lift to 0.83 mm, then retighten the mounting bolts to the specified torque.

6 If the correct lift cannot be achieved the valve timing must be checked.

7 Refer to Chapter 1, Section 9 and remove the valve cover and drivebelt cover.

8 Ensure that the engine is set with No 1 piston at TDC on the compression stroke and then lock the camshaft in position with tool 2065, or a suitable substitute.

9 Check that the lockpin fits in the holes in the pump sprocket and injection pump mounting plate. If the pin cannot be located in both holes the valve timing will have to be adjusted.

10 To adjust the valve timing, slacken the drivebelt tensioner locknut, then slacken the camshaft sprocket retaining nut half a turn and free the sprocket from the taper on the camshaft by tapping it with a plastic hammer. This will allow the sprocket to be rotated independently of the camshaft.

11 Now rotate the injection pump sprocket to align the mark on the sprocket with the mark on the pump mounting plate, see Fig. 3.12, and fit the locking pin to lock the pump sprocket in position.

12 Adjust the tension of the drivebelt as described in Chapter 1, Section 9, then tighten the camshaft sprocket retaining bolt to the specified torque.

13 Remove the camshaft locking tool and the pump sprocket lockpin.

14 Recheck the injection pump timing.

15 Refit the drivebelt cover and valve cover, refer to Chapter 1, Section 9.

9 Injection pump timing (CAV system) – checking and adjusting

1 Follow the instructions given in Section 8, paragraph 1.

2 Clean the area surrounding the inspection cover on the injection pump, then remove the retaining screws and withdraw the cover. Check that the cold start control is fully released.

3 With No 1 piston at TDC as set in paragraph 1, the fixed pointer must be in alignment with the mark which passes through the hole on the rotor driveplate (see Fig. 3.16). If necessary, loosen the pump flange retaining bolts and the mounting bracket bolt at the rear, and turn the injection pump as required.

4 When the adjustment is correct, tighten the bolts and check the timing by turning the engine through two complete revolutions until No 1 piston is at TDC again.

5 Refit the inspection cover and tighten the retaining screws.

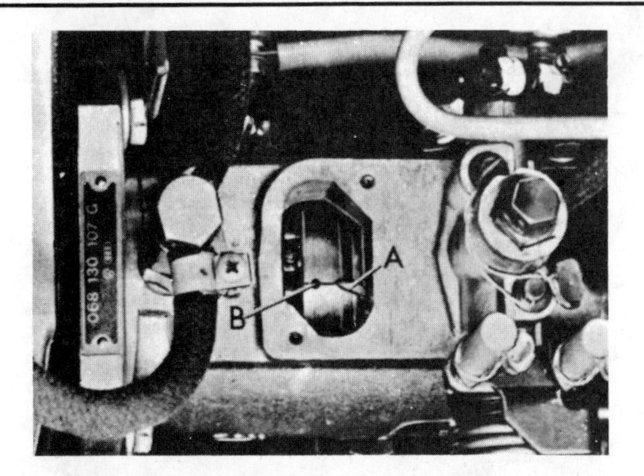

Fig. 3.16 CAV injection pump timing pointer (A) and drilled mark (B) (Sec 9)

10 Accelerator cable – adjustment

Note: *The following paragraphs refer to the Bosch system, but the same principles apply to the CAV system.*

1 Ensure that the ball-pin on the pump lever is pointing upwards and is opposite the line marked on the lever (Fig. 3.17 and photo).

2 Check that the accelerator cable is located in the upper hole of the cable support bracket, indicated by the black arrow in Fig. 3.17 (photo).

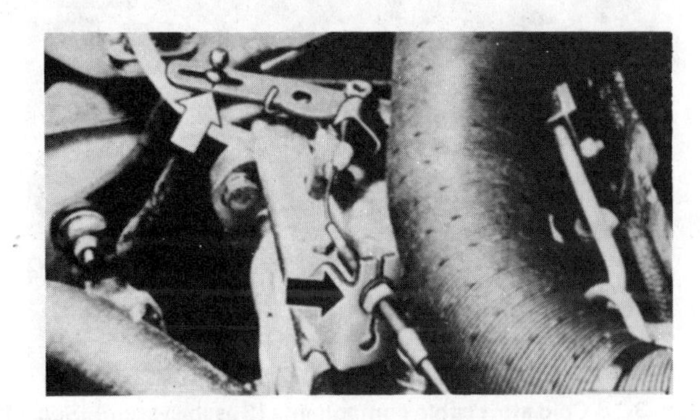

Fig. 3.17 Throttle lever ball-pin mark (white arrow) and accelerator cable location (black arrow) (Sec 10)

10.1 Bosch injection pump, showing the throttle lever

10.2 Injection pump end of the accelerator cable

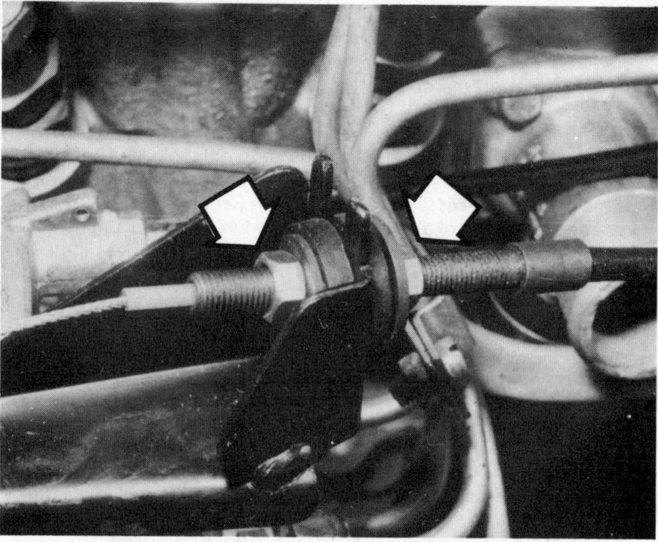

10.3 Accelerator cable adjustment locknuts (arrowed) (Bosch system)

Fig. 3.18 Cold start cable components (Bosch system) (Sec 11)

1 Washer *2 Lockwasher* *3 Pin*

11.3 Bosch injection pump, showing the cold start lever

Fig. 3.19 Idle speed adjusting screw (arrowed) on the Bosch injection pump (Sec 12)

Fig. 3.20 Maximum speed adjusting screw (arrowed) on the Bosch injection pump (Sec 12)

3 Have an assistant depress the accelerator fully, then adjust the cable by turning the locknuts as necessary until the pump lever just contacts the maximum rpm adjusting screw without putting any strain on the cable (photo).
4 With the cable correctly adjusted, ensure that the adjusting nuts are locked against the cable support bracket, and then check the operation of the cable while your assistant operates the accelerator pedal.

11 Cold starting cable – adjustment

Refer to the note at the beginning of Section 10.
1 When the cold starting cable knob on the dashboard is pulled out, this advances the fuel injection pump timing by 2.5° and thus improves starting.
2 When fitting the cable to the pump insert the washer (1 in Fig. 3.18) onto the cable, then fit the outer cable in the bracket wih the rubber bush.
3 Connect the inner cable to the cold start lever by inserting the cable in pin (3) and then fit the lockwasher (2) (photo).
4 Move the lever as far as possible in the direction of the arrow (Fig. 3.18), then while holding the lever in this position, pull the inner cable tight and lock it in the pivot pin by tightening the clamp screw.

12 Idle speed and maximum speed – checking and adjusting

Checking and adjustment of the idle speed and maximum speed requires the use of special equipment, adaptor VW 1324 and tester VW 1267, so this is a job for your local VW garage. For owners who have access to this equipment the checking and adjustment procedure is described below.
1 Connect the test equipment to the engine in accordance with the instructions printed on the equipment.
2 Start the engine and warm it up to normal operating temperature, engine oil temperature at 50°–70°C (122°–158°F).

Idle speed (Bosch system)

3 Check that the idle speed is between 770 and 870 rpm, if not, slacken the idle adjusting screw locknut and screw the adjusting screw in or out as necessary to obtain the correct idle speed.
4 Hold the adjusting screw to prevent it turning and tighten the locknut.

Maximum speed without load (Bosch system)

5 Move the pump lever to the full throttle position and check that the engine speed is between 5400 and 5450 rpm. If necessary adjust by slackening the adjusting screw locknut, arrowed in Fig. 3.20, and turning the adjusting screw to obtain the specified setting.
6 Tighten the adjusting screw locknut.
7 Remove the test equipment.

Idle speed (CAV system)

8 Check that the cold start control is fully released.
9 Increase the engine speed to 2000 rpm several times, then allow it to idle. The idle speed should be between 760 and 860 rpm.
10 If adjustment is necessary, first loosen the locknut on the idle speed stabilising screw (see Fig. 3.21) and back off the screw from the throttle shaft. The screw may be locked with locking wire and a lead seal, and the owner is warned that warranty conditions may be infringed if the seal is disturbed.
11 Loosen the idle speed adjustment locknut and adjust the screw until the engine speed is between 820 and 840 rpm. Tighten the locknut.
12 Turn the idling speed stabilising screw clockwise until the engine speed just starts to rise, then back it off $\frac{1}{2}$ to 1 turn and tighten the locknut.
13 Increase the engine speed to 2000 rpm several times and recheck the idling speed.

Maximum speed without load (CAV system)

14 Start the engine and fully open the throttle; the maximum engine speed should be between 5450 and 5650 rpm. **Note:** *If the setting is too high, do not exceed 5850 rpm.*
15 If adjustment is necessary, loosen the locknut and adjust the stop screw (arrowed in Fig. 3.22) as required. Tighten the locknut.

13 Fuel injectors – removal and refitting

1 Injector failure is usually indicated by knocking in one or more cylinders, engine overheating, loss of power, smoky black exhaust and increased fuel consumption. To locate a faulty injector run the engine at a fast idle (engine at normal operating temperature) and slacken the pipe union on each injector in turn. If there is no drop in engine speed when a pipe is slackened then that injector is faulty.
2 Stop the engine, then disconnect the fuel pipe and leak off pipes from the injector (photo).
3 Unscrew the injector from the cylinder head with a 27 mm ring spanner (photos).
4 Remove the heat shield from the bottom of the injector hole in the cylinder head (photo). Discard the heat shield.
5 Fit a new heat shield and then screw the injector into the cylinder head and tighten it to the specified torque.
6 Reconnect the fuel pipes to the injector and tighten the union nut to the specified torque. Do not overtighten as this may result in fuel leakage if the union nut is distorted.

Fig. 3.21 Idle speed stabilising screw (1) and adjusting screw (2) on the CAV injection pump. Arrow indicates locknut (Sec 12)

Fig. 3.22 Maximum speed adjusting screw (arrowed) on the CAV injection pump (Sec 12)

13.2 Injector and fuel pipe union

13.3A Removing an injector

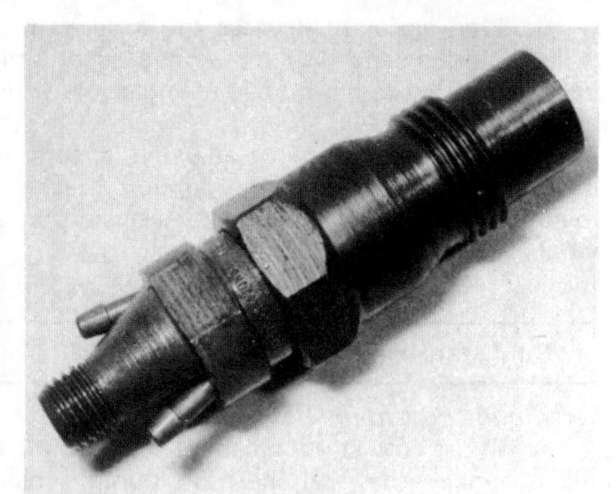

13.3B A Bosch injector, showing leak-off outlets

13.3C An injector nozzle

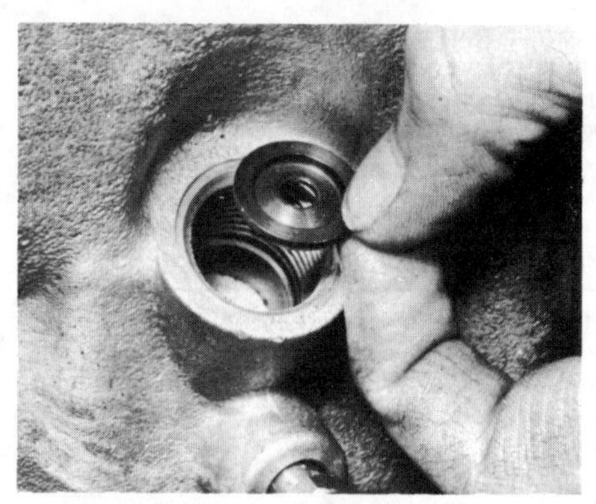

13.4 Removing an injector heat shield

14 Fuel injectors – testing

1 A pressure pump is required for testing the fuel injectors. Fig. 3.23 shows an injector being tested with pressure pump VW 1322. For those owners having access to a fuel injector test pump the testing procedure is described in the following paragraphs.

WARNING: *Never expose your hands to injector spray as working pressure can cause the fuel oil to penetrate the skin.*

2 Connect the injector to the test pump as shown in Fig. 3.23.

Spray testing

3 Close the valve on the pump to isolate the gauge.

4 With rapid short strokes of the test pump lever, 4 to 6 strokes per second, the spray should be even and stop cleanly after each stroke. The injectors must not drip.

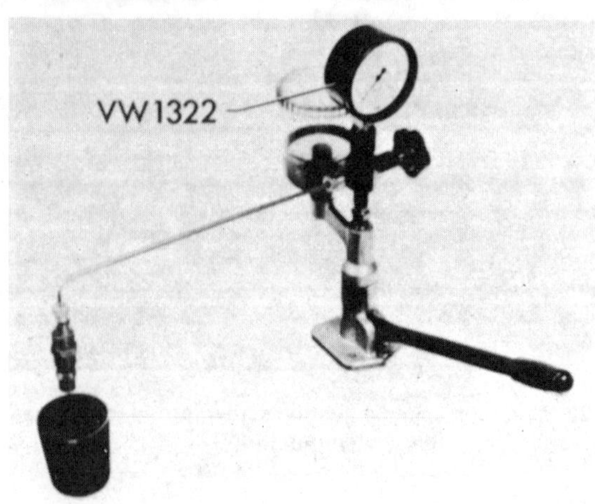

Fig. 3.23 Testing an injector with a pressure pump (Sec 14)

Breaking pressure test

5 Open the valve on the test pump so that the gauge will register.

6 Press the pump lever down very slowly and note the pressure at which the injector works. The specified pressure is 120–130 atm (Bosch) or 120–128 atm (CAV). Adjust, if necessary, by altering the shims as described in Section 15. A thicker shim will increase the breaking pressure – a thinner shim will lower the breaking pressure. Shims are supplied in various thicknesses according to the type of system installed. Increasing the shim thickness by 0.05 mm increases the breaking pressure by approximately 5 atmospheres (5.0 kgf/cm^2) and pro rata.

Leakage test

7 With the pump gauge working, very slowly press down on the pump lever until the gauge is registering approximately 110 atmospheres (110 kgf/cm^2), then hold the pressure for 10 seconds and check that there is no fuel leakage from the tip of the nozzle during this period.

8 Repair or renew a faulty injector.

15 Fuel injectors – servicing

1 The injectors can be dismantled for cleaning, renewal of defective ports and adjustment.

2 Clamp the upper part of the injector in a vice and slacken the lower part, then remove it from the vice, turn it over and clamp the lower part in the vice. This will prevent the internal parts falling out as the upper part is unscrewed.

3 Remove all the parts and clean them in diesel fuel. Keep all the parts belonging to the injector together and do not interchange them with parts from another injector. The needle and nozzle are supplied as a matched set. Always renew the heat shield.

4 Assemble the parts in the order shown in Fig. 3.24 and tighten the upper and lower parts to the specified torque.

5 Test and, if necessary, adjust the breaking pressure, refer to Section 14.

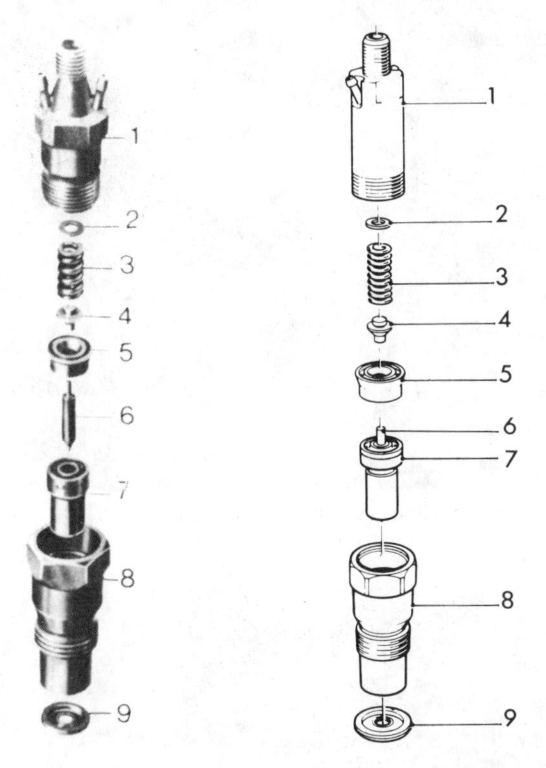

Fig. 3.24 Injector components — Bosch (left) and CAV (right) (Sec 15)

1 *Body*
2 *Shim*
3 *Spring*
4 *Seat*
5 *Nozzle holder*
6 *Needle*
7 *Nozzle*
8 *Nut*
9 *Heat shield*

16 Glow plugs – removal and refitting

1 Provided the glow plugs are working correctly and the engine starts satisfactorily, there is no need to remove the glow plugs except to clean them when the injectors are serviced, normally at 60 000 mile (96 000 km) intervals or every 4 years, whichever occurs first.
2 To remove the glow plugs, first disconnect the battery negative terminal.
3 Unscrew the terminal nuts and remove the bus bar noting that the supply wire is connected to No 4 cylinder (photos).
4 Unscrew the glow plugs and remove them from the cylinder head (photos).
5 Clean any accumulations of carbon from the glow plug tips and apply a little graphite grease to the threads.
6 Refitting is a reversal of removal.

17 Fuel pipes and unions

1 To ensure correct sealing, all fuel pipe unions should only be tightened to the specified torque (photo). Bending of fuel pipes must also be avoided and it is preferable to remove a pipe completely, rather than leave one end attached and bend the other.
2 When fitting fuel distribution pipes to the Bosch injection pump, first tighten the unions into the pump head to the specified torque, then tighten the pipe nuts to the specified torque (photo). The unions and seals should be renewed if a leak occurs after installing them in this manner.
3 Refer to Fig. 3.25 and note that on the Bosch injection pump the inlet union bolt (A) differs from the return union bolt (B). The latter incorporates a restrictor and if fitted in the inlet union will result in unsatisfactory performance, 'hunting', smoky exhaust and incorrect engine maximum speed (photos).

16.3A Glow plug supply wire location

16.3B Glow plug bus bar location

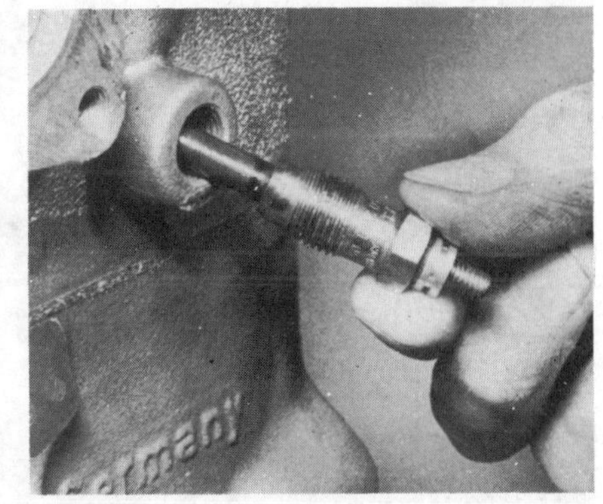

16.4A Removing a glow plug

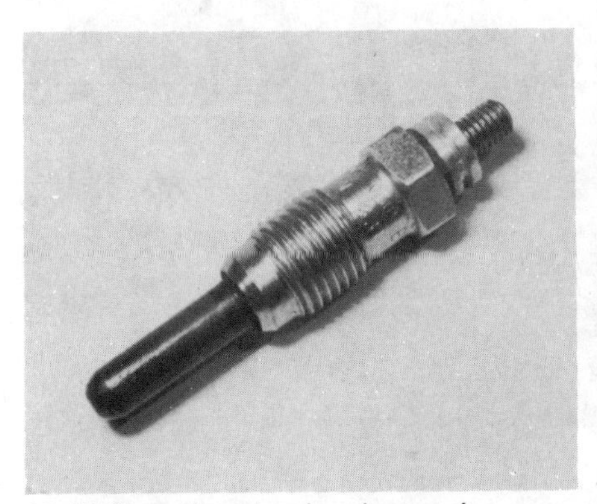

16.4B Glow plug, showing element tip

17.1 Injection pump leak-off outlet union (Bosch system)

17.2 Fuel distribution pipe locations (Bosch system)

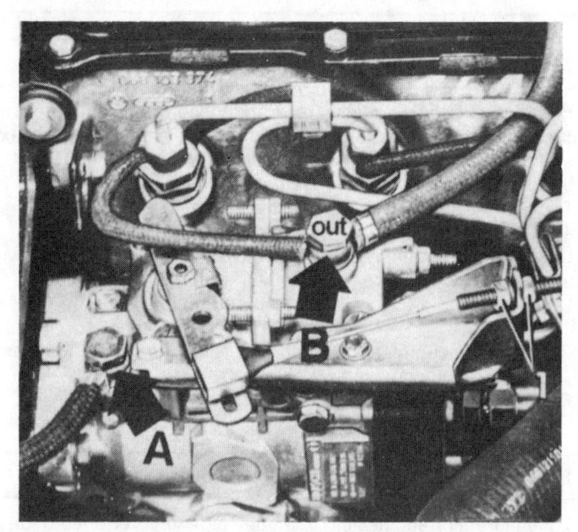

Fig. 3.25 Inlet union (A) and outlet union (B) location on the Bosch injection pump (Sec 17)

17.3A Injection pump inlet union (Bosch system)

17.3B Injection pump inlet union bolt (Bosch system)

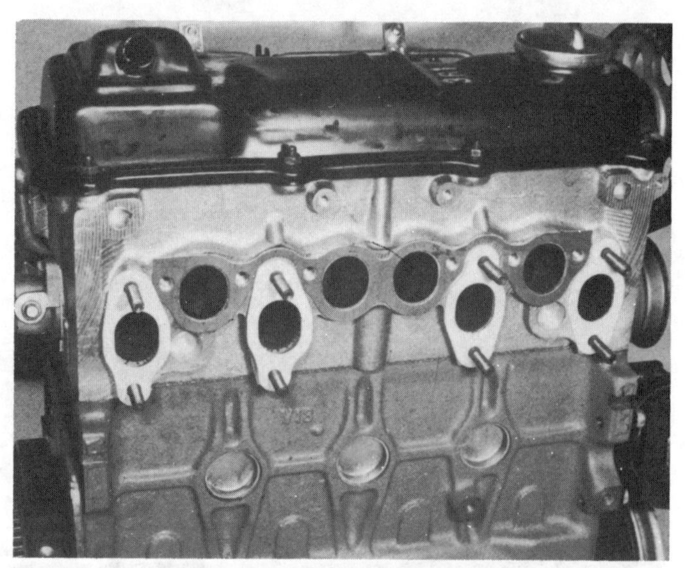
18.1A Cylinder head ports, showing manifold gaskets

18.1B Intake manifold mounting on the exhaust manifold

18.2A Exhaust manifold location

18 Manifolds and exhaust system

1 The intake manifold is bolted to the cylinder head and incorporates four separate induction passages, each communicating with a single cylinder. Always use a new gasket when installing, and tighten the bolts evenly to the specified torque (photos).

2 The exhaust manifold is located just below the intake manifold and is attached to the cylinder head by eight studs. The four gaskets must be renewed whenever the exhaust manifold is removed, and the retaining nuts should always be tightened to the specified torque. The lower end of the exhaust manifold is supported by two brackets (photos).

3 The exhaust system comprises three sections: (a) front pipe and expansion unit which is corrugated and attached to the manifold flange, (b) intermediate pipe and (c) rear section incorporating the silencer. The system is mounted to the body underframe by rubber O-rings (photos).

4 At six monthly intervals, the exhaust system should be examined for leaks and deterioration. Very often a small leak can be repaired by welding, but if the deterioration is extensive the particular section must be renewed. Never ignore a leak, as poisonous gases may seep into

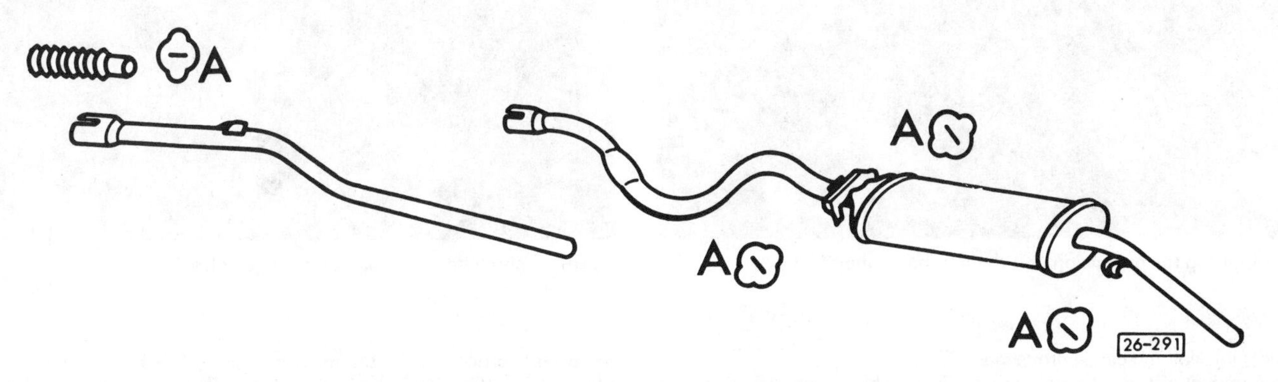

Fig. 3.26 Exhaust system components (Sec 18)

18.2B Right-hand exhaust manifold support bracket

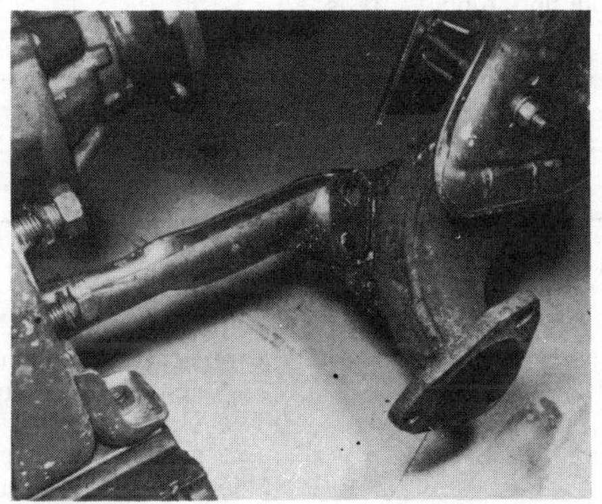

18.2C Left-hand exhaust manifold support bracket

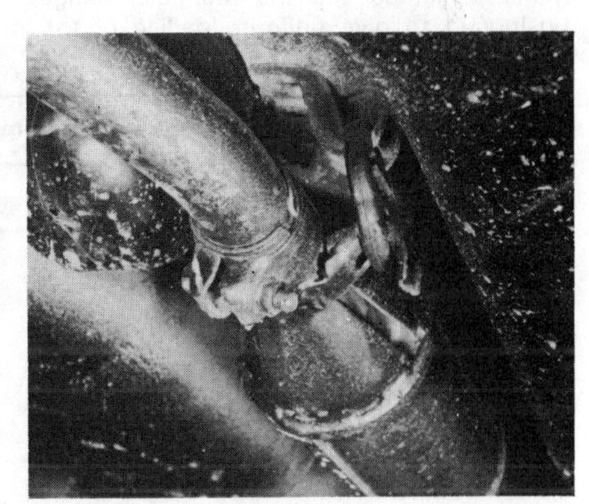

18.3A Centre exhaust mounting and joint

18.3B Rear exhaust mounting

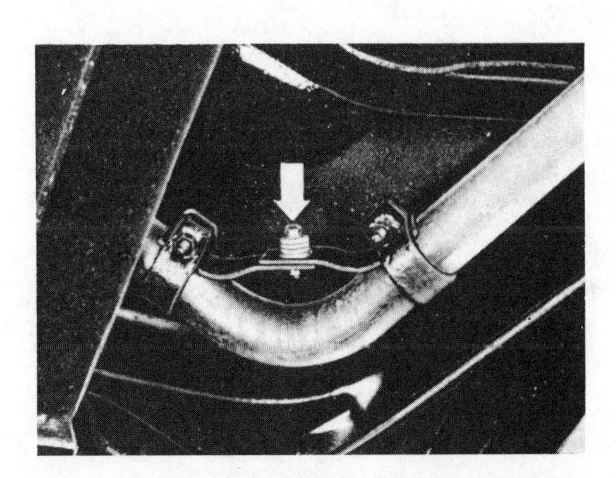

Fig. 3.27 Damping bracket fitted to the rear exhaust section (Sec 18)

18.6A Fitting the expansion unit to the exhaust manifold ...

18.6B ... showing the gasket and flange bolts

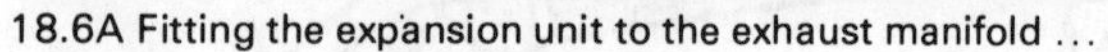

the car interior and cause drowsiness.
5 Removal of the exhaust system is best achieved by disconnecting and removing the rear section first, then removing the front section.
6 Refitting is a reversal of removal, but before tightening the intermediate clamps and the triangular flange, the system should be positioned to give uniform loading of the mounting rubbers. It is important to note that a damping bracket is fitted to the bend in the rear section (Fig. 3.27) and when fitting a new section, the bracket must be transferred from the old section. If it is unserviceable, the bracket must be renewed. Always fit a new gasket between the expansion unit and the exhaust manifold (photos).

19 Fault diagnosis – fuel and exhaust systems

Symptom	Reason/s
Engine does not start	Lack of fuel Battery discharged Stop solenoid faulty or fuse blown Glow plug relay faulty Fuel filter blocked Fuel supply hose trapped Injection timing incorrect Injection pump or injectors faulty
Uneven idling	Incorrect idle speed adjustments Loose fuel pipe or union Rear injection pump bracket broken Air in fuel system Faulty injectors Injection timing incorrect Injection pump faulty
Smoky exhaust	Choked air cleaner element Injection pump adjustments incorrect Faulty injectors Injection timing incorrect Injection pump faulty
High fuel consumption	Choked air cleaner element Leaking fuel pipe or union Return pipes blocked Injection pump adjustments incorrect Faulty injectors Injection timing incorrect Injection pump faulty

Chapter 4 Clutch

Refer to Chapter 11 for specifications and information applicable to later models

Contents

Specifications

Type
Single dry plate, diaphragm spring, cable operated

Clutch pedal free travel
15 mm (5/8 inch)

Driven plate (friction disc)

Maximum out of true (at 179 mm dia)	0·4 mm (0·015 in)
Minimum depth of lining above rivets	0·6 mm (0·023 in)

Pressure plate

Maximum inward taper	0·3 mm (0·012 in)

Torque wrench settings

	kgf m	lbf ft
Pressure plate to crankshaft	7·5	54
Flywheel to pressure plate	2	14
End cover plate	1·5	11

1 General description

Unlike the clutch on most engines, the clutch pressure plate is bolted to the crankshaft flange and the flywheel, which is dish shaped, is bolted to the pressure plate with the friction disc being held between them. This is in effect the reverse of the more conventional arrangement where the flywheel is bolted to the crankshaft flange and the clutch pressure plate bolted to the flywheel (Fig. 4.1).

The release mechanism consists of a metal disc, called the release plate, which is clamped in the centre of the pressure plate by a retaining ring. In the centre of the release plate is a boss into which the clutch pushrod is fitted. The pushrod passes through the centre of the main driveshaft of the gearbox, which is tubular, and is housed in a release bearing situated at the back of the gearbox. A single finger lever presses on this bearing when the shaft to which it is splined is turned by operation of the cable from the clutch pedal. In effect the clutch lever pushes the clutch pushrod, which in turn pushes the centre of the release plate inwards towards the crankshaft. The outer edge of the release plate presses on the pressure plate fingers forcing them back towards the engine and removing the pressure plate friction face from the friction disc, thus disconnecting the drive. When the clutch pedal is released the pressure plate reasserts itself clamping the friction disc firmly against the flywheel and restoring the drive.

It is not possible to dismantle the clutch pressure plate, if it is defective it must be renewed as an assembly.

2 Clutch pedal free travel – adjustment

1 There should be 15 mm (5/8 inch) of free travel downwards on the clutch pedal before the lever on the gearbox begins to turn the withdrawal mechanism shaft. At this point the additional load can be felt at the pedal. This ensures that the withdrawal mechanism is not in contact with the clutch and is not wearing out during normal operation. It also makes certain that the clutch is fully engaged.

2 The easiest way to measure the free play is to use a short wooden strip. Hold it against the pedal with the end on the floor of the car. Mark the position of the top of the clutch pedal. Next make a mark 15 mm (5/8 inch) from the first mark, and replace the strip alongside the pedal as before. Press down the pedal until the end of the free play is felt and make another mark. This mark should coincide with your second mark. If it does not then the cable must be adjusted until it does. 3 mm (1/8 in) either way does not matter. Keep the piece of wood in your tool box for subsequent checks.

3 The cable is adjusted at the gearbox end. At the bracket on the gearbox where the outer cable (cable sheath) ends there are two locknuts. Slacken these, and turn the adjuster. Undo it to shorten the free play, screw it in to increase free play (Fig. 4.2). Two people make the job easier, one to check the measurement, one to do the adjustment.

3 Clutch cable – removal and refitting

1 Slacken off the cable adjuster until it is possible to disconnect the cable from the operating lever on the transmission. Remove the inner and outer cable from the operating lever on the transmission. Remove the inner and outer cable from the bracket on the transmission casing (photos).

2 Working under the facia unhook the cable from the clutch pedal, and then withdraw the cable through the grommet in the bulkhead.

3 Refitting is the reverse of the removal procedure. Adjust the clutch pedal free travel as described in Section 2.

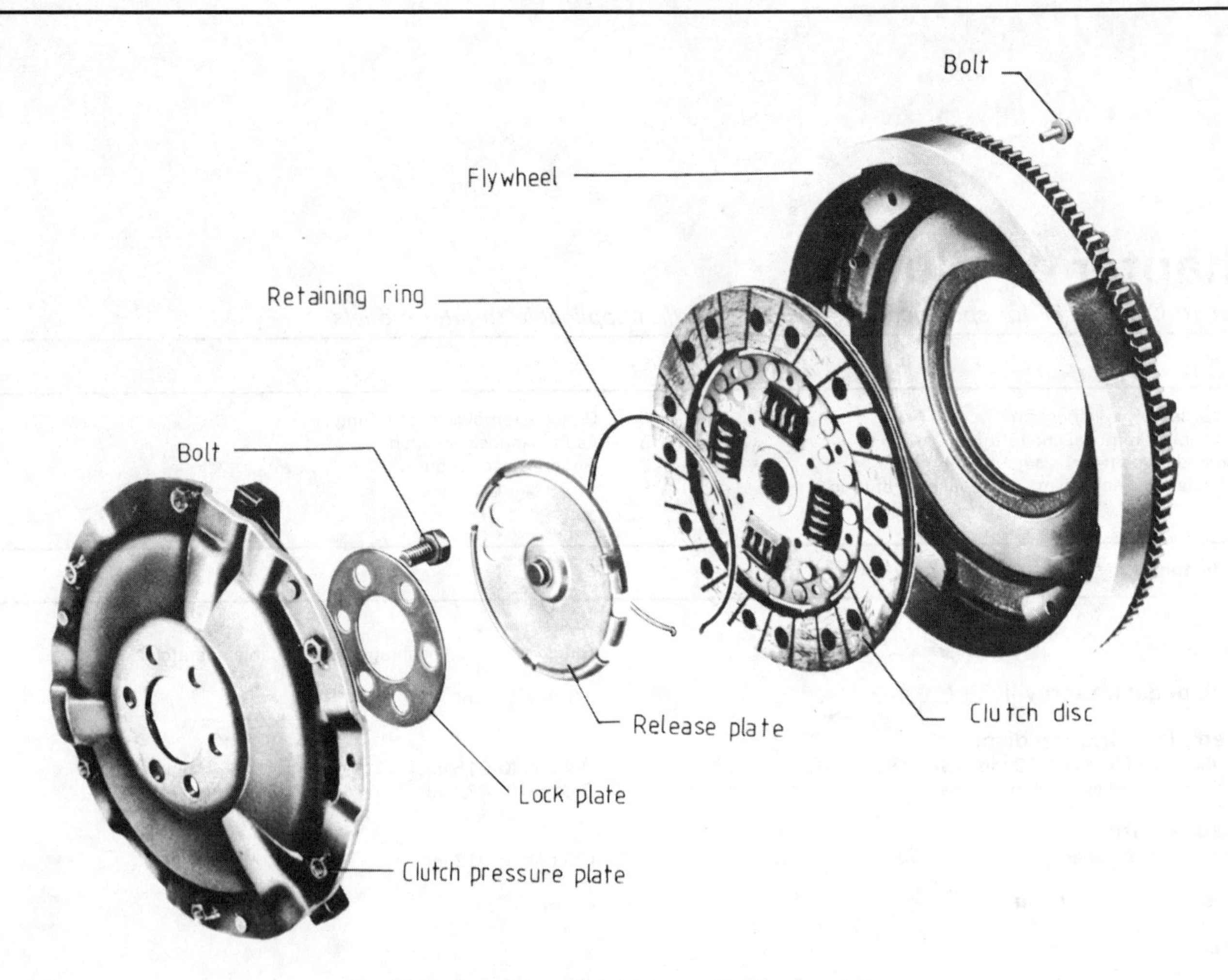

Fig. 4.1 Exploded view of clutch assembly (Sec 1 and 4)

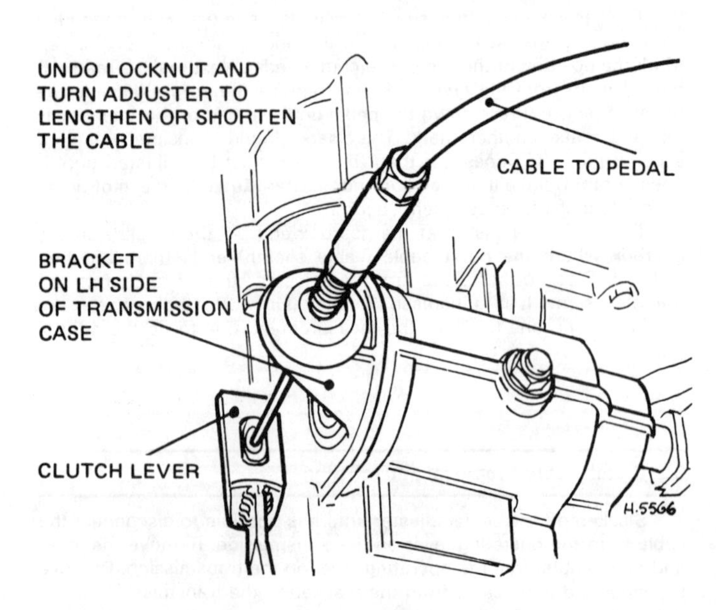

Fig. 4.2 Clutch cable adjustment (Sec 2)

4 Clutch – removal and refitting

1 Support the engine and remove the transmission as described in Chapter 5. Clamp the flywheel to prevent it turning and remove the bolts holding the flywheel to the pressure plate in diagonal fashion. Release each one two or three threads at a time until they are all slack and then take them out. The flywheel and the friction disc may now be removed. Note which way the disc is fitted (Fig. 4.1).

2 Examine the pressure plate surface. If it is clean and free from scoring there is no reason to remove it unless the friction disc shows signs of oil contamination.

3 If the plate surface is defective then it must be removed. Note exactly where the ends of the retaining ring are located (it must be refitted in this way later), and prise the ring out with a screwdriver. The release plate may now be removed. The pressure plate is held to the crankshaft flange by six bolts fitted using a thread locking compound (photos). These will be difficult to remove as they were tightened to a high torque before the locking fluid set, so the plate must be held with a clamp.

4 Refitting of the clutch is the reversal of removal. Use thread locking compound on the bolts securing the plate to the crankshaft flange if they have been removed and tighten them to the specified torque. Make sure that the retaining ring is correctly seated (Fig. 4.3 and photos). Take care that no oil or grease is allowed to get onto the pressure plate or friction surfaces.

5 When refitting the clutch disc make sure the greater projecting boss which incorporates the torsion springs is furthest from the engine, then fit the flywheel over the pressure plate. Fit the securing bolts and tighten them finger-tight only.

3.1A Clutch operating lever and cable

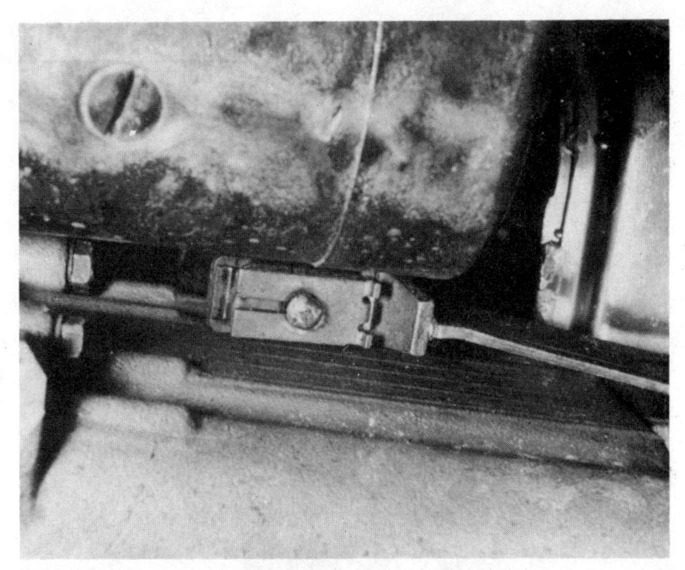

3.1B Clutch inner cable connection to operating lever

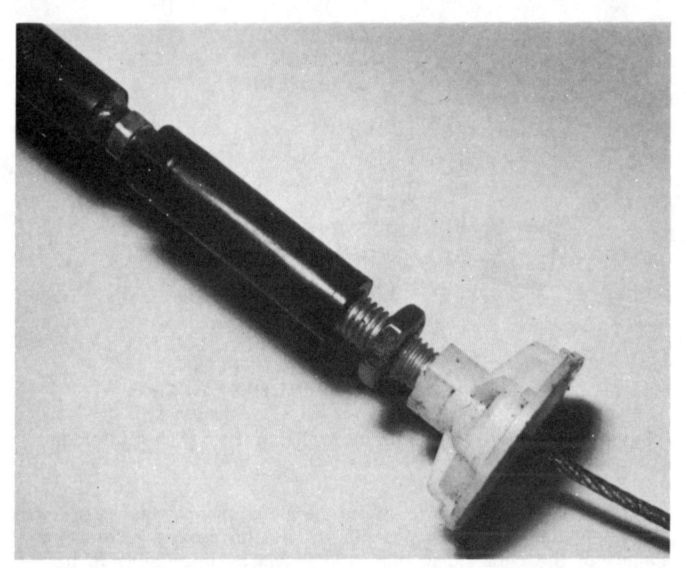

3.1C Clutch adjustment and locknut

4.3A Removing the clutch release plate (retaining ring already extracted)

4.3B Removing the clutch pressure plate

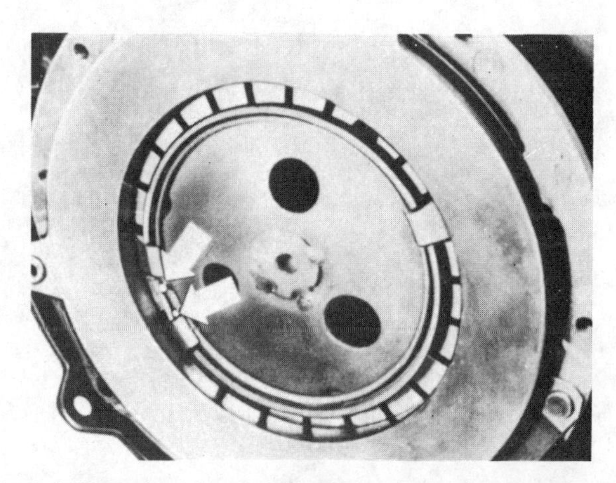

Fig. 4.3 The ends of the retaining ring must rest between the two cut-outs of the release plate as shown – arrowed (Sec 4)

4.4A Refitting the clutch pressure plate

4.4B Flywheel expanding peg for locating pressure plate

Fig. 4.4 Using VW tool 547 to centre the clutch friction disc (Sec 4)

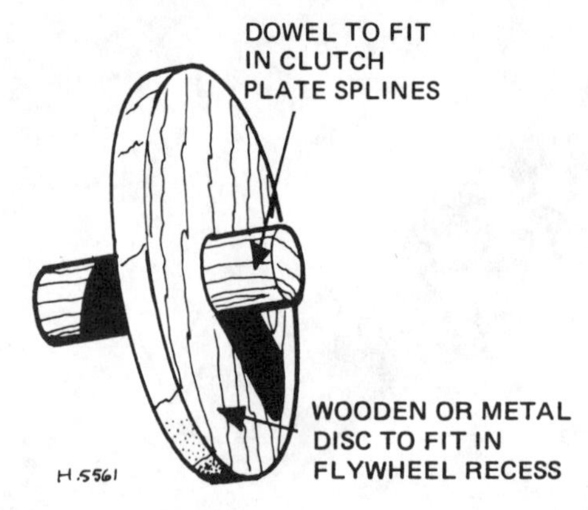

Fig. 4.5 Home-made tool for centralising the clutch friction disc (Sec 4)

4.6 Using a vernier caliper to double check the clutch disc centralisation

6 The next operation is to centre the clutch disc. If this is not done accurately the gearbox mainshaft will not be able to locate in the splines of the clutch disc hub, and it will be impossible to fit the gearbox. The best centralising tool is VW 547 which fits in the flywheel and has a spigot which fits exactly in the centre of the clutch disc hub (Fig. 4.4). If you cannot borrow or hire tool VW 547 then we suggest you make up a tool as shown in Fig. 4.5. Once the clutch disc is centred correctly, tighten the securing bolts in a diagonal sequence to the specified torque and check the centralisation again (photo).

7 When refitting the transmission, put a smear of multi purpose grease on the end of the clutch pushrod at the release plate end.

5 Clutch assembly – inspection

1 The most probable part of the clutch to require attention is the friction disc. Normal wear will eventually reduce its thickness. The lining must stand proud of the rivets by not less than 0.6 mm (0.025 inch). At this measurement the lining is at the end of its life and a new friction plate is needed.

2 The friction disc should be checked for run-out if possible. Mount the disc between the centres of a lathe and measure the run-out at 175 mm diameter. The quoted limit is given in the Specifications. However, this requires a dial gauge and a mandrel. If the clutch has not shown signs of dragging then this test may be passed over, but if it has we suggest that expert help be sought to test the run-out (Fig. 4.6).

3 Examine the pressure plate. There are three important things to check. Put a straight-edge across the friction surface and measure any bow or taper with feeler gauges. This must not exceed the specified value. The fingers of the diaphragm will be rough. Scoring of up to 0.3 mm (0.012 ins) is acceptable. If the surface is rough enough to

Fig. 4.6 Checking the friction disc for run-out (Sec 6)

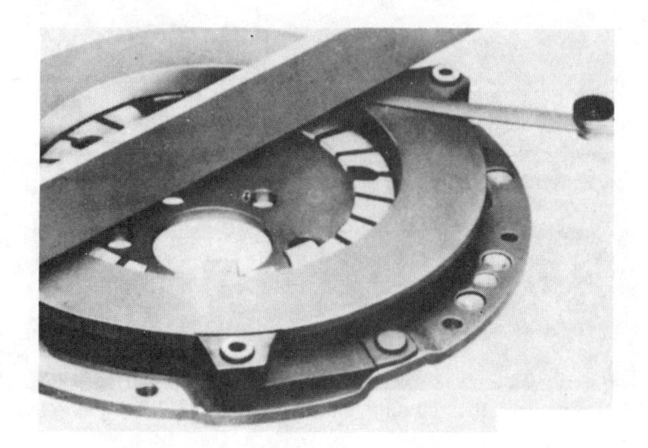

Fig. 4.7 Checking the inward taper of the pressure plate friction surface (Sec 4)

damage the release bearing remove the burrs with a stone. If in doubt consult the VW agent.

4 The rivets which hold the spring fingers in position must be tight. If any of them are loose the pressure plate must be scrapped. Finally, the condition of the friction surface. Ridges or scoring indicate undue wear and unless they can be removed by light application of emery paper it would be better to renew the plate.

5 The flywheel friction surface must be similarly checked.

6 So far the inspection has been for normal wear. Two other types of damage may be encountered. The first is overheating due to clutch slip. In extreme cases the pressure plate and flywheel may have radial cracks. Such faults mean that they require renewal. The second problem is contamination by oil or grease. This will cause clutch slip but probably without the cracks. There will be shiny black patches on the friction disc which will have a glazed surface. There is no cure for this, a new friction disc is required. In addition it is **imperative** that the source of contamination be located and rectified. It will be either the engine rear oil seal, or the gearbox front oil seal, or both. Examine them and renew the faulty ones before fitting a new disc. Failure to do this will mean dismantling the transmission again very shortly. The fitting of new seals is discussed in Chapters 1 and 5.

6 Clutch release mechanism – removal and refitting

1 The clutch release mechanism is accessible after removal of the gearbox end cover plate. Undo the four securing screws and remove the cover plate (photo). The clutch torque shaft with its finger is now visible (Fig. 4.8).

2 The finger is held in place by two circlips and a spring. The two circlips can now be removed and the shaft, finger and return spring removed from the casing (photos).

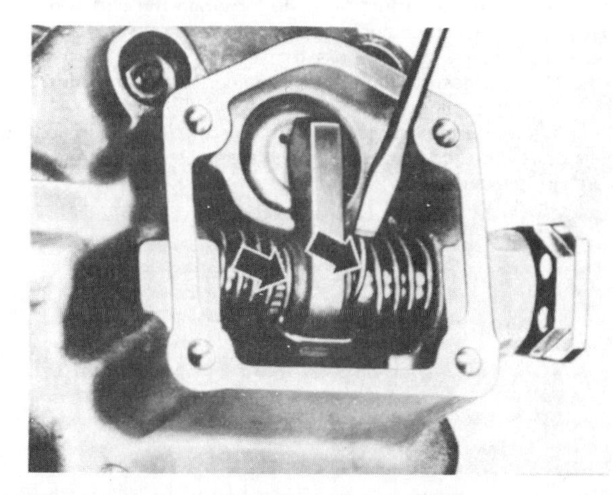

Fig. 4.8 View of clutch release mechanism with gearbox end cover removed. Arrows indicate location of the two circlips (Sec 6)

3 Lift out the clutch release bearing and guide sleeve (photo). The pushrod cannot be withdrawn until the transmission is in the car.

4 Rotate the clutch release bearing: if it is noisy or feels rough it must be renewed. Inspect the shaft oil seal for wear or deterioration

6.1 Removing the gearbox end cover plate

6.2A Removing the clutch release finger retaining circlips

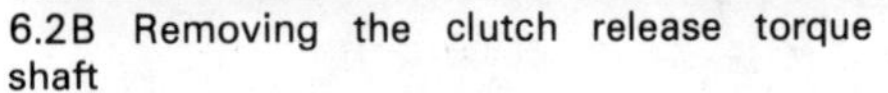

6.2B Removing the clutch release torque shaft

6.3 Removing the clutch release bearing

6.5 Fitted position of the clutch release finger return spring

and renew it, if necessary.

5 Refitting is the reverse of the removal procedure. When fitting the return spring ensure that the bent ends bear against the gearbox casing and the centre part hooks into the clutch finger (photo). Always use a new gasket when fitting the end cover.

7 Fault diagnosis – clutch

There are four main faults to which the clutch and release mechanism are prone. They may occur by themselves or in conjunction with any of the other faults. They are clutch squeal, slip, spin and judder.

Clutch squeal

1 If on taking up the drive or when changing gear, the clutch squeals, this is a good indication of a badly worn clutch release bearing.

2 As well as regular wear due to normal use, wear of the clutch release bearing is much accentuated if the clutch is ridden, or held down for long periods in gear, with the engine running. To minimise wear of this component the car should always be taken out of gear at traffic lights and for similar holdups.

Clutch slip

3 Clutch slip is a self-evident condition which occurs when the clutch pedal free travel is insufficient, the clutch friction plate is badly worn, when oil or grease have got onto the flywheel or pressure plate faces, or when the pressure plate itself is faulty.

4 The reason for clutch slip is that, due to one of the faults listed above, there is either insufficient pressure from the pressure plate, or insufficient friction from the friction plate to ensure solid drive.

5 If small amounts of oil get onto the clutch, they will be burnt off under the heat of clutch engagement, and in the process, gradually darken the linings. Excessive oil on the clutch will burn off leaving a carbon deposit which can cause quite bad slip, or fierceness, spin and judder.

6 If clutch slip is suspected, and confirmation of this condition is required, there are several tests which can be made.

7 With the engine in second or third gear and pulling lightly up a moderate incline sudden depression of the accelerator may cause the engine to increase in speed without any increase in road speed. Easing off on the accelerator will then give a definite drop in engine speed without the car slowing.

8 In extreme cases of clutch slip the engine will race under normal acceleration conditions.

9 If slip is due to oil or grease on the linings a temporary cure can sometimes be effected by squirting carbon tetrachloride into the clutch. The permanent cure is, of course, to renew the clutch friction disc and trace and rectify the oil leak.

Clutch spin

10 Clutch spin is a condition which occurs when the release arm travel is excessive, there is an obstruction in the clutch either on the mainshaft splines or in the operating lever itself, or oil may have partially burnt off the clutch linings and have left a resinous deposit which is causing the clutch disc to stick to the pressure plate or flywheel.

11 The reason for clutch spin is that due to any, or a combination of the faults just listed, the clutch pressure plate is not completely freeing from the centre plate even with the clutch pedal fully depressed.

12 If clutch spin is suspected, the condition can be confirmed by extreme difficulty in engaging first gear from rest, difficulty in changing gear, and very sudden take up of the clutch drive at the fully depressed end of the clutch pedal travel as the clutch is released.

13 Check that the clutch cable is correctly adjusted and, if in order, the fault lies internally in the clutch. It will then be necessary to remove the clutch for examination and to check the gearbox mainshaft.

Clutch judder

14 Clutch judder is a self evident condition which occurs when the gearbox or engine mountings are loose or too flexible, when there is oil on the faces of the clutch friction plate, or when the clutch pressure plate has been incorrectly adjusted during assembly.

15 The reason for clutch judder is that due to one of the faults just listed, the clutch pressure plate is not freeing smoothly from the friction disc, and is snatching.

16 Clutch judder normally occurs when the clutch pedal is released in first gear or reverse gear, and the whole car shudders as it moves backwards or forwards.

Chapter 5 Transmission

Refer to Chapter 11 for specifications and information applicable to later models

Contents

Specifications

Type

Four forward speeds (all synchromesh), one reverse. Final drive integral with main gearbox. Transfer to front wheels by double CV jointed driveshafts

Gear ratios

1st	3·45 : 1
2nd	1·94 : 1
3rd	1·37 : 1
4th	0·969 : 1
Reverse	3·17 : 1
Final drive	3·89 : 1

Oil capacity

1·25 litres (2·2 Imp pints, 2·65 US pints)

Oil type

Hypoid oil marked GL4. SAE 80 or 80W/90

Torque wrench settings

	kgf m	lbf ft
Transmission to engine bolts	5·5	40
Driveshaft flange bolts	4·5	32
Engine mounting bolts	5·5	40
Pinion shaft bearing retainer bolts	4·0	29
Housing flange joint bolts	2·0	14
End cover plate bolts	1·5	11
Selector shaft cover	4·5	32
Selector shaft detent plunger	2·0	14
Reverse gear shaft lock bolt	2·0	14
Mainshaft bearing retaining nuts	1·5	11

1 General description

The transmission assembly is bolted to the rear of the engine. The larger section of the casing, the gearbox housing, is withdrawn from the bearing housing leaving the gear trains and the differential unit in the bearing housing.

The mainshaft can be removed complete, but the pinion shaft must be completely dismantled, while held in the bearing housing, before it can be removed. Only after the removal of the pinion shaft can the differential unit be lifted out. The drive for the speedometer is provided by the helical teeth on the pinion shaft.

The clutch release mechanism is housed in the rear of the gear casing and is accessible after removal of the end cover plate. The release shaft bearing and sleeve can be removed while the gearbox is in the car (see Chapter 4).

The inner selector shaft operates one of the three shift rods. These are controlled by gear detents and interlocks located in the gearbox housing. The 1st/2nd gear shift rod operates a fork which engages with the synchromesh hub for 1st/2nd gear on the pinion shaft. A similar arrangement operates the 3rd/4th gearchange, moving gears on the mainshaft. A third shift rod operates the reverse gearchange by means of a relay lever. The reverse gear is located in the end of the gearbox housing on a short shaft which is pressed into the casing. The shaft does not rotate. Reverse gear has a plain metal bush and no roller or ball bearings.

Power is transferred to the differential unit by a helical gear on the pinion shaft which meshes with the ring gear on the differential unit. The drive flanges can be removed without removing the transmission assembly from the car, see Chapter 8, and the oil seals renewed. For all other repairs to the transmission the assembly must be removed from the car.

2 Transmission removal – general

1 The transmission is removed downwards so the car must be raised from the ground to provide sufficient height for the withdrawal of the gearbox from underneath. The ideal arrangement is to work over a pit, but axlestands or similar suitable supports under the body can be used. Note however that you must be able to rotate the wheels when disconnecting the driveshafts from the transmission drive flanges. Approximately 60 cm (24 inches) clearance is required to withdraw the transmission assembly from under the car.
2 Since the engine will be left unsupported at the rear it is necessary to make provision to take its weight. If a suitable hoist is available this will be a simple matter, but if not, a support similar to that shown in Fig. 5.1 can be made and placed in position with the end supports resting in the channels which house the bonnet sides on top of the wings. Alternatively the engine can be supported from underneath with blocks placed under the sump but this method means that the car cannot be moved while the transmission is out of the car.

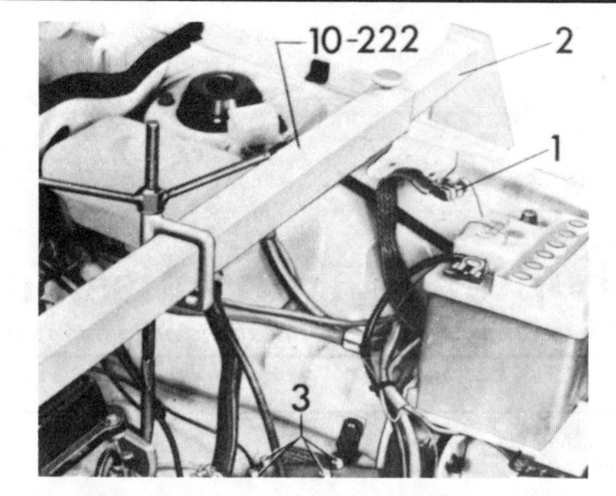

Fig. 5.1 Engine support bar necessary when removing the transmission only (Sec 2)

1 Battery earth strap
2 Engine support bar
3 Left-hand transmission mounting bolts

3 Transmission – removal and refitting

1 Support the engine as described in Section 2.
2 Remove the ground strap from the battery negative terminal.
3 Remove the drain plug and drain the transmission oil into a suitable container.
4 For purposes of this Section, the front of the transmission is regarded as the end nearest the engine. Left and right are as if you are standing at the left-hand side of the car behind the transmission looking towards the engine.
5 Remove the plastic plug from the timing access hole in the bellhousing. Rotate the crankshaft until the cut-out on the flywheel is opposite the pointer on the clutch housing (Fig. 5.2). This lines up the recess in the flywheel with the driveshaft flange and unless this is done the transmission will not come away from the engine.
6 Remove the clamp bolt and lift out the speedometer drive cable and pinion assembly (photos). Plug the hole to prevent any dirt or small items getting into the gearbox.
7 Pull the wiring plug from the reversing light switch. Disconnect the clutch cable, refer to Chapter 4.
8 Undo the three left-hand gearbox mounting bolts and lift away the mounting and the earth strap.
9 Undo the upper gearbox-to-engine bolts and remove them.
10 Remove the retaining clips and disconnect the shift linkage from the selector shaft lever and the relay lever.
11 Disconnect the leads from the starter motor, then remove the securing bolts and lift away the starter.
12 Remove the engine mounting support from the body and the engine.

Fig. 5.2 Removal cut-out location on the flywheel (Sec 3)

3.6A Removing the speedometer drive cable

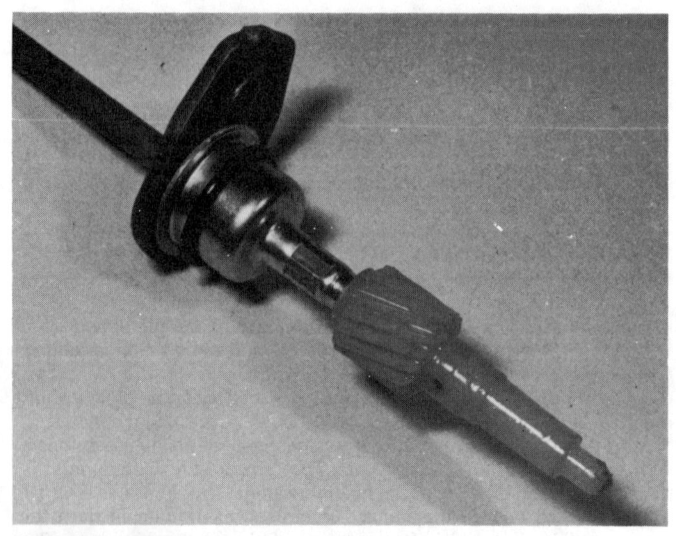

3.6B Speedometer drive cable and driven gear

13 Remove the mounting which attaches the transmission to the car body. It is best to remove it completely from the rubber mounting on the transmission.
14 Undo the socket bolts securing the left-hand driveshaft to the transmission drive flange with an Allen key and remove the CV joint from the flange. Cover the CV joint with a plastic bag and tie the shaft out of the way. Rotate the roadwheel as necessary to get at each bolt in sequence. Repeat this procedure to disconnect the right-hand driveshaft.
15 Just below the driveshaft flange is a nut, (18 in Fig. 5.3), remove this and then the bolts securing the flywheel cover plate and the smaller cover plate on the sump. Remove the small cover plate, the large plate remains on the engine.
16 With the help of an assistant pull the transmission off the locating dowels and lower it to the ground (photo). When removing the transmission keep it level with the engine until the clutch pushrod and gearbox mainshaft are withdrawn from the boss of the clutch friction disc. This will prevent any stress being put on the mainshaft which could damage the shaft.
17 Refitting is the reverse of the removal procedure. When positioning the transmission check that the flywheel recess is in the correct position, see paragraph 5, before trying to slide the assembly into position. As the gearbox mainshaft splines have to enter the clutch friction disc boss it may be necessary to rotate the crankshaft slightly to align the splines. Use a spanner on the crankshaft sprocket centre bolt to turn the crankshaft.
18 Tighten the driveshaft flange bolts, the transmission-to-engine bolts and the mounting bolts to the specified torques. Refill the transmission with the correct grade of oil.
19 Check the gear selector operation and adjust the clutch pedal free travel if necessary, refer to Chapter 4.

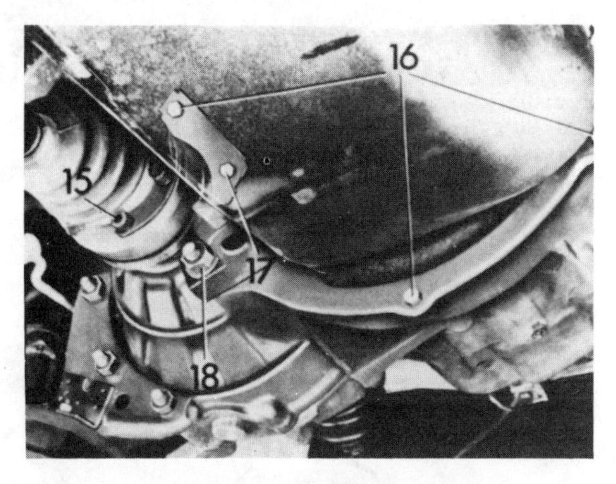

Fig. 5.3 Flywheel cover plate (Sec 3)

15 Driveshaft retaining bolt
16 Large cover plate bolts
17 Small cover plate bolt
18 Nut

3.16 Removing the transmission from the engine

4 Transmission overhaul – general

1 The overhaul of the transmission assembly requires a number of special tools, and adjustments are critical if the job is to be done successfully. For this reason we do not recommend that the home mechanic should attempt a complete overhaul. However, the transmission can be dismantled and the following Sections describe the dismantling and overhaul procedures.
2 The dismantling of the final drive is described only as far as removing the differential unit from the casing. It is considered that the overhaul of the differential unit is not within the capacity of the owner who does not have the necessary jigs and fixtures required for this work and therefore, this aspect of the overhaul procedure has not been included.
3 Before dismantling the transmission, decide first if it is worthwhile from an economic point of view. If it has done a high mileage and is in generally poor condition then the cost of new parts could be more than the cost of an exchange unit.

5 Transmission – separating the housings

1 Remove the clutch pushrod. Undo the four bolts securing the gearbox end cover plate and remove the cover plate. This will give access to the clutch release mechanism. Lift out the clutch release bearing and sleeve.
2 There are two circlips one on each side of the clutch release lever (Fig. 5.5). Later models may only have one circlip. Remove these and slide the operating shaft out of the housing, collecting the return spring and release lever as the shaft is withdrawn. Note that there is a master spline on the shaft and that the release lever will fit on the shaft in one way only.
3 Prise out the plastic cap from the centre of the drive flange and remove the circlip and spring washer. Withdraw the flange using a suitable puller. Fig. 5.4 shows VW tool 391 being used to pull off the flange (photos). There is no need to remove the opposite driveshaft flange.
4 Remove the selector shaft detent plunger and the lockbolt for the reverse gear shaft. Remove the reversing light switch (photos).
5 On the side of the housing below the clutch withdrawal shaft is the cover for the selector shaft. Using a plug spanner, remove this cover and lift off the spring seat, then remove the two detent springs. Later transmissions have only one spring (photo).
6 Withdraw the selector shaft from the housing (photos).

Fig. 5.4 Tool for removing the drive flanges (Sec 5)

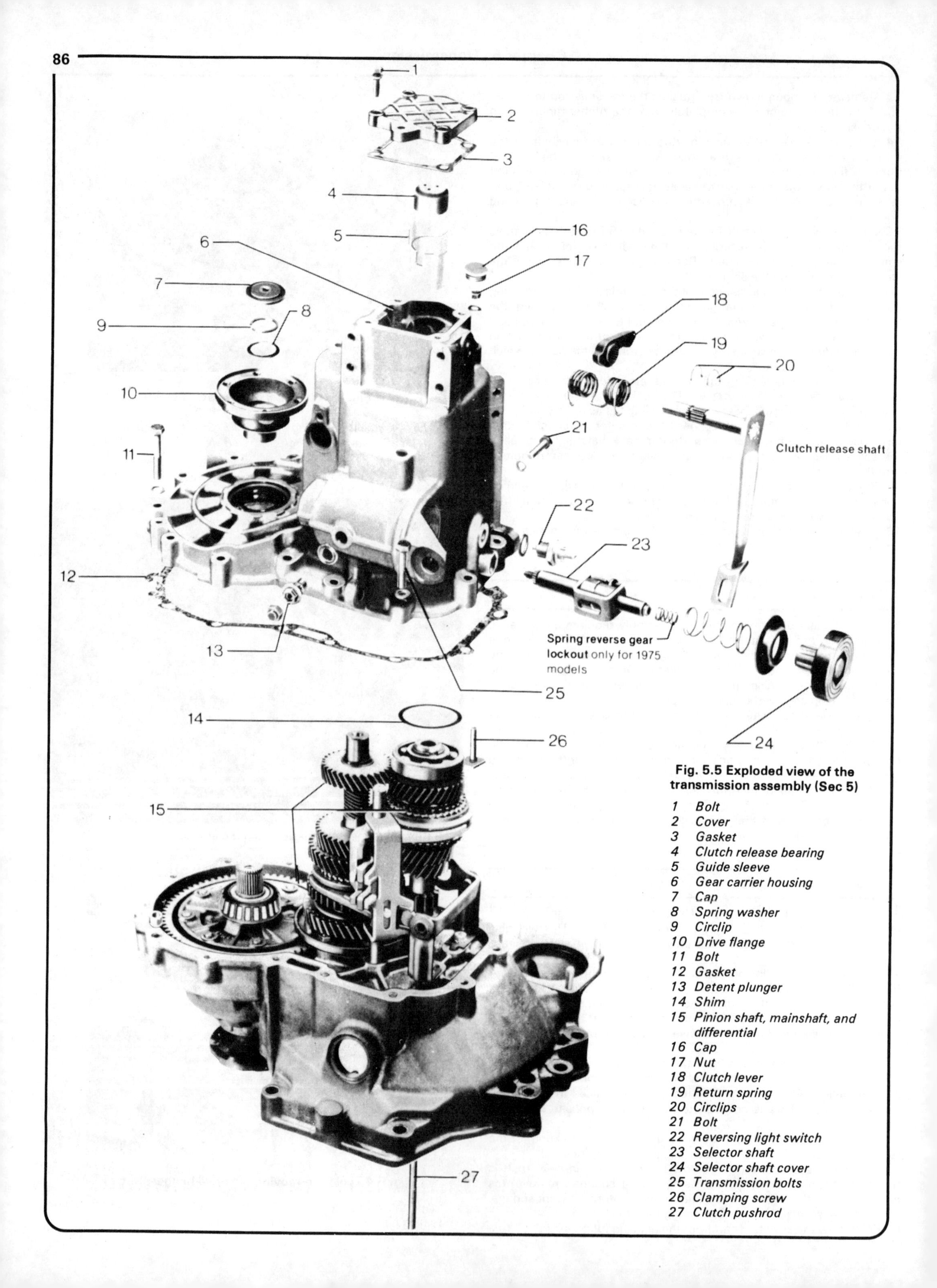

Fig. 5.5 Exploded view of the transmission assembly (Sec 5)

1 Bolt
2 Cover
3 Gasket
4 Clutch release bearing
5 Guide sleeve
6 Gear carrier housing
7 Cap
8 Spring washer
9 Circlip
10 Drive flange
11 Bolt
12 Gasket
13 Detent plunger
14 Shim
15 Pinion shaft, mainshaft, and differential
16 Cap
17 Nut
18 Clutch lever
19 Return spring
20 Circlips
21 Bolt
22 Reversing light switch
23 Selector shaft
24 Selector shaft cover
25 Transmission bolts
26 Clamping screw
27 Clutch pushrod

5.3A Removing a drive flange plastic cap

5.3B Drive flange showing circlip and spring washer

5.3C Withdrawing a drive flange with a puller

5.4A Selector shaft detent plunger

5.4B Removing the selector shaft detent plunger

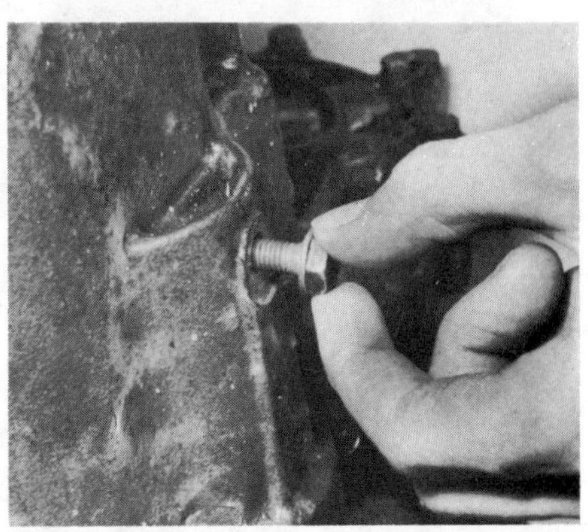
5.4C Removing the reverse gear shaft lockbolt

5.4D Removing the reversing light switch

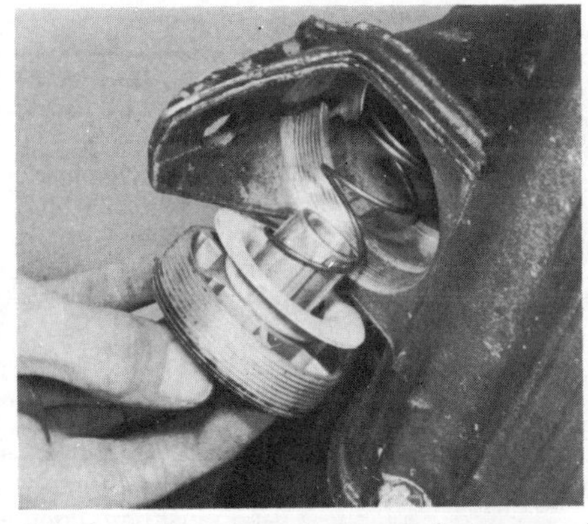
5.5 Removing the selector shaft cover and spring

5.6A Removing the selector shaft

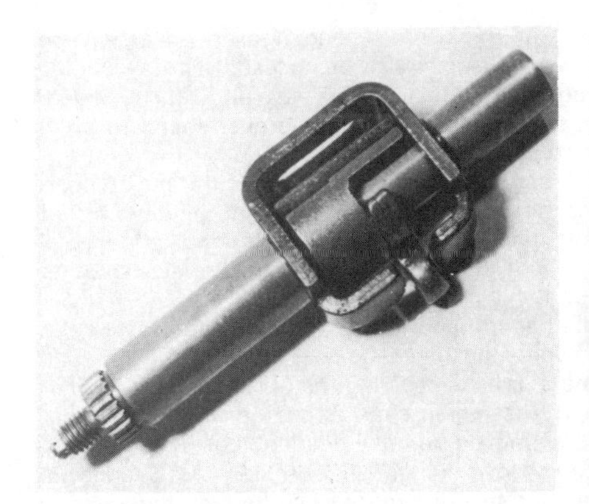
5.6B The gear selector shaft

5.7A Removing the plastic caps to expose the mainshaft bearing clamp retaining nuts

5.7B Showing the three mainshaft bearing clamp retaining nuts

Fig. 5.6 Mainshaft bearing clamp nut locations, showing the plastic caps (arrowed) (Sec 5)

5.9 Removing the gear carrier housing

Fig. 5.7 Separating the transmission casings using the special tool VW 391 (Sec 5)

A 7 mm bolt

7 On the end of the gear carrier housing (where the clutch withdrawal mechanism is located) are two plastic caps. Prise these out and undo the nuts underneath them. There is a third nut inside the housing from which the clutch withdrawal mechanism was removed, this must also be removed (photos). If these nuts are not removed, the mainshaft bearing cannot be pulled out of the casing and the casing will fracture if pressure is applied to draw it off the bearing housing. The three nuts which must be removed are shown in Fig. 5.6, with the plastic caps (arrowed) already removed.
8 Undo and remove the 14 bolts securing the two casings together. Twelve of these bolts are M8 x 50 and two are M8 x 36, note where the shorter bolts are fitted.
9 The casings are now ready for separation. Fig. 5.7 shows VW tool 391 being used. Secure the tool in the holes for the cover plate with two 7 mm bolts and then screw the centre screw down on the top of the mainshaft until is just touches. Fasten a bar or piece of angle iron across the bellhousing in such a manner as to support the end of the mainshaft and then continue to screw in the centre screw of the tool until the casing is pulled away, leaving the mainshaft bearing complete on the mainshaft. Lift away the gear carrier housing (photo). On top of the bearing there may be one or more shims, collect them and label them to ensure that they can be identified at reassembly. The needle bearing for the pinion shaft will remain in the gear carrier housing; it can be removed, if necessary, using a suitable extractor.

6 Mainshaft, pinion shaft and differential – removal

1 Refer to Fig. 5.8. The mainshaft assembly can be removed quite easily, but the pinion shaft assembly must be partially dismantled before the pinion shaft and the differential unit can be removed.
2 Remove the two shift fork shaft circlips and withdraw the shaft from the bearing housing, then lift away the shift fork set (photos).
3 Remove the circlip retaining the 4th speed gear on the pinion shaft, then lift the mainshaft out of its bearing in the bearing housing and at the same time remove the 4th speed gear from the pinion shaft (photos). The mainshaft needle bearing and oil seal will remain in the bearing housing.
4 Remove the circlip retaining the 3rd speed gear on the pinion shaft. This circlip is used to adjust the axial play of the 3rd speed gear and must be refitted in the same position, so label it for identification at reassembly. Remove the 3rd speed gear (photos).
5 Remove the 2nd speed gear and then the needle bearing from over its inner sleeve (photos).
6 To remove the rest of the gears a long hooked puller will be required. Fig. 5.9 shows VW tool 447h being used with a suitable puller (A). Before pulling off the synchro hub/sleeve and 1st speed gear remove the reverse gear by tapping the reverse gear shaft out of its seating, then lift the shaft and gear away (photo).
7 Remove the plastic stop button from the end of the pinion shaft and fit the puller under the 1st speed gear. Note that the pinion shaft bearing retainer has two notches to accommodate the puller legs. Pull the gear and synchro hub off the shaft. Tape the synchro unit together to prevent it coming apart.
8 Remove the needle bearing and thrust washer. Note that the flat side of the washer is towards the 1st speed gear (photos).
9 Remove the four nuts or bolts securing the pinion bearing retainer and lift off the retainer. The pinion shaft is seated in a taper roller bearing, and can now be removed from the bearing housing (photos).
10 Remove the second drive flange as described in Section 5, paragraph 3, and then lift the differential unit out of the bearing housing (photo). We do not recommend trying to overhaul the differential unit, if it is in any way suspect seek advice from your VW agent.

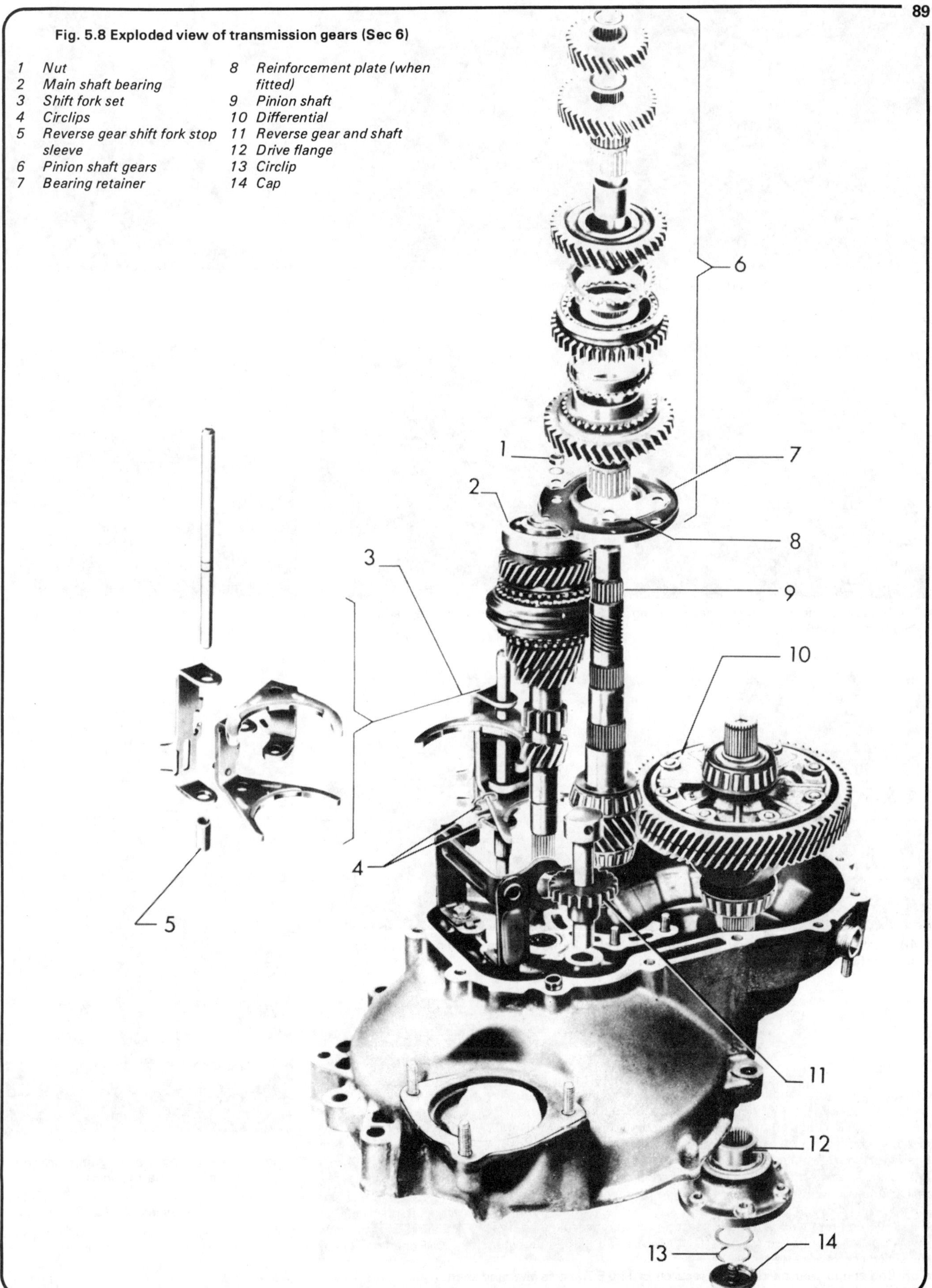

Fig. 5.8 Exploded view of transmission gears (Sec 6)

1 *Nut*
2 *Main shaft bearing*
3 *Shift fork set*
4 *Circlips*
5 *Reverse gear shift fork stop sleeve*
6 *Pinion shaft gears*
7 *Bearing retainer*
8 *Reinforcement plate (when fitted)*
9 *Pinion shaft*
10 *Differential*
11 *Reverse gear and shaft*
12 *Drive flange*
13 *Circlip*
14 *Cap*

6.2A Removing a shift fork shaft circlip (arrowed)

6.2B Removing the shift fork shaft

6.2C Removing the shift fork set

6.3A Removing the 4th speed gear retaining circlip (pinion shaft)

6.3B Removing the mainshaft assembly

6.4A Removing the 3rd speed gear retaining circlip (pinion shaft)

6.4B Removing the 3rd speed gear from the pinion shaft

6.5A Removing the 2nd speed gear from the pinion shaft

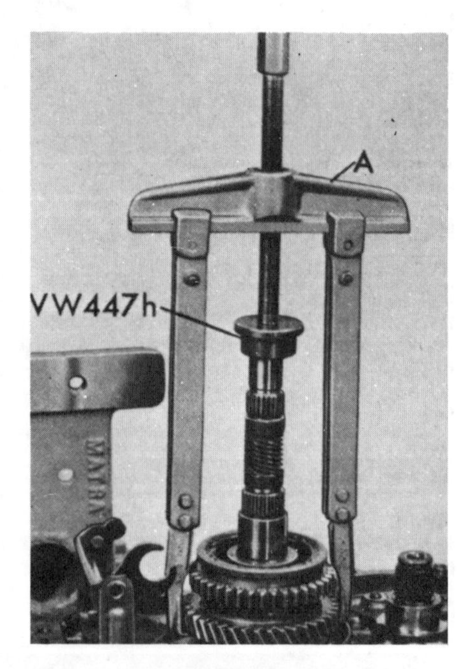

Fig. 5.9 Removing the 1st gear from the pinion shaft (Sec 6)

A VW puller US 1078

6.5B 2nd speed gear needle roller location on the pinion shaft

6.6 Reverse gear and shaft

6.8A 1st speed gear needle roller bearing

6.8B Removing the 1st speed gear thrust washer

6.9A Removing the pinion bearing retainer bolts

6.9B Removing the pinion bearing retainer

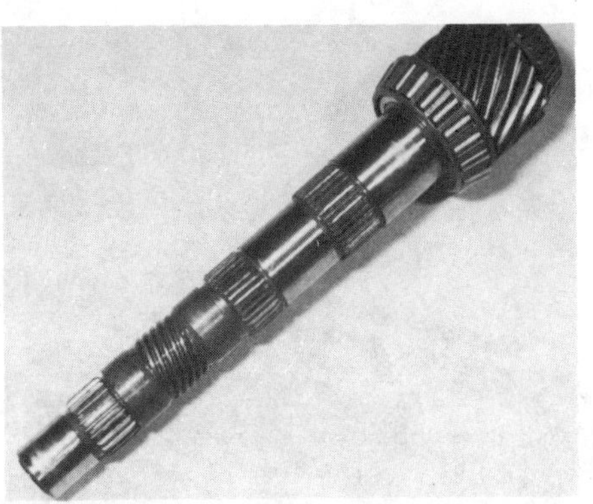
6.9C The pinion or output shaft

6.10 The differential unit on the bearing housing

7 Bearing housing – overhaul

1 Clean the housing using a grease solvent, remove all oil and sludge. Renew both the oil seals (photo). Fill the space between the lips of the seals with multi-purpose grease before fitting. The drive flange oil seal must be driven in as far as it will go. Fig. 5.10 shows VW tool 194 being used, but a piece of suitable diameter steel tubing can also be used.

2 The mainshaft needle bearing may be removed, if necessary, with a suitable extractor. Do not remove the bearing unless it is defective as it is likely to be damaged during removal (photo).

3 If the outer races of the differential bearings are defective then it will be necessary for the casings and the differential unit to be taken to the VW agent for servicing.

4 The starter motor shaft should be tried in the starter bush. If undue wear is apparent, remove the bush with a puller and fit a new one.

7.1 Mainshaft oil seal in the bearing housing

Fig. 5.10 Fitting the drive flange oil seal with the special tool (Sec 7)

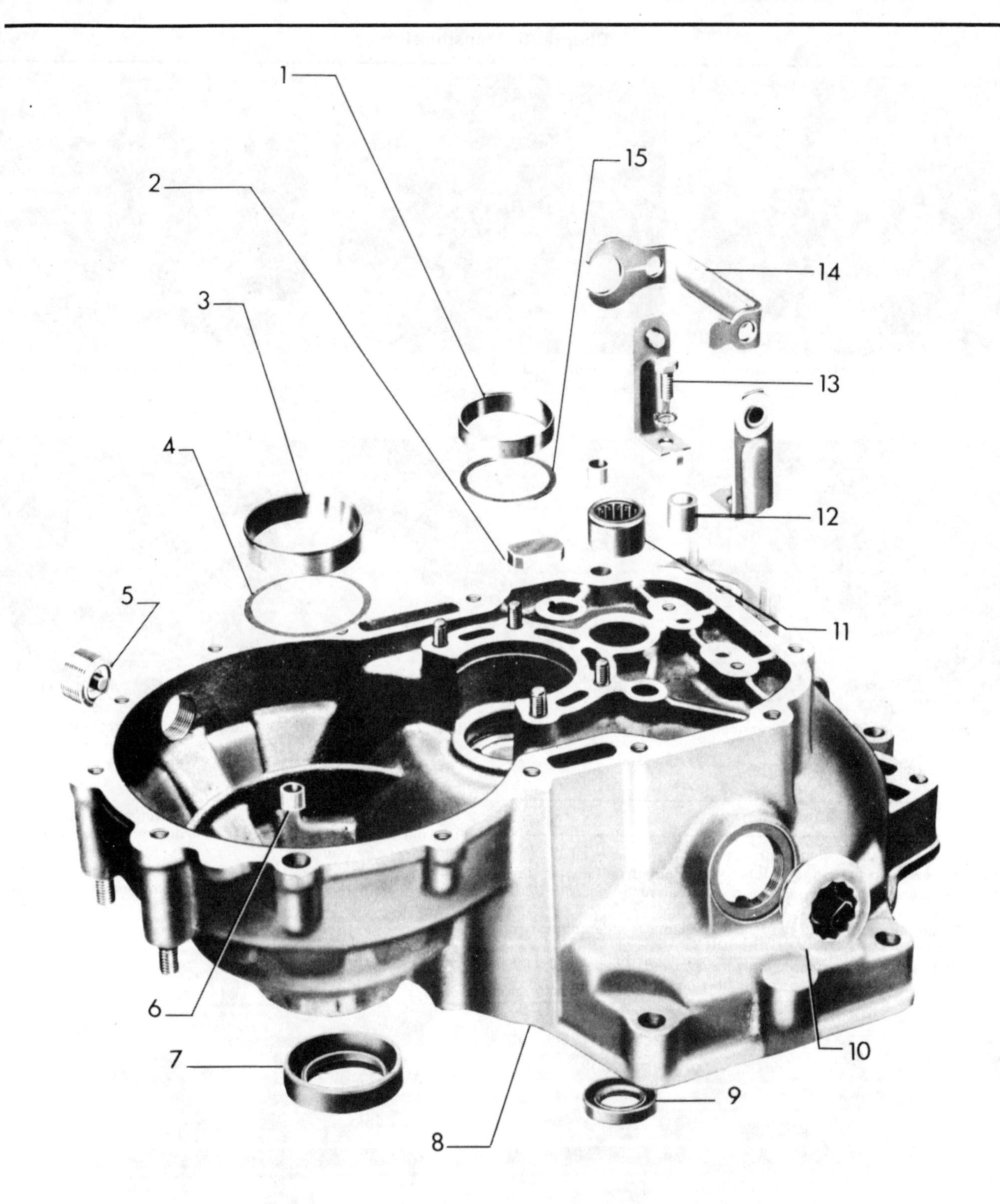

Fig. 5.11 Exploded view of the bearing housing (Sec 7)

1 Pinion bearing outer race
2 Magnet
3 Differential bearing outer race
4 Shim
5 Drain plug
6 Dowel sleeve
7 Drive flange oil seal
8 Bearing housing
9 Mainshaft oil seal
10 TDC sender unit (early models)
11 Mainshaft needle bearing
12 Starter bush
13 Bolt
14 Reverse shift fork
15 Shim

7.2 Mainshaft needle bearing in the bearing housing

8 Gear carrier housing – overhaul

1 There are three oil seals to renew: one for the clutch operating lever, one for the selector shaft, and a large one for the drive flange. In each case prise out the old seal, noting which way round it is fitted. Fill the seal lips with multi-purpose grease and drive the seals squarely into the housing using a suitable mandrel or piece of tubing (photos).

2 The needle bearing for the pinion shaft is difficult to extract unless the correct extractor is available, see Fig. 5.13. The shaft may be tried in the bearing without having to remove the bearing (photo).

3 If the outer race of the final drive is renewed, the complete unit must be taken to the VW agent for setting up with the correct shims (photo).

9 Pinion shaft taper bearings

1 The large and small bearings accurately locate the pinion shaft gear with the crownwheel of the differential. If either bearing is defective then both must be renewed. In the removal process the bearings are destroyed. New ones have to be shrunk on and the shim under the smaller bearing changed for one of the correct size.

2 This operation is quite complicated and requires special equipment for preloading of the shaft and measurement of the torque required to rotate the new bearings. In addition the shim at the top of

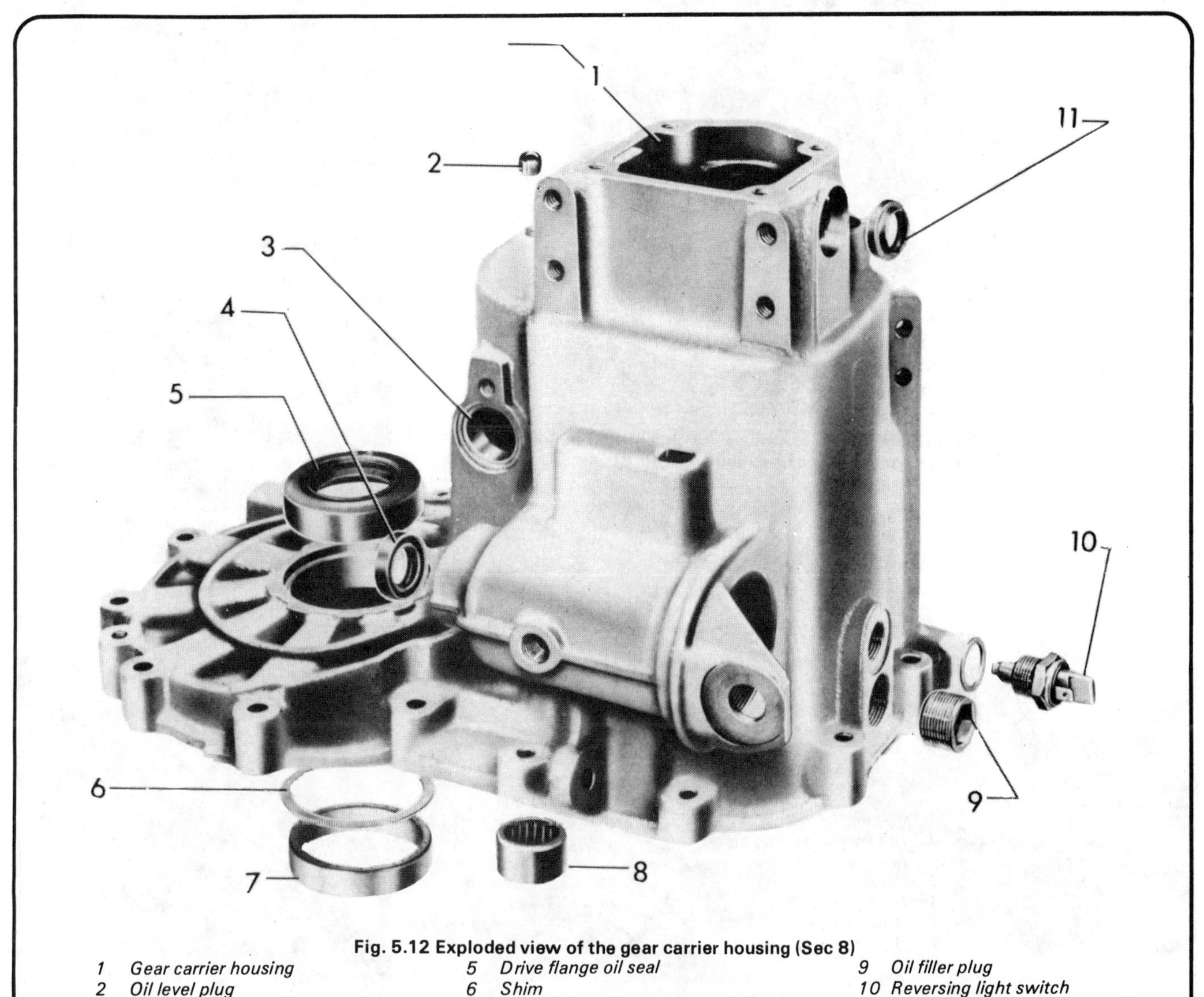

Fig. 5.12 Exploded view of the gear carrier housing (Sec 8)

1 Gear carrier housing
2 Oil level plug
3 Speedometer drive aperture
4 Selector shaft oil seal
5 Drive flange oil seal
6 Shim
7 Differential bearing outer race
8 Mainshaft needle bearing
9 Oil filler plug
10 Reversing light switch
11 Clutch operating lever oil seal

8.1A Clutch operating lever oil seal

8.1B Selector shaft oil seal

8.1C Drive flange oil seal location in the gear carrier housing

Fig. 5.13 Using an extractor (A) to remove the pinion shaft needle bearing from the gear carrier housing (Sec 8)

8.2 Pinion shaft needle bearing location in the gear carrier housing

8.3 Differential bearing outer race location in the gear carrier housing

10.2A Removing the 4th speed gear from the mainshaft

10.2B Removing the 4th speed gear needle roller bearing

10.4A 3rd speed gear needle roller bearing location on the mainshaft

the mainshaft and the axial play at the circlip of the 3rd speed gear on the pinion shaft will be affected. This will mean selection of a new shim and circlip. There are six different thicknesses of circlip. Therefore it is recommended that if these bearings require renewal, the work should be left to your VW agent.

10 Mainshaft – dismantling and reassembly

1 Refer to Fig. 5.14. Remove the ball bearing retaining circlip and then, supporting the bearing under the inner race, press the shaft out of the inner race. VW tool 402 can be used to remove the bearing but a suitable tool can be made from a piece of steel. At assembly the bearing is pressed into the gear carrier housing and the shaft pressed into the race.

2 Remove the thrust washer, 4th speed gear, and the needle bearing together with the synchro ring from the mainshaft (photos).

3 Remove the circlip, then support the 3rd speed gear and press the mainshaft through the 3rd/4th synchro hub. Tape the synchro unit together to prevent it from coming apart.

4 Remove the needle bearing to complete the dismantling of the shaft (photos).

5 Should the clutch pushrod be loose in the mainshaft the bush may be driven out of the end of the shaft and a new bush and oil seal fitted. If the shaft is to be renewed then the tolerances will be affected and this means going back to the agent with his special tools and gauges. The problem is the play between the 2nd speed gear on the pinion shaft and 3rd speed gear on the mainshaft when both shafts are fitted. This must be 1.0 mm (0.040 in). Adjusting this also requires a new shim on the top of the ball bearing between the bearing and the gear carrier housing.

6 If gears on either shaft are to be renewed then the mating gear on the other shaft must be renewed as well. They are supplied in pairs only.

7 The inspection of the synchro units is dealt with in Section 11.

8 When reassembling the mainshaft, lightly oil all the parts.

9 Fit the 3rd speed gear needle bearing and the 3rd speed gear. Press on the 3rd/4th gear synchro hub and fit the retaining circlip. When pressing on the synchro hub and sleeve, turn the rings so that the keys and grooves line up. The chamfer on the inner splines of the hub must face 3rd gear (photos).

10 On transmission up to N. 18065, fit the thrust washer and 4th speed gear. Later transmissions have a ball bearing with a wider inner race and the thrust washer is not fitted. If the old type bearing is being replaced with a new type, the 4th gear thrust washer must be left out.

11 The mainshaft ball bearing should now be pressed into the gear carrier housing. Ensure that the same shim/s removed at dismantling are refitted between the bearing and the casing. The bearing is fitted in the housing with the closed side of the ball bearing cage towards the 4th speed gear. Insert the clamping screws and tighten the clamping

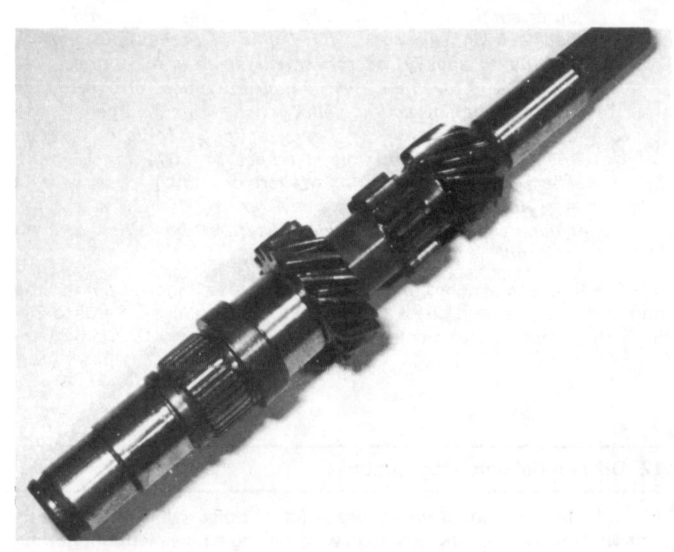

10.4B The mainshaft completely dismantled

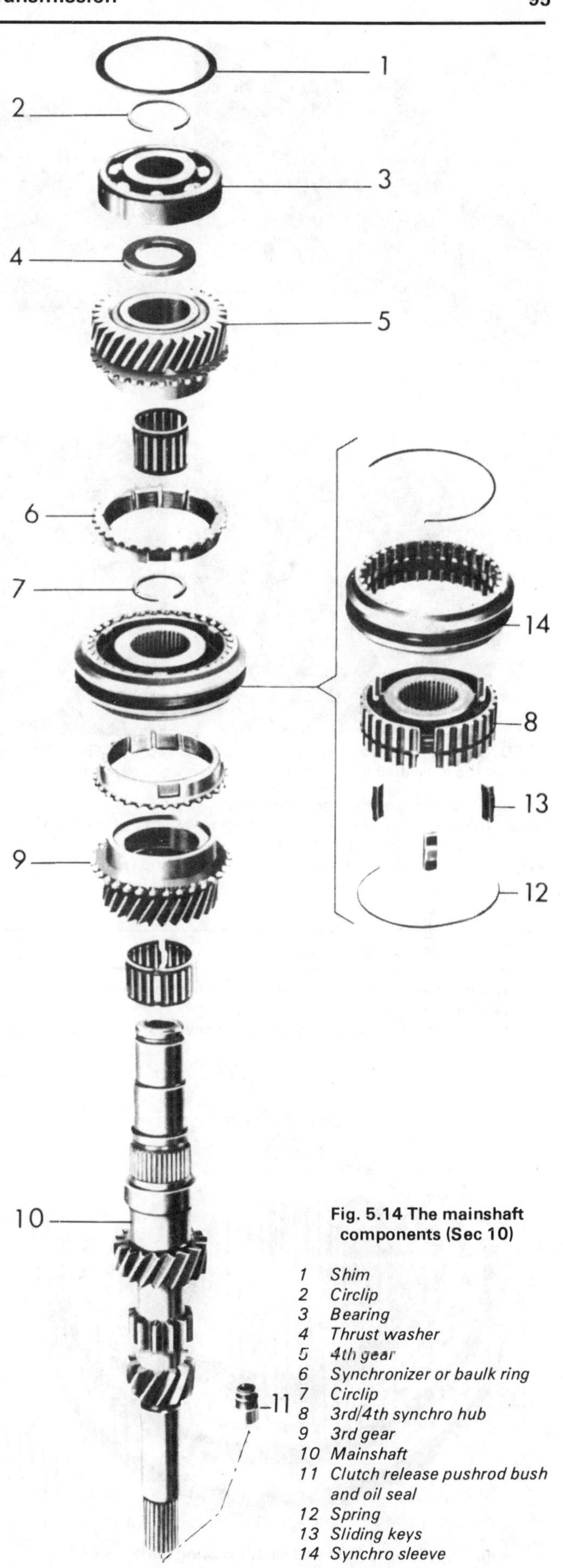

Fig. 5.14 The mainshaft components (Sec 10)

1 *Shim*
2 *Circlip*
3 *Bearing*
4 *Thrust washer*
5 *4th gear*
6 *Synchronizer or baulk ring*
7 *Circlip*
8 *3rd/4th synchro hub*
9 *3rd gear*
10 *Mainshaft*
11 *Clutch release pushrod bush and oil seal*
12 *Spring*
13 *Sliding keys*
14 *Synchro sleeve*

10.9A Fitting the 3rd speed gear to the mainshaft

10.9B 3rd speed gear baulk ring location

10.9C Fitting the 3rd/4th synchro assembly to the mainshaft

10.9D 3rd/4th synchro hub retaining circlip fitted on the mainshaft

10.11A Fitting the mainshaft ball bearing into the gear carrier housing

10.11B Correct location of mainshaft ball bearing outer race clamps

screw nuts to the specified torque (photos).

Note: The endplay will have to be adjusted if either of the bearings, the thrust washer or mainshaft have been renewed, so the help of a VW agent with the necessary special tools and gauges will be required.

11 Synchroniser units – inspection

1 Synchroniser units are supplied as an assembly, parts should not be interchanged.

2 When synchro baulk rings are being renewed, it is advisable to fit new blocker bars (sliding keys) and retaining springs in the hub. This will ensure that full advantage is taken of the new, unworn cut-outs in the rings.

3 The splines on the inner hub and outer sleeve are matched, either by selection on assembly or by wear patterns during use. Those matched on assembly have etched lines on the inner hub and outer sleeve so that they can be easily realigned. For those with no marks, scribe a line on the hub and sleeve before dismantling (see Fig. 5.15) to ensure correct alignment at reassembly.

4 When examining for wear there are two important features to check:

(a) *The fit of the splines. With the keys removed, the inner and outer sections of the hub should slide easily with minimum backlash or axial rock. The degree of permissible wear is difficult to specify, no movement at all is exceptional, yet excessive movement will affect operation and result in jumping out of gear. If in doubt consult your VW agent*

(b) *Selector fork grooves and selector forks should not exceed the maximum permissible clearance of 0.012 in (0.3 mm). The wear can be either on the fork or in the groove, so try a new fork in the existing sleeve groove first to see if the clearance is reduced considerably. If not then a new synchro assembly is needed.*

5 The fit of the synchro ring on the gear is also important. Press the ring onto the gear and check the gap with feeler gauges. Refer to Fig. 5.16, the dimension (a) must not be less than 0.5 mm (0.020 in).

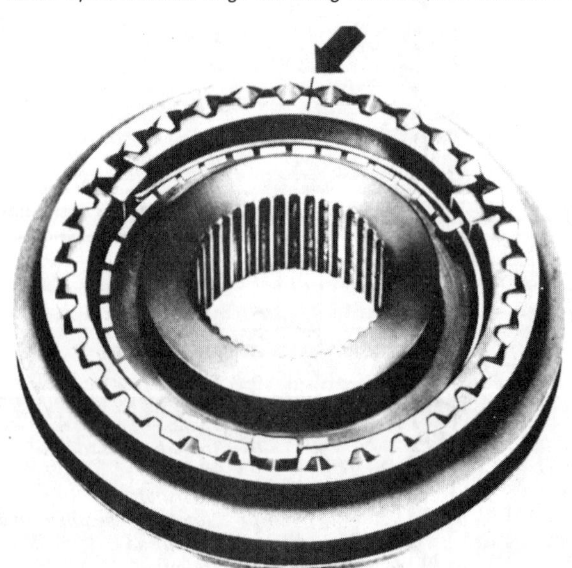
Fig. 5.15 Synchro hub and sleeve mating marks (Sec 11)

12 Differential unit – inspection

1 A faulty differential will cause a lot of noise while the car is in motion. However, it may go a long way making a noise without getting any worse. If you decide it needs renewing then renew it as a unit.

2 There are several problems. If a new differential is fitted then a new crownwheel to an old shaft is inviting noise. It may be possible for

Fig. 5.16 Synchro ring wear limit dimension (a) (Sec 11)

the agent to build the old differential if only the taper bearings are at fault, and he may be able to fit a new crownwheel to your differential if only the crownwheel is damaged. It is worth asking, but most likely you will need a replacement differential.

13 Differential, pinion shaft and mainshaft – reassembly in bearing housing

1 Refit the differential unit in the casing. Using tool VW 391, fit the drive flange to the bearing housing and fit the spring washer, retaining circlip and cap.

2 Check that the mainshaft ball bearing is correctly fitted in the gear carrier housing, plastic cage towards the casing, and that the bearing clamp bolt nuts are tight.

3 Fit the pinion shaft complete with its taper bearings into the bearing housing, so that the pinion gear meshes with the crownwheel (photo).

4 Fit the bearing retaining plate and the four securing bolts (photo). Fit the 1st speed gear thrust washer with its flat side up (towards the 1st gear). Fit the needle roller cage.

13.3 Fitting the pinion shaft into the bearing housing

5 Slide the 1st speed gear over the needle bearing. Warm the synchro hub a little and press it into position. The hub will slide on if heated to 120°C (250°F) and it can then be tapped into position. Make sure that the grooves are in line with the shift keys in the 1st/2nd synchro to avoid damage to the baulk ring on assembly. The shift fork groove in the operating sleeve should be nearer 2nd gear and the groove on the hub nearer 1st gear (photos).

6 The inner race for the 2nd speed gear needle bearing must be fitted next and pressed down as far as it will go.

7 Fit the reverse idler gear and shaft with the shaft aligned as shown

13.4 Tightening the pinion shaft bearing retaining plate bolts

13.5A Assembling the 1st speed gear to the pinion shaft

13.5B 1st speed gear baulk ring location

13.5C Aligning baulk ring slots with the sliding keys in the 1st/2nd synchro assembly

13.5D Fitting the 1st/2nd synchro assembly to the pinion shaft

13.5E Using a metal tube to drive the 1st/2nd synchro assembly onto the pinion shaft

in Fig. 5.17. Use a plastic hammer to drive the shaft into the casing.

8 Fit the 2nd speed gear needle bearing on the pinion shaft and the 2nd gear with the shoulder downwards.

9 Warm the 3rd speed gear and press it down over the splines with the collar thrust face towards the 2nd gear.

10 Fit the 3rd gear retaining circlip and using feeler gauges measure the play between the gear and the circlip. It must be less than 0.008 in (0.20 mm). If it is more, a thicker circlip must be fitted. The following table gives the sizes available:

Part no	Thickness (mm)	Thickness (in)	Colour
020 311 381	*2·5*	*0·098*	*brown*
020 311 381 A	*2·6*	*0·102*	*black*
020 311 381 B	*2·7*	*0·106*	*bright*
020 311 381 C	*2·8*	*0·110*	*copper*
020 311 381 D	*2·9*	*0·114*	*yellow*
020 311 381 E	*3·0*	*0·118*	*blue*

11 At this stage the mainshaft must be fitted in position on the bearing housing. Slide it into the needle bearing in the casing and fit the shift forks in the operating sleeves. Insert the retaining circlips. Fit the reverse gear shift fork (photos).

12 Fit the 4th speed gear and its retaining circlip on the pinion shaft. Finally insert the stop button for the pinion shaft needle bearing in the end of the pinion shaft.

13 The bearing housing and shafts are now assembled ready for the assembly of the gear carrier housing to the bearing housing, which is described in the next Section (photo).

Fig. 5.17 Reverse idler gear shaft alignment (Sec 13)

14 Transmission – reassembling the housings

1 Check that the reverse gear shaft is in the correct position, see Fig. 5.17, and set the gear train in neutral. Fit a new gasket on the bearing housing flange.

13.11A Fitting the shift forks

13.11B Fitting the reverse shift fork pivot posts

13.11C Reverse shift fork assembly

13.11D Reverse shift fork to gear assembly

13.13 Bearing housing assembly ready for fitting of the gear carrier housing

2 Lower the gear carrier housing over the shafts onto the bearing housing flange, checking that the pinion shaft is aligned with the pinion shaft needle bearing in the casing, and drive the mainshaft into its bearing, using a suitable mandrel on the inner race. A piece of suitable diameter steel tube can be used. Ensure that the mainshaft is supported on a block of wood when driving the mainshaft into the bearing.
3 Insert the 14 bolts which secure the two housings together and tighten them to the specified torque.
4 Fit the circlip over the end of the mainshaft, working through the release bearing hole. Insert the clutch pushrod into the mainshaft. Ensure that the circlip is properly seated, then fit the clutch release bearing and sleeve assembly (photos).
5 Fit the clutch release shaft and lever. Ensure that the spring is hooked over the lever in the centre and that the angled ends rest against the casing. The shaft can be inserted in the lever in one position only. Fit the two circlips, one each side of the lever.
6 Position a new gasket on the end of the housing and fit the end cover plate and four securing screws. Tighten the screws to the specified torque.
7 Lubricate the selector shaft and insert it into the casing. When it is in position, fit the spring and screw in the shaft cover with a plug spanner, tightening it to the specified torque.
8 Fit the selector shaft detent plunger. This has a plastic cap. Only if the housing, selector shaft or plunger are faulty and new ones are required should the plunger need adjusting. If necessary, adjust as follows:

(a) Refer to Fig. 5.18. Slacken the locknut and screw in the adjusting sleeve until the lockring lifts off the adjusting sleeve
(b) Screw the adjusting sleeve out until the lockring just contacts the sleeve
(c) Check that the lockring lifts as soon as the shaft is turned. Tighten the locknut and fit the plastic cap

9 Fit the reverse gear shaft lockbolt and the reversing light switch.

14.4A Fitting the circlip over the end of the mainshaft

14.4B Inserting the clutch pushrod into the mainshaft

15 Gearshift linkage – adjustment

1 Align the holes in the lever housing cover plate with the holes in the bearing plate, A in Fig. 5.19, and check that the threaded holes B are in the middle of the slots. If they are not, turn the bearing plate through 180°. Tighten the lever bearing.
2 Slacken the shift rod clamp bolt (Fig. 5.20) to permit the selector lever to move on the shift rod. Pull the boot off the lever housing and push it back out of the way (photo).
3 Refer to Fig. 5.21. Ensure that the shift finger is in the centre of

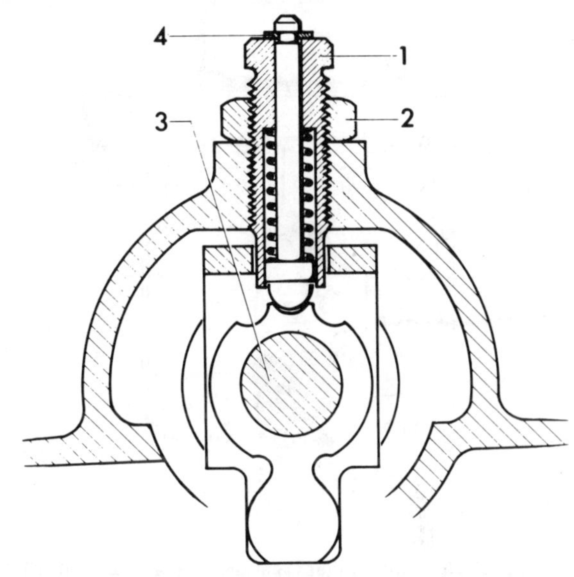

Fig. 5.18 Cross-sectional view of the selector shaft detent plunger (Sec 14)

1 Adjusting sleeve
2 Locknut
3 Selector shaft
4 Spring clip

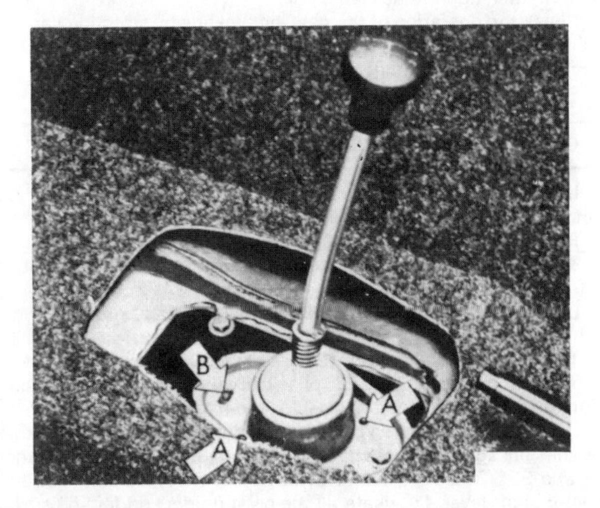

Fig. 5.19 Gear lever housing cover. Arrows show alignment holes (Sec 15)

Fig. 5.20 Gear shift rod clamp bolt (arrowed) (Sec 15)

15.2 Shift rod clamp bolt and rubber boot

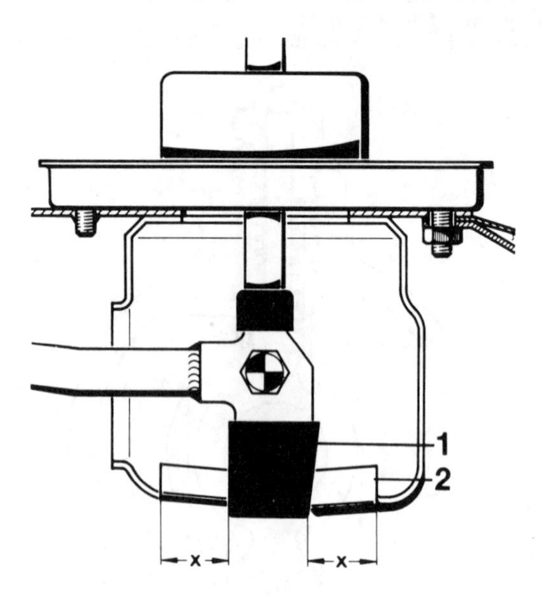

Fig. 5.21 Gear shift finger (1) and stop plate (2) (Sec 15)

Fig. 5.22 Gear shift rod adjusting dimension (a) (Sec 15)

the stop plate and that dimension X is the same on both sides.

4 Now adjust the shift rod so that dimension a in Fig. 5.22 is 0.75 in (20 mm), then tighten the clamp bolt.

5 Check the operation of the gearshift linkage by selecting each gear in turn. If the linkage is spongy or binding check the adjustment of the selector shaft detent plunger, refer to Section 14, paragraph 8.

16 Gearchange lever and linkage – removal and refitting

1 Unscrew the gearchange lever knob. Release the rubber gaiter from the gear lever housing and pull it off the lever.

2 Release the boot from under the lever housing, slide it forward on the shift rod and remove the clevis pin securing the shift rod to the gearchange lever (photo).

3 Undo the two lever bearing assembly securing bolts then lift out the bearing and lever together.

4 To remove the linkage, first refer to Fig. 5.23. When disconnecting the front and rear selector rods, prise back the ends of the plastic clips before pulling the rods off the selector and relay levers (photos).

5 Refitting is the reverse of the removal procedure. When fitting the gearshift linkage rod note that the ends are bent at different angles, 90° and 95°. The 95° end, arrowed in Fig. 5.23, is connected to the selector shaft lever. Lubricate all the pivot points and friction surfaces with grease. Adjust the gearshift linkage as described in Section 15 (photos).

16.2 Gear lever boot location

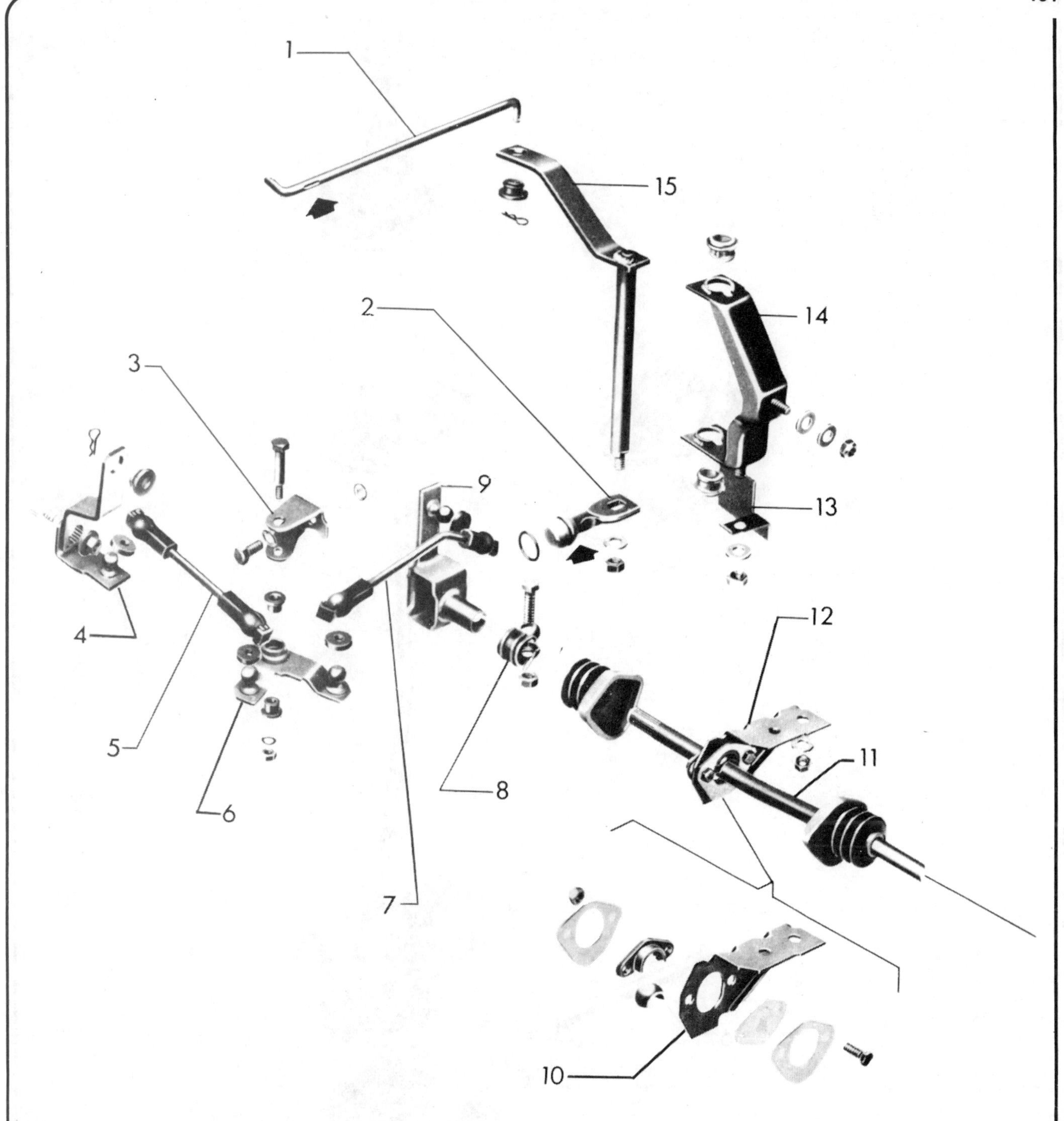

Fig. 5.23 Exploded view of the gearchange linkage (Sec 16)

1 Linkage rod with 95° angled end (arrowed)
2 Lever
3 Relay lever bracket
4 Selector shaft lever
5 Front selector rod
6 Relay lever
7 Rear selector rod
8 Shift rod clamp
9 Selector lever
10 Bearing plate
11 Shift rod
12 Shift rod bearing assembly
13 Protective plate
14 Relay shaft bracket
15 Relay shaft

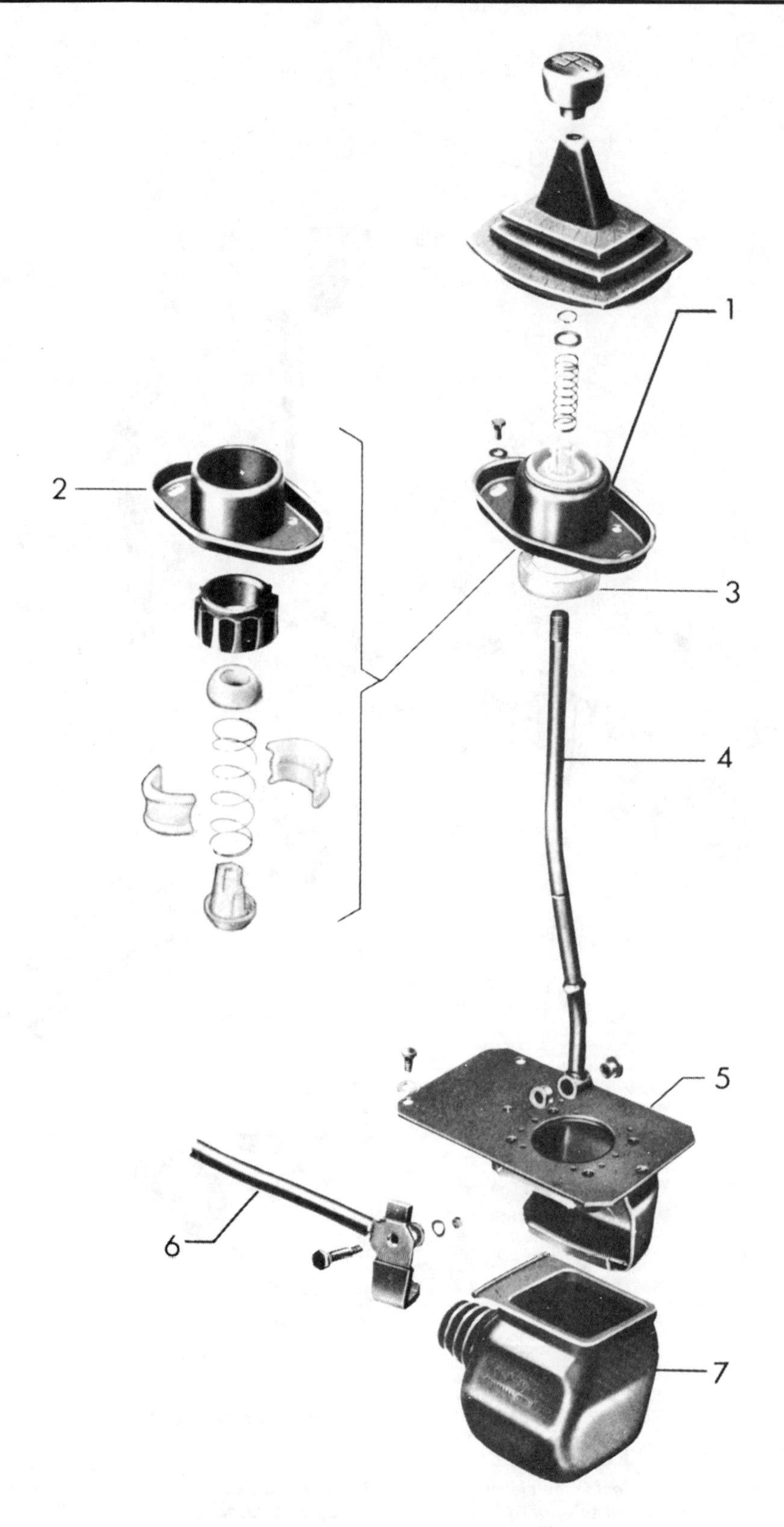

Fig. 5.24 Exploded view of the gear lever (Sec 16)

1 Gear lever bearing assembly
2 Gear lever bearing plate
3 Spacer
4 Gearshift lever
5 Gear lever housing
6 Shift rod
7 Gear lever boot

16.4A Removing the gearchange linkage crank pivot bolt

16.4B Removing the gearchange linkage crank and bushes

16.4C Removing the gearchange linkage crank bracket

16.5A The gearchange relay shaft location

16.5B The gearchange relay shaft and connecting link

16.5C Selector shaft lever location

17 Fault diagnosis – transmission

It is sometimes difficult to decide whether all the effort and expense of dismantling the transmission is worthwhile to cure a minor irritant such as a whine or where the synchromesh can be beaten by a rapid gearchange. If the noise gets no worse consideration should be given to the time and money available, for the elimination of noise completely is almost impossible, unless a complete new set of gears and bearings is fitted. New gears and bearings will still make a noise if fitted in mesh with old ones.

Symptom	Reason/s
Synchromesh not giving smooth change	Worn baulk rings or synchro hubs
Jumps out of gear on drive or over-run	Weak detent spring Worn selector forks Worn synchro hubs
Noisy rough whining and vibration	Worn bearings, chipped or worn gears
Gear difficult to engage	Clutch fault Gearshift mechanism out of adjustment

Chapter 6 Braking system, wheels and tyres

Refer to Chapter 11 for specifications and information applicable to later models

Contents

Specifications

Type ...... Hydraulic, servo-assisted. Dual circuit diagonally connected. Front – disc brakes. Rear – drum brakes

Handbrake ...... Cable operated, acting on rear brakes

Front brakes (disc)

Caliper piston diameter	44 mm (1·73 in)
Disc diameter	239 mm (9·4 in)
Disc thickness (new)	12 mm (0·47 in)
Disc thickness (min) after refinishing	10·5 mm (0·41 in)
Pad thickness (new)	14 mm (0·55 in)

Rear brakes (drum)

Drum internal diameter	180 mm (7·08 in)
Drum internal diameter (max) after refinishing	181 mm (7·13 in)
Wheel cylinder bore diameter	14·29 mm (0·56 in)
Lining thickness:	
Riveted	5 mm (0·20 in)
Wear limit	2·5 mm (0·097 in)
Bonded	4 mm (0·16 mm)
Wear limit	1·5 mm (0·060 in)
Lining width	30 mm (1·2 in)

Master cylinder bore diameter ...... 20·64 mm (0·81 in)

Brake servo diameter ...... 177·8 mm (7·0 in)

Wheels and tyres

Wheels	$4\frac{1}{2}$J x 13 or 5J x 13 with 45 mm offset	
Tyres	Radial ply 145SR or 155SR	
Tyre pressures	**Front**	**Rear**
Half load	1·7 bar (24 lbf/in²)	1·7bar (24 lbf/in²)
Full load	1·8 bar (26 lbf/in²)	2·2 bar (31 lbf/in²)

Torque wrench settings

Torque wrench settings	kgf m	lbf ft
Front hub nut	24·0	174
Caliper to wheel bearing housing	6·0	43
Disc to hub	0·7	5
Splash plate to hub	0·8	6
Backplate to stub axle	3·0	22
Wheel nuts	9·0	65
Servo to adapter	1·0	14
Master cylinder to servo	1·5	11
Brake hoses to caliper/wheel cylinder	1·0	7
Brake pipes to master cylinder	1·0	7

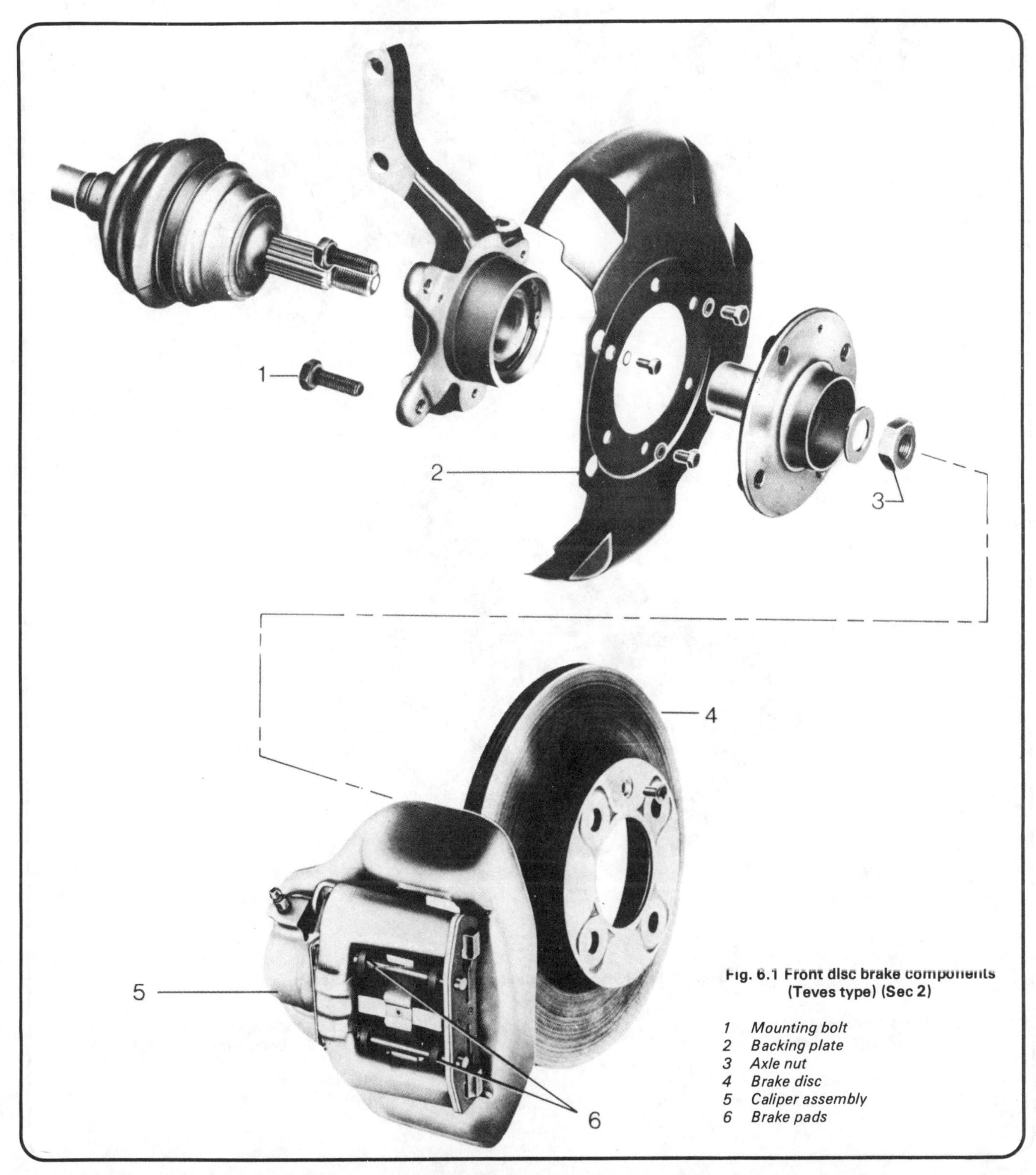

Fig. 6.1 Front disc brake components (Teves type) (Sec 2)

1 *Mounting bolt*
2 *Backing plate*
3 *Axle nut*
4 *Brake disc*
5 *Caliper assembly*
6 *Brake pads*

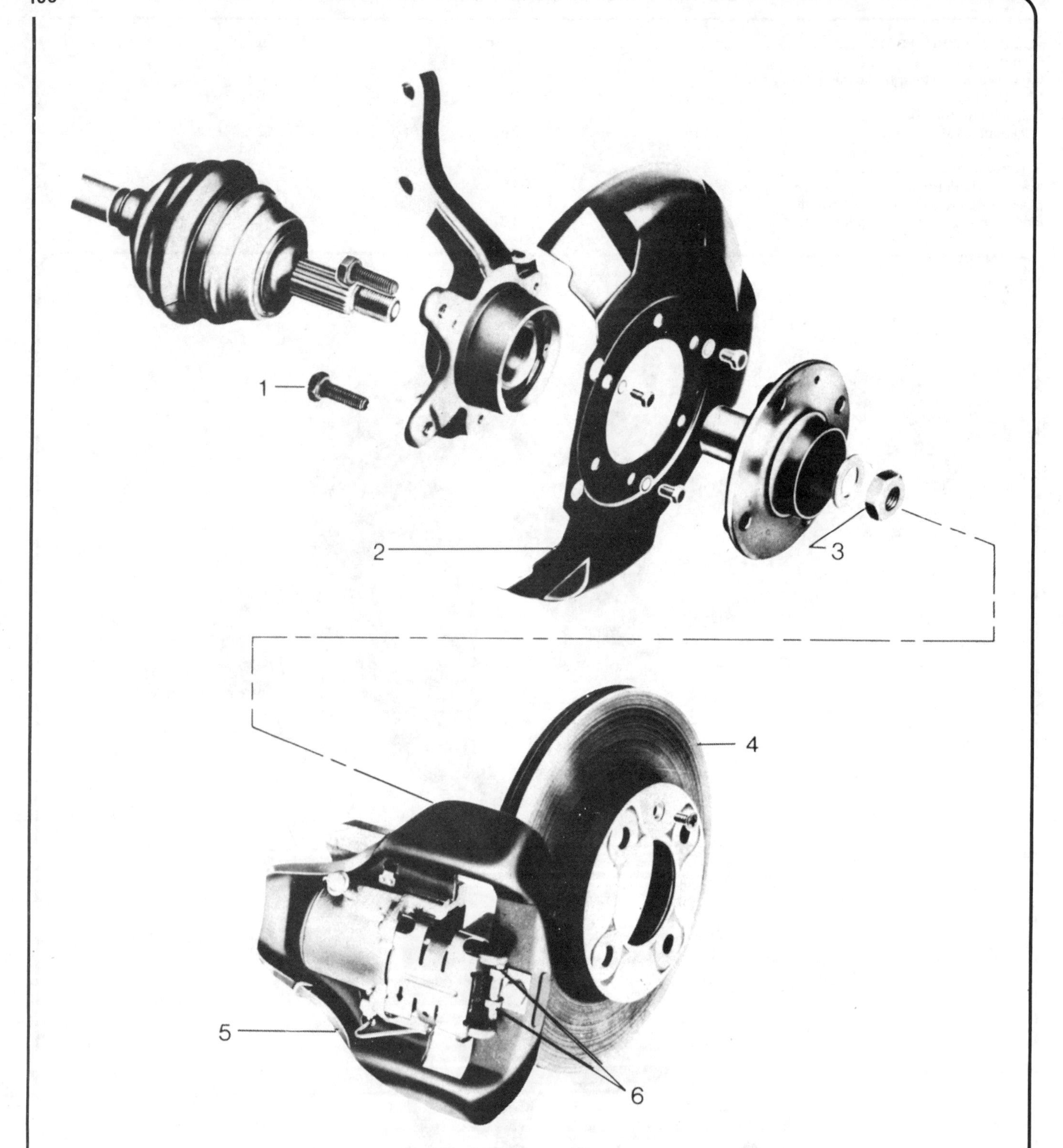

Fig. 6.2 Front disc brake components (Girling type) (Sec 2)

1	*Mounting bolt*	3	*Axle nut*	5	*Caliper assembly*
2	*Backing plate*	4	*Brake disc*	6	*Brake pads*

1 General description

The main braking system is operated hydraulically. The vacuum powered servo is fitted between the brake pedal and the master cylinder to reduce the pressure required at the brake pedal to operate the brakes. A pressure regulator is fitted to the rear brakes.

The system is fail-safe, with dual hydraulic circuits connecting the brakes diagonally: the right front wheel and the left rear wheel are supplied by one circuit and the left front wheel and the right rear wheel by the other circuit. If one circuit fails there will still be one front and one rear brake working. If the pressure in either of the circuits is too low a warning light on the instrument panel will light up to warn the driver.

The front wheels are fitted with floating caliper disc brakes. Two makes of caliper may be found, those made by Girling and those made by Teves. The latter are stamped 'Ate'. The rear wheels are fitted with drum brakes. The handbrake is cable operated, and applies the rear brakes only. The equalizer bar is located at the bottom of the handbrake lever.

A vacuum pump, gear driven from the intermediate shaft, is mounted on the front side of the engine in the position normally occupied by the distributor on petrol (gasoline) engines and supplies vacuum for the operation of the servo. This is necessary because of the low inlet manifold vacuum of diesel engines.

2 Front brakes – general description

1 One of two types of floating caliper is used on the cars covered by this manual: one type is made by Teves (Ate) and the other by Girling. Refer to Figs. 6.1 and 6.2 to see the differences.
2 The Teves caliper has a fixed mounting frame bolted to the stub axle and a floating frame held in position on the mounting frame and able to slide on the mounting frame in a direction 90° to the face of the disc. Fixed to the floating frame is a hydraulic cylinder with piston and seals. The friction pads are held in the mounting frame by pins and a spreader spring.
3 When the brake pedal is pressed the piston is forced against the inner (direct) brake pad pushing it against the disc. The reaction causes the floating frame to move away until the floating frame presses against the outer (indirect) pad, pushing that one against the disc. Further pressure holds the disc between the two pads.
4 The Girling, commonly known as the Girling A type single cylinder caliper, works on a different principle. The cylinder is fixed to the stub axle and contains two pistons. The caliper body which carries the pads is free to slide along grooves in the cylinder body. One piston pushes the direct pad against the disc and the other piston pushes the floating caliper so that the indirect pad is forced against the disc.
5 Despite the obvious structural differences the two calipers are interchangeable but must be changed as pairs (ie, two Teves or two Girlings but not one of each).
6 The brake pads on both types of caliper are identical.

3 Disc pads (Teves caliper) – inspection and renewal

1 Brake pads wear more quickly than drum brake shoes. They should be checked for wear at the intervals specified in Routine Maintenance. The thickness of the pad (including the backing plate) must not be less than the dimension given in the Specifications. If the pads wear below this thickness then damage to the brake disc may result. Measure the pad thickness through the holes in the roadwheels, see Figs. 6.3 and 6.4.
2 To remove the pads, jack-up the front of the car and remove the wheels. Using a suitable drift tap out the pins securing the disc pads (photos). On some cars there may be a wire securing clip fitted round these pins. This should be pulled off. If there is no clip the pins will have sleeves.
3 Remove the spreader and pull out the inner, direct pad. If you are going to put the pads back then they must be marked so that they go back in the same place. Use a piece of wire with a hook on the end to pull the pad out. Now lever the caliper over so that there is space between the disc and the outer pad, ease the pad away from the caliper onto the disc and lift it out (photo).
4 Clean out the pad holder and check that the rubber dust cover is not damaged. Insert the outer pad and fit it over the projection on the caliper (photo). It will be necessary to push the piston in to insert the inner pad. This will cause the master cylinder reservoir to overflow unless action is taken to prevent it. Either draw some fluid out of the reservoir or slacken the bleeder screw (photo).
5 Push the piston in and check that the edges of the raised face of the piston are at 20° to the face of the caliper (photo). Make a gauge

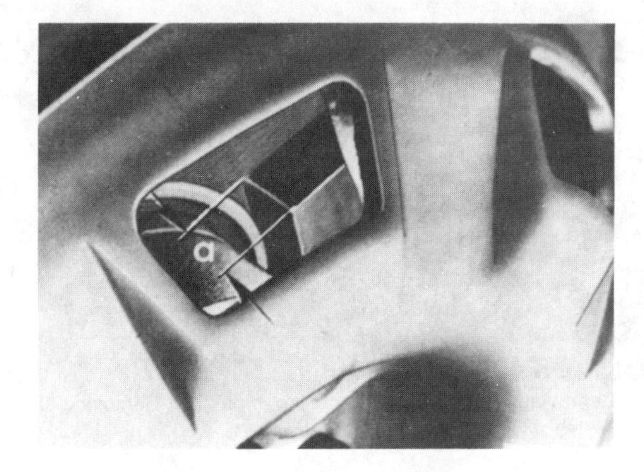

Fig. 6.3 Brake disc pad thickness dimension (a) including backplate not to be less than 6 mm ($\frac{1}{4}$ in) (Sec 3)

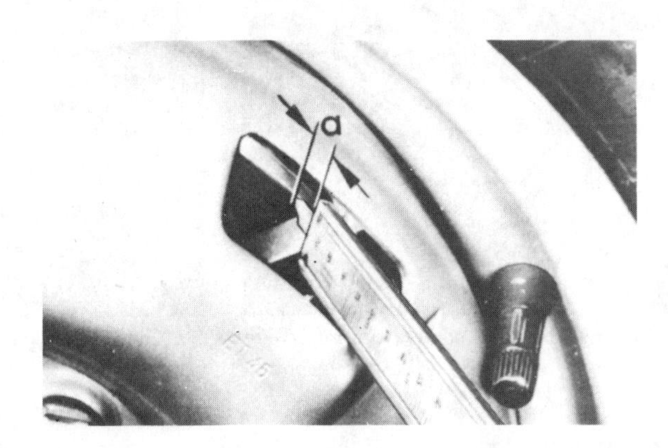

Fig. 6.4 Using vernier calipers to check the brake disc pad thickness (a) (Sec 3)

3.2A Disc pad pin retaining clip location (Teves type)

3.2B Removing the disc pad retaining pins (Teves type)

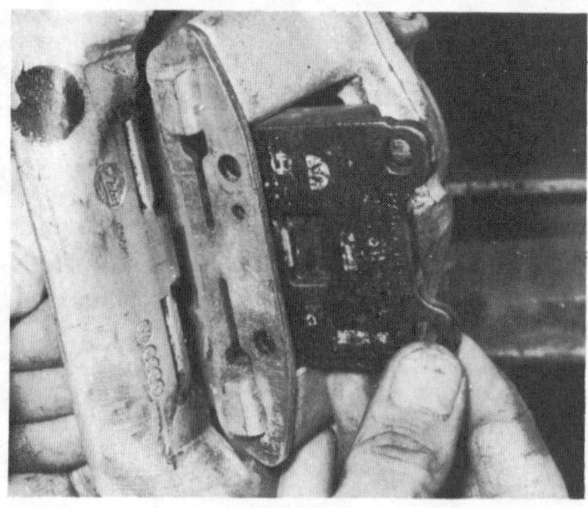
3.3 Removing an outer disc pad (Teves type)

3.4A Caliper projection engaged in outer disc pad recess (Teves type)

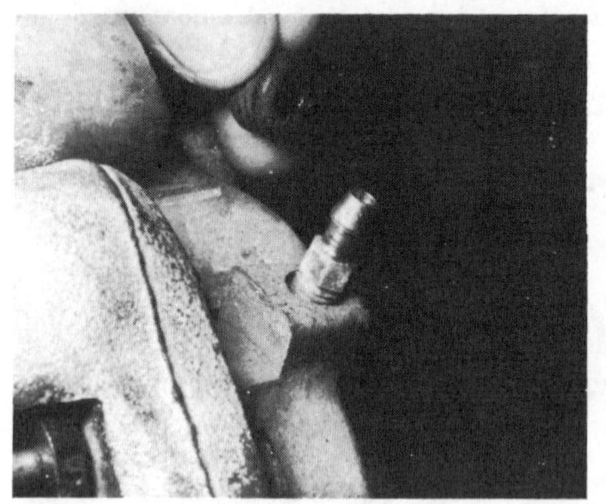
3.4B Caliper bleeder screw location (Teves type)

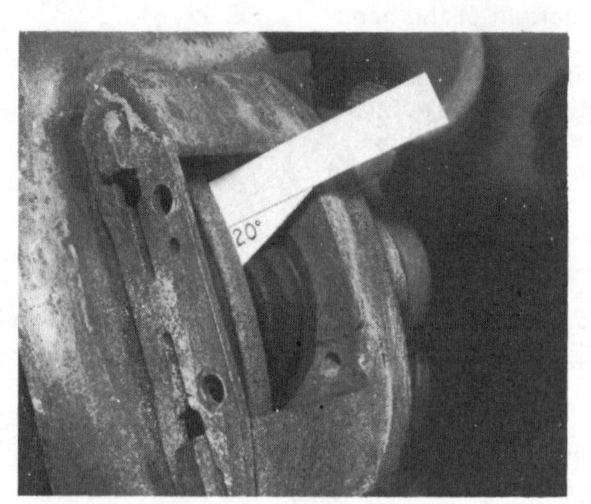

3.5 Check the angle of the piston projections

3.6 Disc pad spreader and retaining pins assembled to the caliper (Teves type)

out of cardboard if necessary. If the angle is more or less, turn the piston until the angle is correct.
6 Insert the inner pad, fit the spreader and the pins (photo). Fit a new locking wire if the type of pin requires it.
7 Do not forget to shut the bleed screw if it was opened as soon as the piston has been forced back. We do not like this method because there is a chance of air entering the cylinder and we do not like spare brake fluid about on the caliper while working on the friction pads.
8 Work the footbrake a few times to settle the pistons. Now repeat the job for the other wheel.
9 Replace the roadwheels, lower to the ground and take the car for a test run.

4 Disc pads (Girling caliper) – inspection and renewal

1 In general, the method is the same as for the Teves caliper (Section 3) with the following differences.
2 Lever off the pad spreader spring with a screwdriver, and pull out the pins with pliers after removing the screw which locks them in position.
3 Remove and refit the pads as with the Teves caliper. Fit the pins and the locking screw. A repair kit for brake pads will include new pins and retainer so use them.
4 The pad spreader spring is pressed on. The arrow must point in the direction of rotation of the disc when the car is travelling forward.

5 Calipers, pistons and seals (Teves) – removal, inspection and refitting

1 Support the front of the car on stands and remove the wheels.
2 If it is intended to dismantle the caliper then the brake hoses must be removed, plugged and tied out of the way.
3 Mark the position of the brake pads and remove them if the caliper is to be dismantled.
4 Remove the two bolts holding the caliper to the stub axle (photo). This should not be done while the caliper is hot.
5 Withdraw the caliper from the car and take it to a bench.
6 Refer to Fig. 6.5. A repair kit should be purchased for the overhaul of the caliper.
7 Ease the floating frame away from the mounting frame. Press the brake cylinder and guide spring off the floating frame. The cylinder may now be dismantled.
8 Remove the retaining ring and the rubber boot. The next problem is to remove the piston which is probably stuck in the piston seal. The obvious way is to blow it out using air pressure in the hole which normally accommodates the hydraulic pressure hose. However, be

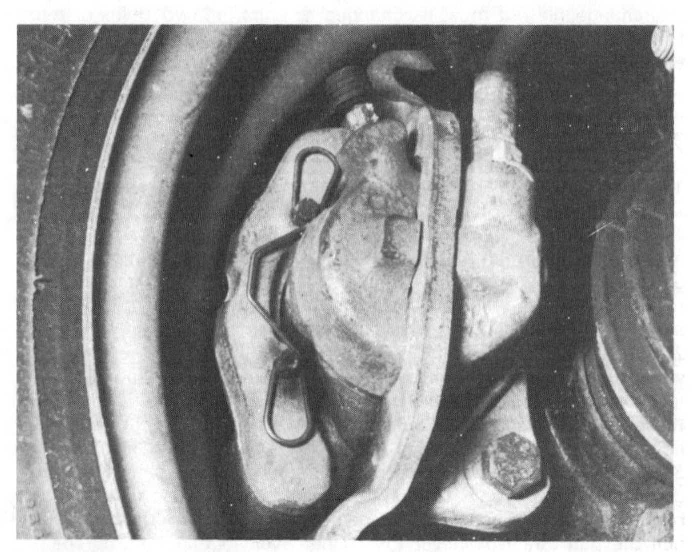
5.4 Caliper lower retaining bolt location (Teves type)

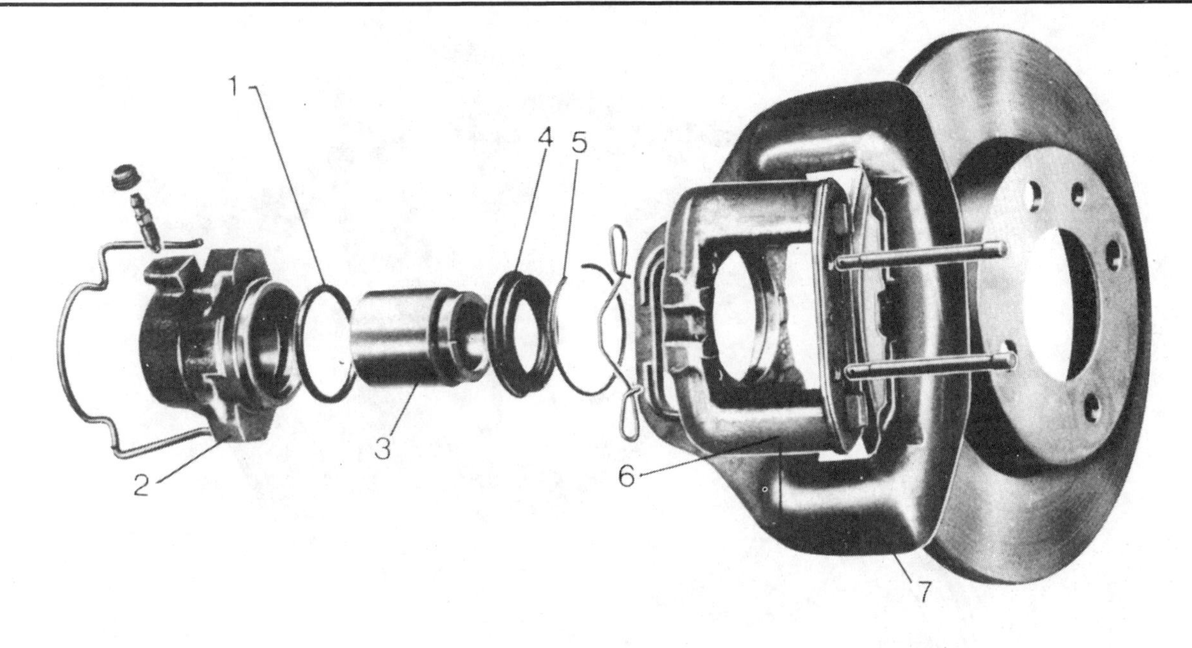

Fig. 6.5 Exploded view of the brake caliper (Teves type) (Sec 5)

1 *O-ring seal*
2 *Cylinder*
3 *Piston*
4 *Dust cap*
5 *Circlip*
6 *Mounting frame*
7 *Floating frame*

careful. The piston may be stuck in the seal but when it does come out it will come quickly. Fit the cylinder in a vice with a piece of wood arranged to act as a stop for the piston. If you do not, as most people do not, have a ready supply of compressed air use a foot pump. Low air pressure is all that is normally required.

9 When the piston is out of the cylinder clean carefully the bore of the cylinder and the piston with brake fluid or methylated spirit.

10 It is difficult to define wear on the piston. When it has been cleaned it should have a mirror finish. Scratches or dull sections indicate wear. The inside of the bore must be clean with no scratches or distortion. If there is any doubt renew the whole unit. In any case renew the seal and dust excluder. Dip the cylinder in clean brake fluid. Coat the piston and seal with brake rubber grease (Ate) and press the piston and seal into the cylinder. Install new dust excluder and its retaining ring. Fit a new locating spring and knock the cylinder onto the floating frame with a brass drift.

11 Refit the floating frame to the mounting frame and set the piston recess at an inclination of 20° to the lower guide surface of the caliper (where the pad rests), as described in Section 3.

12 Bolt the mounting frame onto the stub axle and fit the pads. Refit the wheel and lower to the ground.

13 Top-up the brake fluid reservoir and bleed the brakes as described in Section 23.

14 Road test the car to check the operation of the brakes.

6 Calipers, pistons and seals (Girling) – removal, inspection and refitting

1 Refer to Fig. 6.6. The general rules for overhaul are the same as for the Teves (Section 5) but with the complication of the extra piston.

2 Remove the cylinder and floating mounting from the stub axle and press them apart. Again use air pressure to dislodge the pistons and be careful that they do not come out suddenly. Clean the pistons and bore carefully. Use a Girling kit of seals and parts. Wash the cylinder and pistons with clean brake fluid and reassemble.

3 In this case the pistons and cylinder are bolted to the front suspension. The caliper sliding frame must move easily over the cylinder casting. Tighten the holding bolts to the specified torque.

7 Disc pad wear indicators – general

1 A built-in method of indicating to the driver that the brake pads require renewing is fitted on cars for the USA and certain other territories. Refer to Fig. 6.7. The chamfer (arrow B in the illustration) on both sides of the brake disc has a lug about 25 mm (1 in) long flush with the face of the disc. The pads have a thin extension piece (A) on the bottom which is coated with wear resistant material. When the pads are worn to the permissible limit the extension on the pad makes contact with the lug on the disc. This causes a vibrating or pulsating effect at the foot pedal, thus indicating to the driver that pad renewal is a matter of urgency.

2 Discs and brake pads of this type may be fitted to cars not so originally fitted.

8 Front brakes – squeaking pads

If the pads squeal or squeak excessively, relief from this problem may be obtained by removing and cleaning the pads and holders and then applying sparingly a substance known as 'Plastilube'. This substance **must not** be applied to friction surfaces but to the ends, sides and back of the pad. It should also be applied to the pins and spreader and the sliding surfaces of the floating caliper.

9 Rear brakes – general description

1 One of two types of drum brake is found on the Golf/Rabbit models. The early models are equipped with manually adjusted brakes; from 1979 self-adjusting brakes are fitted.

2 With the manually adjusted type, the shoes are held against the connecting link by a U-spring. The shoes are adjusted through a hole in the backplate, using a screwdriver to rotate the brake shoe adjuster.

3 The construction of the self-adjusting brakes is similar to the type found on earlier models except that the U-spring is replaced by two coil springs and the adjusting mechanism consists of a pushrod, wedge and springs. Refer to Fig. 6.9. A single operation of the brake

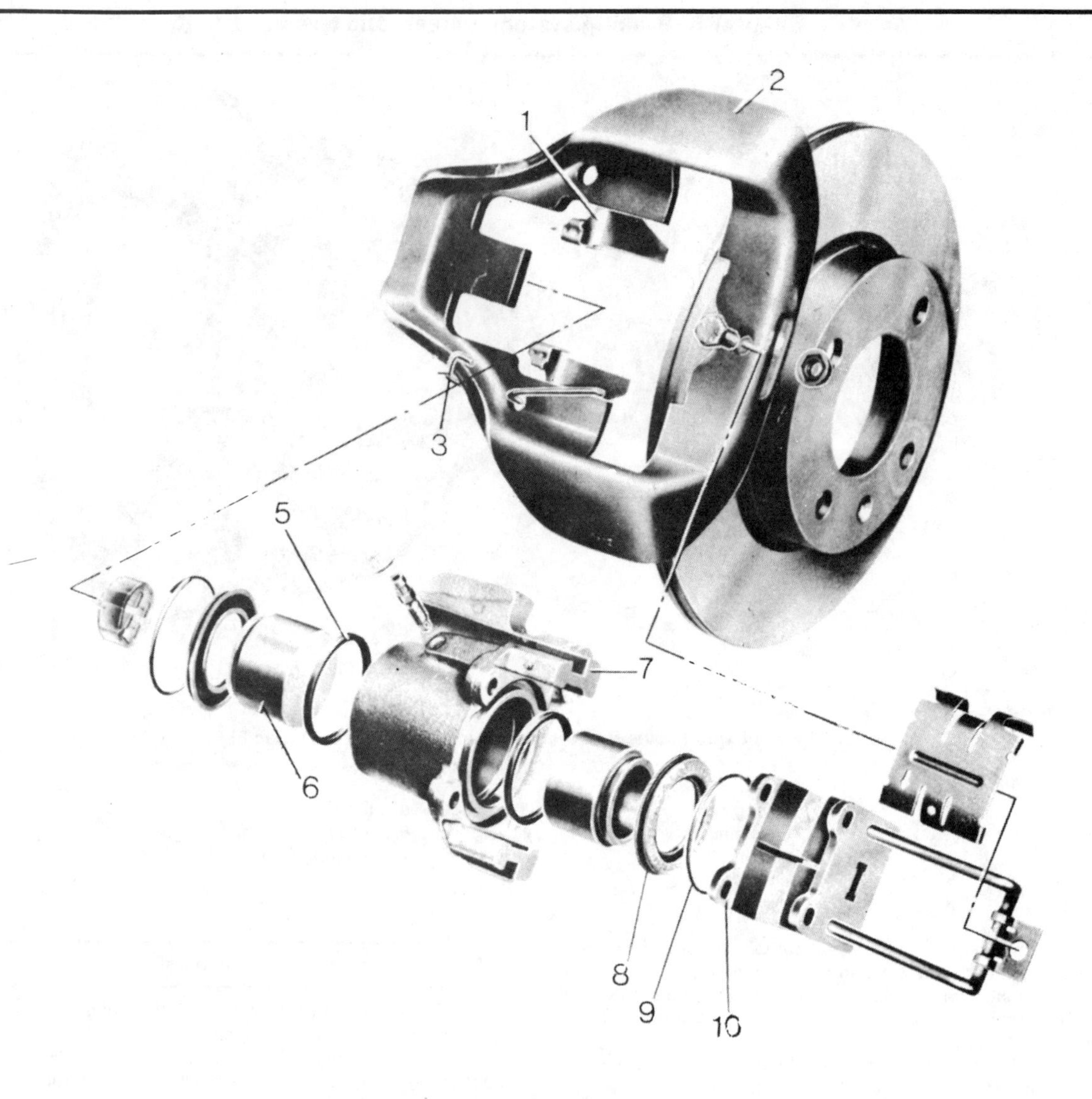

Fig. 6.6 Exploded view of the brake caliper (Girling type) (Sec 6)

1 Clip
2 Frame
3 Retaining spring
4 Support
5 Sealing ring
6 Piston
7 Cylinder housing
8 Dust cap
9 Circlip
10 Brake pads

Fig. 6.7 Brake pad wear indicator system (Sec 7)

For A and B see text

pedal sets a fixed clearance between the brake drum and the brake shoes. As the linings wear and shoe travel becomes greater than the fixed clearance, the wedge is moved downwards by the arrangement of the pushrod and the springs. This movement of the wedge alters the effective length of the pushrod and thus automatically adjusts the position of the brake shoes in relation to the brake drum.

10 Rear brakes (manually adjusted type) – adjustment of shoes and handbrake

Note: Adjustment of the handbrake on models with self-adjusting rear brakes is only necessary if the handbrake lever or cables have been removed and refitted, or the cables have stretched after a high mileage, refer to Section 12.

Brake shoe adjustment

1 Chock the front wheels, then jack-up the rear of the car and support it on axle stands or other suitable supports.
2 Find the adjuster access hole on the backplate. It is just below the wheel cylinder bleed screw. Remove the rubber plug from the hole.
3 Move the handbrake lever to the fully off position. Depress the lever on the brake pressure regulator valve (Section 21) towards the valve body. This will release any pressure in the right-hand pipeline.
4 Through the access hole in the left-hand rear brake backplate a toothed wheel is visible. Using a screwdriver, lever the adjuster wheel teeth upwards (handle of the screwdriver down) until the brake linings bind on the drum. Now back-off (work the screwdriver the other way) the adjuster until the roadwheel will rotate without the drum dragging on the brake shoes. Refit the rubber plug in the adjuster access hole.
5 Repeat this procedure to adjust the other rear brake, but note that this time the adjuster works in the reverse direction, (lever the teeth of the adjuster downwards to bring the shoes in contact with the brake drum).

Handbrake adjustment

6 After the footbrake is correctly adjusted the handbrake may be adjusted, if it fails to lock the rear wheels after the control lever is pulled over three or four notches. Remove the handbrake lever boot securing screws and pull up the boot. Pull the handbrake lever up two notches from the fully off position. Tighten each of the adjusting nuts a little at a time, keeping the equalizer bar level, until neither of the rear wheels can be turned by hand. Release the handbrake lever and check that the rear wheels rotate freely. Tighten the locknuts.
7 If the equalizer bar is positioned at an acute angle after the correct adjustment has been achieved then one of the cables has been stretched and must be renewed, refer to Section 25.

11 Rear brakes (manually adjusted type) – removal and refitting

1 Slacken the roadwheel bolts. Chock the front wheels, jack-up the rear of the car and support it on axle stands or other suitable supports. Release the handbrake lever.
2 Take off the hub cap, remove the split-pin and the locking ring. Remove the hub nut and pull the brake drum off the stub axle. The thrust washer and wheel bearing will come off with the drum. Set the drum on one side for inspection (Section 13). **Note:** It may be necessary to back-off the brake shoes to remove the brake drum, refer to Section 10.
3 The brake assembly is now exposed. **Do not** depress the brake pedal from now on. Remove the large U-spring and unhook the two springs at the bottom of the brake shoes. Take off the leaf springs from the steady pins and remove the shoes from the backplate. It will be necessary to unhook the handbrake cable to remove the shoes. The adjuster rod will come away with the shoes. The backplate can be removed, if necessary, by undoing the four securing bolts and lifting away the backplate and stub axle.
4 Overhaul of the brake drum and shoes is dealt with in Section 13

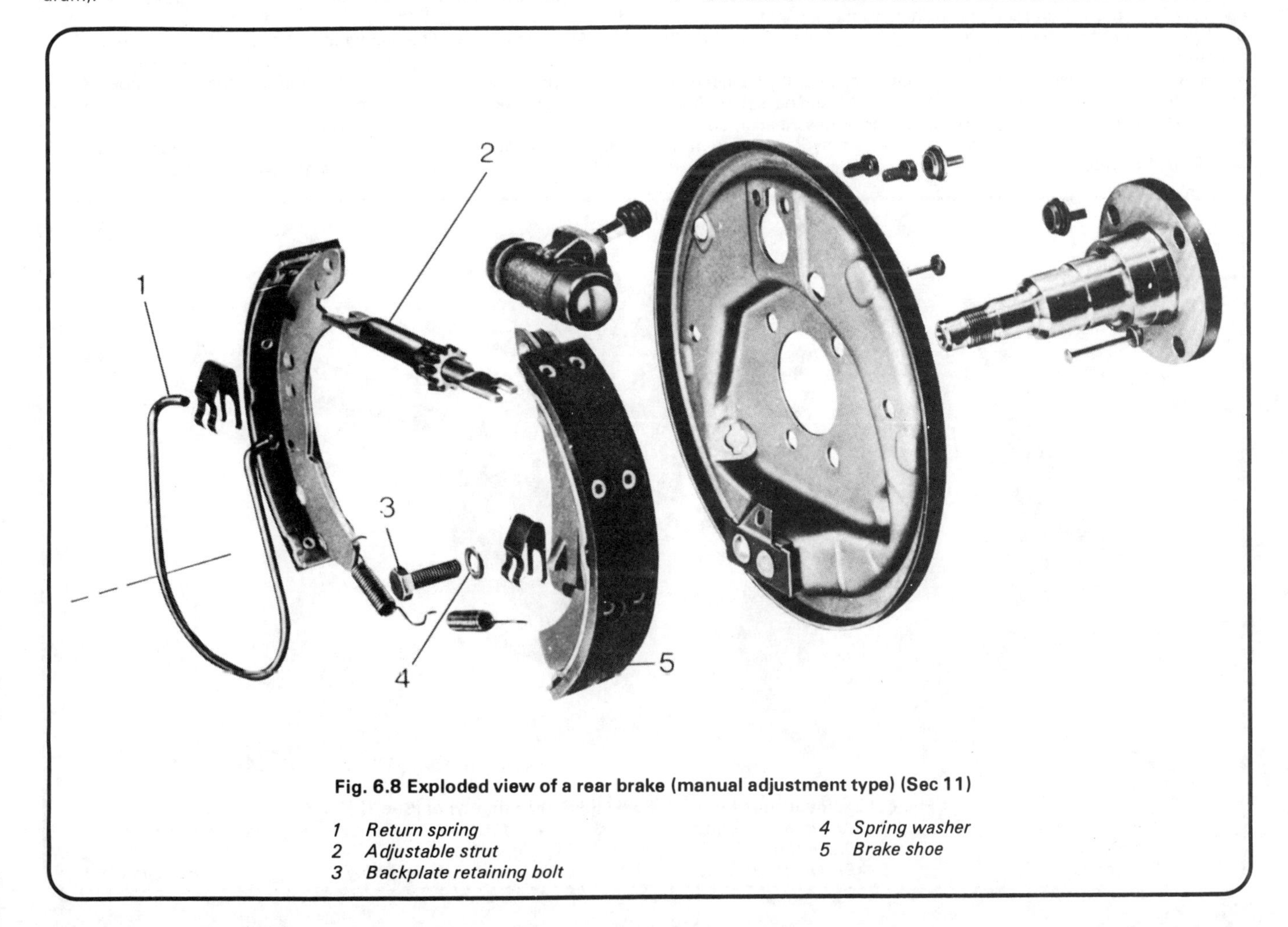

Fig. 6.8 Exploded view of a rear brake (manual adjustment type) (Sec 11)

1 Return spring
2 Adjustable strut
3 Backplate retaining bolt
4 Spring washer
5 Brake shoe

and the wheel cylinder in Section 14.
5 Refitting is the reverse of the removal procedure. Fit the handbrake cable to the trailing shoe lever, locate the shoes on the backplate and fit the steady pins. Fit the bottom springs. Lever the shoes apart and fit the adjuster rod. Finally fit the U-spring.
6 Refit the drum and bearing. Adjust the bearing as described in Chapter 9. Refit the roadwheel and adjust the brakes, refer to Section 10. If the wheel cylinder was removed, bleed the hydraulic system as described in Section 23.

12 Rear brakes (self-adjusting type) – removal and refitting

1 Refer to Fig. 6.9. Repeat the operations described in Section 11, paragraphs 1 and 2, but note that the self-adjusting mechanism will have kept the lining right up to the drum and the adjuster must be backed off before the drum can be removed. To do this insert a screwdriver through a roadwheel bolt hole and push the adjusting wedge right up, then withdraw the brake drum (photo). **Do not** depress the brake pedal from now on.
2 Using a pair of pliers remove the washers retaining the shoe steady springs by depressing them and rotating them through 90° and remove the springs (photo).
3 Pull the bottom ends of the shoes out of the support and unhook the lower return spring (photo).
4 Pull back the spring on the handbrake cable and disconnect the cable from the operating lever.
5 Using pliers, unhook the wedge spring and upper return spring, then lift away the brake shoes (photo).
6 The brake wheel cylinder can be removed, if necessary, as described in Section 14.
7 The backplate can be removed, if necessary, by undoing the four securing bolts and lifting away the backplate and stub axle.
8 To disconnect the pushrod from the brake shoe, clamp the shoe in a vice, unhook the locating spring and lift away the pushrod.
9 Overhaul of the brake drum and shoes is dealt with in Section 13 and the wheel cylinder in Section 14.
10 Refitting is the reverse of the removal procedure. When inserting the wedge, the lug on the wedge must be towards the backplate. After fitting the brake drum adjust the wheel bearing as described in Chapter 9. If the wheel cylinder was removed, bleed the hydraulic system as described in Section 23.

Note: The brakes are automatically adjusted when the footbrake pedal is depressed. Do this several times.
11 Refit the roadwheels and lower the car to the ground. Road test the car to check the operation of the brakes.

13 Brake shoes and drums – inspection and renovation

1 One of two types of shoe is used. One type has the lining riveted to the brake shoe and the other has the lining bonded to the shoe. When the lining thickness is worn to less than the wear limit specified at the beginning of this Chapter the shoes must be renewed.
2 The linings must be free from oil or hydraulic fluid contamination, have smooth surfaces and be free from scoring. If any of these defects are present the shoes must be renewed and the cause of the defect discovered and rectified.
3 The four shoes on the rear axle must be renewed at the same time. It is not recommended that DIY owners should attempt to reline the riveted type of brake shoe. It is much better to obtain new shoes.
4 Drums must be smooth and unscored. Any scoring or grooves on the inside diameter must be removed by machining in a lathe. The diameter after machining must not exceed the maximum specified.
5 The drum must also be inspected for cracks, ovality or other damage.

14 Wheel cylinder – removal, overhaul and refitting

1 The wheel cylinder is fastened to the backplate by two screws. It may be removed only after the brake shoes are lifted away from it. It is not necessary to dismantle the shoes from the backplate.
2 Disconnect the hydraulic hose from the cylinder and plug the hose. The screws holding the cylinder to the plate may be difficult to extract. Brake fluid is a good penetrating oil so moisten the threads with a little fluid.
3 An exploded view of a rear wheel cylinder is shown in Fig. 6.10.
4 To dismantle, remove the dust caps and blow the pistons out with low air pressure. Be careful when doing this that the pistons do not come out too quickly. Muffle the cylinder with a large piece of rag.
5 Wash the bores and pistons with clean brake fluid and examine for mirror finish and scratches. If there is any blemish discard the complete unit and fit a new one. Always fit new seals and dust caps. Coat

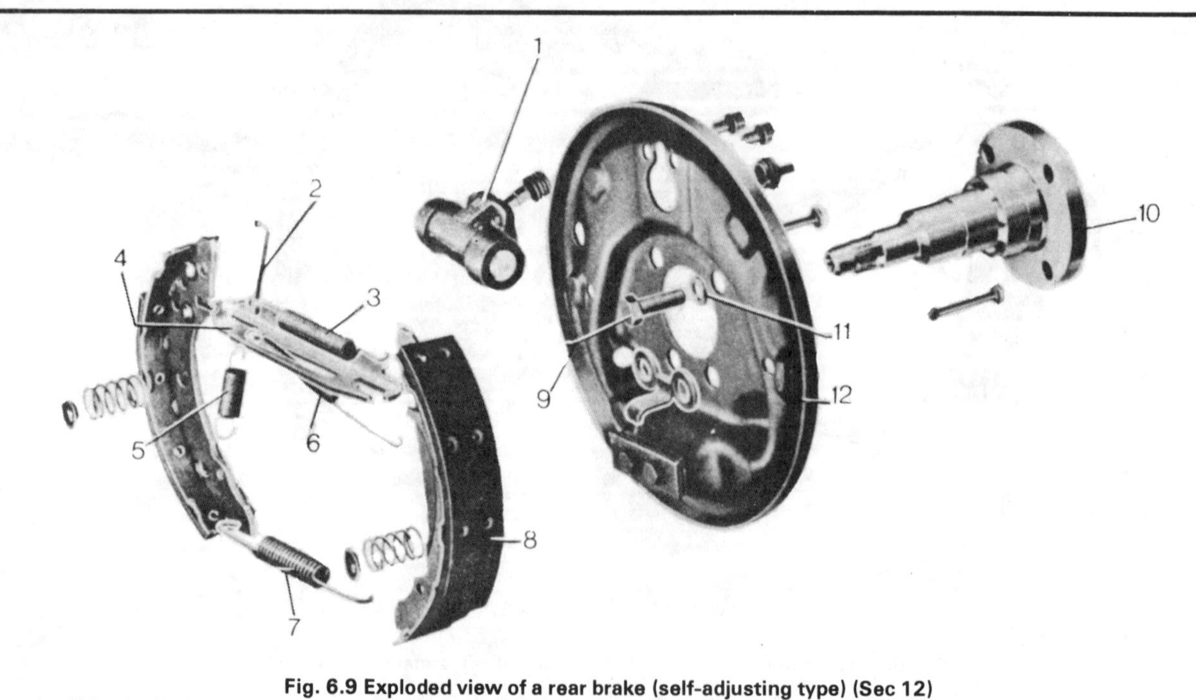

Fig. 6.9 Exploded view of a rear brake (self-adjusting type) (Sec 12)

1 Wheel cylinder
2 Wedge
3 Spring
4 Pushrod
5 Spring
6 Upper return spring
7 Lower return spring
8 Brake shoe
9 Backplate retaining bolt
10 Stub axle
11 Washer
12 Backplate

12.1 Removing the rear brake drum

12.2 Removing the brake shoe steady springs

12.3 Lower return spring location on rear brakes (self-adjusting type) and handbrake cable and operating lever location

12.5 View of the rear brakes showing spring and strut locations (self adjusting type)

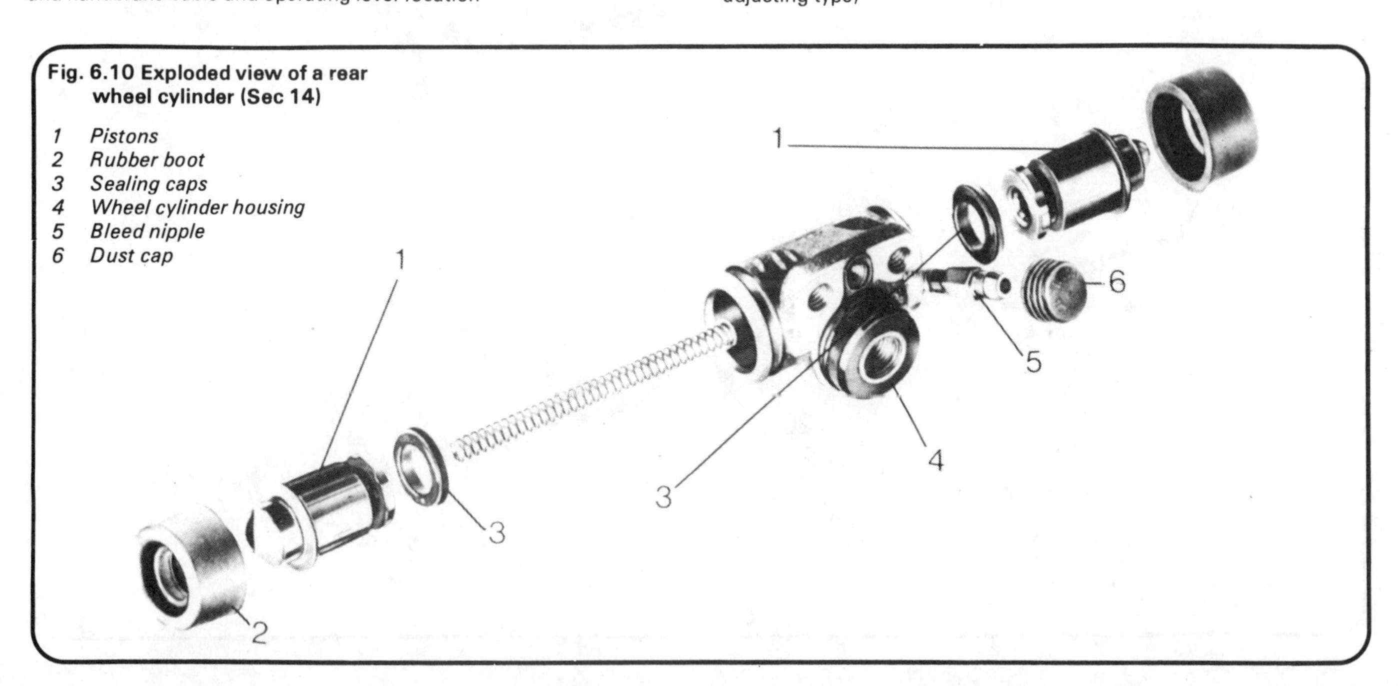

Fig. 6.10 Exploded view of a rear wheel cylinder (Sec 14)

1 *Pistons*
2 *Rubber boot*
3 *Sealing caps*
4 *Wheel cylinder housing*
5 *Bleed nipple*
6 *Dust cap*

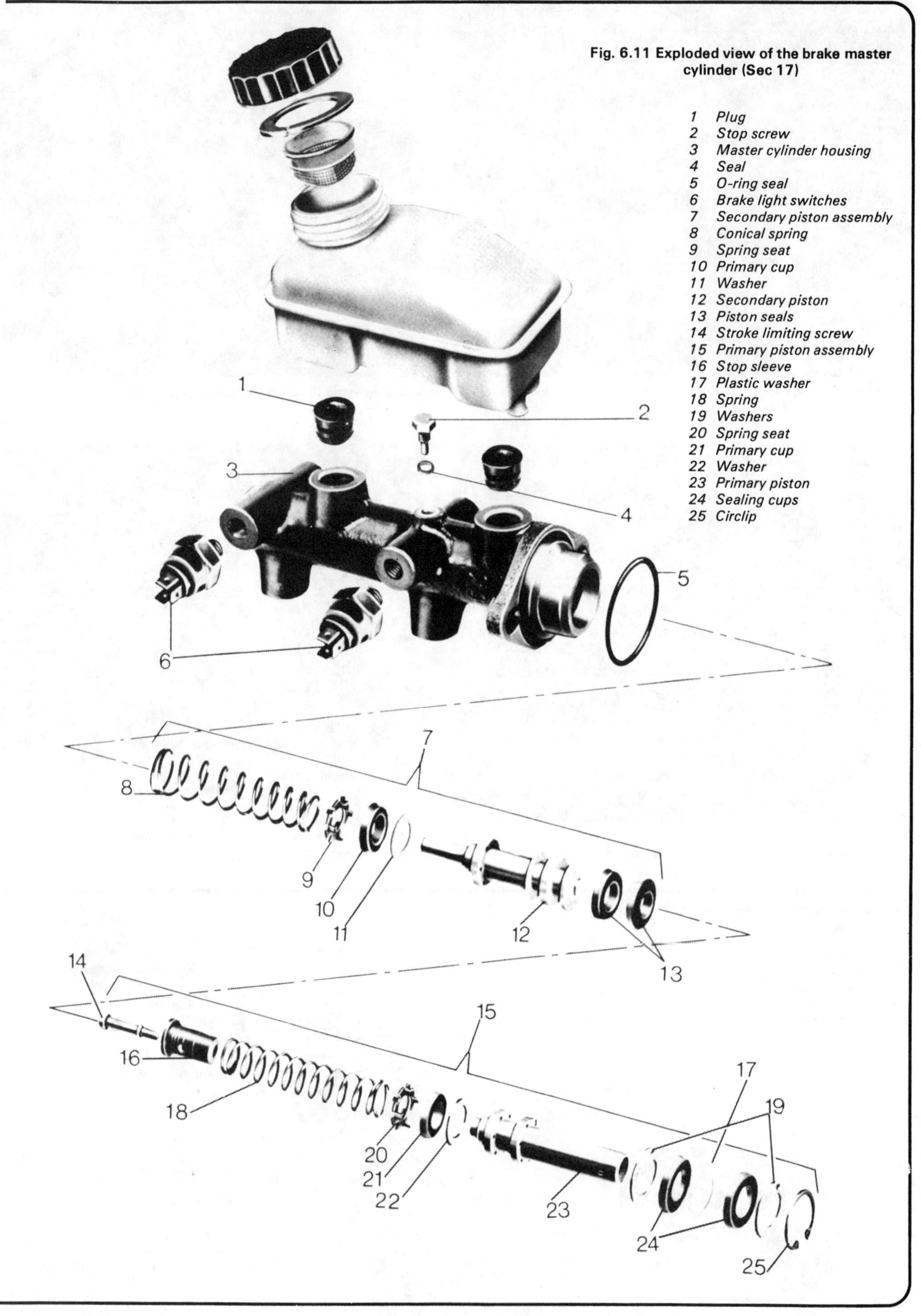

Fig. 6.11 Exploded view of the brake master cylinder (Sec 17)

1 Plug
2 Stop screw
3 Master cylinder housing
4 Seal
5 O-ring seal
6 Brake light switches
7 Secondary piston assembly
8 Conical spring
9 Spring seat
10 Primary cup
11 Washer
12 Secondary piston
13 Piston seals
14 Stroke limiting screw
15 Primary piston assembly
16 Stop sleeve
17 Plastic washer
18 Spring
19 Washers
20 Spring seat
21 Primary cup
22 Washer
23 Primary piston
24 Sealing cups
25 Circlip

them with a little brake rubber grease before fitting and make sure the seals are the right way round.

6 Refitting is the reverse of the removal procedure. Bleed the hydraulic system as described in Section 23 and road test the car to check the operation of the brakes.

15 Brake hydraulic system – description and operation

1 The master cylinder has two pistons, the front or secondary supplying pressure to the left-hand front wheel and the right-hand rear wheel. The rear, or primary, piston supplies pressure to the right-hand front wheel and the left-hand rear wheel. Inspection of the unit will show four pipes leading from the main casting to the wheel cylinders and calipers.

2 Brake fluid is supplied from the fluid reservoir, the white plastic container which feeds both circuits through the master cylinder.

3 When the brake pedal is pressed the pushrod moves the primary piston forward so that it covers the port to the fluid reservoir. Further movement causes pressure to build up between the two pistons exerting pressure on the secondary piston which also moves forward covering the port to the fluid reservoir. The pressure now builds up in the pipe.

4 If the pipes of the secondary piston circuit fracture or the system fails in some way the secondary piston will move forward to the end of the piston compressing the conical spring and sealing the outlet port to the left front and right rear brakes. The primary piston circuit will continue to operate.

5 Failure of the primary piston circuit causes the piston to move forward until the stop sleeve contacts the secondary piston and the primary piston simultaneously when pressure will again be applied to the secondary piston circuit.

6 The various springs and seals are designed to keep the pistons in the right place when the system is not under pressure.

7 Pressure switches are screwed into the cylinder body and operate the stoplights and warning lamps.

16 Master cylinder – removal and refitting

1 The master cylinder is mounted on the front of the brake servo unit in the engine compartment (photo).

2 Disconnect the battery earth strap and then the wiring from the brake light and warning light switches on the master cylinder.

3 Disconnect the pipes from the master cylinder and collect the brake fluid, which will drain out, in a suitable container. Take care to avoid brake fluid spilling on paintwork.

4 Undo the nuts securing the master cylinder to the servo unit and lift the master cylinder off the mounting studs. Discard the O-ring located between the master cylinder and the servo.

5 Refitting is the reverse of the removal procedure. Always fit a new O-ring between the master cylinder and the servo unit. Ensure that the wiring and pipes are reconnected to their original locations. Tighten the master cylinder securing nuts and the brake lines to the specified torque.

6 Fill the brake fluid reservoir with fresh fluid (never re-use fluid removed from the system) and bleed the hydraulic system as described in Section 23.

17 Master cylinder – overhaul

1 Clean the outside of the cylinder thoroughly and then set it on a clean area. Work only with clean hands. Obtain an overhaul kit and a quantity of fresh brake fluid in a clean container. Always use all the parts supplied in the overhaul kit.

2 Refer to Fig. 6.11. Undo and remove the brake stoplight and warning light switches.

3 Remove the fluid reservoir by pulling it upwards out of the rubber grommets. Remove the grommets.

4 Remove the stop screw and seal from the top of the cylinder.

5 Push the piston back slightly and then remove the retaining circlip. The contents of the cylinder should now slide out if the cylinder is tilted. If they do not, apply a low air pressure and blow them out. As the parts are removed place them in the order in which they should be refitted.

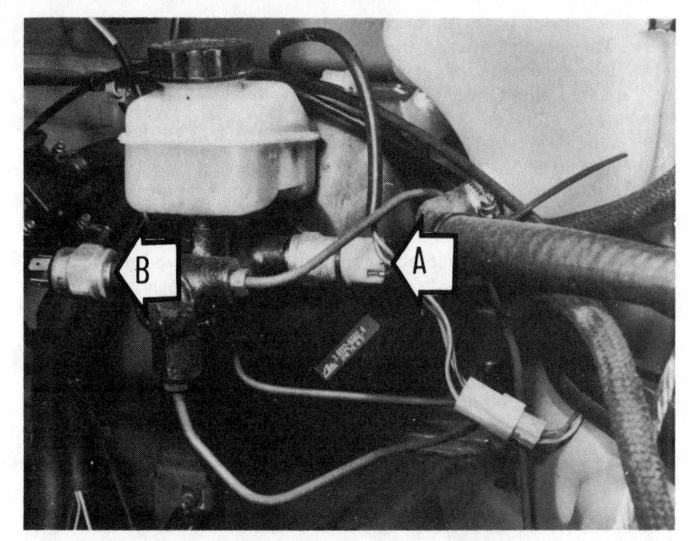

16.1 Brake master cylinder location, showing primary brake circuit stop and warning light switch (A) and secondary brake circuit stop and warning light switch (B)

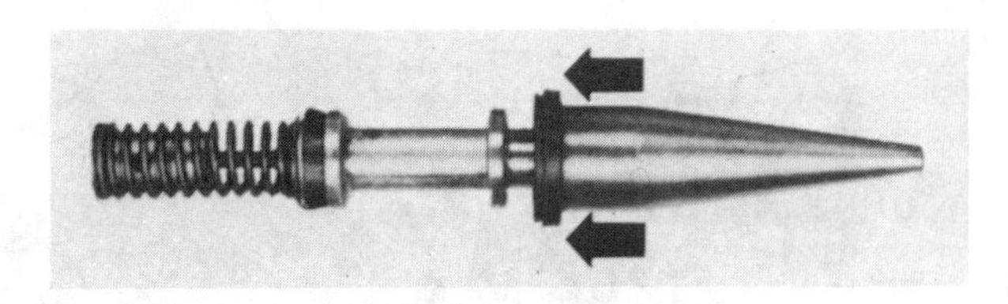

Fig. 6.12 Using a tapered mandrel to fit the master cylinder piston seals (Sec 17)

6 Examine the bore of the cylinder. If it is scored or rusty the cylinder must be renewed. It is possible to hone out slight marks if you have access to the right equipment, but our advice is to fit a new cylinder.

7 Clean the master cylinder with brake fluid or methylated spirit only, never use paraffin or petrol, and ensure that all passageways and holes are clear.

8 The piston seals are not easy to fit. VW make a special taper mandrel, see Fig. 6.12, which just fits on the ends of the piston and then the seal is eased on over the taper. The seals can be fitted without this tool, but take care not to damage them as they slide over the piston lands. Use the fingers only – never a metal tool.

9 Assemble the two pistons with their associated parts. During assembly wet all the parts with fresh brake fluid. Ensure that the primary cups and piston seals are fitted correctly, see Fig. 6.11.

10 Hold the cylinder vertically and insert the complete secondary piston from underneath.

11 Now insert the primary piston and retaining circlip. Fit the stop screw. It may be necessary to move the secondary piston to fit the stop screw fully.

12 Insert the brake fluid reservoir locating grommets in the top of the master cylinder and then fit the fluid reservoir.

18 Servo unit – testing, repair, removal and refitting

1 If the brakes seem to need more pedal pressure than normal a check of the servo is indicated. The brake servo pump should also be checked.

2 First of all check that the hose connecting the vacuum pump to the servo unit is securely attached and in good condition. Then with the engine running disconnect the hose from the servo unit and check that there is suction at the end of the hose. If there is no suction the vacuum pump is defective, refer to Sections 19 and 20.

3 If the vacuum hose and vacuum pump are in order and the servo is still not assisting the brakes the trouble is either a leaking servo diaphragm or a defect in the master cylinder. Check that the sealing ring between the master cylinder and the servo is not leaking.

4 The repairs possible to the servo are very limited. Refer to Fig. 6.13. The O-ring between the servo and the master cylinder should always be renewed whenever the units are separated. At the rear end of the servo, the filter and damping washer may be removed from the sleeve and renewed, note that the slots in the damping washer and filter should be offset 180° on assembly. Apart from this the servo, which is a sealed unit, cannot be serviced further.
5 To remove the servo, first remove the master cylinder as described in Section 16 and plug the hydraulic lines. Disconnect the vacuum hose from the servo.
6 Remove the clevis pin connecting the servo pushrod to the brake pedal (LHD models) or to the brake pedal to servo link rod (RHD models) and unscrew the fork end (photo). Remove the nuts securing the servo unit to the mounting bracket and lift away the servo.
7 Refitting is the reverse of the removal procedure. Bleed the hydraulic system as described in Section 23 and adjust the brake pedal pushrod, refer to Section 26.

19 Brake servo vacuum pump – general description, removal and refitting

1 As the intake manifold vacuum on diesel engines is much less than that on petrol (gasoline) engines, a vacuum pump is needed to create enough vacuum for the efficient operation of the brake servo unit.
2 The diaphragm type of pump used is mounted on the location used by the distributor on spark ignition engines and is driven by a gear on the intermediate shaft.
3 Slacken the hose clips and disconnect both hoses from the pump. Disconnect the oil feed at the union on the pump body (photos).
4 Undo the clamp plate securing bolt, remove the clamp plate and then withdraw the pump from the engine block (photos). Discard the O-ring.
5 Refitting is the reverse of the removal procedure. Ensure that the end of the pump drive engages correctly with the oil pump shaft. Tighten the clamp plate securing bolt to the specified torque.

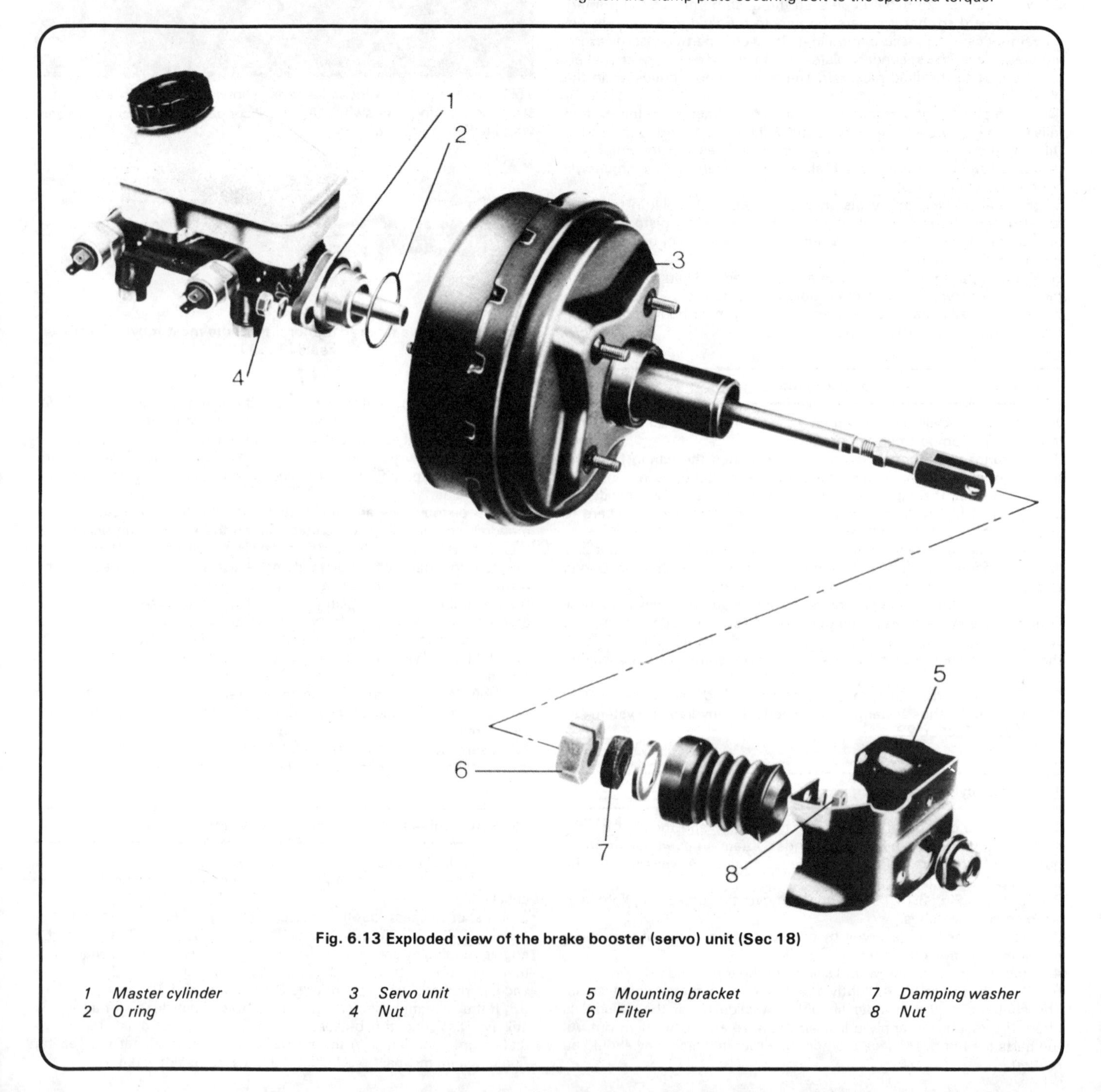

Fig. 6.13 Exploded view of the brake booster (servo) unit (Sec 18)

1 Master cylinder
2 O ring
3 Servo unit
4 Nut
5 Mounting bracket
6 Filter
7 Damping washer
8 Nut

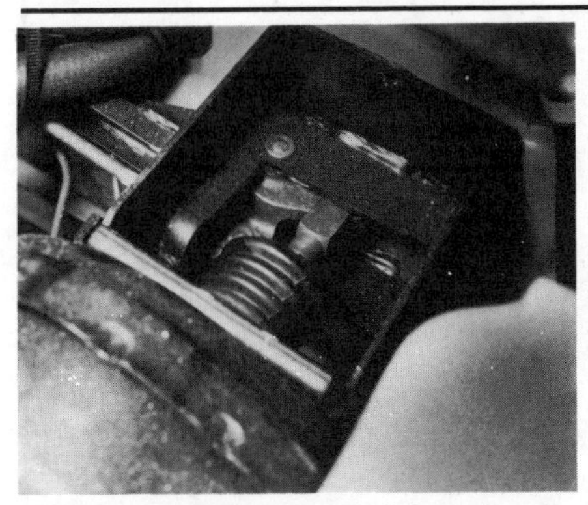

18.6 Brake servo bell crank linkage location (left-hand side)

19.3A Brake servo vacuum pump (exhauster) lower hose connection

19.3B Brake servo vacuum pump (exhauster) upper hose connection and non-return valve

19.3C Brake servo vacuum pump (exhauster) oil feed pipe location

19.4A Brake servo vacuum pump (exhauster) clamp plate

19.4B Removing the brake servo vacuum pump (exhauster)

20 Vacuum pump – overhaul

1 Obtain a vacuum pump repair kit before dismantling the pump. Refer to Fig. 6.15.
2 Make an alignment mark on both parts of the pump body to ensure that they can be reassembled in their original position. Undo the eight securing bolts and separate the top section from the lower part of the body.
3 Undo the diaphragm assembly securing nut and remove the diaphragm and its backing plates. Remove the O-ring.
4 Undo the four securing screws and remove the endplate and gasket.
5 Undo the two securing screws and remove the cover plate, springs, spring seats, valves and washers.
6 Wash all the parts in a cleaning solvent and inspect them for wear and damage. Examine the diaphragm for deterioration or splits. If in doubt, renew it.
7 Fit a new O-ring on the pushrod.
8 Fit the diaphragm and backing plates with the moulded part of the diaphragm towards the top part of the body and the holes in line with the holes in the pump body.
9 Coat the threads of the diaphragm assembly securing nut with a thread locking compound. Fit the nut and tighten it to the specified torque.
10 Assemble the top part loosely to the main body.
11 Working through the endplate opening remove the circlip, arrowed in Fig. 6.14 from the driveshaft, then push the shaft back until the pushrod is free of the driveshaft and the tension is released from the diaphragm.
12 Push the pushrod against the stop in the direction of the arrow in Fig. 6.16 and hold it in this position while tightening the screws securing the top part to the body in a diagonal sequence, a few threads at a time. Push the driveshaft into position and fit the thrust washer and circlip on the end of the shaft.
13 Refit the washers, valves, spring seats and springs. When fitting the valves the spring seats must be towards the housing. Using a new gasket, refit the valve cover plate.
14 Using a new gasket, refit the pump body endplate.

Fig. 6.14 Brake vacuum pump driveshaft circlip (arrowed) (Sec 20)

21 Brake pressure regulator – description

1 This device is fitted to some models to limit the hydraulic pressure in the rear wheel hydraulic cylinders and so limit the braking force. In effect it should prevent the rear wheels being locked solid by the brakes and so avoid skidding.
2 It is fitted just in front of the rear axle, to which it is attached by a spring. When the brakes are applied sharply the front of the car sinks

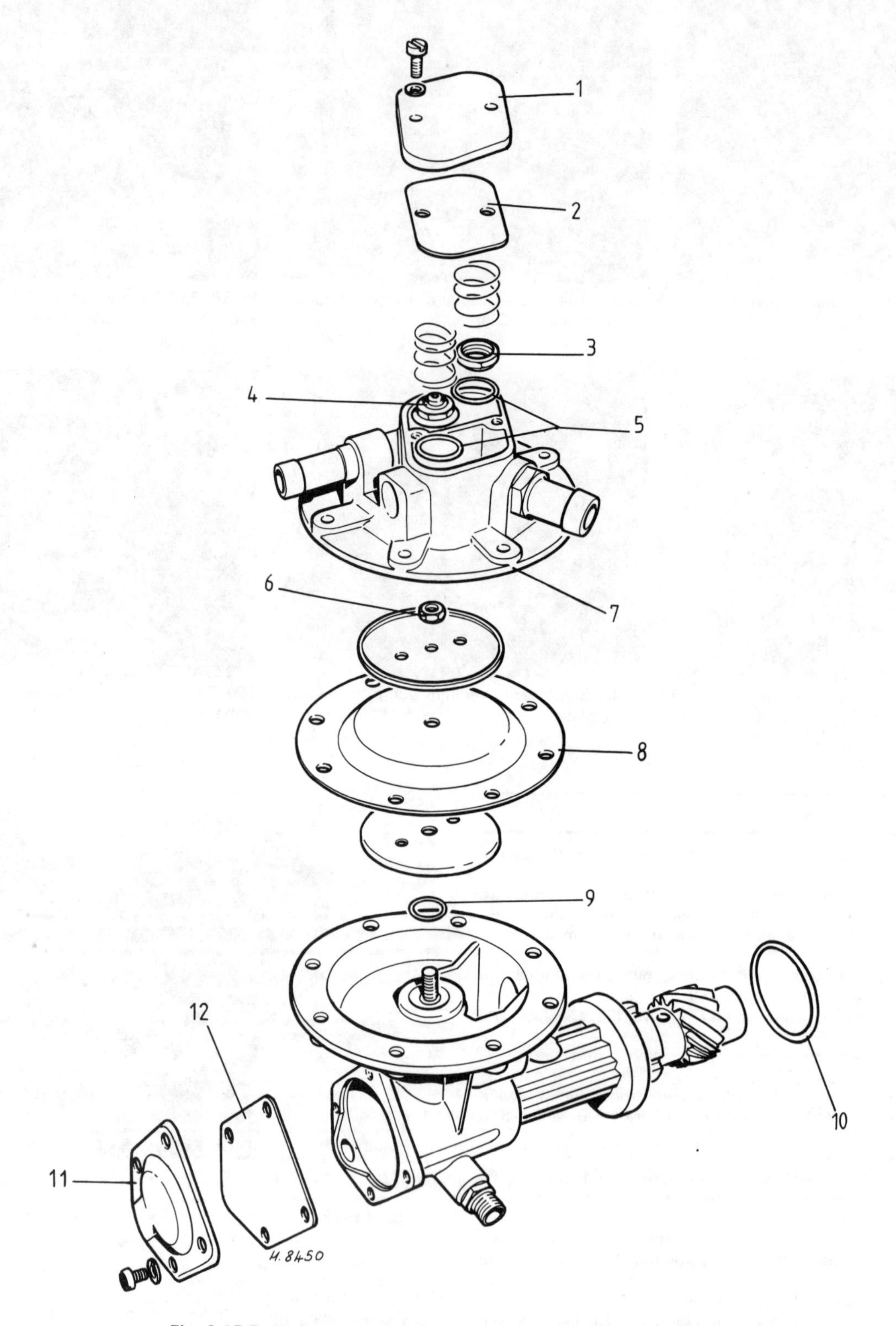

Fig. 6.15 Exploded view of the brake vacuum pump (exhauster) (Sec 20)

1 Cover
2 Gasket
3 Inlet valve
4 Outlet valve
5 Sealing washer
6 Nut
7 Upper body
8 Diaphragm
9 O-ring
10 O-ring
11 Cover
12 Gasket

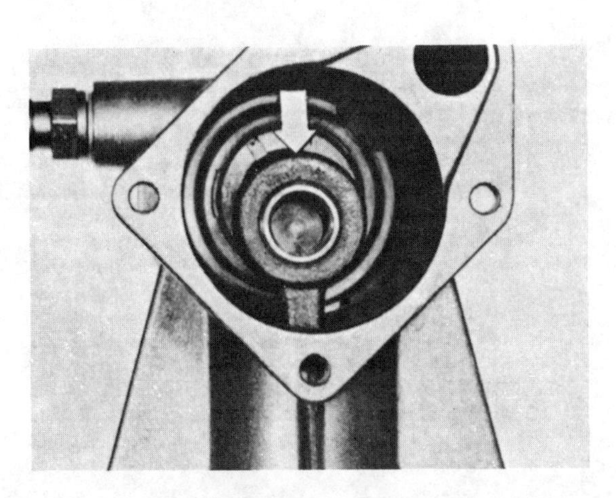

Fig. 6.16 Arrow shows direction to move pushrod of brake vacuum pump to pre-tension the diaphragm (Sec 20)

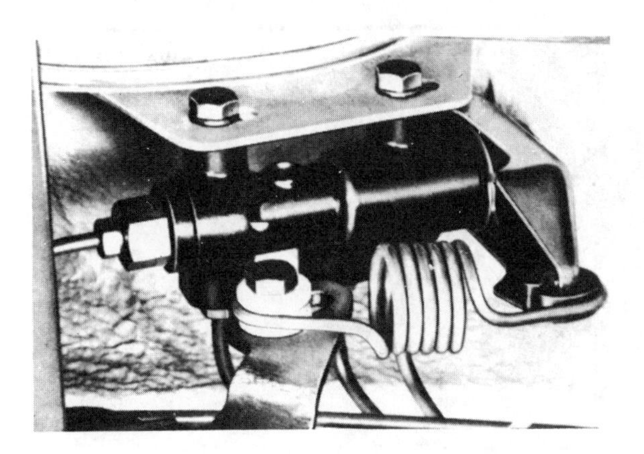

Fig. 6.17 Brake pressure regulator (Sec 21)

and the rear rises a little. This causes the rear axle beam to pivot slightly in its mountings and the pressure on the spring is reduced. This enables a spring inside the valve to restrict the hydraulic pressure available to the rear brakes.

3 When testing and adjusting the pressure regulator it is necessary to position the car body accurately and then to measure the pressures at the front and rear brake bleed nipples when the brakes are applied sharply. The tension of the spring between the regulator and the rear axle beam is then adjusted so that the correct pressures are obtained.

4 This requires special equipment to set the position of the rear suspension springs, and meters to measure up to 100 kgf/cm^2 (1420 lbf/in^2). This job must be left to the VW agent who will have the necessary equipment. Inaccurate setting may result in rear wheel skids or reduced braking efficiency.

5 When bleeding the hydraulic brake system of cars fitted with a pressure regulator, the lever of the regulator should be pressed towards the rear axle.

22 Hydraulic pipes and hoses – inspection and renewal

1 The magnitude of the pressure in the hydraulic lines is not generally realized. The test pressures are 100 kgf/cm^2 (1420 lbf/in^2) for the front brakes. These pressures are with the braking system cold. The temperature rise in the drums and discs for an emergency stop from 60 mph is as much as 80°C (176°F), and during a long descent may reach 400°C (752°F). The pressure must be even further raised as the temperature of the brake fluid in the cylinder rises. The normal pressure in the hydraulic system when the brakes are not in use is negligible. The pressure builds up quickly when the brakes are applied and remains until the pedal is released. Each driver will know how quick the build up is when equating it to the speed of his own reaction in an emergency brake application.

2 Recent research in the USA has shown that brake line corrosion may be expected to lead to failure after only 90 days exposure to salt spray such as is thrown up when salt is used to melt ice or snow. This in effect makes a four year old vehicle automatically suspect. It is possible to use pipe made of copper alloy used in marine work called 'Kunifer 10' as a replacement. This is much more resistant to salt corrosion, but as yet it is not a standard fitting.

3 All this should by now have indicated that pipes need regular inspection. The obvious times are in the autumn before the winter conditions set in, and in the spring to see what damage has been done.

4 Trace the routes of all the rigid pipes and wash or brush away accumulated dirt. If the pipes are obviously covered with some sort of underseal compound do not disturb the underseal. Examine for signs of kinks or dents which could have been caused by flying stones. Any instances of this means that the pipe section should be renewed, **but before actually taking it out read the rest of this Section.** Any unprotected sections of pipe which show signs of corrosion or pitting on the outer surface must also be considered for renewal.

5 Flexible hoses, running to each of the front wheels and from the underbody to each rear wheel should show no external signs of chafing or cracking. Move them about to see whether surface cracks appear. If they feel stiff and inflexible or are twisted they are nearing the end of their useful life. If in any doubt renew the hoses. Make sure also that they are not rubbing against the bodywork.

6 Before attempting to remove a pipe for renewal it is important to be sure that you have a replacement source of supply within reach if you do not wish to be kept off the road for too long. Pipes are often damaged on removal. If an official agency is near you may be reasonably sure that the correct pipes and unions are available. If not, check first that your local garage has the necessary equipment for making up the pipes and has the correct metric thread pipe unions available. The same goes for flexible hoses.

7 Where the couplings from rigid to flexible pipes are made there are support brackets and the flexible pipe is held in place by a U clip which engages in a groove in the union. The male union screws into it. Before getting the spanners on, soak the unions in penetrating fluid as there is always some rust or corrosion binding the threads. Whilst this is soaking in, place a piece of plastic film under the fluid reservoir cap to minimise loss of fluid from the disconnected pipes. Hold the hexagon on the flexible pipe coupling whilst the union on the rigid pipe is undone. Then pull out the clip to release both pipes from the bracket. For flexible hose removal this procedure will be needed at both ends. For a rigid pipe the other end will only involve unscrewing the union from a cylinder or connector. When you are renewing a flexible hose, take care not to damage the unions of the pipes that connect into it. If a union is particularly stubborn be prepared to renew the rigid pipe as well. This is quite often the case if you are forced to use open ended spanners. It may be worth spending a little money on a special pipe union spanner which is like a ring spanner with a piece cut out to enable it to go round the tube.

8 If you are having the new pipe made up, take the old one along to check that the unions and pipe flaring at the ends are identical.

9 Refitting of the hoses or pipes is a reversal of the removal procedure. Precautions and care are needed to make sure that the unions are correctly lined up to prevent cross threading. This may mean bending the pipe a little where a rigid pipe goes into a fixture. Such bending must not, under any circumstances, be too acute or the pipe will kink and weaken.

10 When fitting flexible hoses take care not to twist them. This can happen when the unions are finally tightened unless a spanner is used to hold the end of the flexible hose and prevent twisting.

11 If a pipe is removed or a union slackened, thereby admitting air into the hydraulic system, then the system must be bled as described in Section 23.

23 Brake hydraulic system – bleeding

1 First locate the bleed nipples on all four brakes. The rear wheel bleed nipples are at the back of the drum at the centre of the hydraulic cylinder. A small dust cap covers the nipple. This will probably be

covered with mud. Clean the mud from the back of the drum, wipe the dust cap and the area around it with a clean rag and the operation may start. The front brake bleed nipples are on the inside surface of the caliper.

2 As fluid is to be pumped out of the system make sure you have plenty of new clean fluid. It must conform to SAE recommendation J1703, J1703R but better still, get the official VW fluid. If other than hydraulic fluid is used the whole system may become useless through failure of piston seals. Top-up the fluid reservoir generously, and keep topping it up at intervals throughout the whole job (photo). On cars fitted with a pressure regulator valve, refer to Section 21, paragraph 5.

3 Start with the rear right-hand wheel. A piece of rubber or plastic hose 4 mm ($\frac{5}{32}$ in) inside diameter and about 600 mm (2 feet) long is required. Fit this over the bleed nipple and immerse the other end in a jar or bottle with some clean brake fluid in it. Fix the hose so that the end of it cannot come out of the brake fluid, and stand the bottle on the ground in a secure place.

4 You will need a helper whose job is to depress the brake pedal when required (but see Section 24, paragraph 6). It is as well to rehearse the operation before opening the bleed nipple valve. Open the valve about one turn and depress the pedal slowly to the floor of the vehicle. As soon as it is on the floor close the bleed valve **before** the pedal is released. Now release the pedal slowly. Brake fluid and air bubbles should have passed down the tube into the bottle. Repeat the operation until no further air bubbles are observed. Make the final tightening at the end of the last down-stroke. Check the fluid reservoir

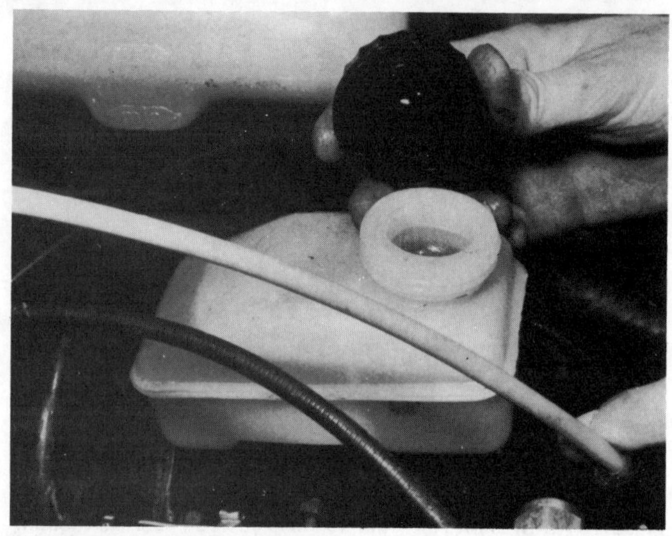

23.2 Removing the brake fluid reservoir filler cap

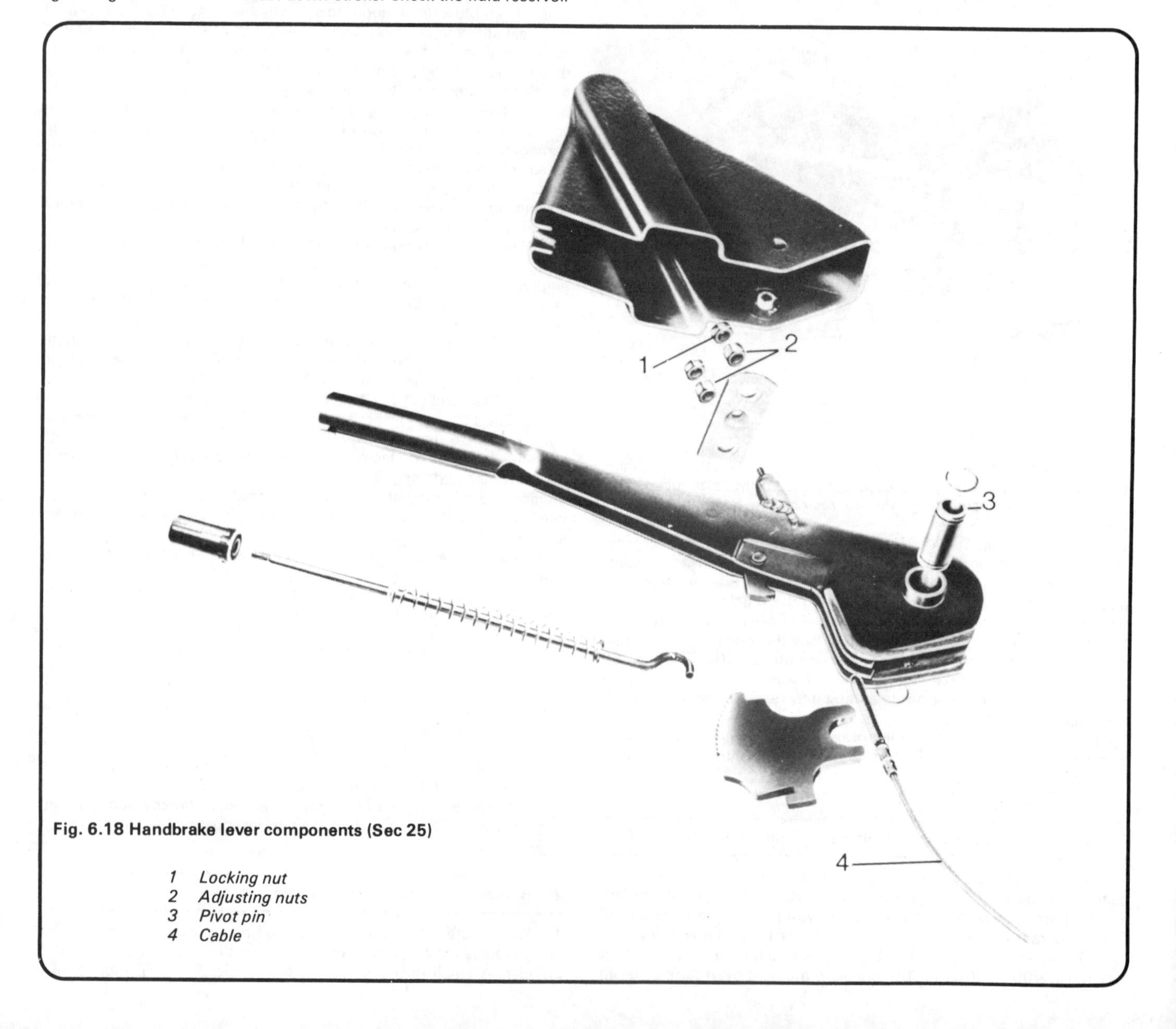

Fig. 6.18 Handbrake lever components (Sec 25)

1 *Locking nut*
2 *Adjusting nuts*
3 *Pivot pin*
4 *Cable*

level after every two strokes and top-up if necessary.
5 When you are satisfied that the rear right-hand brake line is clear of air bubbles then wipe down the brake and proceed to the next task; in order left rear, right front caliper, left front caliper.
6 After each session clean down the brakes with care and wash your hands. Brake fluid is poisonous and it is a splendid paint remover. Use a soapy solution and wash any paintwork that has been splashed.
7 Finally, the brake fluid in the jar should be discarded. Never use it for topping-up the brake fluid reservoir, for this always use fresh brake fluid.

24 Brake hydraulic system – changing fluid

1 The brake fluid is hygroscopic, which means that if it is allowed to come into contact with the open air it will absorb moisture. If it does then when the brakes get hot the water will boil and the brakes will not work properly – at the moment when they are needed most.
2 VW recommend that the fluid should be changed every two years– they give a variety of reasons – but the fact that they do recommend it should be reason enough.
3 The change over is simple to do. First of all clean the rear drums, particularly by the bleed nipples, and give the front calipers the same treatment.
4 Connect all four bleed nipples to plastic pipe and suitable containers, then open all four bleed nipples and get a helper to pump the brake pedal until no further fluid comes out.
5 Close the nipples, fill the system with clean brake fluid and then bleed the system as in Section 23.
6 The use of a proprietary pressure brake bleeder will enable the operations described in Section 23 and 24 to be carried out without an assistant.

25 Handbrake lever and cables – removal and refitting

1 To remove the lever, the plastic boot must be removed from the handbrake lever. Refer to Fig. 6.18. Slacken off the locking nut and remove it and the adjusting nut from each cable. Lift off the equalizer bar. Remove the clip from the pivot pin and push the pin out of the housing. The handbrake lever will now lift out leaving the cables in place.
2 The cables run through conduits under the car and then through clips to the rear brake backplate and to the shoe operating lever. It is necessary to remove the brake drum, as described in Sections 11 and 12, to unhook the cable from the operating lever. The cable can then be withdrawn from the conduit.
3 Refitting is the reverse of the removal procedure. Adjust the handbrake as described in Section 10 for manually adjusted rear brakes. The procedure for adjusting the handbrake cable on models with self-adjusting rear brakes (only necessary when the lever and/or cables have been renewed) is the same as for manually adjusted rear brakes.

26 Brake and clutch pedal assembly – removal and refitting

LHD models – refer to Fig. 6.19

1 Both pedals are fitted on the same shaft. Disconnect the clutch cable from the clutch pedal, refer to Chapter 4.
2 Disconnect the brake pedal pushrod clevis from the pedal by removing the clevis pin. Unhook the pedal return spring.
3 Remove the pedal shaft retaining clip, withdraw the shaft and collect the brake and clutch pedals.
4 Refitting of the pedals is the reverse of the removal procedure. After the pedals are fitted check the clutch pedal free play, refer to Chapter 4, and adjust the brake pedal pushrod as follows:

(a) Slacken the clevis locknut and adjust the position of the clevis on the pushrod to obtain a dimension of 206 mm (8.1 in) between the brake servo housing and the clevis hole, indicated by A in Fig. 6.19
(b) Fit the clevis pin in the second hole in the brake pedal which is 51 mm (2.0 in) from the centre of the pedal bush

RHD models – refer to Fig. 6.20

5 On RHD models the brake pushrod is connected to the servo unit through left and right-hand bell cranks which are connected by a link rod (photo).
6 The removal and refitting of the pedals is basically the same as for LHD models but note that the pushrod clevis pin and clip are different and that the bell cranks and connecting link are mounted to the brackets attached to the bulkhead.
7 When refitting the assembly, adjust the length of the brake pushrod by positioning the clevis at each end of the pushrod so that dimension B in Fig. 6.20 is 172 mm (6·8 in) and then tighten the locknuts.
8 When fitting the link rod between the bell cranks, adjust the length of the rod so that it is free of play and then tighten the locknuts.
9 Lubricate all friction surfaces inside the car with MOS_2 grease and all the linkage joints in the engine compartment with multi-purpose grease.

27 Tyres – inspection and maintenance

1 These can be an expensive problem if neglected. The most important matter is to keep them inflated at the correct pressure. A table of pressures is given at the start of the Chapter. If the vehicle is driven over rough ground, or over glass in the road, the tread should be inspected to see whether stones or glass fragments are lodged in the tread. These should be removed forthwith.
2 A careful study of the tread once a week will pay dividends. Tyres should wear evenly right across the tread but they rarely do, they are usually replaced because of misuse. A table is given showing some of the main troubles. Study it and watch your tyre treads.

Wear description	Probable cause
Rapid wear of the centre of the tread all round the circumference	*Tyre overinflated*
Rapid wear at both edges of the tread, wear even all round the circumference	*Tyre underinflated*
Wear on one edge of the tyre	
(a) Front wheels only	*Steering geometry needs checking*
(b) Rear wheels only	*Check rear suspension for damage*
Scalloped edges, wear at the edge at regular spacing around the tyre	*Maybe wheel out of balance, or more likely wear on the wheel bearing housing or steering balljoint*
Flat or rough patches on the tread	*Caused by harsh braking Check the brake adjustment*
Cuts and abrasions on the wall of the tyre	*Usually done by running into the kerb*

3 If the tyre wall is damaged the tyre should be removed from the wheel to see whether the inside of the cover is also damaged. If not the dealer may be able to repair the damage.
4 Tyres are best renewed in pairs, putting the new ones on the front and the older ones on the rear. Do not carry a badly worn spare tyre, and remember to keep the spare properly inflated.
5 The Specification to this Chapter gives the recommended grade of tyre to fit the wheels. We suggest you keep to this recommendation. However, if you wish to alter the arrangement there are two golden rules to remember:

(a) Do not fit a radial and a crossply on the same axle. In many countries this is illegal. The braking characteristics are different and you will be inviting an accident
(b) If you have two radials and two crossplys, the radials MUST be on the rear wheels. This again is bound up with braking characteristics

Fig. 6.19 Brake pedal components for LHD models (Sec 26)

1 Servo unit
2 Pushrod
3 Bracket
4 Pushrod yoke
5 Bushes
6 Clevis pin
A = 206 mm (8.1 in)

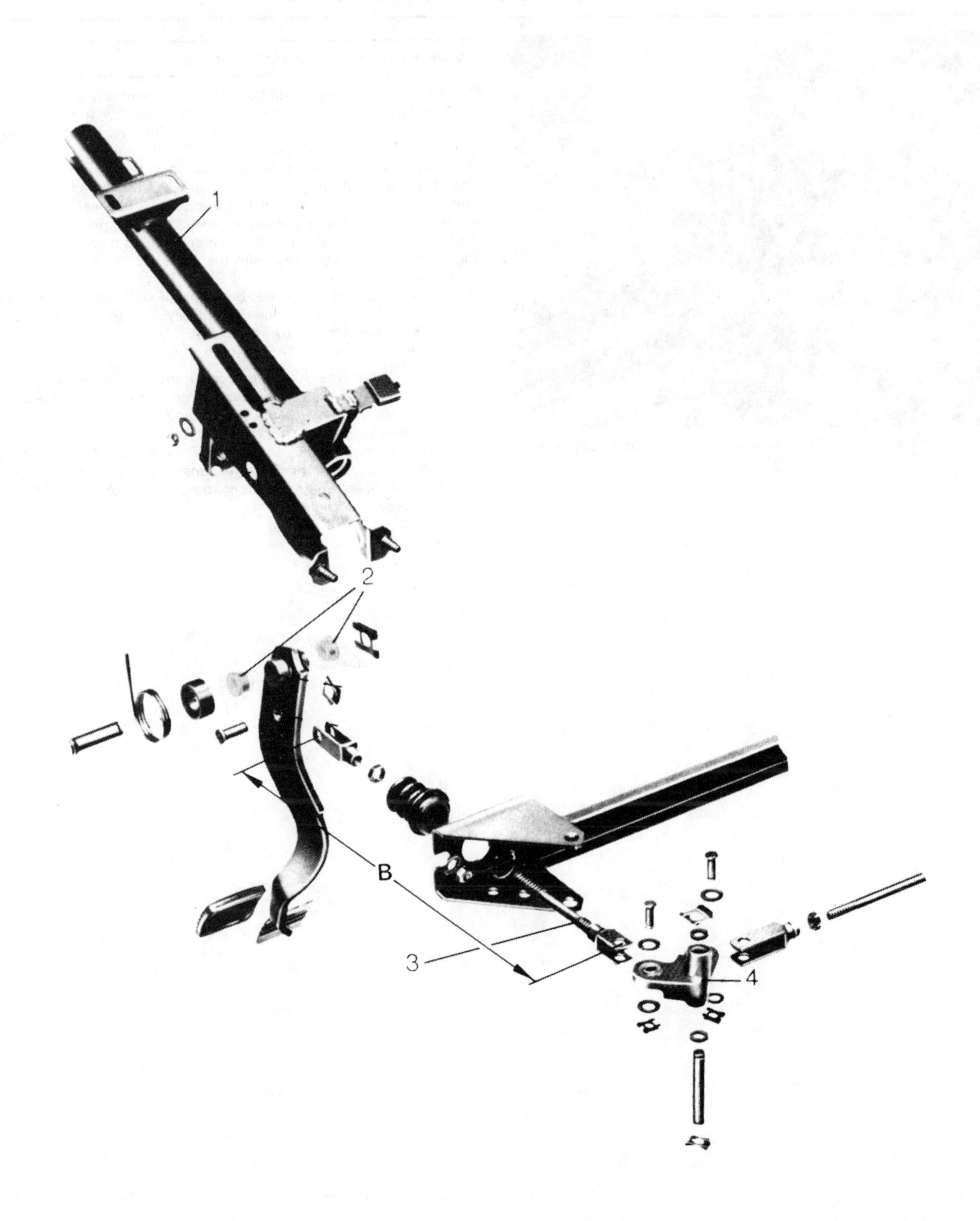

Fig. 6.20 Brake pedal components for RHD models (Sec 26)

1 Steering column and pedal bracket
2 Bushes
3 Adjustable pushrod
4 Right-hand side bell crank
B = 172 mm (6.8 in)

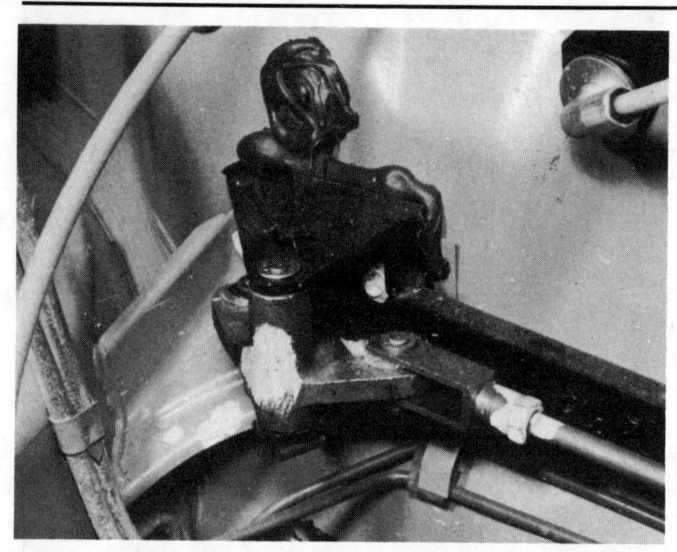

26.5 Brake pedal to servo bell crank linkage (right-hand side) on RHD cars

28 Wheels – inspection and balancing

1 Tyres and wheels should be balanced dynamically when new tyres are fitted. Mark the position of the balance weights and note the size of them. They have been known to fly off. This will affect the steering in the case of the front wheel.

2 If you suspect a wheel is out of balance, jack-up that wheel and spin it gently. Mark the bottom position when it comes to rest. Spin it several times more and note the bottom position each time. If the wheel comes to rest in the same position each time then the wheel and tyre are definitely out of balance and should be taken for balancing. It is best to do this test with the wheel on the rear axle where it may spin more freely.

3 Even though the wheel may be balanced statically (as in paragraph 2) it may still be out of balance dynamically. The only way to check for certain is to have it tested by a specialist.

There are two simple checks to do on the wheel:

(a) Jack the wheel off the ground and spin the wheel. Place a heavy weight so that it almost touches the rim and watch the clearance between the weight and rim as the wheel spins to see that the rim is not distorted

(b) Examine the holes through which the fixing bolts are fitted. If the bolts are allowed to get loose while the car is in service these holes may be elongated. Check the rim for cracks between bolt holes

29 Fault diagnosis – braking system

Before diagnosing faults in the brake system check that irregularities are not caused by any of the following faults:

1. *Incorrect mix of radial and crossply tyres*
2. *Incorrect tyre pressures*
3. *Wear in the steering mechanism, suspension or shock absorbers*
4. *Misalignment of the bodyframe*

Symptom	Reason/s
Pedal travels a long way before the brakes operate	Seized adjuster on rear shoes or shoes require adjustment Disc pads worn past limit
Stopping ability poor, pedal pressure firm	Linings, pads, discs or drums worn, contaminated, or wrong type One or more caliper piston or rear wheel hydraulic cylinder seized Loss of vacuum in servo
Car veers to one side when brakes are applied	Brake pads on one side contaminated with oil Hydraulic pistons in calipers seized or sticking Wrong pads fitted
Pedal feels spongy when brakes are applied	Air in the hydraulic system Spring weak in master cylinder
Pedal travels right down with no resistance	Fluid reservoir empty Hydraulic lines fractured Seals in master cylinder head failed
Brakes overheat or bind when car is in motion	Compensating port in master cylinder blocked Reservoir air vent blocked Pushrod requires adjustment Brakes shoes return springs broken or strained Caliper piston seals swollen Unsuitable brake fluid
Brakes judder or chatter and tend to grab	Linings worn Drums out of round Dirt in drums or calipers Discs run-out of true excessive
Brake shoes squeak (rear brake)	Dirt in linings Backplates distorted Brake shoe return springs broken or distorted Brake linings badly worn
Disc pads squeak (front brakes)	Wrong type of pad fitted Pad guide surfaces dirty Spreader spring deficient or broken Pads glazed Lining on pad not secure
Foot pedal must be pressed harder in one position only	Groove in master cylinder pushrod due to wear at sealing cups. Air entering vacuum side of servo
Very high pedal pressure required to operate brakes, linings found to be in good condition and correctly adjusted	No servo assistance Vacuum pump defective

Chapter 7 Electrical system

Refer to Chapter 11 for specifications and information applicable to later models

Contents

Specifications

System type ..... 12 volt, negative earth

Battery rating ..... 63 amp-hours

Alternator

Make	Bosch or Motorola
Maximum output	55 amp/770W
Stator resistance:	
Bosch	0·14 to 0·154 ohms
Motorola	0·15 to 0·17 ohms
Rotor resistance:	
Bosch	3·4 to 3·7 ohms
Motorola	3·9 to 4·3 ohms
Maximum slip ring ovality	0·03 mm (0·001 in)
Brush protrusion:	
New	10·0 mm (0·4 in)
Minimum	5·0 mm (0·2 in)
Charging voltage	12·5 to 14·5 volts

Starter motor

Type	Pre-engaged
Output	1·1 kW (1·5 hp) or 1·5 kW (2·0 hp)
Brush length – minimum	13·0 mm (0·5 in)
Commutator minimum diameter	38·5 mm (1·52 in)
Armature endfloat	0·1 to 0·15 mm (0·004 to 0·006 in)

Bulbs	**Type No**	**VW part No (USA)**
Sealed beam headlamps	6014	ZVP 118114
Front direction indicator/parking lamps	1034	ZVP 118034
Side marker lamps	1816	ZAP 118816
Rear direction indicator lamps	1073	ZVP 118073
Stop lamps	1073	ZVP 118073
Tail lamps	67	ZVP 118067
Reversing lamps	1073	ZVP 118073

Rear number plate lamp	1816	ZAP 118816
Control illumination	—	N 177512
Instrument illumination	—	N 177512
Indicator and warning lamps	—	N 177512
Interior lamp	211	ZVP 118211

Fuses

For fuse ratings and circuits protected, see Section 33

Torque wrench settings

	kgf m	lbf ft
Starter mounting bracket to transmission	1·6	11
Starter to bracket nuts	0·6	4
Alternator pulley nut	4·0	29
Alternator mounting nuts and bolts	2·0	14
Windscreen wiper arm nut	0·4 to 0·6	3 to 4
Tailgate wiper arm nut	0·4 to 0·6	2·4 to 4·2

1 General description

The major electrical system components comprise a 12 volt battery, of which the negative terminal is earthed, a Bosch or Motorola alternator which is mounted on the left front side of the engine and is driven by a belt from the crankshaft pulley, and a starter motor which is mounted on the right-hand rear of the engine.

The battery supplies current for starting and for the lighting and other electrical circuits. It provides a reserve of electricity when the current consumed by the electrical equipment exceeds that being produced by the alternator. Normally an alternator is able to meet any demand placed upon it.

When fitting electrical accessories to cars with a negative earth system, it is important, if they contain silicone diodes or transistors, that they are connected correctly otherwise serious damage may result to the components concerned. Items such as radios, tape recorders etc, should all be checked for correct polarity.

It is important that both battery leads are disconnected if the battery is to be boost charged. If body repairs are to be carried out using electric arc welding equipment, the alternator must be disconnected otherwise serious damage can be caused to the diodes and regulator. Whenever the battery is disconnected it must always be reconnected with the negative terminal earthed.

2 Battery – removal and refitting

1 The battery is mounted on the left-hand side at the front of the engine compartment. Remove the earth strap (negative) and the positive cable terminal. The battery is held in position by a clamp which fits over a rim at the base. A 13 mm socket spanner preferably with an extension, is needed to undo the clamp nut (photo).

2 Lift the battery out and clean the battery platform. Any sign of corrosion should be neutralized with an alkali solution. Ammonia or ordinary baking powder will do the job. If the corrosion has reached the metal, scrape the paint away to give a bright surface and repaint right away.

3 Refitting is the reverse of removal. Smear the terminals with a little petroleum jelly. **Do not** use grease.

3 Battery – maintenance and inspection

1 Normal weekly battery maintenance consists of checking the electrolyte level of each cell to ensure that the separators are covered by 6 mm ($\frac{1}{4}$ in) of electrolyte. If the level has fallen, top up the battery using distilled water only (photo). Do not overfill. If a battery is overfilled or any electrolyte spilt, immediately wipe away the excess as electrolyte attacks and corrodes any metal it comes into contact with very rapidly.

2 As well as keeping the terminals clean and covered with petroleum jelly, the top of the battery, and especially the top of the cells, should be kept clean and dry. This helps prevent corrosion and ensures that the battery does not become partially discharged by leakage through dampness and dirt.

3 Once every three months remove the battery and inspect the battery securing bolts, the battery clamp plate, tray, and battery leads for corrosion (ie. white fluffy deposits on the metal which are brittle to touch). If any corrosion is found, clean off the deposits with ammonia and paint over the clean metal with an anti-rust/anti-acid paint.

4 At the same time inspect the battery case for cracks. If a crack is found, clean and plug it with one of the proprietary compounds marketed for this purpose. If leakage through the crack has been

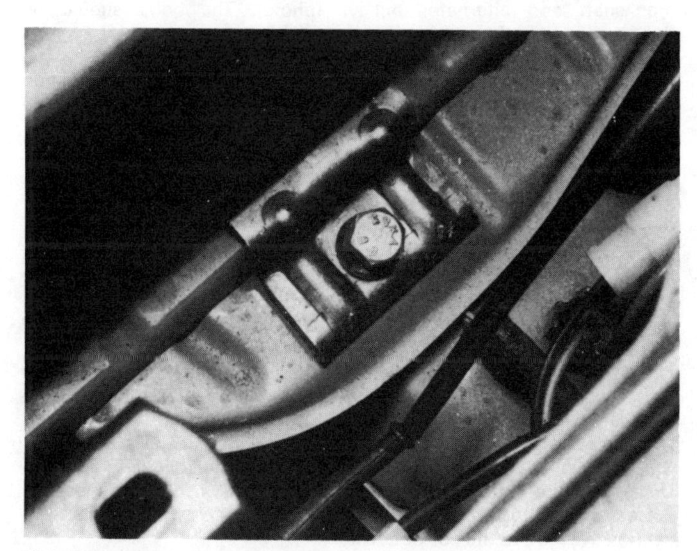

2.1 The battery retaining clamp

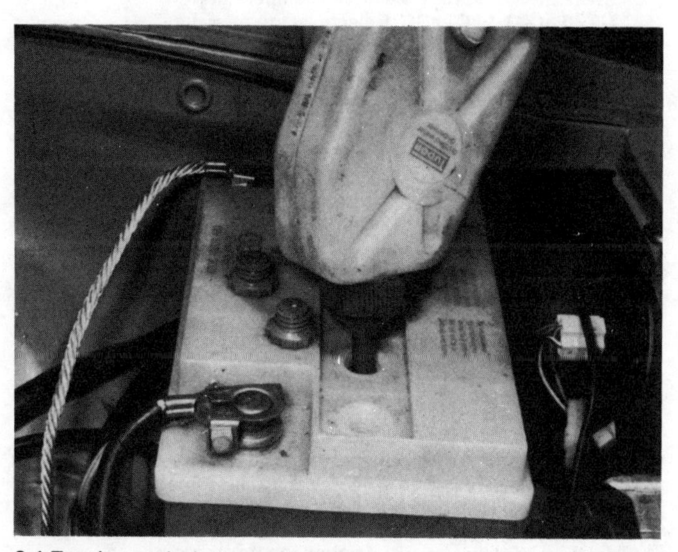

3.1 Topping up the battery with distilled water

excessive then it will be necessary to refill the appropriate cell with fresh electrolyte as detailed later. Cracks are frequently caused to the top of battery cases by pouring in distilled water, in the middle of winter, *after* instead of *before* a run. This gives the water no chance to mix with the electrolyte and so the former freezes and splits the battery case.

5 If topping-up the battery becomes excessive and the case has been inspected for cracks that could cause leakage, but none are found, the battery is being overcharged and the voltage regulator will have to be checked.

6 With the battery on the bench at the three monthly interval check, measure the electrolyte specific gravity with a hydrometer to determine the state of charge and condition of the electrolyte. There should be very little variation between the different cells and if a variation in excess of 0·025 is present it will be due to either:

(a) Loss of electrolyte from the battery caused by spillage or a leak, resulting in a drop in the specific gravity of the electrolyte when the deficiency was replenished with distilled water instead of fresh electrolyte

(b) An internal short circuit caused by buckling of the plates or a similar malady pointing to the likelihood of total battery failure in the near future

7 The specific gravity of the electrolyte for fully charged conditions at the electrolyte temperature indicated, is listed in Table A. The specific gravity of a fully discharged battery at different temperatures of the electrolyte is given in Table B.

8 Specific gravity is measured by drawing up into the body of a hydrometer sufficient electrolyte to allow the indicator to float freely. The level at which the indicator floats indicates the specific gravity.

Table A

Specific gravity – battery fully charged
1·268 at 38°C or 100°F electrolyte temperature
1·272 at 32°C or 90°F electrolyte temperature
1·276 at 27°C or 80°F electrolyte temperature
1·280 at 21°C or 70°F electrolyte temperature
1·284 at 16°C or 60°F electrolyte temperature
1·288 at 10°C or 50°F electrolyte temperature
1.292 at 4°C or 40°F electrolyte temperature
1·296 at –1·5°C or 30°F electrolyte temperature

Table B

Specific gravity – battery fully discharged
1·098 at 38°C or 100°F electrolyte temperature
1·102 at 32°C or 90°F electrolyte temperature
1·106 at 27°C or 80°F electrolyte temperature
1·110 at 21°C or 70°F electrolyte temperature
1·114 at 16°C or 60°F electrolyte temperature
1·118 at 10°C or 50°F electrolyte temperature
1·122 at 4°C or 40°F electrolyte temperature
1·126 at – 1·5°C or 30°F electrolyte temperature

4 Battery – charging

1 In winter time when heavy demand is placed upon the battery such as when starting from cold and much electrical equipment is continually in use, it is a good idea occasionally to have the battery fully charged from an external source at the rate of 4 to 6 amps. Always disconnect it from the car electrical circuit when charging.

2 Continue to charge the battery at this rate until no further rise in specific gravity is noted over a four hour period.

3 Alternatively, a trickle charger, charging at the rate of 1·5 amps, can safely be used overnight. Disconnect the battery from the car electrical circuit before charging or you will damage the alternator.

4 Specially rapid 'boost' charges which are claimed to restore the power of the battery in 1 to 2 hours can cause damage to the battery plates through over-heating.

5 While charging the battery note that the temperature of the electrolyte should never exceed 100°F (37.8°C).

6 Make sure that your charging set and battery are set to the same voltage.

5 Battery – electrolyte replenishment

1 If the battery is in a fully charged state and one of the cells maintains a specific gravity reading which is 0·025 or lower than the others, and a check of each cell has been made with a voltage meter to check for short circuits (a four to seven second test should give a steady reading of between 1·2 and 1·8 volts), then it is likely that electrolyte has been lost from the cell with the low reading at some time.

2 Top-up the cell with a solution of 1 part sulphuric acid to 2·5 parts of water, obtainable ready mixed from a service station. If the cell is already fully topped-up draw some electrolyte out of it with a battery hydrometer

3 Continue to top-up the cell with the freshly made electrolyte and to recharge the battery and check the hydrometer readings.

6 Alternator – safety precautions

1 The alternator has a negative earth circuit. Be careful not to connect the battery the wrong way or the alternator will be damaged.

2 **Do not** run the alternator with the output wire disconnected.

3 When welding is being done on the car the battery and the alternator output cable should be disconnected.

4 If the battery is to be charged in the car, both the leads of the battery should be disconnected before the charging leads are connected to the battery.

5 Do not use temporary test connections which may short circuit accidentally. The fuses will not blow, the diodes will burn out.

6 When replacing a burnt out alternator clear the fault which caused the burn out first or a new alternator will be needed a second time.

7 Alternator – drivebelt adjustment

1 The alternator is driven by a belt from the crankshaft pulley. The belt also drives the water pump.

2 The alternator has two lugs on its casing. A bolt threaded through these is mounted in a bracket bolted to the cylinder block (photo). This bolt forms the hinge on which the alternator is mounted. The head of the bolt is a hollow (socket) hexagon which is accessible through a hole in the timing belt cover. This bolt must be slackened before the alternator belt tension may be adjusted.

3 On the bottom of the alternator is yet another lug through which a bolt is fitted to a slotted strap. The strap is hinged on the cylinder block (photo).

4 Thus the alternator may be rotated about the hinge bolt to tighten the drivebelt. The tension in the drivebelt is correct when the belt may be depressed with a thumb as shown in Fig. 7.1 halfway between the crankshaft and alternator pulleys (photo). The bolts should be tightened to hold the alternator in this position.

5 It may be difficult to slacken the socket head bolt. If so, remove the strap, and then remove the bracket with alternator complete. Undo the wiring plug, remove the belt and then take the alternator away. The socket head bolt may then be held in a vice and undone that way.

8 Alternator – testing

1 There is a way of testing the alternator in the car, but it requires a lot of expensive equipment and does not provide much conclusive evidence. Refer to Fig. 7.2. The following are required. A battery cut-out switch, a variable resistance capable of consuming up to 500 watts, an ammeter reading 0–30 amps, a voltmeter reading 0–20v, and a tachometer. A special adaptor is required to enable a tachometer to be used on a Diesel engine. In its absence an estimate of engine speed may suffice.

2 The battery cutout switch is illustrated in Fig. 7.3.

3 Connect up as shown in the diagram in the following manner. Disconnect the battery earth strap and the positive cable. Connect the cut-out switch to the battery positive terminal and then connect the car positive lead to the cut-out switch.

4 So far no interference with the normal circuit. Now arrange an alternative one to take the place of the battery. From the battery cut-out switch connect the variable resistance and ammeter in series to

7.2 The alternator with drivebelt removed, showing the mounting bracket and upper pivot bolt

7.3 Alternator lower mounting and adjusting strap

7.4 The alternator and water pump drivebelt

Fig. 7.1 Checking point for alternator drivebelt tension (Sec 7)

a = 10 – 15 mm ($\frac{3}{8}$ – $\frac{9}{16}$ in)

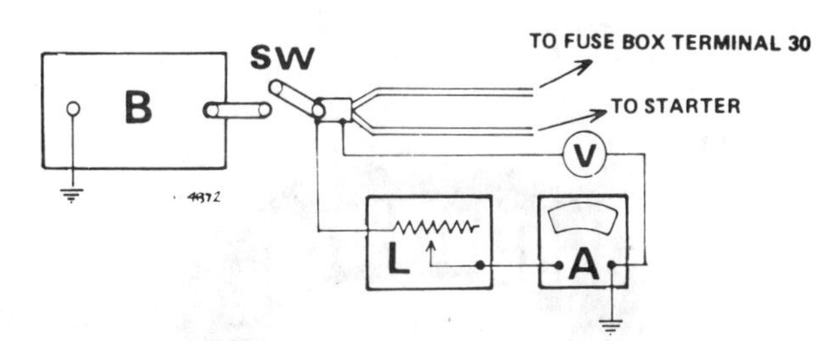

Fig. 7.2 Circuit diagram for testing alternator (Sec 8)

B Battery
SW Battery cut-out switch
L Variable resistance
A Ammeter (0–30 amps)
V Voltmeter (0–20 volts)

the chassis (earth) of the car. Arrange a voltmeter to measure the voltage between the battery cut-out switch and earth. Reconnect the battery earth strap.

5 Start the engine and run it up to 2800 rpm. Set the variable resistance so that the ammeter reading is between 20 and 30 amps. Now open the battery cut-out switch, that is, cut the battery out of the circuit so that the current flows only through the resistance. Alter the resistance to bring the current back to 25 amps. Now read the voltmeter. It should read between 12.5 and 14·5 volts.

6 If the voltmeter reading is outside these limits close the cut-off switch, stop the engine and replace the alternator regulator with a new one (or a borrowed one). Repeat the test. If the desired 12·5 to 14·5 volts is obtained then the old regulator was faulty. If not then the alternator is faulty and must be changed. It seems a lot to do for little reward but the only other way is to take the alternator to an official agent for testing.

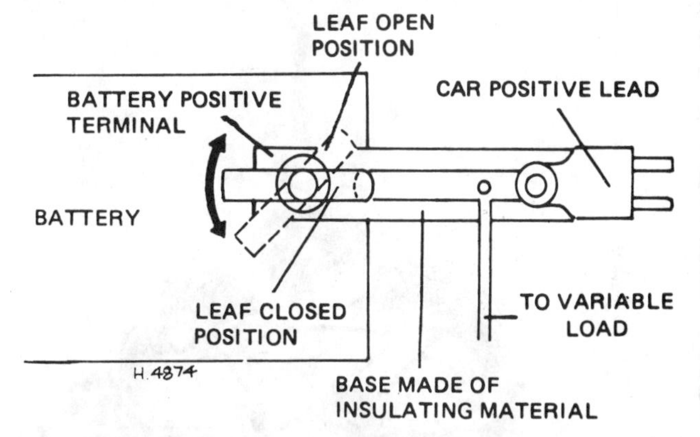

Fig. 7.3 Diagrammatic arrangement of the battery cut-out switch (Sec 8)

9 Alternator (Bosch K1 14V) – overhaul

1 The regulator is fitted into the alternator housing. Remove a small screw and it may be removed. Refer to Fig. 7.4 (photo).

2 Inside it will be seen the two slip ring brushes. These must be free in the guides and protrude at least 5 mm (0·2 in). The new brush protrusion is 10 mm (0·4 in). The brushes may be renewed by unsoldering the leads from the regulator, fitting new brushes and resoldering the leads (Fig. 7.5).

9.1 Alternator wiring plug and regulator locations (Bosch type)

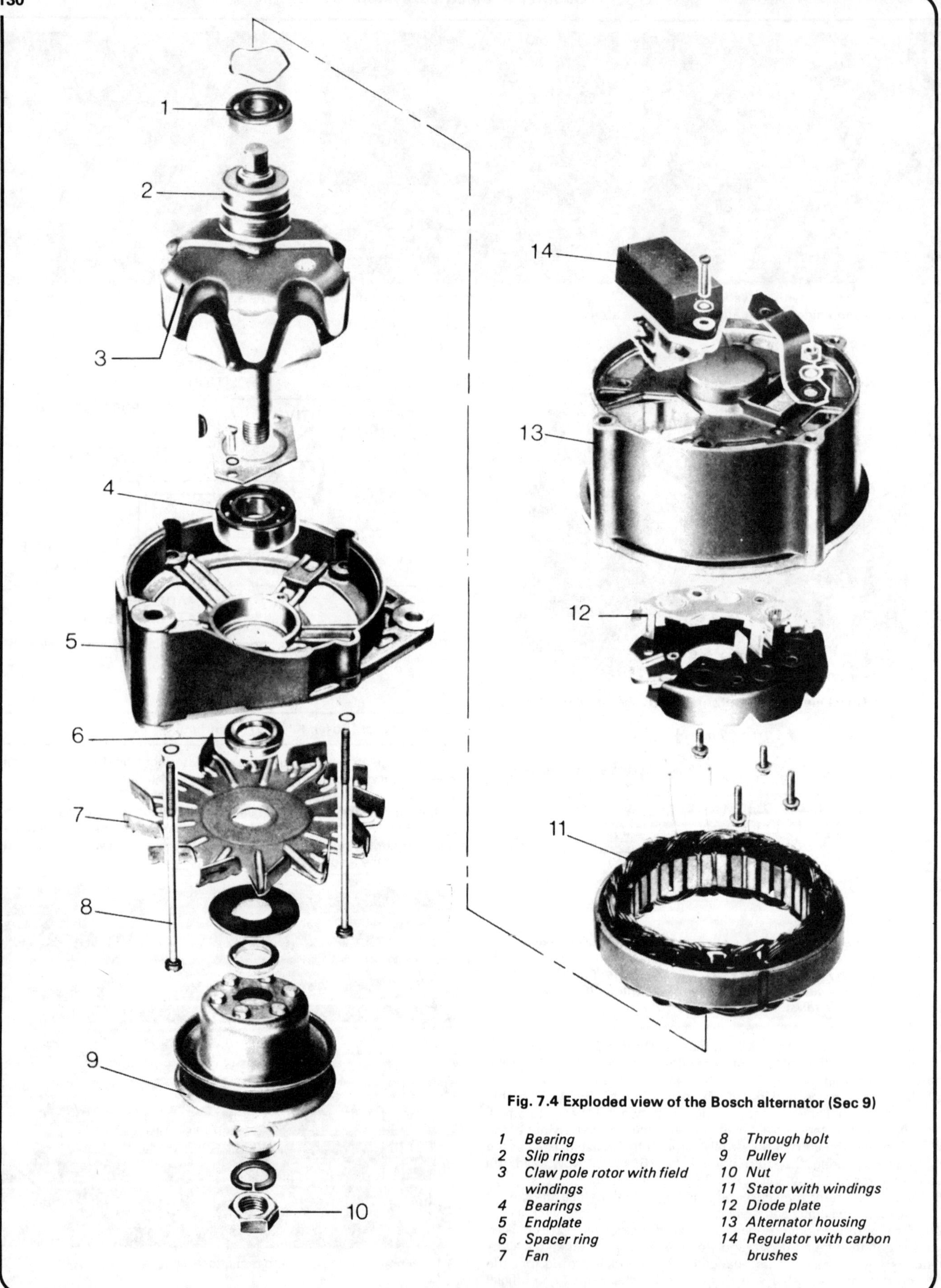

Fig. 7.4 Exploded view of the Bosch alternator (Sec 9)

1 Bearing
2 Slip rings
3 Claw pole rotor with field windings
4 Bearings
5 Endplate
6 Spacer ring
7 Fan
8 Through bolt
9 Pulley
10 Nut
11 Stator with windings
12 Diode plate
13 Alternator housing
14 Regulator with carbon brushes

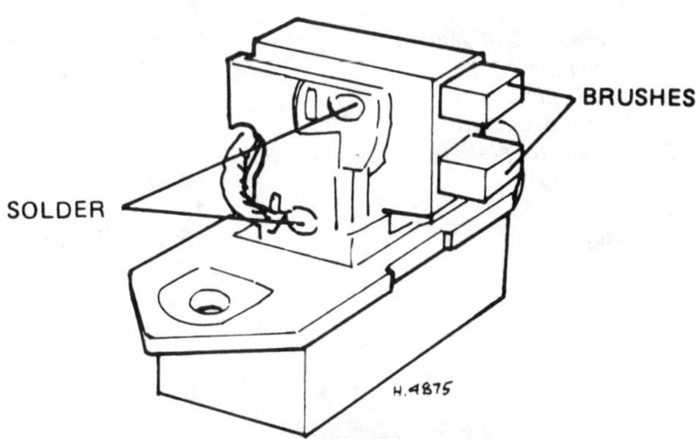

Fig. 7.5 Alternator brush leads are soldered to the regulator (Sec 9)

3 Undo the pulley nut and remove the pulley, the spacer ring, the large washer and the fan. Note which way the fan fits to make assembly easier. There is an arrow showing the direction of rotation.

4 Remove the bracket from the housing which held the wiring plug and if not already removed, take away the regulator.

5 Undo the housing bolts and separate the components. The armature will stay in the endplate and the housing bearing will stay on the shaft. Have a good look at the various components. Clean off all the dust using a soft brush and then wipe clean. Any smell of burnt carbon or signs of overheating must be investigated. Check the slip rings for burning, scoring and ovality. You will have had reason to check the bearings before dismantling, but have a further look now. At this point you must make up your mind whether to do the repair yourself, or whether to take the alternator to a specialist. If you have the tools and the skill, it is possible to renew the bearings, renew the diode carrier complete, clean up the slip rings and to fit a new rotor or stator. It is not possible to repair the winding, renew individual diodes, renew the slip rings or repair the fan.

6 Dealing with the rotor first. The rotor may be removed from the endplate by using a mandrel press. Then take the screws out of the cover over the endplate bearing and press the bearing out of the frame. The slip ring end bearing may be pulled off using an extractor on the inner race. If you pull on the outer race the bearing will be scrapped. Renew the bearings if necessary.

7 The slip rings may be cleaned up with fine glasspaper (not emery paper).

8 Test the rotor electrically. Check the insulation resistance between the slip rings and the shaft. This must be infinity. If it is not there is a short circuit and the armature must be renewed. Get an auto-electrical specialist to confirm your findings first. Check the resistance of the winding. Measure this between slip rings. It should be as specified. If there is an open circuit or high resistance, then again the rotor must be renewed.

9 The stator and the diode carrier are connected by wires. Make a simple circuit diagram so that you know which wire goes to which diode and then unsolder the connections. This is a delicate business as excess heat will destroy the diode and possibly the winding. Grip the wire as close as possible to the soldered joint with a pair of long nosed pliers and use as small a soldering iron as possible.

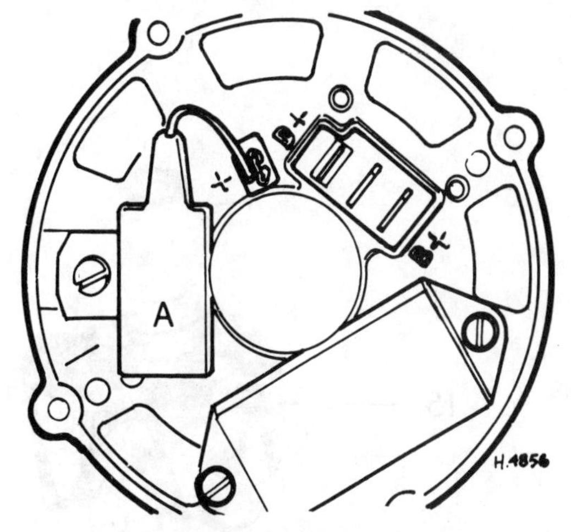

Fig. 7.6 Anti-surge condenser (A) fitted to the alternator (Sec 9)

10 The stator winding may now be checked. First check that the insulation is sound. The resistance between the leads and the frame must be infinity. Next measure the resistance of the winding. It should be as specified. A zero reading means a short circuit, and of course a high or infinity reading, an open circuit.

11 The diode carrier may now be checked. Each diode should be checked in turn. Use a test lamp or an ohmmeter. Current must flow only one way; ie, the resistance measured one way must be high and the other way (reverse the leads), low. Keep the current down to 0·8 milliamps and do not allow the diode to heat up. If the resistance both ways is a high one, then the diode is open circuited, a low one, short circuited. If only one diode is defective the whole assembly (diode plate) must be renewed.

12 Reconnect the stator winding to the diode circuit, again be careful not to overheat the diode, and reassemble the stator and diode carrier to the housing.

13 A new diode carrier, or a new stator may be fitted, but be careful to get the correct parts.

14 Assembly is the reverse of dismantling. Be careful to assemble the various washers correctly.

15 It has been found that voltage surge in the electrical system damages the alternator diodes. If this happens when requesting repair of the diode plate ask for and fit a condenser (part no 059 035 271) to prevent this occurring again (Fig. 7.6).

10 Alternator (Motorola type) – overhaul

1 Refer to Fig. 7.7 and 7.8. It will be seen that although the construction is basically the same as the Bosch generator the Motorola differs considerably in detail.

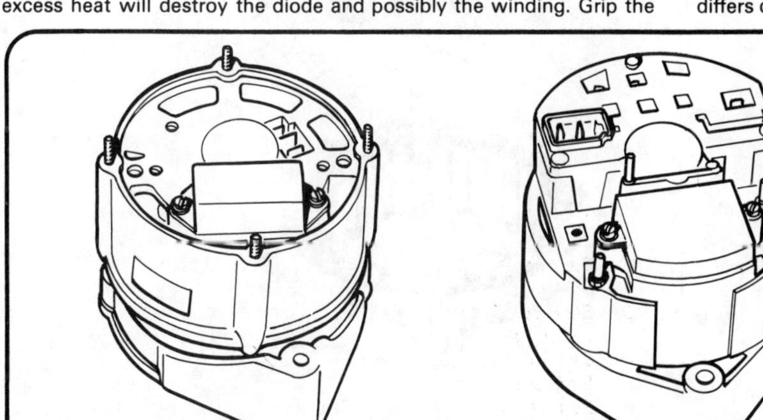

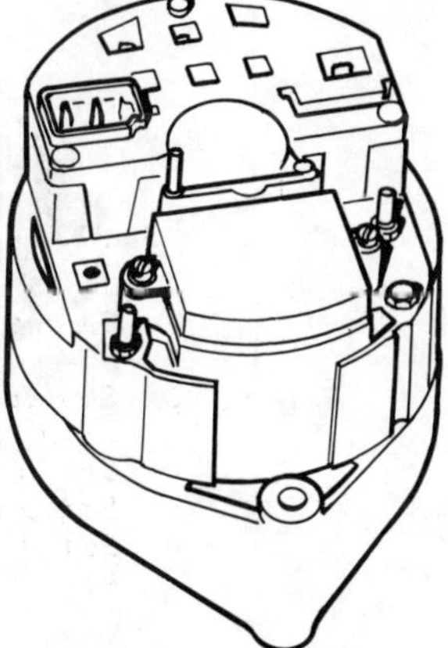

Fig. 7.7 Comparison of Bosch (left) and Motorola (right) alternators (Sec 10)

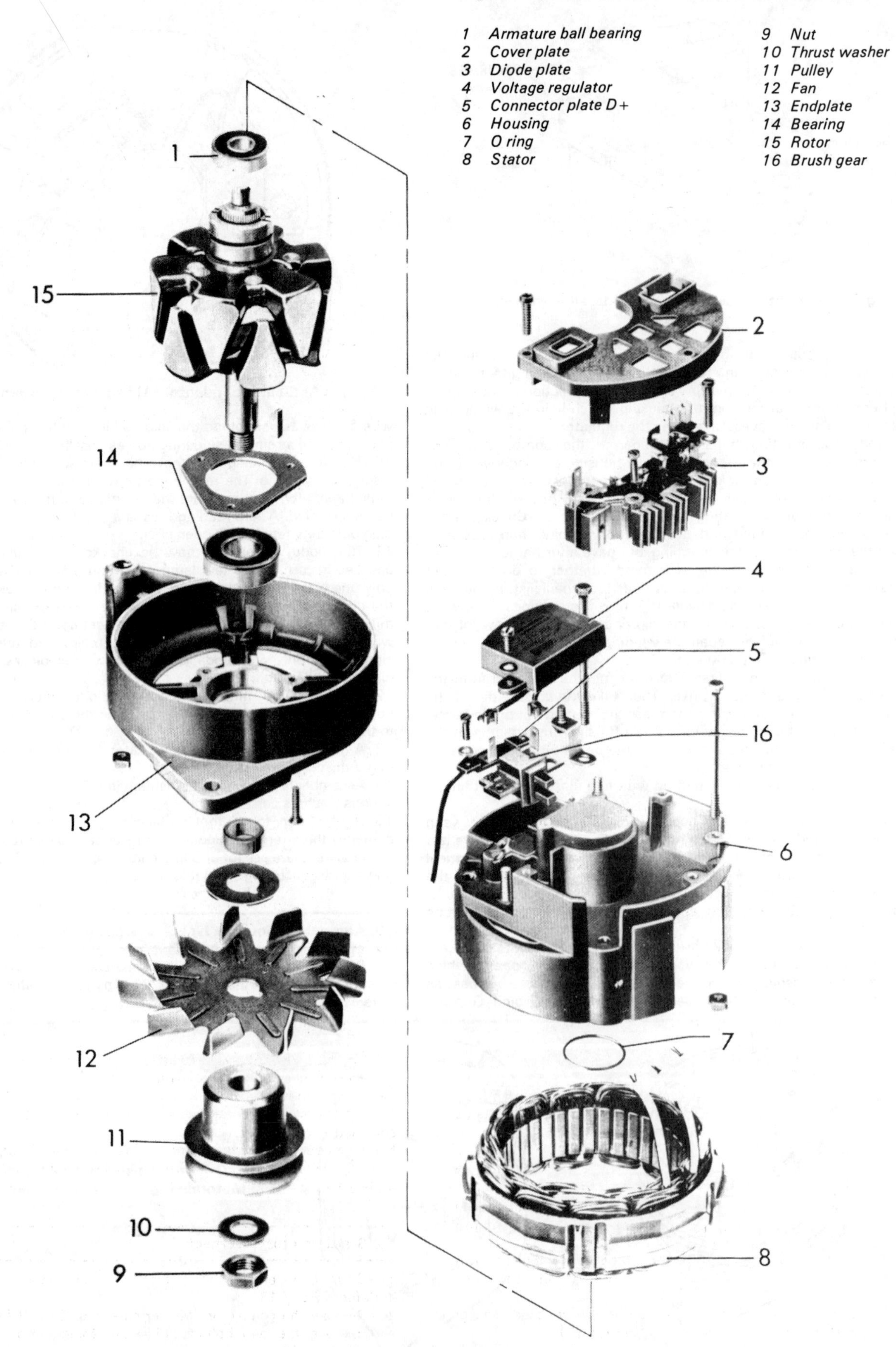

Fig. 7.8 Exploded view of the Motorola alternator (Sec 10)

1 *Armature ball bearing*
2 *Cover plate*
3 *Diode plate*
4 *Voltage regulator*
5 *Connector plate D+*
6 *Housing*
7 *O ring*
8 *Stator*
9 *Nut*
10 *Thrust washer*
11 *Pulley*
12 *Fan*
13 *Endplate*
14 *Bearing*
15 *Rotor*
16 *Brush gear*

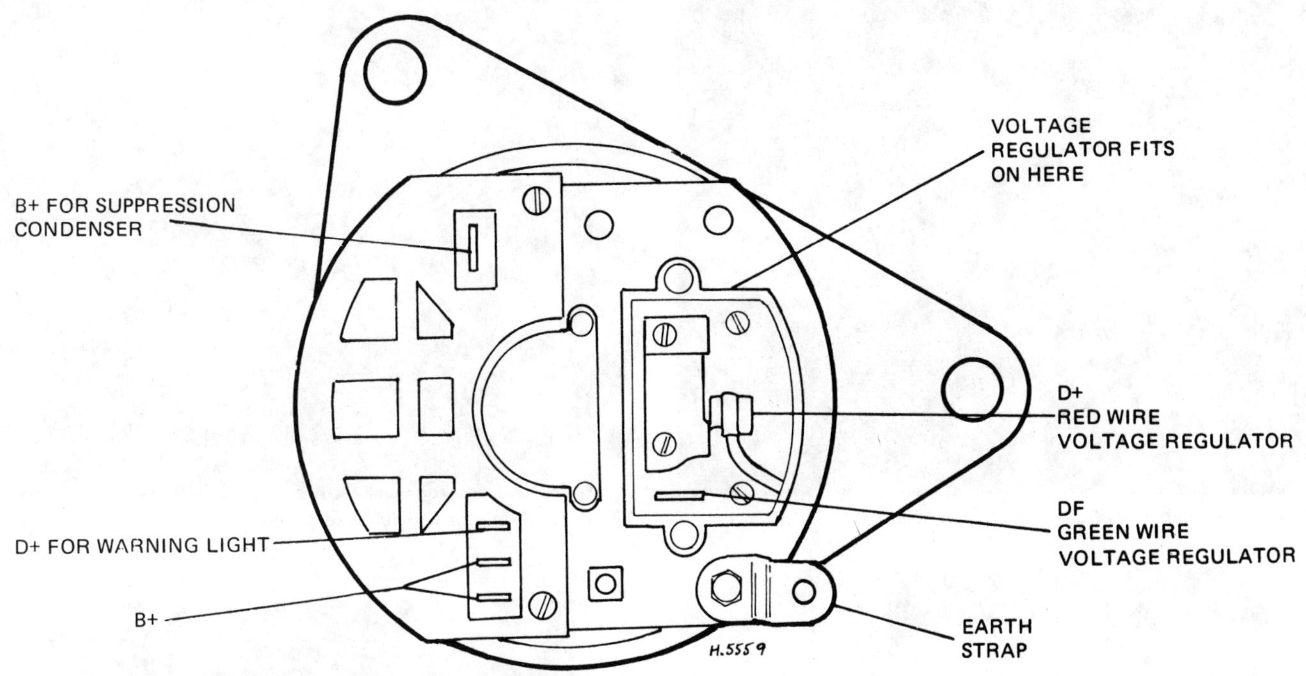

Fig. 7.9 Diagrammatic view of the Motorola alternator endplate with regulator removed (Sec 10)

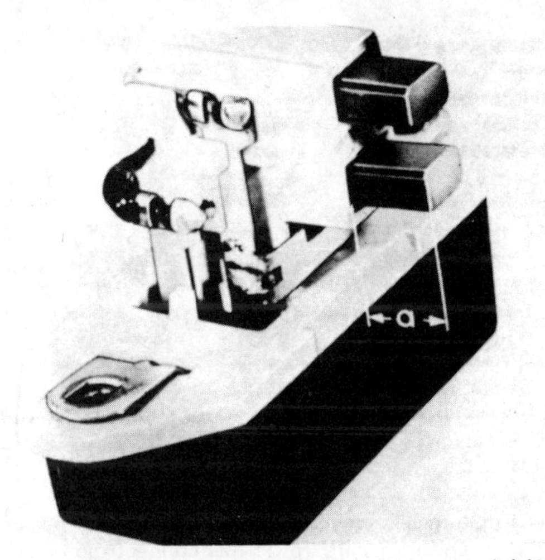

Fig. 7.10 Alternator brush checking dimension (a) (Sec 10)

2 The stator and rotor have the same form as those of the Bosch but the cover, housing and diode plate are of different construction. The earth strap is bolted to the cover, not the hinge as in the case of the Bosch.

3 The same principles apply for overhaul. The rotor should be checked for earth short circuit and continuity. The resistance between slip rings must be as specified, ovality of slip rings must be within limits. Bearings may be drawn off with a puller and renewed if necessary.

4 The stator may be disconnected from the diode plate and the winding tested for open and short circuit. The resistance should be as specified.

5 Once isolated the diode plate may be tested, as in Section 9, and a new one fitted if required. It is not recommended that any attempt be made to renew individual diodes.

6 The routing of the D+ wire inside the cover is important. It must be fitted in the two sets of clips provided or it will interfere with the armature.

7 The voltage regulator connections must be checked. The green wire goes to DF and the red wire to D+ (see Fig. 7.9).

8 The connections on the cover must be checked carefully. A diagram is given for information.

11 Starter motor – testing on engine

1 If the starter motor fails to operate then check the condition of the battery by turning on the headlamps. If they glow brightly for several seconds and then gradually dim, the battery is in an uncharged condition.

2 If the headlamps continue to glow brightly and it is obvious that the battery is in good condition, then check the tightness of the earth lead from the battery terminal to its connection on the body frame particularly, and the other battery wiring. Check the tightness of the connections at the rear of the solenoid. Check the wiring with a voltmeter for breaks or short circuits.

3 If the wiring is in order check the starter motor for continuity using a voltmeter.

4 If the starter motor drive fails to release when the ignition key is released from the start position (indicated by a continuous whirring noise) then the solenoid spring is broken or the drive teeth are jammed in the ring gear. Remove the starter motor for examination.

12 Starter motor – removal and refitting

1 Disconnect the earth strap from the battery negative terminal.

2 Disconnect the heavy cable from the battery and pull the leads off the two solenoid terminals (photos).

3 Undo and remove the bolts securing the starter motor to the bellhousing and withdraw the starter motor from the housing.

4 To refit, insert the starter into the transmission bellhousing and locate the armature shaft in the bush in the housing.

5 Fit the starter motor securing bolts and tighten them to the specified torque.

6 Reconnect the wiring to the starter solenoid. Route the battery cable as shown in Fig. 7.12, with the cables pointing towards the rear of the car.

13 Starter motor – overhaul

1 Clean the exterior of the starter motor before starting to dismantle it. Refer to Fig. 7.13.

2 Remove the connector strip terminal nut (D) and from the other end remove the two bolts holding the solenoid to the mounting bracket. Now lift the solenoid pull rod so that it is clear of the operating lever and remove the solenoid.

12.2A Battery cable location on the starter solenoid

12.2B Starter motor with leads removed

Fig. 7.11 Starter motor mounting plate (1) and retaining nuts (2) (Sec 12)

Fig. 7.12 Correct position of starter motor supply cable (Sec 12)

3 At the front end of the starter is a cap held by two screws. Remove this and under it there is a shaft with a circlip and bush. Remove the clip.

4 Now remove the through bolts and remove the cover.

5 The brush gear is now visible. Lift the brushes out of the holder and remove the brush holder. The starter body holding the field coils may now be separated from the endplate. This will leave the armature still in the mounting bracket.

6 To remove the mounting bracket from the drive end of the shaft, first push back the stop ring with a suitable tube so that the circlip underneath may be released from its groove. It is not possible to remove the mounting bracket and pinion from the shaft.

7 Finally remove the operating lever pin from the mounting bracket and remove the pinion assembly.

8 Clean and examine the pinion, shaft and lever and inspect for wear. If possible run the armature between centres in a lathe and check that the shaft is not bent. Check the fit of the drive pinion on the shaft. Check that the pinion will revolve in one direction only (one way clutch) and that the teeth are not chipped.

9 Examine the commutator. Clean off the carbon with a rag soaked in petrol. Minor scoring may be removed with fine glass paper. Deep scoring must be removed by machining in a lathe. Commutator copper is harder than the commercial grade, and requires the lathe tool to be ground differently. Unless you have had instruction on machining commutators we suggest that the skimming and under-cutting be left to the expert. The minimum diameter for the commutator is given in the Specifications.

10 Test the armature electrically. Check the insulation between the armature winding and the shaft. To do this connect the negative terminal of the ohmmeter to the shaft and place the positive probe on each commutator segment in turn.

11 Burning on the commutator is usually a sign of an open circuited winding. If you have access to a 'growler' have the armature checked for short circuits.

12 Inspect the field windings for signs of abrasion or stiff and damaged insulation, particularly where the leads leave the coil. Check the field coil for short circuit to the pole piece and for open circuits. Renew if necessary.

13 The brushes must be at least of the specified length and must slide easily in the holder. There are two schools of thought about brush renewal. One says that the entire field coil must be replaced or the brush plate with the armature current brushes. The VW/Audi method is somewhat different.

14 Isolate the brushes, pull them out of the holders and hold them away from the winding. Crush the old brush with a powerful pair of pliers until the lead is free from the brush. Clean the end of the lead and prepare it for soldering. The new brush, obtainable from official agents, is drilled and has a tinned insert. Push the end of the lead into the drilling and splay it out, then using silver solder, solder the brush to the lead.

15 If it is your first attempt at soldering it could be better to get expert help. Use a large soldering iron (250 watts plus), do not let any of the solder creep along the wire and file off any surplus. Do not let the lead get too hot, or damage will occur to the field coils. Use a flat pair of

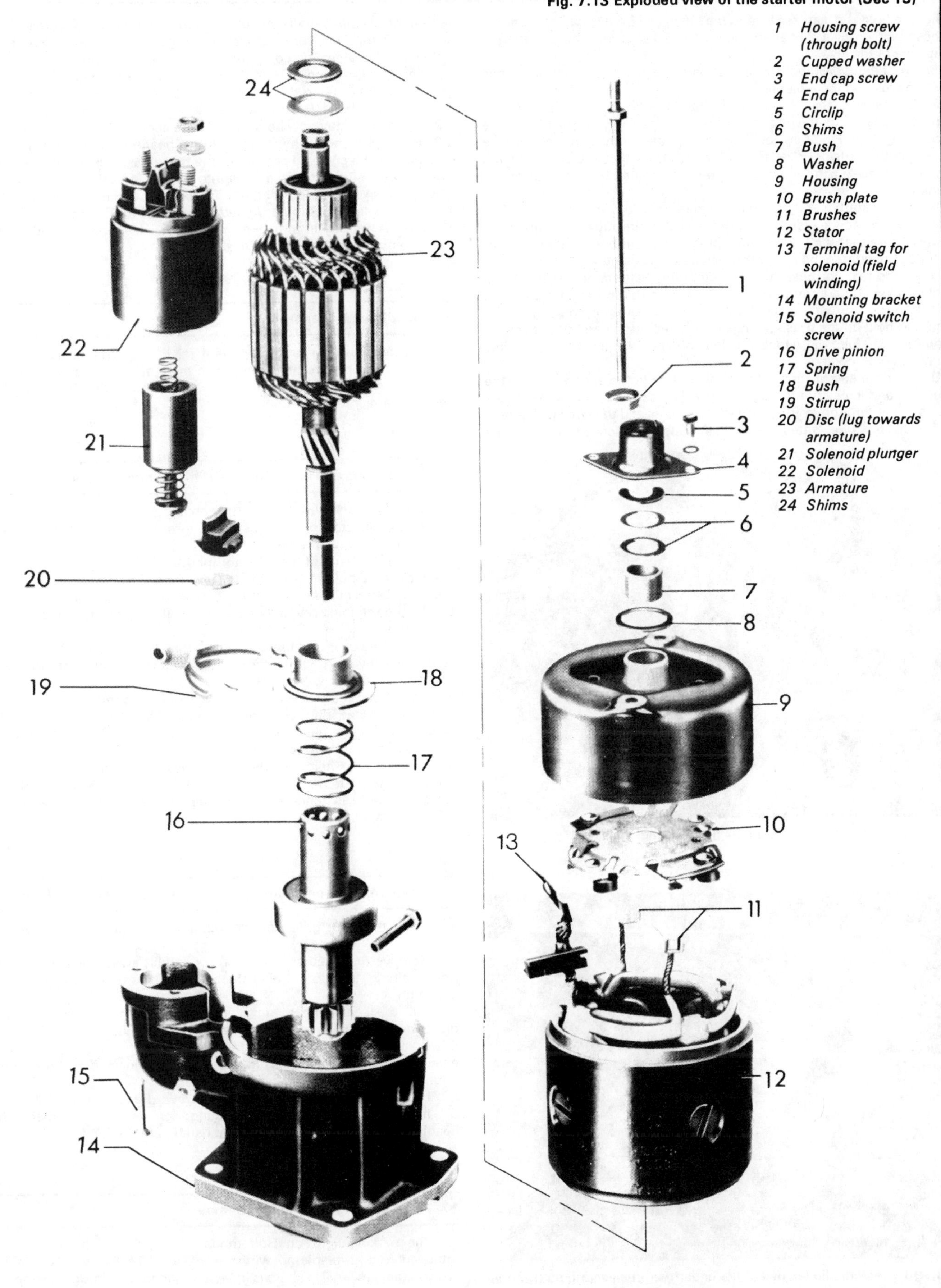

Fig. 7.13 Exploded view of the starter motor (Sec 13)

1 *Housing screw (through bolt)*
2 *Cupped washer*
3 *End cap screw*
4 *End cap*
5 *Circlip*
6 *Shims*
7 *Bush*
8 *Washer*
9 *Housing*
10 *Brush plate*
11 *Brushes*
12 *Stator*
13 *Terminal tag for solenoid (field winding)*
14 *Mounting bracket*
15 *Solenoid switch screw*
16 *Drive pinion*
17 *Spring*
18 *Bush*
19 *Stirrup*
20 *Disc (lug towards armature)*
21 *Solenoid plunger*
22 *Solenoid*
23 *Armature*
24 *Shims*

pliers to hold the lead as close to the brush as possible while soldering. These will act as a a heat sink and will also stop the solder getting in the core of the lead.

16 One final word about brushes. Check that you can get new ones before crushing the old ones.

17 Assembly is the reverse of dismantling. Fit the drive pinion and operating lever to the mounting bracket. Fit the drive pinion to the armature shaft. Fit a new lock ring (circlip) and fit the stop ring (groove towards the outside) over the lock ring. Check that the stop ring will revolve freely on the shaft. It fits on the armature shaft outside the pinion.

18 Fit the starter body over the armature to the mounting bracket. See that the tongue on the body fits in the cut-out of the mounting bracket and that the body seats properly on the rubber seating. Smear a little joint compound round the joint before assembly.

19 Fit the two washers onto the armature shaft and install the brush holder over the commutator. In order to get the holder in place with the brushes correctly assembled cut two lengths of wire and bend them to hold up the brush springs while the brushes are fitted over the commutator (photos). Once the four brushes are in place the wires may be withdrawn.

20 Wipe the end of the shaft and oil it, then fit the endcover onto the housing and install the through-bolts. Again seal the joint, and seal the ends of the through-bolts. Now refit the shims and the circlips. If a new armature has been fitted the endplay must be checked. It should be within the specified limits and is adjusted by fitting appropriate shims.

21 Check that the solenoid lead grommet is in place and refit the solenoid. Use a sealing compound on the joint faces, move the pinion to bring the operation lever to the opening and reconnect pullrod. Seat the solenoid firmly on the mounting bracket in the sealing compound and install the bolts. Reconnect the wire to the starter body (D).

22 The starter may now be refitted to the car.

23 The pinion end of the shaft fits into a bearing in the clutch housing and this can be checked only when the transmission is dismantled. The commutator end of the shaft fits into a bearing bush in the endplate. The old bush may be pressed out if necessary and a new one pressed in. The endplate should be dipped into hot oil for five minutes before the bush is pressed in to give a shrink fit. Grease the bush with multi-purpose grease before fitting the shaft.

13.19A Using two clips to hold the brush spring on assembly

13.19B The wire clip (arrowed) fits under the brush spring so that the brush may be lifted easily to enter the plate over the commutator

14 Headlamps – general

The fixing and beam adjustment for the Golf/Rabbit is generally the same. The main differences are in the type of bulb and method of renewal. The Golf is fitted with renewable bulbs while the Rabbit has sealed beam units.

15 Headlamps and bulbs (Golf) – removal and refitting

1 To remove the bulb, pull off the connector and cover (photo).

2 Press the retaining ring in slightly, turn it anti-clockwise to release it and then lift away the ring.

3 Lift the bulb out of the reflector (photo).

4 Fit a new bulb in the reflector, ensuring that the lug on the bulb plate is located in the recess at the bottom of the reflector.

5 Fit the retaining ring in position, press it in and turn it clockwise to lock in position.

6 Refit the cover and the electrical connector. Ensure that the cover seats correctly on the reflector or water will find its way into the lamp and a new reflector will be required.

7 To remove the reflector it is necessary first to remove the grille by undoing the securing clips and screws. The screws securing the headlamp (not the beam adjusting screws, see Fig. 7.16) can now be removed and the lamp drawn out from the front (photos). The frame and reflector are held together by clamps which must be turned to separate the parts. It is better to renew them as an assembly. If the lamp assembly is renewed adjust the headlamp beam adjustment, refer to Section 23.

16 Headlamp sealed beam unit (Rabbit) – removal and refitting

1 Undo the front grille securing clips and screws and remove the grille.

2 Remove the three securing screws and lift away the sealed beam unit retaining ring (do not alter the setting of the two headlamp beam adjusting screws).

3 Pull the sealed beam unit forward out of the support ring and disconnect the wiring plug from the back of the unit.

4 Connect the plug to a new unit then fit the unit in the support ring ensuring that the three glass lugs are correctly located in the support ring.

5 Fit the retaining ring and securing screws. Refit the front grille.

6 If the headlamp beam adjusting screws are disturbed, the headlamp beam will have to be re-adjusted, refer to Section 23.

17 Front parking light – bulb renewal

1 The parking light on Golf models is inside the headlamp and attached to the headlamp wiring, see Fig. 7.14 (photo). On Rabbit models the parking light bulb is located in the front turn signal lamp.

2 Turn the bulb to the left and pull it out.

3 Fit the new bulb in the holder and turn it to the right.

15.1 Removing the headlamp rear cover

15.3 Removing the headlamp bulb

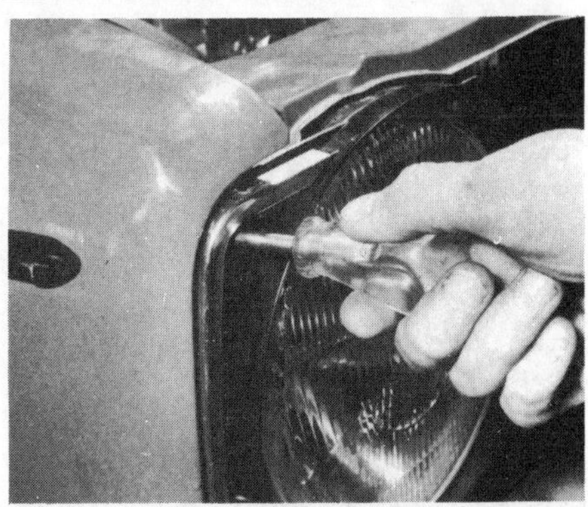
15.7A Removing the headlamp surround ...

15.7B ... lamp retaining screw ...

15.7C ... and headlamp complete

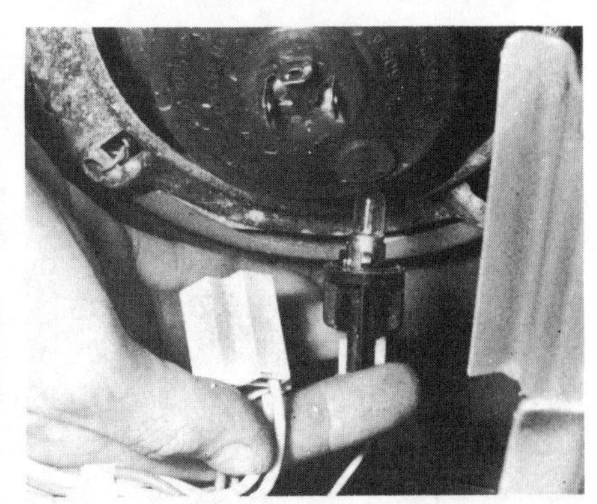
17.1 Removing the parking light bulb

Fig. 7.14 Headlamp and parking light components (Sec 17)

1 Cover
2 Retaining ring
3 Headlamp bulb
4 Reflector
5 Parking light bulb

18 Front turn signal lamp – bulb renewal

1 Remove the two screws securing the lens to the bumper bar (photo). The bulb can now be pressed into the holder, turned to the left and removed.
2 Fit the new bulb. When refitting the lens ensure that the gasket is in good condition and seated correctly. Do not overtighten the lens securing screws or the lens may crack.

19 Light cluster (rear) – bulb renewal

1 Open the tailgate and remove the plastic cover by pressing the lugs outwards (photo).
2 Remove the bulb holder and extract the bulbs as necessary by pressing them in and turning them anti-clockwise.
3 If required, the lens can be removed by unscrewing the retaining nuts (photos).
4 Refitting is a reversal of removal.

20 Interior light – bulb renewal

1 Using a wide bladed screwdriver, prise the lamp from the roof, taking care not to damage the headlining (photo).
2 Remove the festoon type bulb from the spring contacts.
3 Fit the new bulb. Refit the lamp by inserting the switch end first and pressing the remaining end in until the spring clip is fully engaged.

21 Number plate light – bulb renewal

1 Remove the two screws and lift off the lens and cover (photo).
2 Depress the bulb,turn it anti-clockwise, and remove it.
3 Fit the new bulb and refit the lens, taking care not to overtighten the retaining screws.

22 Side markers (USA) – bulb renewal

1 Remove the two retaining screws and pull the unit away from the front wing to expose the rubber cover.
2 Pull off the cover, press the lugs outward, and remove the bulb holder and bulb.
3 Refitting is a reversal of removal.

18.1 Removing the front turn signal lens

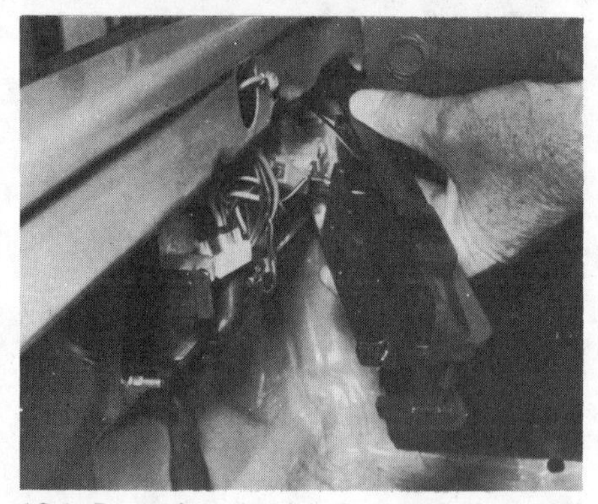
19.1 Removing the interior cover from the rear light cluster

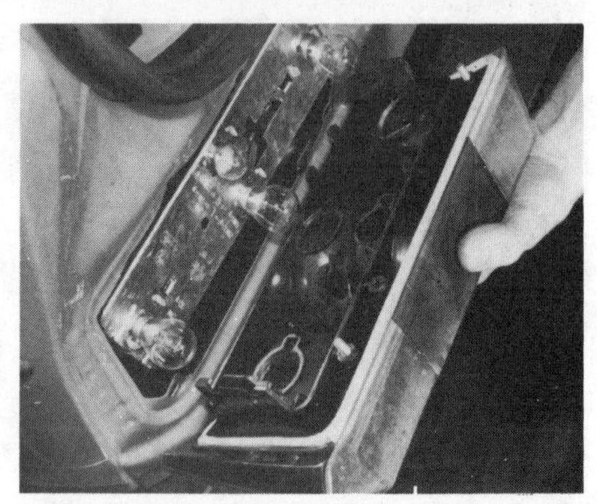
19.3A Removing the rear light cluster lens ...

19.3B ... and bulbholder

20.1 The interior lamp and bulb

21.1 Removing the number plate lamp lens and cover

Fig. 7.15 Side marker lens retaining screws (arrowed) (Sec 22)

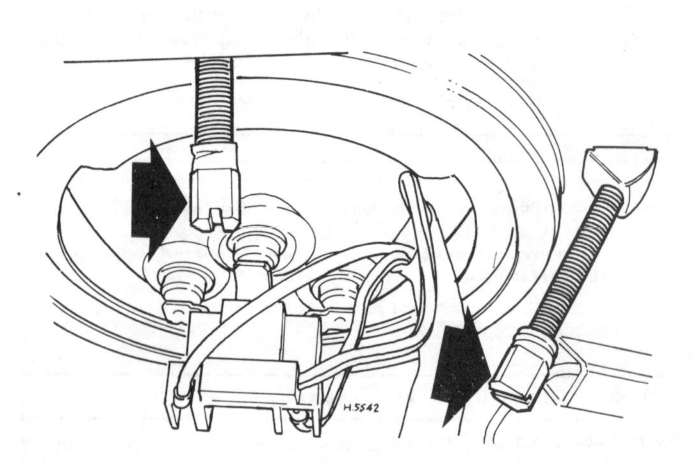

Fig. 7.16 Beam adjustment screws (arrowed) on the Rabbit (Sec 23)

23 Headlamps – beam adjustment

1 Refer to Fig. 7.16 and note the location of the two adjusting screws.
2 It is recommended that the headlamp beams are adjusted by your dealer or service staion. In an emergency use the following procedure.
3 Having ascertained how the beam is to be aimed on dip, whether it must swing to the right or left when dipped, set the car on level ground about 20 feet (6 metres) from a vertical wall or door. The vehicle tyre pressures should be correct and there should be the equivalent of the driver's weight in the driving seat.
4 Mark with relation to the centre-line of the car and the height of the lamps from the ground, the equivalent positions of the lamps on the wall. Using this as a datum, mark the area to which the lights should be directed on dipped beam. Cover one light and switch on. Using the adjusting screws direct the beam to the area required. Cover that light and repeat with the other one. When the beams are correctly focussed on dip the main beams will automatically be correct.

24 Direction indicators and emergency flashers – general

1 The direction indicators are controlled by the left-hand column switch.
2 A switch on the facia board operates all four flashers simultaneously and although the direction indicators will not work when the ignition is switched off the emergency switch over-rides this and the flasher signals continue to operate.
3 All the circuits are routed through the relay on the console and its fuse.
4 If the indicators do not function correctly, a series of tests may be done to find which part of the circuit is at fault.
5 The most common fault is in the flasher lamps, defective bulbs, and dirty or corroded contacts. Check these first, then test the emergency switch. Remove it from the circuit and check its operation. If the switch is in good order refit it and again turn on the emergency

lights. If nothing happens then the relay is not functioning properly and it should be renewed. If the lights function on emergency but not on operation of the column switch then the wiring and column switch are suspect (see Section 25).

25 Steering column switches – removal and refitting

1 There are four switches on the steering column:

(a) *The horn switch pad in the centre of the steering wheel*
(b) *The turn signal/headlight dip and flasher lever on the left of the column*
(c) *The windscreen wiper/washer control lever on the right of the column which, on later models, incorporates the rear window wash/wipe control*
(d) *The steering lock/starter switch on the lower right of the column (see Section 26)*

2 Before commencing dismantling remove the earth strap from the battery.

3 To remove the horn pad simply pull it upwards. It takes a very strong pull, but it does come off that way. Remove the wire connection. Undo the nut and remove the steering wheel.

4 Two crosshead screws hold the two halves of the cover for the column switches together. Remove these and take the lower half of the cover away. The various components of the switches are now visible.

5 From below the switch pull out the three multipin plugs. There is one for each lever switch and one for the starter switch. They will go back only one way so there is no need to mark them. Unscrew the retaining screws which hold the wiper/turn signal switch assemblies in place, and lift the complete switch away.

6 The action of the lever switches may now be tested. If the switch does not move decisively and there is slackness in the linkage, then the switch must be renewed.

7 Assembly is the reverse of removal. When refitting the steering

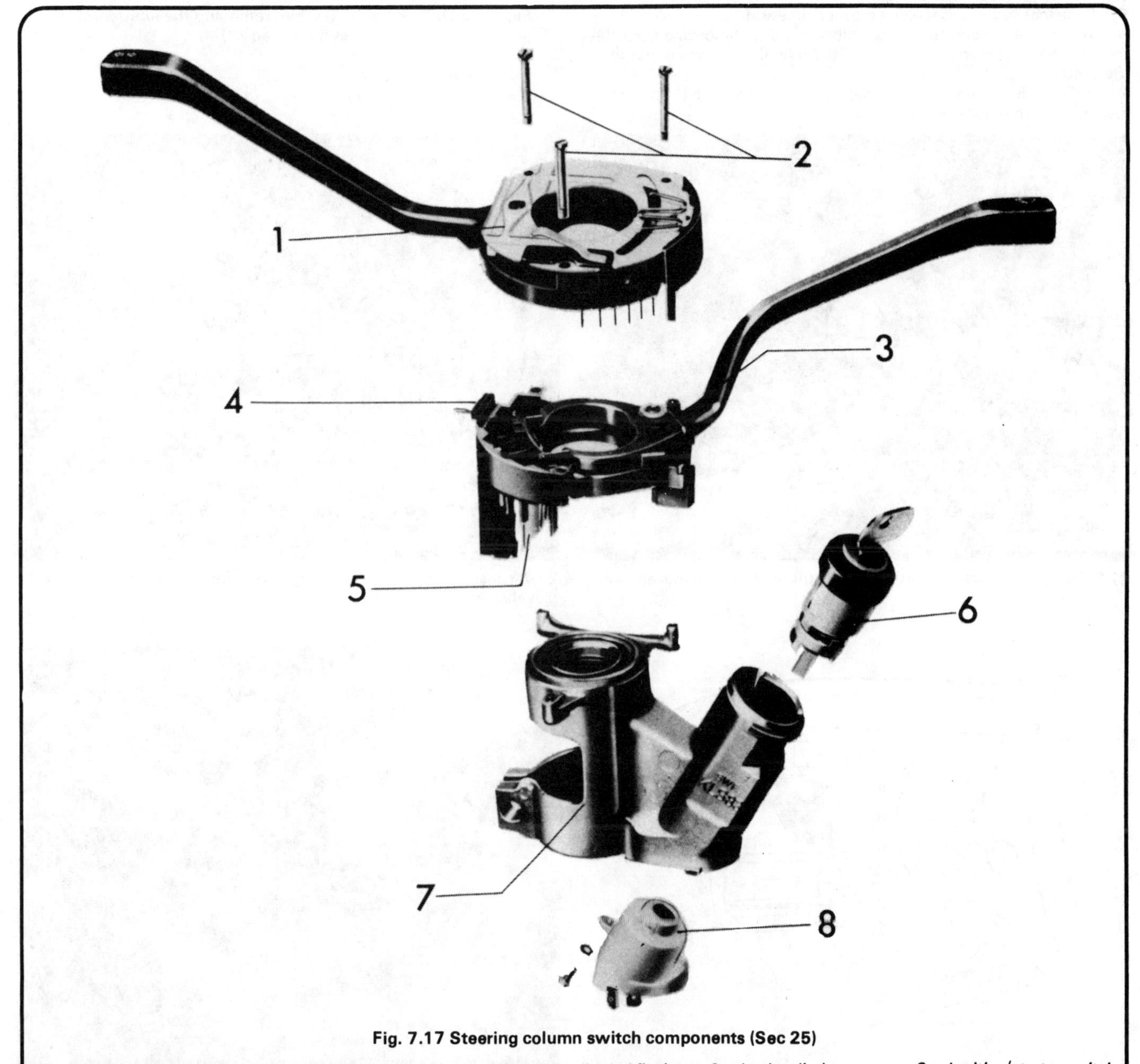

Fig. 7.17 Steering column switch components (Sec 25)

1 Turn signal switch
2 Retaining screws
3 Windscreen wiper switch
4 Wedge
5 Headlight dip and flasher switch
6 Lock cylinder
7 Steering lock housing
8 Ignition/starter switch

wheel be careful to align the cancelling lug in the right place or the switch levers will be damaged. The roadwheels should be in the straight ahead position, the turn signal switch in the neutral position and the lug to the right.

26 Ignition and steering lock switch – removal and refitting

1 Disconnect the battery negative terminal.
2 Remove the steering wheel and switches as described in Section 25.
3 On early Rabbit models pull out the locking plate with a pair of pliers, insert and turn the ignition key clockwise approximately one thickness of the key, then withdraw the lock and key.
4 On all other models it is necessary to drill a 3·0 mm (0·11 in) diameter hole in the lock housing at the location shown in Fig. 7.18. Do not drill intothe lock cylinder. Insert a length of stout wire through the hole and depress the spring tensioned retaining peg; the cylinder can now be removed with the ignition key.
5 If the lock cylinder is faulty it must be renewed.
6 To remove the switch, first withdraw the switch housing from the top of the steering column. Unscrew the retaining screw and remove the switch.
7 Refitting is a reversal of removal, but with the early type, peen the locking plate into position with a hammer.

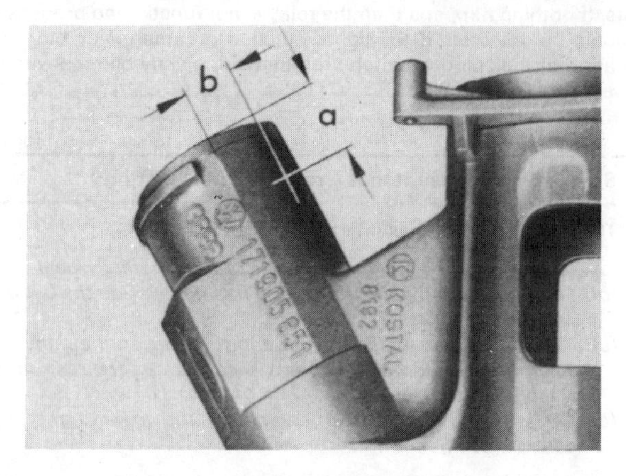

Fig. 7.18 Drilling position when removing the ignition/starter switch (Sec 26)

a = 12.0 mm (0.472 in)
b = 10.0 mm (0.394 in)

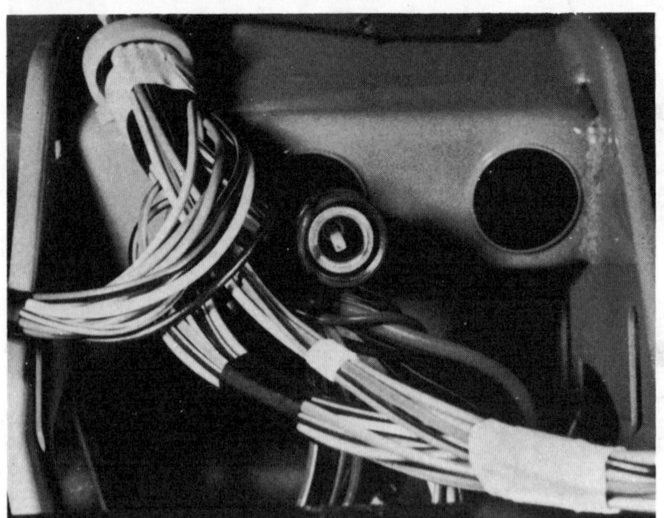

27.5 Speedometer cable location with instrument panel removed

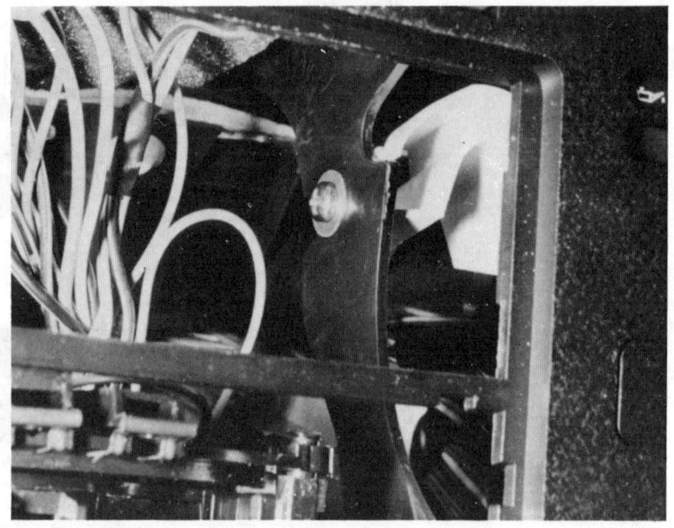

27.7A Instrument panel retaining screw location on RHD cars with glovebox and grille removed

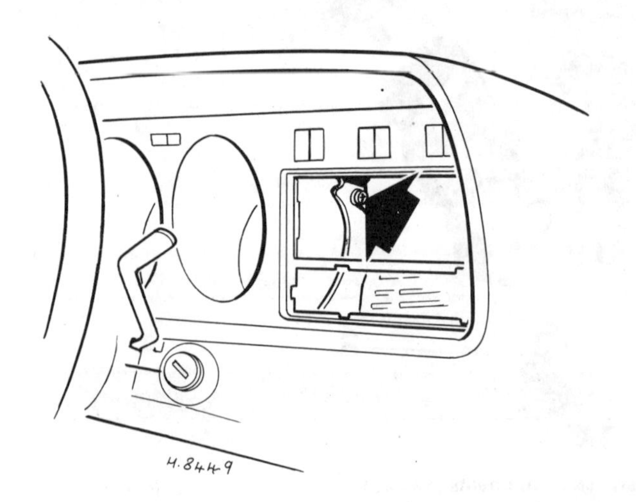

Fig. 7.19 Instrument panel retaining screw (arrowed) on LH drive cars (Sec 27)

27.7B Removing the instrument panel and switches

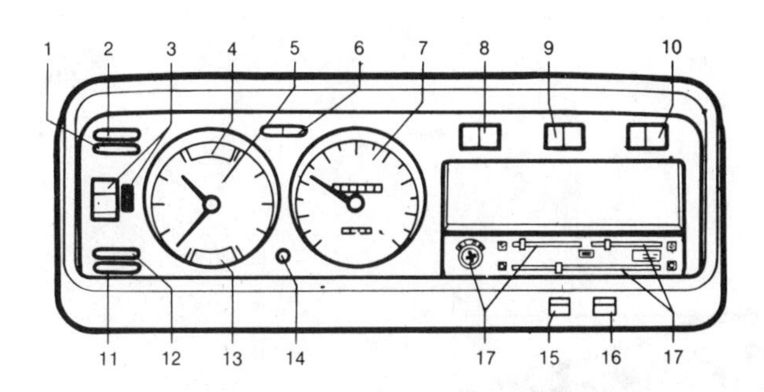

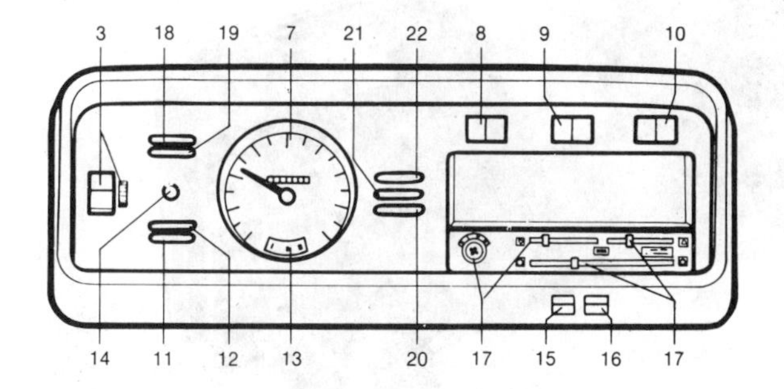

Fig. 7.20 The two types of instrument panel (Sec 27)

1 Oil pressure warning lamp
2 Glow plug warning lamp
3 Light switch
4 Temperature gauge
5 Clock
6 Main beam, turn signal, alternator warning lamps
7 Speedometer
8 Heated rear window switch
9 Hazard light switch
10 Fog light switch
11 Brake warning lamp
12 Vacant for accessory
13 Fuel gauge
14 Vacant for accessory
15 Vacant for accessory
16 Vacant for accessory
17 Heater and vent levers, fan switch
18 Turn signal warning lamp
19 Main beam warning lamp
20 Glow plug warning lamp
21 Alternator and coolant temperature warning lamps
22 Oil pressure warning lamp

27 Instrument panel – removal and refitting

1 Disconnect the battery negative terminal.
2 Remove the steering wheel. This is not essential but will facilitate subsequent operations.
3 Pull off the heater control knobs and remove the trim plate.
4 Prise off the radio control knobs, then reach under the radio and depress the two side clips. Ease the radio out from the front of the facia and disconnect the earth and supply wires, speaker plug, and aerial plug.
5 Reach up behind the instrument panel and disconnect the speedometer cable (photo).
6 Remove the glovebox (if fitted).
7 Remove the retaining screw from the rear of the instrument panel (Fig. 7.19) and withdraw the assembly from the facia (photos).
8 Reference to Fig. 7.20 will show that there are two types of instrument panel and therefore the wiring arrangement will depend on the type fitted. The switches can be removed by opening the dovetails and pulling them out; there is no need to disconnect the wires.
9 With the instrument panel removed,the instruments, warning bulbs, and printed circuit can be dismantled from the casing.
10 Refitting is a reversal of removal.

28 Instrument panel – inspection and testing

1 The printed circuit should be examined, preferably using a magnifying glass. Any defects will show up as breaks or short circuits in

28.1 Rear view of the instrument panel

Fig. 7.21 Instrument panel 12 point multiplug terminals (Sec 28)

1 Turn signal warning lamp
2 Main beam warning lamp
3 Clock
4 Earth
5 Instrument lighting
6 Live (+) wire
7 Fuel gauge sender
8 Temperature gauge sender
9 Alternator warning lamp
10 Oil pressure warning lamp
11 Glow plug lamp
12 Vacant

which case the printed circuit must be renewed (photo).
2 The various circuits can be tested for continuity using a test lamp and leads or a meter. Figs. 7.21 and 7.23 show the multiplug connection terminals and the applicable circuits.

29 Switches (lighting) – removal,testing and refitting

1 Disconnect the battery negative terminal.
2 Remove the instrument panel as described in Section 27.
3 Move up the dovetail clips and withdraw the switch from the panel (photo).
4 The switches cannot be repaired and if one is proved faulty after testing with a meter or test lamp, it must be renewed.
5 Refitting is a reversal of removal.

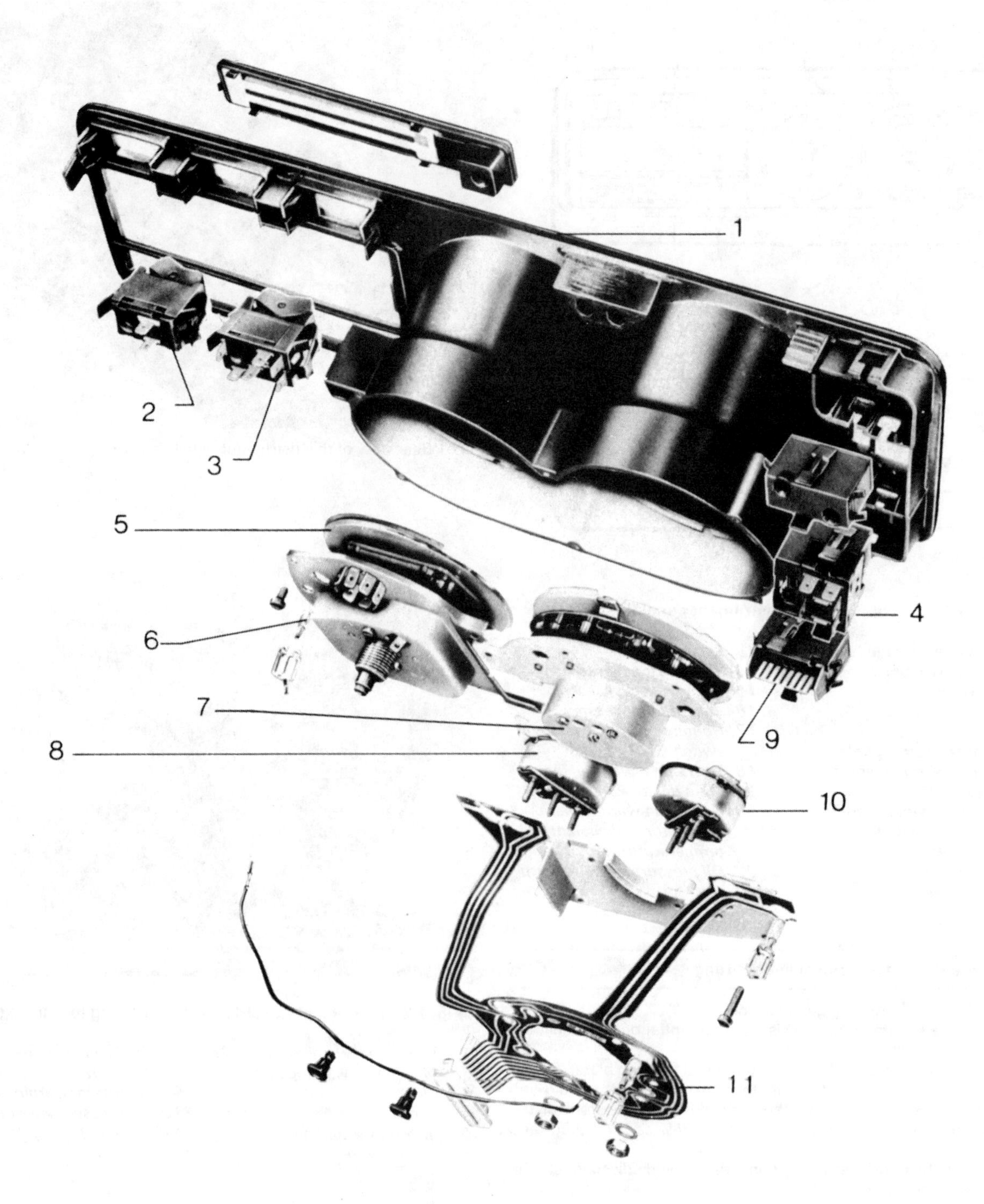

Fig. 7.22 Instrument panel components (Sec 28)

1 *Surround*
2 *Heated rear window switch*
3 *Hazard flasher switch*
4 *Light switch*
5 *Speedometer*
6 *Bulb*
7 *Clock (illustration shows tachometer)*
8 *Fuel gauge*
9 *Dual brake circuit warning light and seat belt light*
10 *Coolant temperature gauge*
11 *Printed circuit*

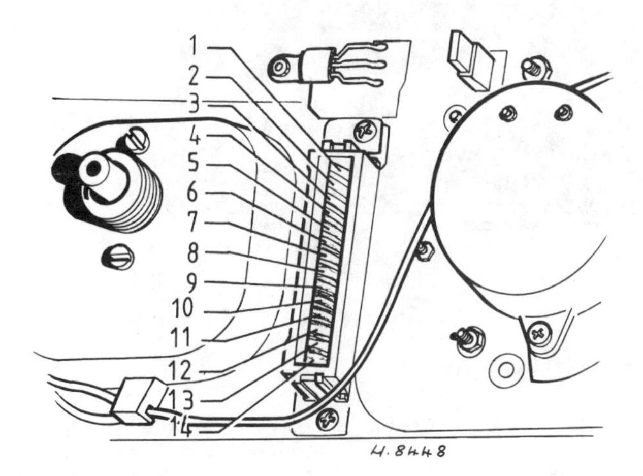

Fig. 7.23 Instrument panel 14 point multiplug terminal (Sec 28)

1 *Vacant*
2 *Turn signal warning lamp*
3 *Main beam warning lamp*
4 *Clock*
5 *Earth*
6 *Instrument lighting*
7 *Live (+) wire*
8 *Fuel gauge sender*
9 *Coolant temperature or warning switch*
10 *Alternator warning lamp*
11 *Oil pressure warning lamp*
12 *Glow plug warning lamp*
13 *Vacant*
14 *Vacant*

29.3 Switches, showing the locating lugs

30 Temperature gauge – testing

1 If the gauge is giving innaccurate readings, first test the voltage stabilizer for correct output.
2 If the gauge is receiving the correct voltage from the stabilizer but is still innaccurate, then either the gauge or the sender is faulty, or alternatively the wiring may have a fault.
3 A quick test is to pull the wire off the sender and using a 12v 6 watt bulb as a series resistance and with the ignition switched on touch the bulb centre pin to earth momentarily. Have someone watch the gauge. It should move right across the scale. It is essential to have the bulb in series with the wire or you may damage the gauge. If there is no reading then press a bit harder with your contact. Still no reading, then either the wiring or the gauge are at fault. Check the fuse, if there is one in the circuit for your vehicle. If there is still no reading then the gauge must be extracted for testing separately.
4 If there is a reading then the gauge is not at fault but the sender is suspect. To test the gauge for accuracy disconnect the sender and fit resistances in its place. Ordinary radio resistances will do. A 47 ohm resistance should indicate that the engine is hot. Replace this with a 270 ohm resistor and the needle should register 'cold'. If the gauge does not pass this test then it is not accurate and should be renewed. If the gauge passes the test, but still refuses to indicate that the engine is getting warm, then the sender unit should be renewed.

31 Fuel gauge and sender unit – testing

1 Read paragraph 1 of Section 30 before proceeding.
2 The handbook states that when the needle of the fuel gauge reaches the reserve mark there is about one gallon of fuel in the tank. It is important therefore that the gauge is accurate.
3 There is a very good way of checking the reserve mark. Buy one gallon of fuel, empty the fuel tank, put in the one gallon and check the reading against the reserve mark. If it is possible to arrange to fill the tank from empty to full, one gallon at a time, a piece of thin cardboard fitted to the face of the gauge glass may be marked at each gallon and you will know exactly how much fuel is in the tank at all times.
4 If the gauge is not reading then a simple test is required. A 47 ohm resistor and a 0–12v meter are needed. Underneath the car at the rear on the right-hand side of the fuel tank is the combined fuel hose and fuel gauge wires entry. Switch on the ignition and measure the voltage across the terminals. If there is no voltage pull off the terminal of the wire to the gauge and check the voltage between that terminal and earth. If there is now a reading the sender unit is at fault. Check that the other wire is in fact connected securely to earth. If there is still no reading then either the wiring or the gauge is faulty. Check that all the fuses are in order. To decide whether the gauge or the wiring is at fault the instrument panel must be removed.
5 If the sender unit is faulty then the tank must be drained. Remove the hose clip and pull off the hose. Disconnect the battery earth strap and disconnect the wires from the sender unit. Turn the locking plate until the unit can be withdrawn from the tank, and then remove the fuel pipe and sender unit. There is little that can be done to the sender unit, it is best to take it to an expert, who may be able to repair it or supply a new one.
6 To test the accuracy of the gauge, if it is working, disconnect the gauge wires at the sender unit and insert a 47 ohm resistance between them. Switch on the ignition and the gauge should register full. Substitute a 100 ohm and then a 220 ohm resistor in turn and the needle should be on the second mark and then the reserve mark respectively.

32 Fuse and relay panel – removal and refitting

1 Disconnect the battery negative terminal.
2 Unscrew the crosshead retaining screws and withdraw the panel to expose the multiplug connectors.
3 Note the location of the wiring, then disconnect it and remove the panel.
4 Figs. 7.24 and 7.25 show the various terminals on the panel which will be of assistance when tracing a fault.

Fig. 7.24 Front view of the fuse and relay panel (Sec 32)

Socket	Contact	Relay terminal	Internal connections
J – dipswitch relay	*1*	*–*	*J2*
	2	*56*	*G10, D17*
	3	*–*	*J5, fuses S1 and S2*
	4	*56A*	*G8, fuses S3 and S4*
	5	*56B*	*J3*
	6	*–*	*H1 to H7 (terminal 30)*
	7	*S*	*E8*
K – heated rear window relay	*8*	*86*	*D10, L15, M18, N23 (terminal 31)*
	9	*30*	*H1 to H7 (terminal 30)*
	10	*87*	*Fuse S5*
	11	*85*	*D7*
L – Electric fuel pump 1977 on (spark ignition engines only)	*12*	*86*	*A3*
	13	*30*	*H1 to H7 (terminal 30)*
	14	*87*	*A8*
	15	*31*	*D10, K8, M18, N23 (terminal 31)*
	16	*15*	*Fuses S8 and S9*
M – intermittent wiper washer switch relay	*17*	*15*	*Bridge for wiper motor on cars without intermittent switch*
	18	*31*	
	19	*53S*	
	20	*S1*	
	21	*53M*	
N – hazard flasher relay	*22*	*49A*	*D1, E3*
	23	*31*	*D10, L15, K8, M18 (terminal 31)*
	24	*+49*	*D12*
	25	*C*	*D6*

33 Fuses and relays – general

1 The fuses are located in the fusebox beneath the left-hand side of the dashpanel. Access to them is gained by lifting the transparent cover (photo).

2 The function of each fuse is given in the following table; the fuse number is as given on the fusebox lid.

Fuse	Amps	Circuit
1	8	Low beam, left
2	8	Low beam, right
3	8	High beam, left and warning light
4	8	High beam, right
5	16	Heated rear window
6	8	Stop lights, hazard flasher system
7	8	Interior light, cigarette lighter, radio
8	8	Turn signals and warning light
9	16	Reversing lights, horn, fuel cut-off valve
10	8	Fresh air fan
11	8	Wipers, washer pump intermittent relay
12	8	Number plate lights
13	8	Right parking light, tail light, and side marker (USA)
14	8	Left parking light, tail light, and side marker (USA)
15	16	Radiator cooling fan

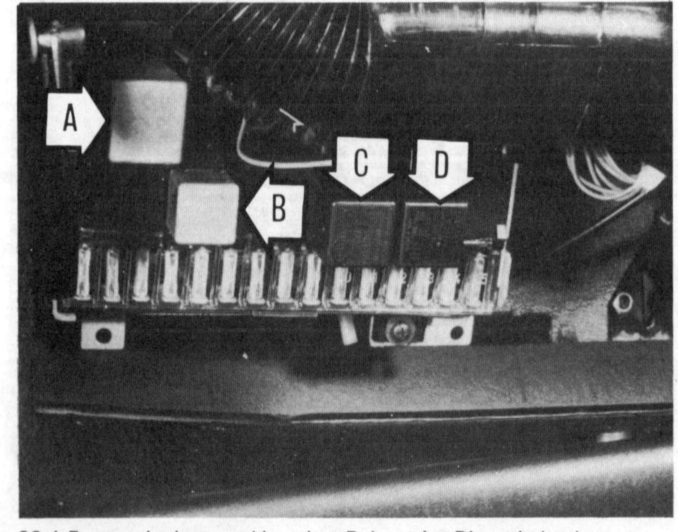

33.1 Fuse and relay panel location. Relays: A – Dip switch relay, B – Heated rear window, C – Intermittent wiper/washer, D – Hazard warning

Fig. 7.25 Rear view of the fuse and relay panel (Sec 32)

A – Front wiring harness socket
B – Test diagnosis socket (early models), vacant (late models)
C – Front wiring harness socket
D – Dashboard wiring harness socket
E – Dashboard wiring harness socket
F – Rear wiring harness socket
G1 – to G6 via fuse S15
G2 – to E7 and M20
G3 – to ignition/starter switch terminal 15
G4 – to fuse S12
G5 – to alternator terminal D+
G6 – to G1 via fuse S15
G7 – to ignition/starter switch terminal X
G8 – to headlight dip relay terminal 56A
G9 – to oil pressure switch
G10 – to light switch terminal 56
H1 to H7 – terminal 30

Additional fuses of 8 amp rating are also located separately above the main fusebox for the rear window wiper/wash system and fog lights (where fitted).
3 To remove a fuse, depress the upper contact spring and lift the fuse out.
4 Always renew a fuse with one of similar rating, and never renew it more than once without finding the source of the trouble.
5 Fit the fuse with the metal strip facing outward. Check that the contacts are free of any corrosion and clean them up with emery tape if necessary.
6 The 50 amp glow plug fuse is contained in a separate fusebox located in the engine compartment on the bulkhead (photo).
7 Relays are plugged into the upper part of the fuse panel and their functions are identified in the photograph 33.1.

34 Horn – testing, removal, and refitting

1 The horn is located on the front crossmember in front of the engine (photo).
2 If the horn fails to operate, first check that the relevant fuse is intact and is firmly fitted between the contacts.
3 Using a test lamp and leads or a voltmeter, check that current is reaching the horn from the supply wire (usually yellow and black) with the ignition switched on. If not, the supply wire is faulty.
4 If current is reaching the supply wire, use the same method to check that current is reaching the horn output terminal. If not, the horn is faulty. If it is, but the horn fails to operate, have an assistant operate the horn push and turn the adjusting screw (if fitted) until the horn operates satisfactorily.
5 If the horn still fails to operate after making the previous tests, connect a wire between the output terminal and earth. If the horn now operates, a fault exists in the horn push or wiring. Use the same method as previously described to trace the fault.
6 To remove the horn, first disconnect the battery negative terminal.
7 Disconnect the horn input and output wires.
8 Unbolt the horn from the mounting bracket.
9 Refitting is a reversal of removal.

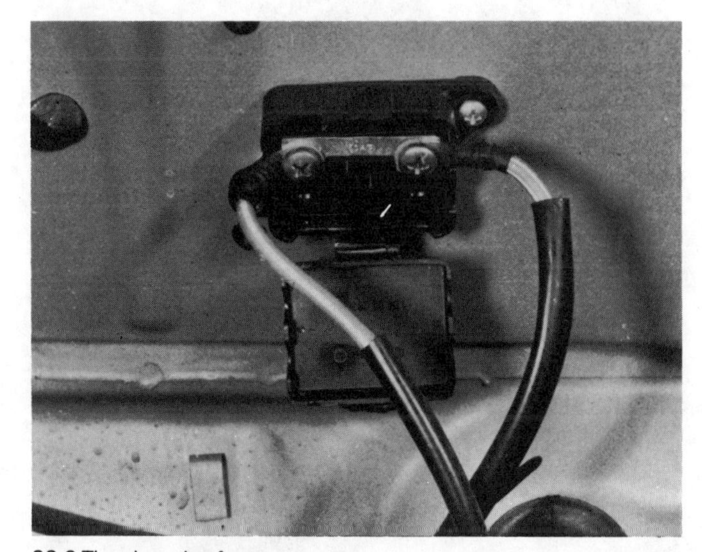

33.6 The glow plug fuse

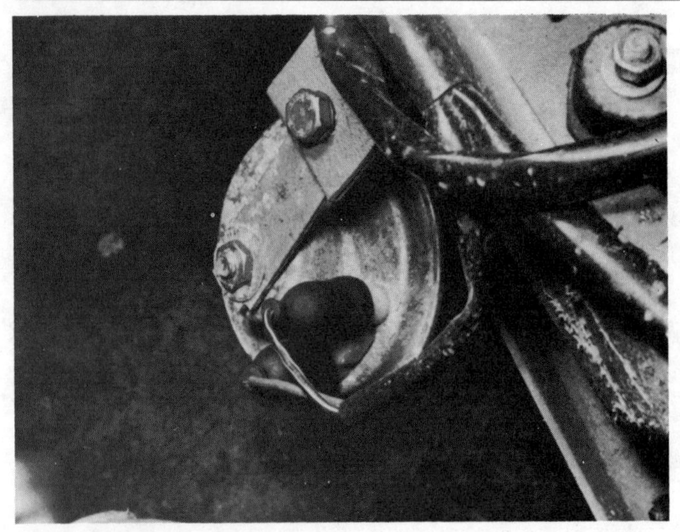

34.1 Location of the horn

35 Speedometer and speedometer drive – removal and refitting

1 Disconnect the battery negative terminal.
2 Remove the instrument panel as described in Section 27.
3 Remove the voltage stabilizer (where fitted).
4 Unscrew the two retaining screws and remove the speedometer head from the rear of the instrument panel.
5 Refitting is a reversal of removal, but make sure that the cardboard washers are located beneath the retaining screw heads. If these are left out, damage will occur due to earthing; replacements can be made out of ordinary cardboard if necessary.
6 To remove the speedometer cable, remove the retaining screw at the transmission end and withdraw the drivegear (photo).
7 With the cable disconnected from the speedometer head pull it through the grommet hole and remove the cable from the car.
8 Refitting is a reversal of removal, but make sure that the cable is positioned with the minimum amount of curving. The cable must be attached to the plastic retainer near the bulkhead in orde- to prevent it touching the clutch cable. Do not grease the cable at the head end as this may cause the needle to stick.

35.6 Speedometer cable retaining clamp location on the transmission (arrowed)

36 Radio – fitting guidelines

1 The fitting of a radio to the Golf presents no problems. It may be fitted to the dashboard, or centre console. A centre console with provision for storing small articles may be purchased and the radio installed in that.
2 VW provide a choice of radios to suit the taste and pocket. Pushbutton or manual control is available, and a radio with a built-in cassette tape player, speakers, a lockable aerial and suppression kits is also supplied, with full fitting instructions.
3 There does not seem to be any point in going elsewhere as fitting a set which is not tailored to suit will mean cutting the dashboard about and building separate supports which will get in the way of things in an already crowded space.
4 The console fitting for the radio is particularly attractive as the storage space is somewhat limited, and putting the radio in the dashboard involves pressing out the small glovebox and discarding it.
5 It is important to get advice from the local experts as to the best set for your requirements. VHF gives patchy reception in built-up or hilly areas and if you do a lot of long distance work, a set which functions well at home may be unsuitable in other areas.
6 If 'electrical noise' appears then it is probably due to some unit of the electrical system which is badly worn. Sparking at the alternator slip rings, a defective voltage regulator, fan motor or blower motor may be responsible. The only way to find out is to isolate each item until the noise disappears and then overhaul that item.
7 It is recommended that a small fuse be inserted in the supply cable. A two amp one is sufficient.

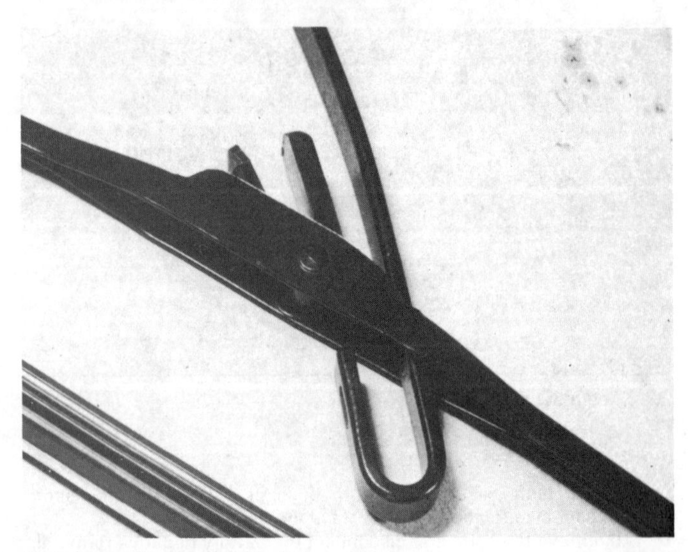

37.1 Removing a wiper blade from the arm

37 Wiper blades and arms – renewal

1 To remove a wiper blade, depress the clip with a screwdriver and slide the blade from the arm (photo). On some models it may be possible to renew the wiper rubber separately. Where this type is fitted, use pliers to squeeze the support rail, then withdraw the rubber.
2 To remove a wiper arm,first prise out the cover with a screwdriver then unscrew the retaining nut (photo). Lever the arm off the shaft being careful not to damage the paintwork.
3 When refitting a wiper arm and blade assembly, make sure that the gap between the blade and the windscreen lower edge is as given in Fig. 7.26; the wiper motor must also be in its parked position. The tailgate wiper should be positioned as shown in Fig. 7.28.

38 Windscreen wiper mechanism – removal and refitting

1 Remove the wiper blades and arms as described in Section 37.
2 Disconnect the battery negative terminal, then remove the boot from each spindle and unscrew the retaining nuts.
3 Unplug the connector from the motor terminals.
4 The wiper motor and frame are held in position by a nut on a bracket (photo). Remove this nut and lift the motor and frame away

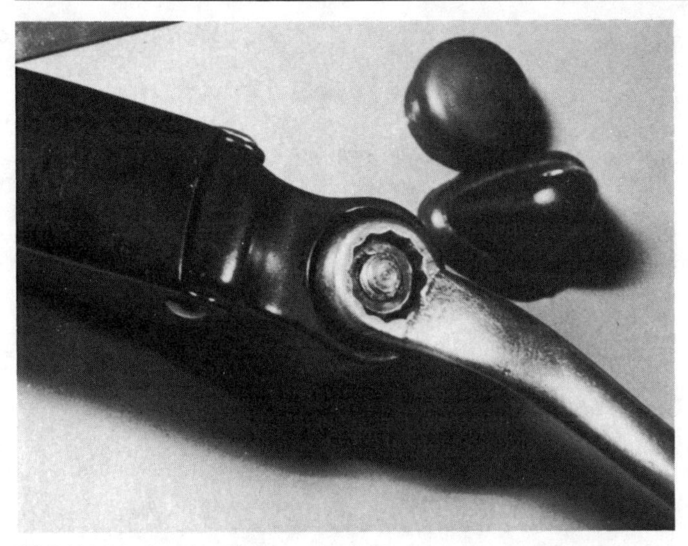

37.2 Unscrewing the wiper arm retaining nut

from the car. The mechanism may now be examined at leisure.

5 The motor may be removed from the frame by undoing two nuts. Alternatively, the motor may be removed from the frame while the frame is still in position but we do not see the point in this as having got so far one might as well have a look at the frame joints as well.

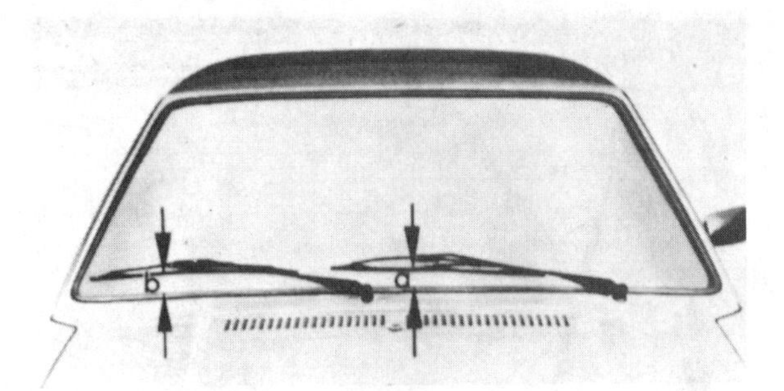

Fig. 7.26 Wiper blade parked position dimensions (Sec 37)

a = 35.0 mm (1.38 in)
b = 65.0 mm (2. 56 in)

Note: LHD shown, RHD is mirror image

6 Check that all the levers and pins are secure, not worn and are well lubricated.

7 Do not remove the crank from the motor unless the motor is to be replaced with a new one. If this is to be done, connect the multipin plug to the motor before installing the motor, switch on and let the motor run for four minutes. Switch off and the motor will stop in the parking position. Fit the crank in the way shown in Figs. 7.29 and 7.30.

8 Repair is by renewal. Whatever is wrong with the electrical components may only be cured by fitting new ones. The renewal of the column switch is discussed in Section 25. The brush gear is not a

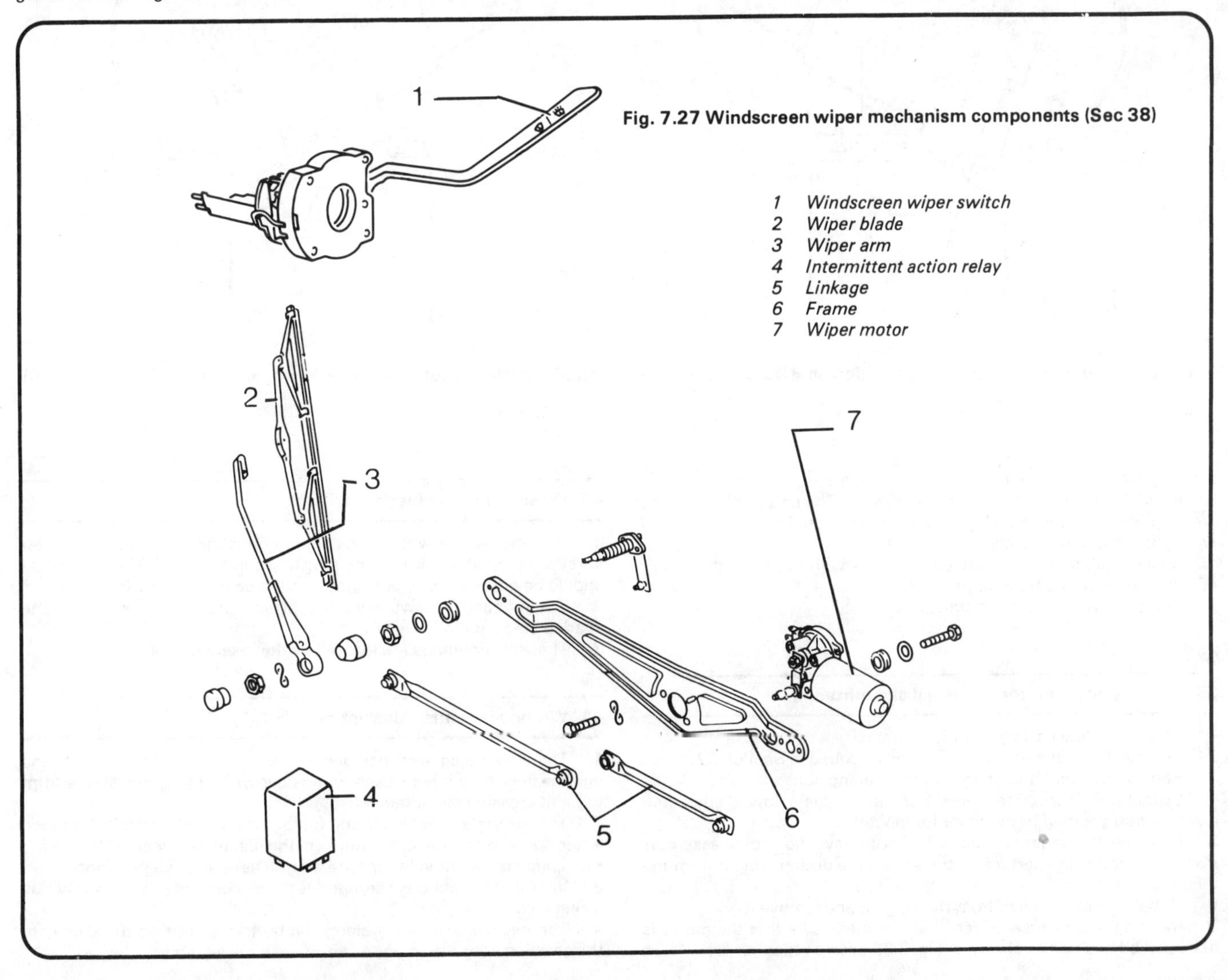

Fig. 7.27 Windscreen wiper mechanism components (Sec 38)

1 Windscreen wiper switch
2 Wiper blade
3 Wiper arm
4 Intermittent action relay
5 Linkage
6 Frame
7 Wiper motor

Fig. 7.28 Tailgate wiper blade parked position (Sec 37)

Distance from top of blade to weatherstrip = 30 mm (1.18 in)

38.4 Windscreen wiper motor location

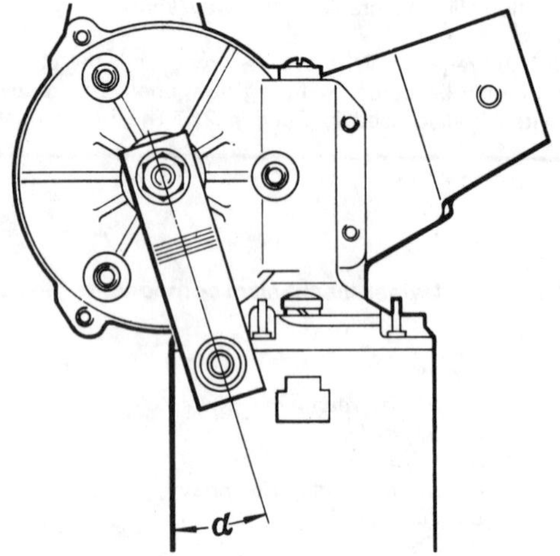

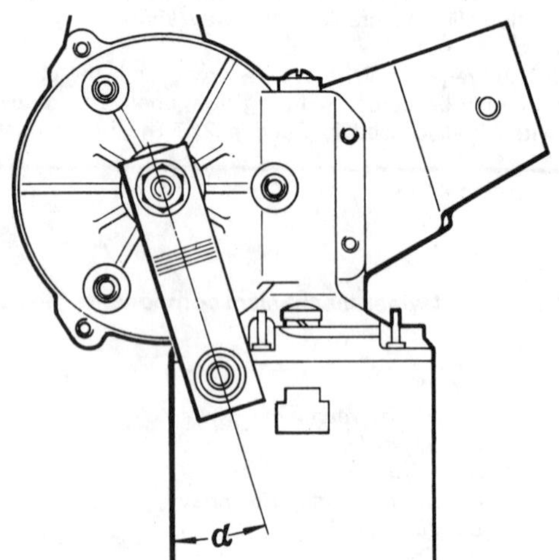

Fig. 7.29 Wiper motor crank parked position on RHD cars (Sec 38)

Angle = 20°

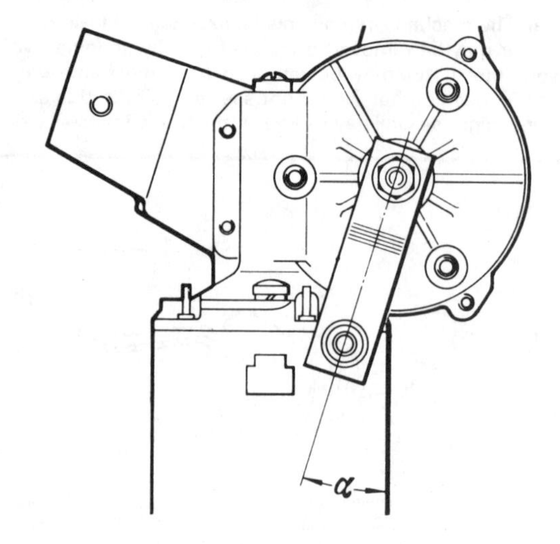

Fig. 7.30 Wiper motor crank parked position on LHD cars (Sec 38)

Angle = 20°

service part so the motor must be renewed if faulty.

9 Inspect the linkage for wear or corrosion. The parts of the linkage are renewable, if required. However, the links are different on left-hand and right-hand drive vehicles.

10 If electrical 'noise' has caused problems with the radio the motor may be replaced by a fully suppressed one.

11 Refitting is a reversal of removal.

39 Tailgate wiper motor – removal and refitting

1 Disconnect the battery negative terminal.

2 Remove the wiper arm and blade as described in Section 37.

3 Remove the boot and unscrew the retaining nut.

4 Extract the inspection panel from the tailgate lower edge and disconnect the wiring from the motor (photo).

5 Unscrew the mounting bolts and withdraw the motor assembly from the inspection aperture, at the same time disengaging it from the mechanism.

6 Unbolt the mechanism from the tailgate and remove it.

7 Refitting is a reversal of removal, but make sure that the motor is in its parked position as shown in Fig. 7.32 before fitting it.

40 Washer jets – adjustment

1 Periodically the washer jets should be adjusted so that the water spray is directed toward the points shown in Fig. 7.33. The tailgate jet should be adjusted to direct spray into the centre of the wiped area.

2 Where headlight washers are fitted, direct the spray into the shaded area shown in Fig. 7.34.

3 The washer pumps are located by the reservoirs (photo).

41 Wiring diagrams – desciption

1 The wiring diagrams included at the end of this Chapter are of the current flow type where each wire is shown in the simplest line form without crossing over other wires.

2 The fuse/relay panel is at the top of the diagram and the combined letter/figure numbers appearing on the panel terminals refer to the multiplug connector in letter form and the terminal in figure form.

3 Internal connections through electrical components are shown by a single line.

4 The encircled numbers along the bottom of the diagram indicate the earthing connecting points as given in the leg end.

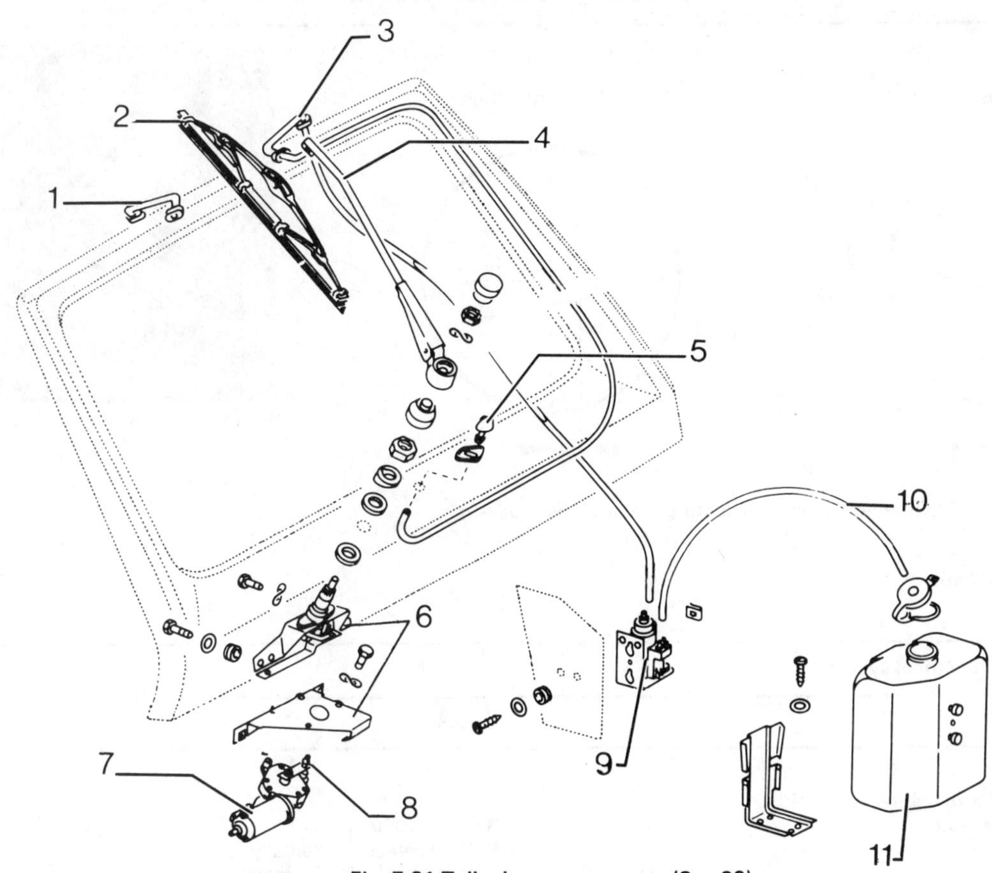

Fig. 7.31 Tail wiper components (Sec 39)

1 *Wiring grommet*
2 *Wiper blade*
3 *Water hose grommet*
4 *Wiper arm*
5 *Jet*
6 *Spindle and frame*
7 *Wiper motor*
8 *Crank*
9 *Washer pump*
10 *Hose*
11 *Reservoir*

39.4 Tailgate wiper motor with inspection cover removed

Fig. 7.32 Tailgate wiper motor crank parked position (Sec 39)

Fig. 7.33 Windscreen washer jet adjusting dimensions (Sec 40)

a = 170 to 270 mm (6.7 to 10.6 in)
b = 275 to 325 mm (10.8 to 12.8 in)
c = 305 to 355 mm (12.0 to 14.0 in)
d = 230 to 330 mm (9.0 to 13.0 in)

Note: LHD shown, RHD is mirror image

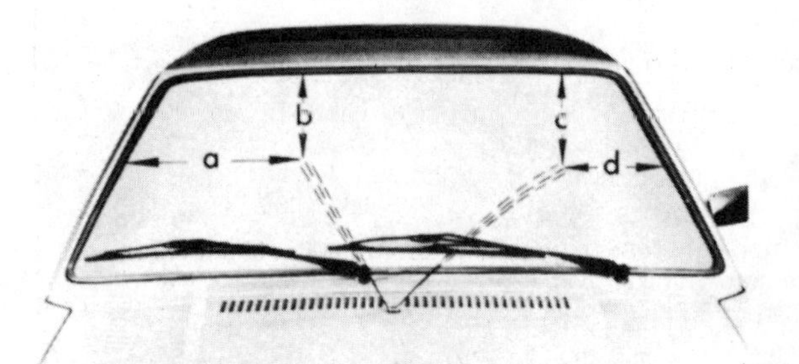

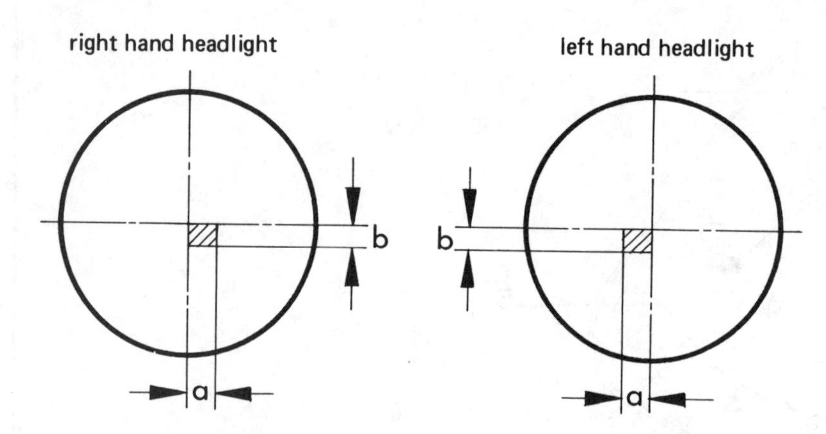

Fig. 7.34 Headlight washer adjusting dimensions (Sec 40)

a = 15 mm (0.6 mm)
b = 10mm (0.4 mm)

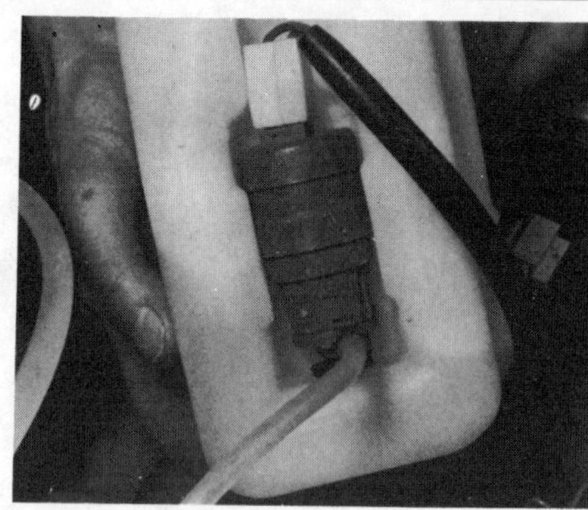

40.3 Windscreen washer reservoir and pump

42 Fault diagnosis – electrical system

Symptom	Reason/s
Starter motor fails to turn engine	
No electricity at starter motor	Battery discharged Battery defective internally Battery terminal leads loose or earth lead not securely attached to body Loose or broken connections in starter motor circuit Starter motor switch or solenoid faulty
Electricity at starter motor: faulty motor	Starter brushes badly worn, sticking, or brush wires loose Commutator dirty, worn, or burnt Starter motor armature faulty Field coils earthed
Starter motor turns engine very slowly	
Electrical defects	Battery in discharged condition Starter brushes badly worn, sticking or, brush wires loose Loose wires in starter motor circuit
Starter motor operates without turning engine	
Mechanical damage	Faulty solenoid Pinion or flywheel gear teeth broken or worn
Starter motor noisy or excessively rough engagement	
Lack of attention or mechanical damage	Pinion or flywheel gear teeth broken or worn Starter motor retaining bolts loose
Battery will not hold charge for more than a few days	
Wear or damage	Battery defective internally Electrolyte level too low or electrolyte too weak due to leakage Plate separators no longer fully effective Battery plates severely sulphated
Insufficient current flow to keep battery charged	Alternator belt slipping Battery terminal connections loose or corroded Alternator not charging properly Short in lighting circuit causing continual battery drain Regulator unit not working correctly
Ignition light fails to go out, battery runs flat in a few days	
Alternator not charging	Fan belt loose and slipping, or broken Brushes worn, sticking, broken, or dirty Brush springs weak or broken
Regulator or cut-out fails to work correctly	Regulator incorrectly set Open circuit in wiring of regulator unit

Fuel gauge

Symptom	Reason/s
Fuel gauge gives no reading	Fuel tank empty! Electric cable between tank sender unit and gauge broken or loose Fuel gauge case not earthed Fuel gauge supply cable interrupted Fuel gauge unit broken
Fuel gauge registers full all the time	Electric cable between tank unit and gauge earthed

Horn

Symptom	Reason/s
Horn operates all the time	Horn push either earthed or stuck down Horn cable to horn push earthed
Horn fails to operate	Blown fuse Cable or cable connection loose, broken or disconnected Horn has an internal fault
Horn emits intermittent or unsatisfactory noise	Cable connections loose Horn incorrectly adjusted

Lights

Symptom	Reason/s
Lights do not come on	If engine not running, battery discharged Light bulb filament burnt out or bulbs broken Wire connections loose, disconnected or broken Light switch shorting or otherwise faulty
Lights come on but fade out	If engine not running battery discharged
Lights give very poor illumination	Lamp glasses dirty Reflector tarnished or dirty Lamps badly out of adjustment Incorrect bulb with too low wattage fitted Existing bulbs old and badly discoloured Electrical wiring too thin not allowing full current to pass
Lights work erratically – flashing on and off, especially over bumps	Battery terminals or earth connection loose Lights not earthing properly Contacts in light switch faulty

Wipers

Symptom	Reason/s
Wiper motor fails to work	Blown fuse Wire connections loose, disconnected, or broken Brushes badly worn Armature worn or faulty Field coils faulty
Wiper motor works very slowly and takes excessive current	Commutator dirty, greasy, or burnt Spindle binding or damaged Armature bearings dry or unaligned
Wiper motor works slowly and takes little current	Brushes badly worn Commutator dirty, greasy, or burnt Armature badly worn or faulty
Wiper motor works but wiper blades remain static	Wiper motor gearbox parts badly worn

SYMBOL	EXPLANATION	Sample use
	ALTERNATOR WITH DIODE RECTIFIERS	Golf alternator
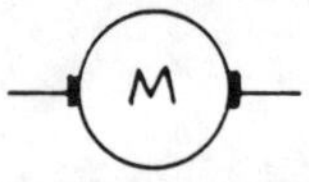	MOTOR	Fan motor, radiator fan
	EXTERNAL WIRING WIRE 10 mm^2 sectional area	Wiring diagram
	WIRE JUNCTION FIXED (SOLDERED)	Relay plate, dash board printed circuit
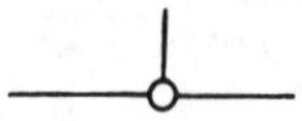	WIRE JUNCTION SEPARABLE	Screw on terminals and eyelets
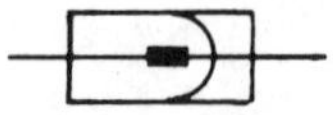	PLUG, SINGLE OR MULTIPIN	T10 written by the side means explanation in the key
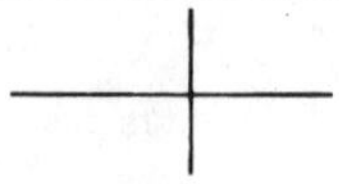	WIRE CROSSING, NOT JOINED	Wiring diagram
	GROUND, OR EARTH	Wiring diagram
	SWITCH CLOSED	Wiring diagram
	SWITCH OPEN	Wiring diagram
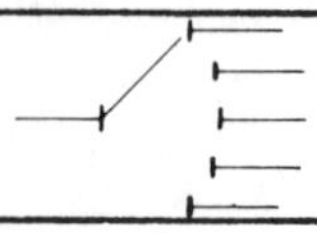	MULTI CONTACT SWITCH	Wiring diagram
 1 2 3	SWITCH, MANUALLY OPERATED	Wiring diagram
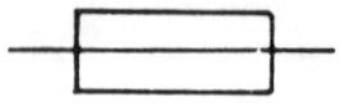	FUSE	
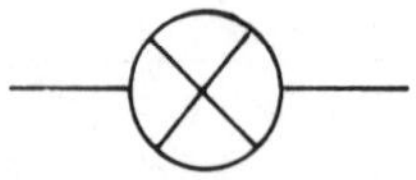	BULB	
 H.5550	CONDENSER	

Fig. 7.35A Current flow diagram symbols

SYMBOL	EXPLANATION	Sample use
	TRANSFORMER, IRON CORE	
	DIODE	Alternator
	TRANSISTOR	Voltage regulator
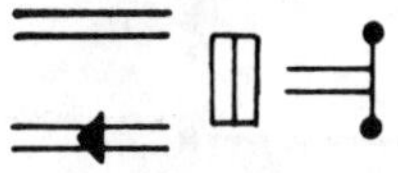	MECHANICAL CONNECTION MECHANICAL CONNECTION SPRING LOADED	Double switch Oil pressure switch
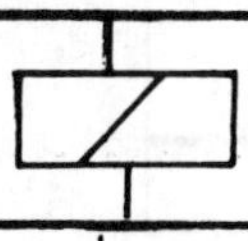	RELAY, COIL	
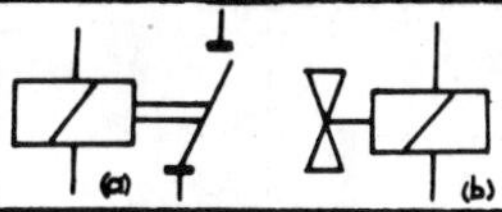 (a) (b)	RELAY, ELECTRO MAGNETIC	Headlamp
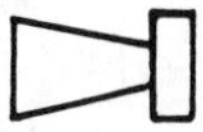	HORN	
	RESISTOR	
	POTENTIOMETER	
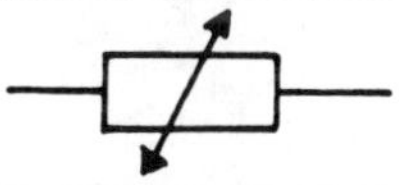	THERMAL RESISTOR AUTOMATIC REGULATING	Temperature sender
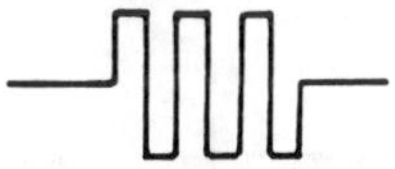	HEATING ELEMENT	Rear window heater
	BATTERY 12 volt	
	MEASURING GAUGE	Fuel gauge Temperature gauge
	SUPPRESSION WIRE	
	CLOCK	
	MECHANICAL PRESSURE SWITCH	Door switch for interior light

Fig. 7.35B Current flow diagram symbols (continued)

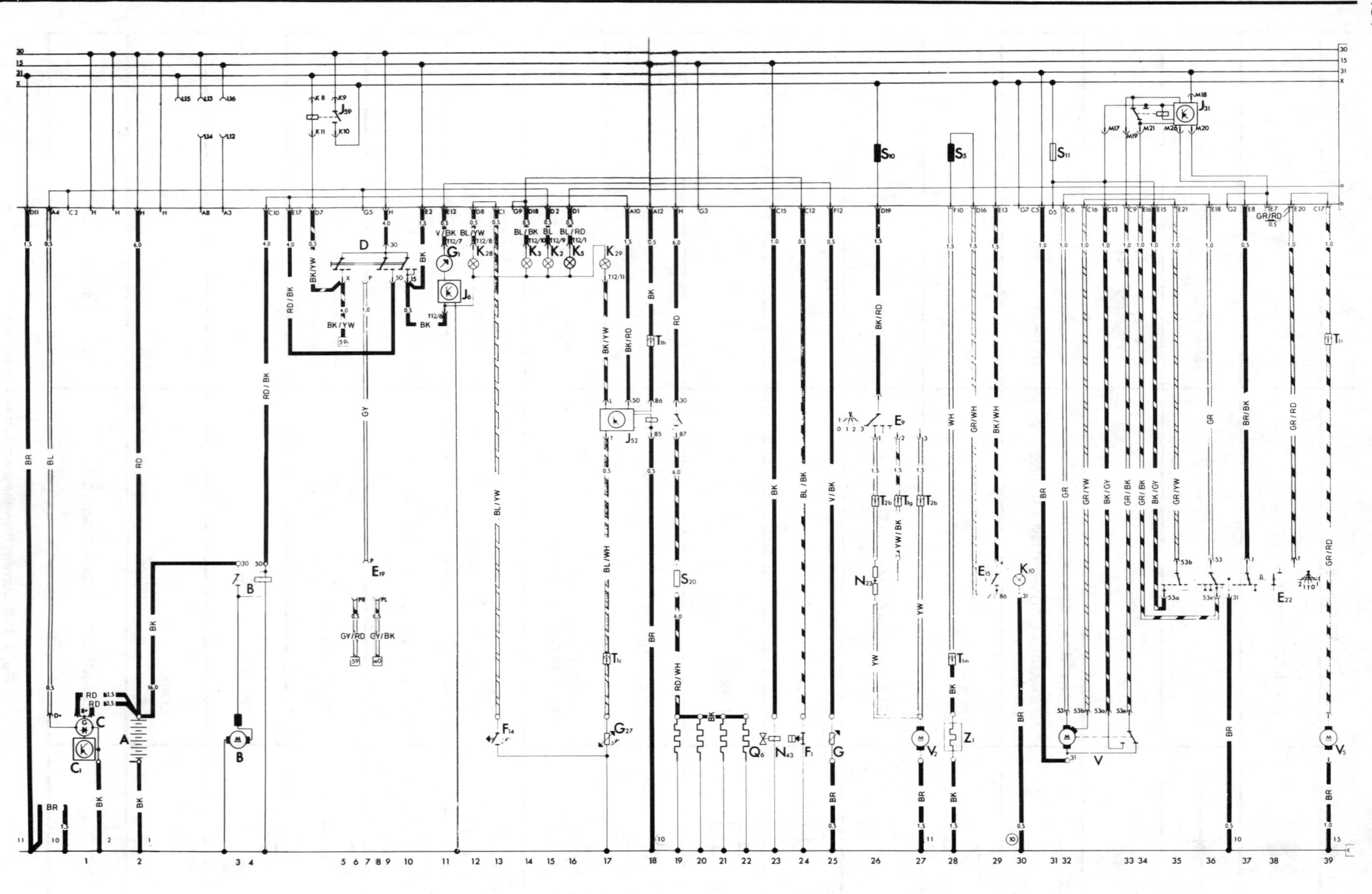

Fig. 7.36A Wiring diagram for 1978 Golf LD (see page 156 for key and page 175 for colour codes)

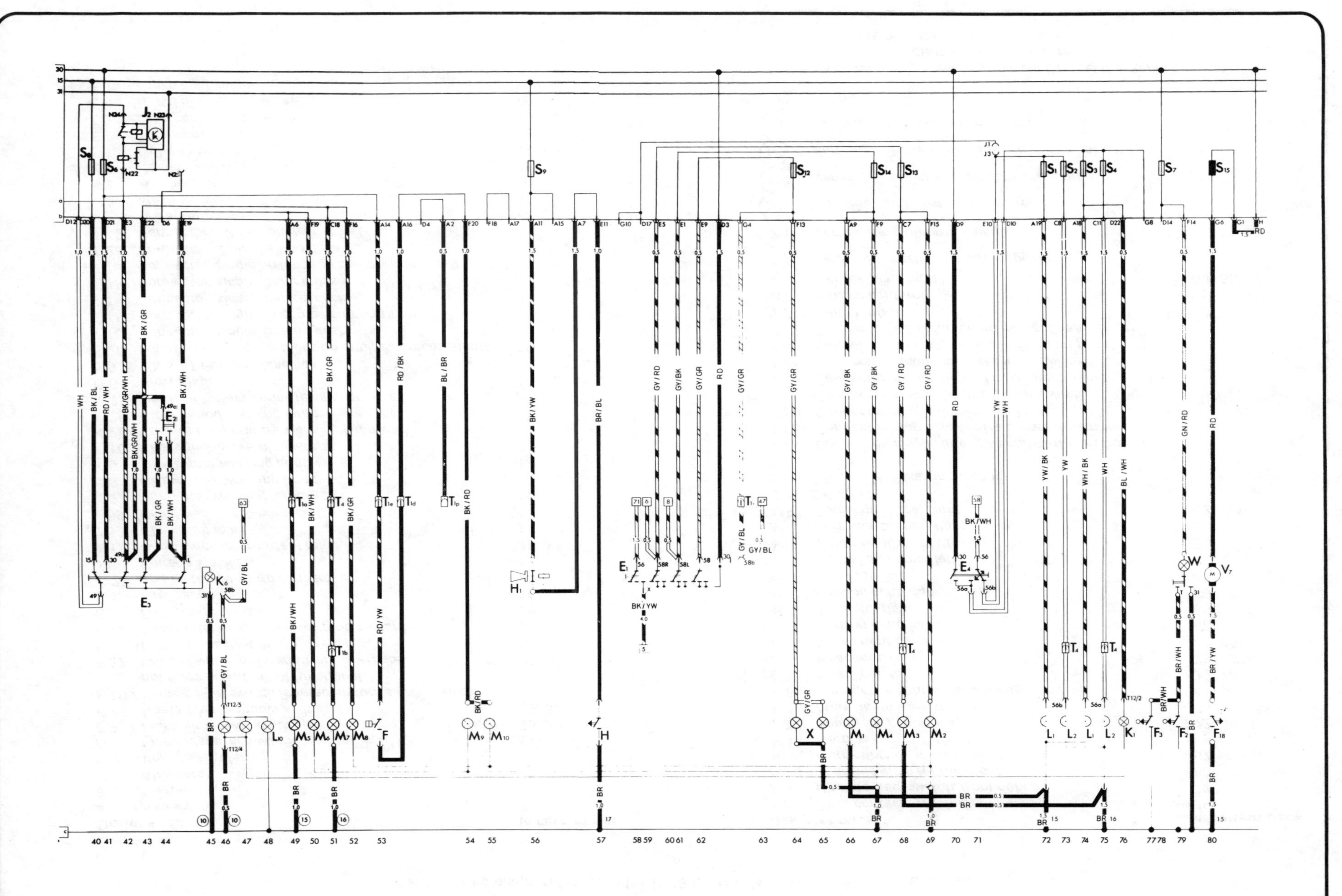

Fig. 7.36B Wiring diagram for 1978 Golf LD (continued) (see page 156 for key and page 175 for colour codes)

Key to wiring diagram Fig. 7.36 for 1978 Golf LD (see page 175 for colour codes)

Designation		in current track
A	*Battery*	2
B	*Starter*	3, 4
C	*Alternator*	1
C 1	*Voltage regulator*	1
D	*Ignition/starter switch*	5–10
E 9	*Fresh air blower switch*	26, 27
E 15	*Heated rear window switch*	29, 30
E 19	*Parking light switch (contact in turn signal switch not fitted – switch not functional)*	7, 8, 9
E 22	*Wiper switch for intermittent operation*	35–38
F 1	*Oil pressure switch*	24
F 14	*Coolant temperature – warning switch**	13
G	*Fuel gauge sender*	25
G 1	*Fuel gauge*	11
G 27	*Engine temperature sender**	17
J 6	*Voltage stabilizer*	11
J 31	*Wash-wipe intermittent relay*	35–36
J 52	*Glow plug relay*	17–19
J 59	*Relief relay for X-contact*	5
K 2	*Alternator warning lamp*	15
K 3	*Oil pressure warning lamp*	14
K 5	*Turn signal warning lamp*	16
K 10	*Rear window warning lamp*	30
K 28	*Coolant temperature warning lamp*	12
K 29	*Glow period warning lamp*	17
N 33	*Fresh air blower series resistance*	37
Q 6	*Glow plugs*	19–22
S 5, S 10, S 11	*Fuses in fuse box*	
S 20	*Fuse (50 A) for glow plugs (in engine compartment)*	19
T 1c	*Connector, single, behind dash*	
T 1f	*Connector, single, in engine compartment*	
T 1g	*Connector, single, behind dash*	
T 1h	*Connector, single, behind dash*	
T 1m	*Connector, single, in luggage compartment, left*	
T 2b	*Connector, 2 pin, behind dash*	
T 12	*Connector, 12 pin, on dash insert*	
V	*Wiper motor*	32–34
V 2	*Fresh air blower*	27
V 5	*Washer pump*	39
Z 1	*Heated rear window*	28
E 1	*Lighting switch*	58–62
E 2	*Turn signal switch*	44
E 3	*Emergency light switch*	40–46
E 4	*Headlight dip/flasher switch*	70–71
F	*Brake light switch*	53

Designation		in current track
F 2	*Door contact switch, front left*	79
F 3	*Door contact switch, front right*	77
F 18	*Thermoswitch for radiator fan*	80
H	*Horn control*	57
H 1	*Horn*	56
J 2	*Emergency light relay*	42–44
K 1	*Main beam warning lamp*	76
K 6	*Emergency light warning lamp*	45
L 1	*Headlight, left*	72, 74
L 2	*Headlight, right*	73, 75
L 10	*Instrument panel light*	46, 48
M 1	*Parking light, left*	66
M 2	*Tail light, right*	69
M 3	*Parking light, right*	68
M 4	*Tail light, left*	67
M 5	*Turn signal, front left*	49
M 6	*Turn signal, rear left*	50
M 7	*Turn signal, front right*	51
M8	*Turn signal, rear right*	52
S 1 – S 4, S 6 – S 9, S 12 – S 15	*Fuses in fusebox*	
T 1a	*Connector, single, in engine compartment front left*	
T 1b	*Connector, single, in engine compartment front right*	
T 1d	*Connector, single, in brake light switch*	
T 1e	*Connector, single, on brake light switch*	
T 1i	*Connector, single, behind instrument panel*	
T 1p	*Connector, single, on brake light switch*	
T 4	*Connector, 4 point, behind instrument panel*	
T 12	*Connector, 12 point, on instrument panel insert*	
V 7	*Radiator fan*	80
W	*Interior light front*	79
X	*License plate light*	64–65
(1)	*Earth wire, battery – body – engine*	
(2)	*Earth wire, generator – engine*	
(10)	*Earthing point, instrument panel*	
(11)	*Earthing point, steering column support*	
(15)	*Earthing point, engine compartment, front left*	
(16)	*Earthing point, engine compartment, front right*	
(17)	*Earth wire, steering box*	

* *F 14 and G 27 in one housing, On Golf LD the coolant temperature warning switch F 14 is replaced by a coolant temperature indicator sender.*

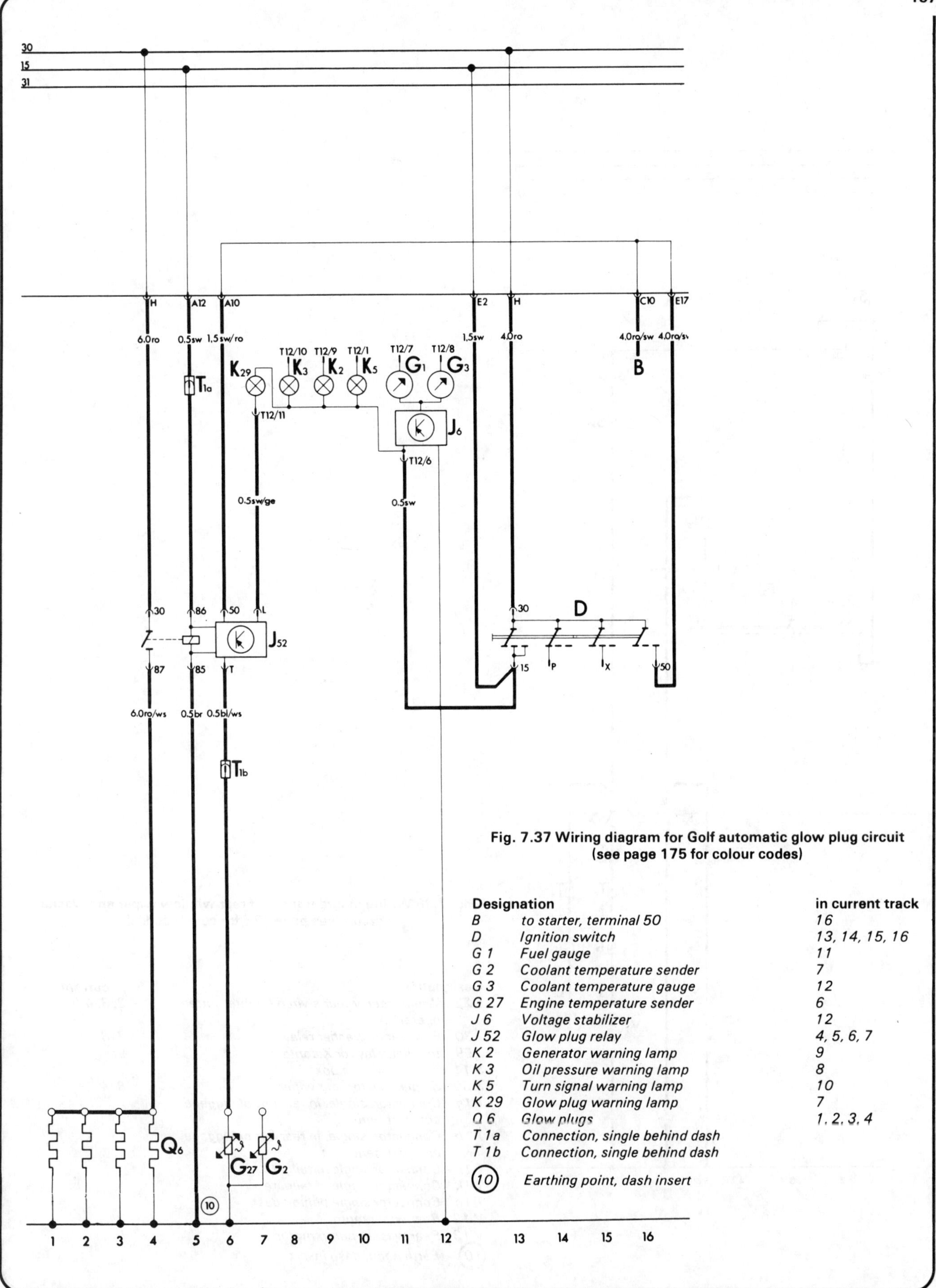

Fig. 7.37 Wiring diagram for Golf automatic glow plug circuit (see page 175 for colour codes)

Designation		in current track
B	*to starter, terminal 50*	*16*
D	*Ignition switch*	*13, 14, 15, 16*
G 1	*Fuel gauge*	*11*
G 2	*Coolant temperature sender*	*7*
G 3	*Coolant temperature gauge*	*12*
G 27	*Engine temperature sender*	*6*
J 6	*Voltage stabilizer*	*12*
J 52	*Glow plug relay*	*4, 5, 6, 7*
K 2	*Generator warning lamp*	*9*
K 3	*Oil pressure warning lamp*	*8*
K 5	*Turn signal warning lamp*	*10*
K 29	*Glow plug warning lamp*	*7*
Q 6	*Glow plugs*	*1, 2, 3, 4*
T 1a	*Connection, single behind dash*	
T 1b	*Connection, single behind dash*	
(10)	*Earthing point, dash insert*	

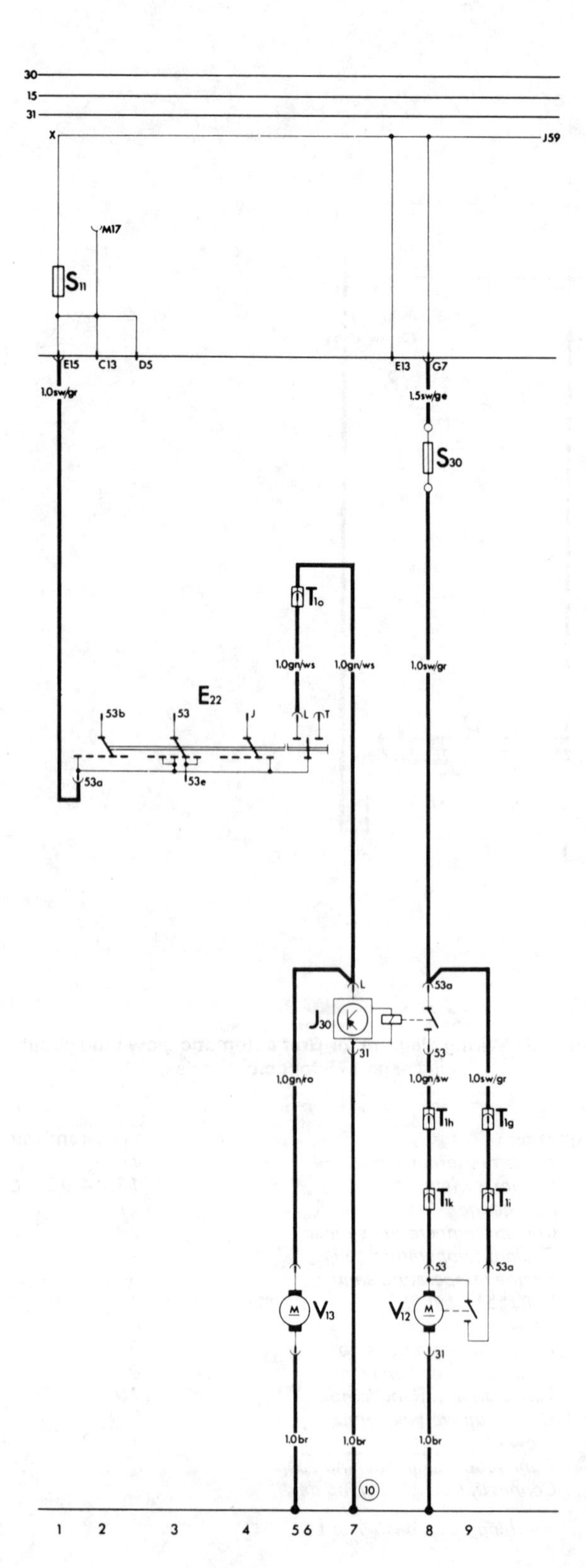

Fig. 7.38 Wiring diagram for Golf rear window wiper and washer circuit (see page 175 for colour codes)

Designation		in current track
E 22	*Windscreen wiper switch for intermittent operation*	*2, 3, 4,6*
J 30	*Rear wiper-washer relay*	*7, 8*
J 59	*to relief relay for X contact*	*9*
S 11	*Fuse 11 in fuse box*	*1*
S 30	*Single fuse for rear wiper*	*8*
T 1g	*Connector, single, in rear left of luggage compartment*	
T 1h	*Connector, single, in rear left of luggage compartment*	
T 1i	*Connector single in tailgate*	
T 1k	*Connector single in tailgate*	
T 1o	*Connector single behind dash*	
V 12	*Rear wiper motor*	*8, 9*
V 13	*Rear washer pump motor*	*5*
(10)	*Earth point, dash insert*	

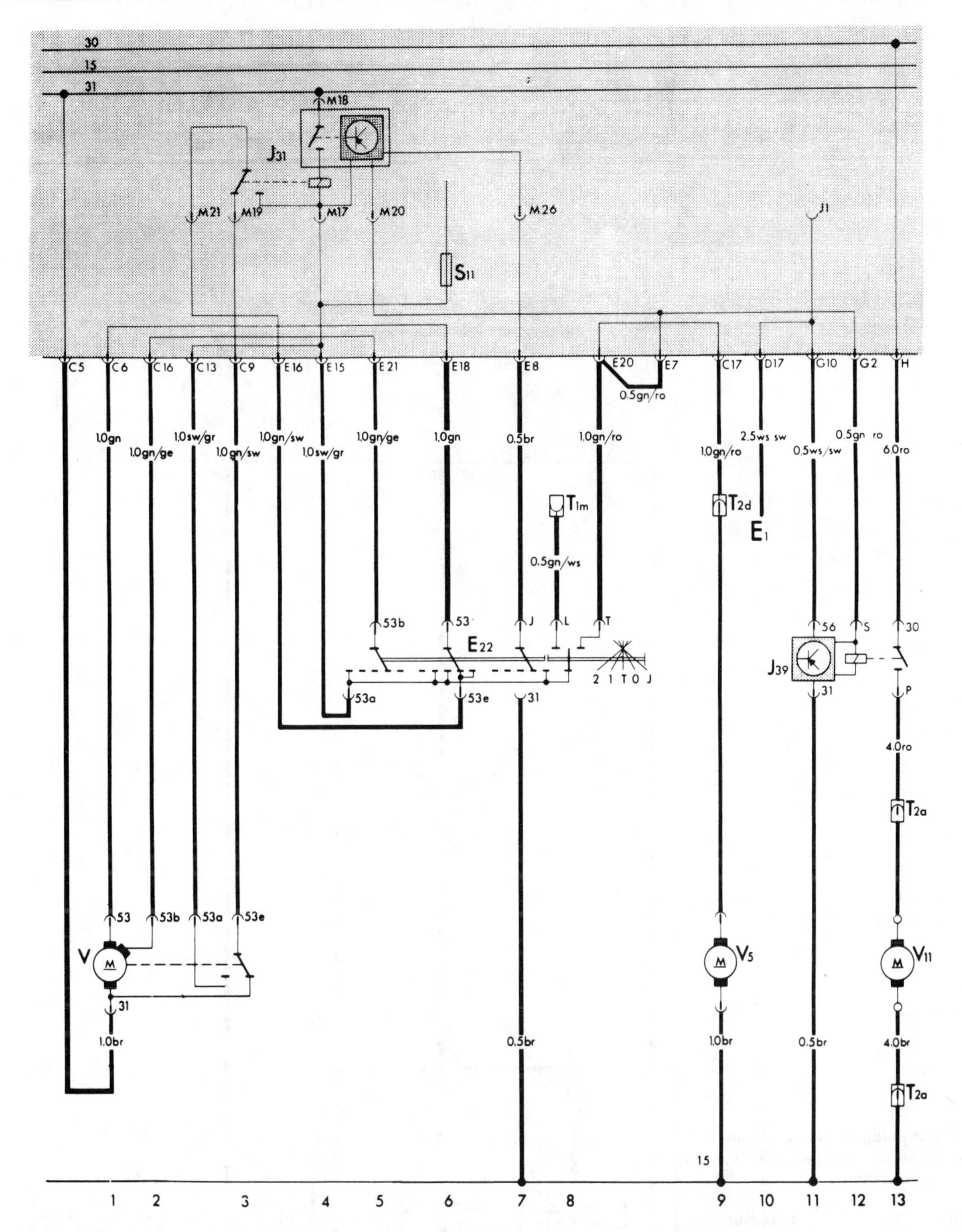

Fig. 7.39 Wiring diagram for Golf headlight washer intermittent facility (see page 175 for colour codes)

Designation		**in current track**
E 1	*to lighting switch*	*10*
E 22	*Wiper switch for intermittent operation*	*5 – 8*
J 31	*Wash/wipe intermittent relay*	*3 – 5*
J 39	*Headlight washer system relay*	*11 – 15*
S 11	*Fuse in fuse box*	*6*
T1m	*Connector, single, behind instrument panel*	
T2a	*Connector, two point, near the headlight washer pump*	
T2d	*Connector, two point, in engine compartment*	
V	*Windscreen wiper motor*	*1*
V5	*Windscreen washer pump motor*	*9*
V11	*Headlight washer pump*	*13*
(15)	*Earthing point in engine compartment*	

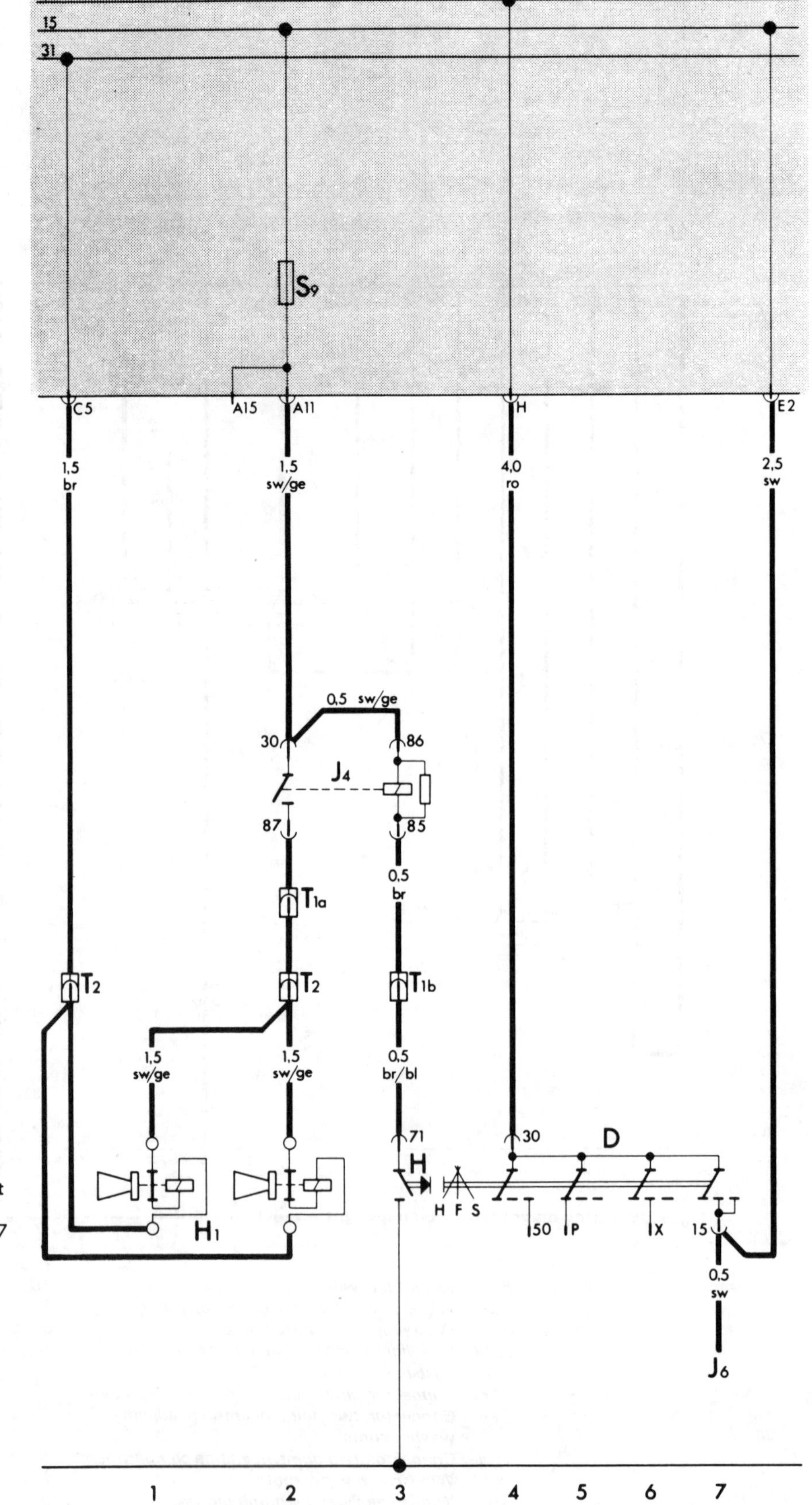

Fig. 7.40 Wiring diagram for Golf dual tone horn system (see page 175 for colour codes)

Designation		**in current track**
D	*Ignition/starter switch*	*4, 5, 6, 7*
H	*Horn plate*	*3*
H 1	*Horn*	*1, 2*
J 4	*Duel tone horn relay*	*2, 3*
J 6	*to voltage stabiliser*	*7*
S 9	*Fuse in fuse box*	*2*
T 1a	*Connector, single, behind instrument panel*	
T 1b	*Connector, single, behind instrument panel*	
T 2	*Connector, two point, near the horns*	

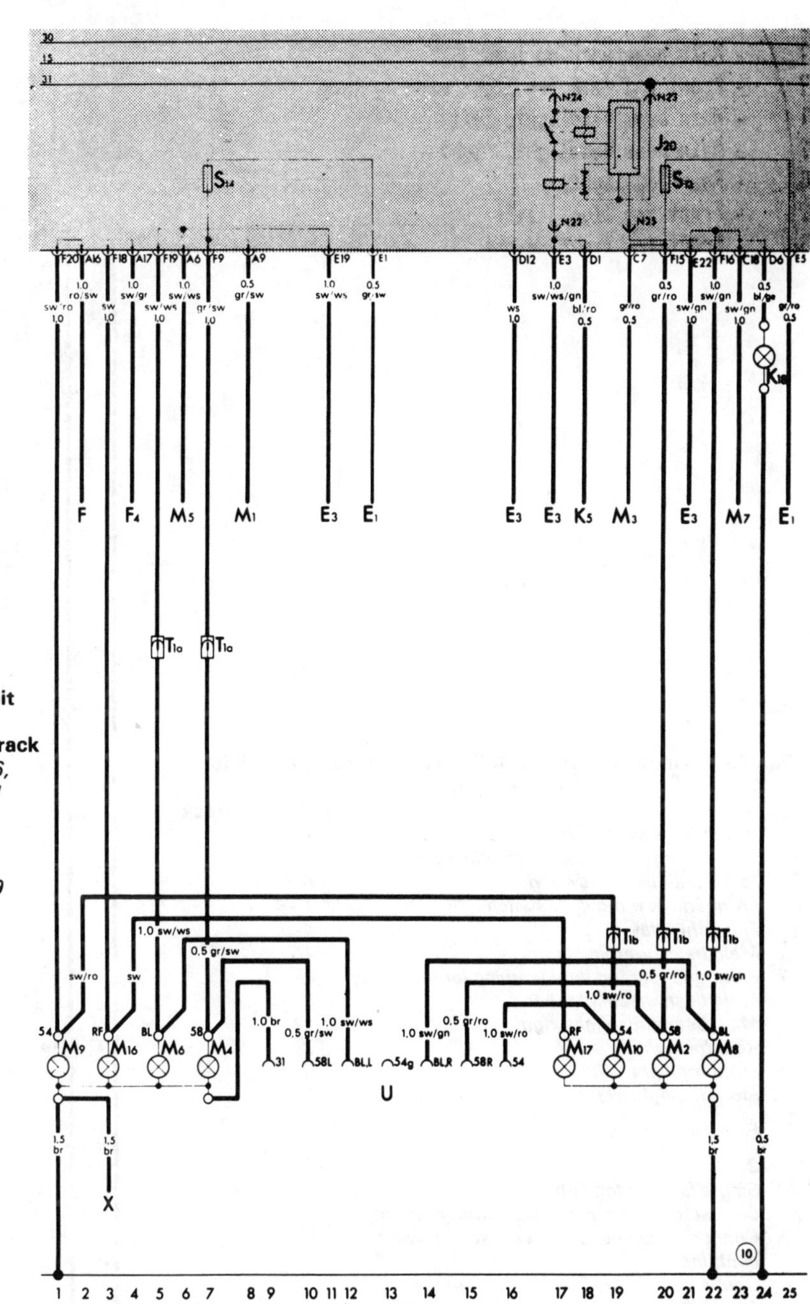

Fig. 7.41 Wiring diagram for Golf trailer circuit (see page 175 for colour code)

Designation		in current track
E3	*to emergency light switch*	*11, 16, 17, 21*
F	*to brake light switch*	*2*
F4	*to reversing light switch*	*4*
J20	*Turn signal/emergency light relay for trailer*	*17, 19*
K 5	*to turn signal warning lamp*	*18*
K 18	*Warning lamp for trailer*	*24*
M 1	*Side light, left*	*8*
M 2	*Tail light, right*	*20*
M 3	*Side light, right*	*19*
M 4	*Tail light, left*	*7*
M 5	*Turn signal, front left*	*8*
M 6	*Turn signal, rear left*	*5*
M 7	*Turn signal, front right*	*23*
M 8	*Turn signal, rear right*	*22*
M 9	*Brake light, left*	*1*
M 10	*Brake light, right*	*19*
M 16	*Reversing light, left*	*3*
M 17	*Reversing light, right*	*17*
S 13 / *S 14*	*Fuses in fuse box*	*7, 20*
T 1a	*Connector, single, in luggage compartment rear left*	
T 1b	*Connector, single, in luggage compartment rear right*	
U	*Trailer socket*	*9, 10*
X	*to number plate lights*	*3*
(10)	*Earthing point, instrument panel*	

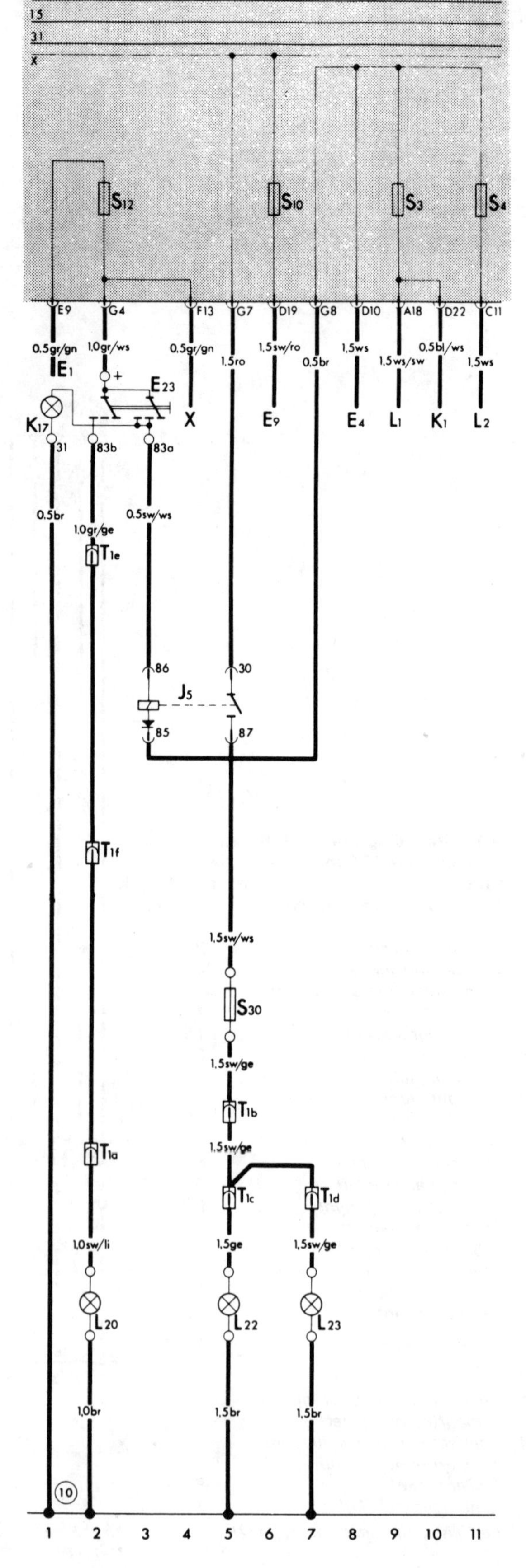

Fig. 7.42 Wiring diagram for Golf fog lights (see page 175 for colour codes)

Designation		**in current track**
E 1	*to lighting switch, terminal 58*	*1*
E 4	*to dimmer/flasher switch, terminal 56a*	*8*
E 9	*to fresh air blower switch*	*6*
E 23	*Front and rear fog light switch*	*1 - 4*
J 5	*Fog light relay*	*4, 5*
K 1	*Main beam warning lamp*	*10*
K 17	*Front and rear fog light warning lamp*	*1*
L 1	*Main beam headlight, left*	*9*
L 2	*Main beam headlight, right*	*11*
L 20	*Rear fog light*	*2*
L 22	*Front fog light, left*	*5*
L 23	*Front fog light, right*	*7*
S 3, 4	*Fuses in fuse box*	
S10, 12		*5*
S 30	*Single fuse for fog lights*	
T 1a	*Connector, single, in luggage compartment*	
T 1b	*Connector, single, under windscreen washer container*	
T 1c	*Connector, single, in engine compartment, front left*	
T 1d	*Connector, single, in engine compartment, right*	
T 1e	*Connector, single, behind instrument panel*	
T 1f	*Connector, single, behind instrument panel*	
X	*to number plate lights*	*3*
(10)	*Earthing point, instrument cluster*	

Key to wiring diagram Fig. 7.43 for 1977 Rabbit (see page 175 for colour codes)

Note: *Rabbit de Luxe petrol version shown. Use diagram in conjunction with wiring diagram Fig. 7.44 but note the following deletions and additions:*

Deletions
1 *Current track 3 to 10 (CIS equipment)*
2 *Current track 10 to 16 (Ignition system)*
3 *Diagnosis system*
4 *EGR components*

Additions
1 *Glow plug system*
2 *Safety belt warning system*
3 *Warning light terminal 15 to D20 now via glow plug relay J52 terminal 86*

Description		in current track
A	Battery	2
B	Starter	10-12
C	Alternator	1
C 1	Regulator	1
D	Ignition/starter switch	18-21
E 1	Light switch	68-70
E 2	Turn signal switch	51
E 3	Emergency flasher switch	47-50, 52, 53, 55
E 4	Headlamp dimmer/flasher sw.	66, 67
E 9	Fresh air fan switch	83
E 15	Rear window defogger switch	86
E 20	Instrument panel lights sw.	71
E 22	Washer/wiper intermitt. sw.	91-94
E 24	Safety belt switch, left	19
E 25	Safety belt switch, right	19
F	Brake light switch	38, 39
F 1	Oil pressure switch	30
F 2	Door sw./buzzer, front left	45
F 3	Door sw./buzzer, front right	43
F 4	Backup light switch	34
F 9	Park. brake warning light switch	28
F 18	Radiator fan thermo switch	96
F 26	Thermo-time sw. for start. valve	8,9
F 27	Elapsed EGR mileage switch	29
G	Fuel guage sending unit	32
G 1	Fuel gauge	22
G 2	Coolant temperature sending unit	31
G 3	Coolant temperature gauge	23
G 6	Electrical fuel pump	7
G 7	Connections for TDC sensor	24
H	Horn button	37
H 1	Horn	36
J 2	Emergency flasher relay	49,50
J 6	Voltage stabilizer	22
J 9	Rear window defogger relay	85
J 17	Electrical fuel pump relay	6-10
J 31	Washer/wiper intermitt. relay	89,90
J 34	Safety belt warning relay	17-21
K 1	Headlight high beam warning light	65
K 2	Alternator charging warning light	25
K 3	Oil pressure warning light	24
K 5	Turn signal warning light	26
K 6	Emergency flasher warning light	54
K 7	Dual circ. brake/safety belt warn. light	27,28
K10	Rear window defog. warning light	86
K 22	EGR warning light	29
L 1	Left headlight, high/low beam	61,63
L 2	Right headlight, high/low beam	62,64
L 10	Instrument panel light	71,72
L 15	Ashtray light	52
L 16	Heater lever light	55
L 28	Cigarette lighter light	50
M 1	Parking light, front left	76
M 2	Tail light, right	81
M 3	Parking light, front right	79
M 4	Tail light, left	78
M 5	Turn signal, front left	57
M 6	Turn signal, rear left	58
M 7	Turn signal, front right	59
M 8	Turn signal, rear right	60
M 9	Brake light, left	40
M 10	Brake light, right	41
M 11	Side marker lights, front	75,80
M 12	Side marker lights, rear	77,82
M 16	Backup light, left	35
M 17	Backup light, right	36
N	Ignition coil	14
N 6	Ballast resistor wire	14
N 9	Control pressure regulator	3
N 17	Cold start valve	8
N 21	Auxiliary air regulator	5
N 23	Speed control resistors for fresh air fan	83
O	Ignition distributor	14,15
P	Spark plug connectors	15,16
Q	Spark plugs	15,16
S	Fuses on fuse/relay panel S1, S2, S3, S4, S5, S6, S7, S8, S10, S11, S12, S13, S14, S15	
S 9	Fuse	33
S 20	Fuse (16A) on fuse/relay panel (on relay J 17)	4
Wire connectors		
T	engine compartment	
T 1a	single, behind dashboard	
T 1b	single, behind dashboard	
T 1c	single, behind dashboard	
T 1d	single, behind dashboard	
T 1e	single, trunk rear left	
T 1f	single, behind dashboard	
T 1g	Single, behind dashboard	
T 1h	single, behind dashboard	
T 1i	single, engine compartment, front left	
T 1k	single, engine compartment, front right	
T 1l	single, trunk, rear left	
T 1m	single, trunk, rear right	
T 2a	double, behind dashboard	
T 2b	double, behind dashboard	
T 2c	double, engine comp., front left	
T 2d	double, engine comp., front right	
T 2e	double, behind dashboard	
T 3a	3 point, behind dashboard	
T 4	4 point, behind dashboard	
T 12	12 point, on instrument cluster	
T 20	Diagnosis socket	22
U 1	Cigarette lighter	46
V	Windshield wiper motor	88,89
V 2	Fresh air fan	84
V 5	Windshield washer pump	95
V 7	Radiator cooling fan	96
W	Interior light	42
X	License plate light	73,74
Y	Clock	69
Z 1	Rear window defogger element	85
(1)	Ground strap, batt./body, engine	
(2)	Ground strap alternator to engine	
(10)	Ground connectors, instrument cluster	
(11)	Ground conn., steer. col. support	
(15)	Ground connectors, engine comp. front left	
(16)	Ground connectors, engine comp. front right	

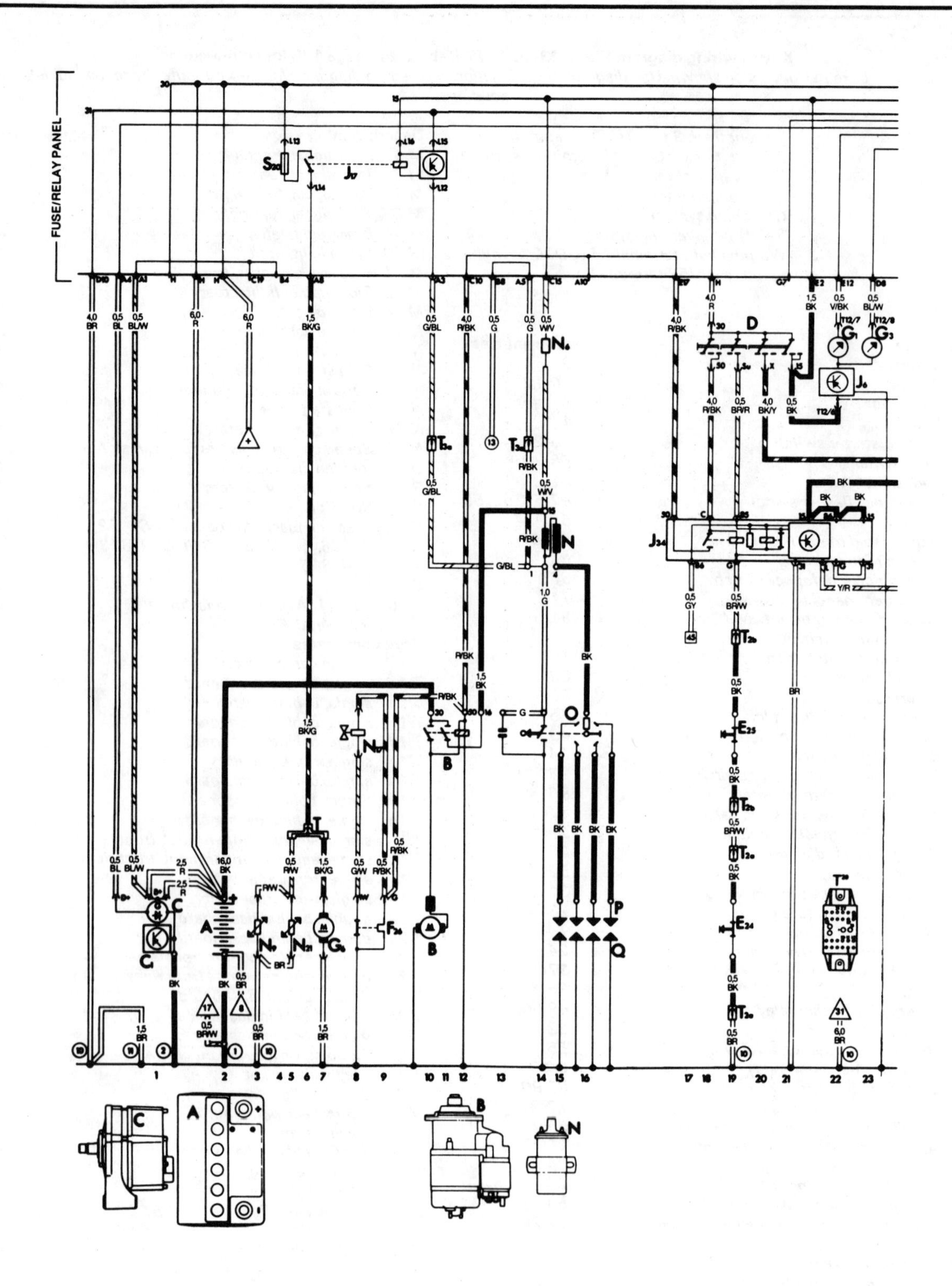

Fig. 7.43A Wiring diagram for 1977 Rabbit (see page 163 for key and page 175 for colour codes)

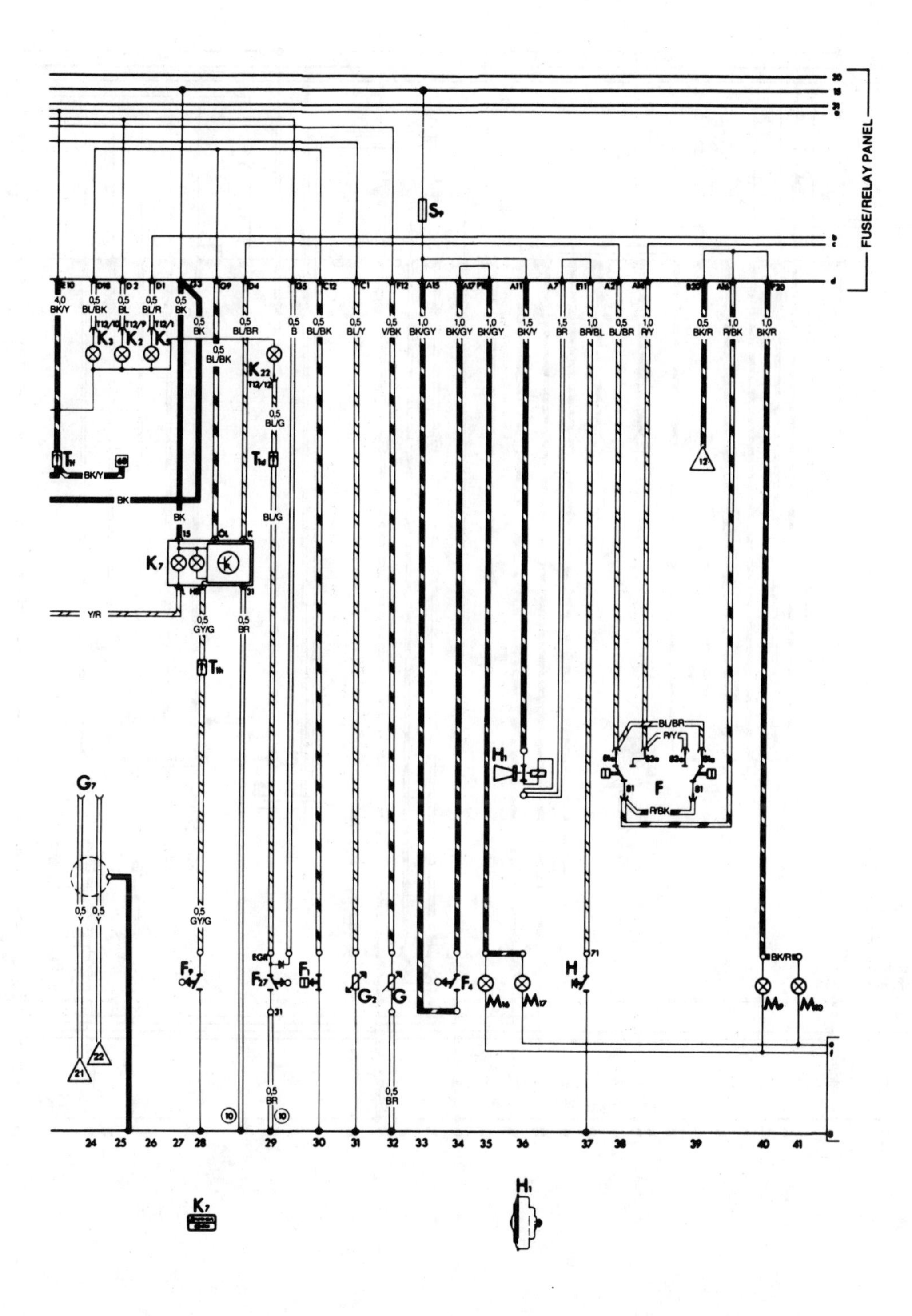

Fig. 7.43B Wiring diagram for 1977 Rabbit (continued) (see page 163 for key and page 175 for colour codes)

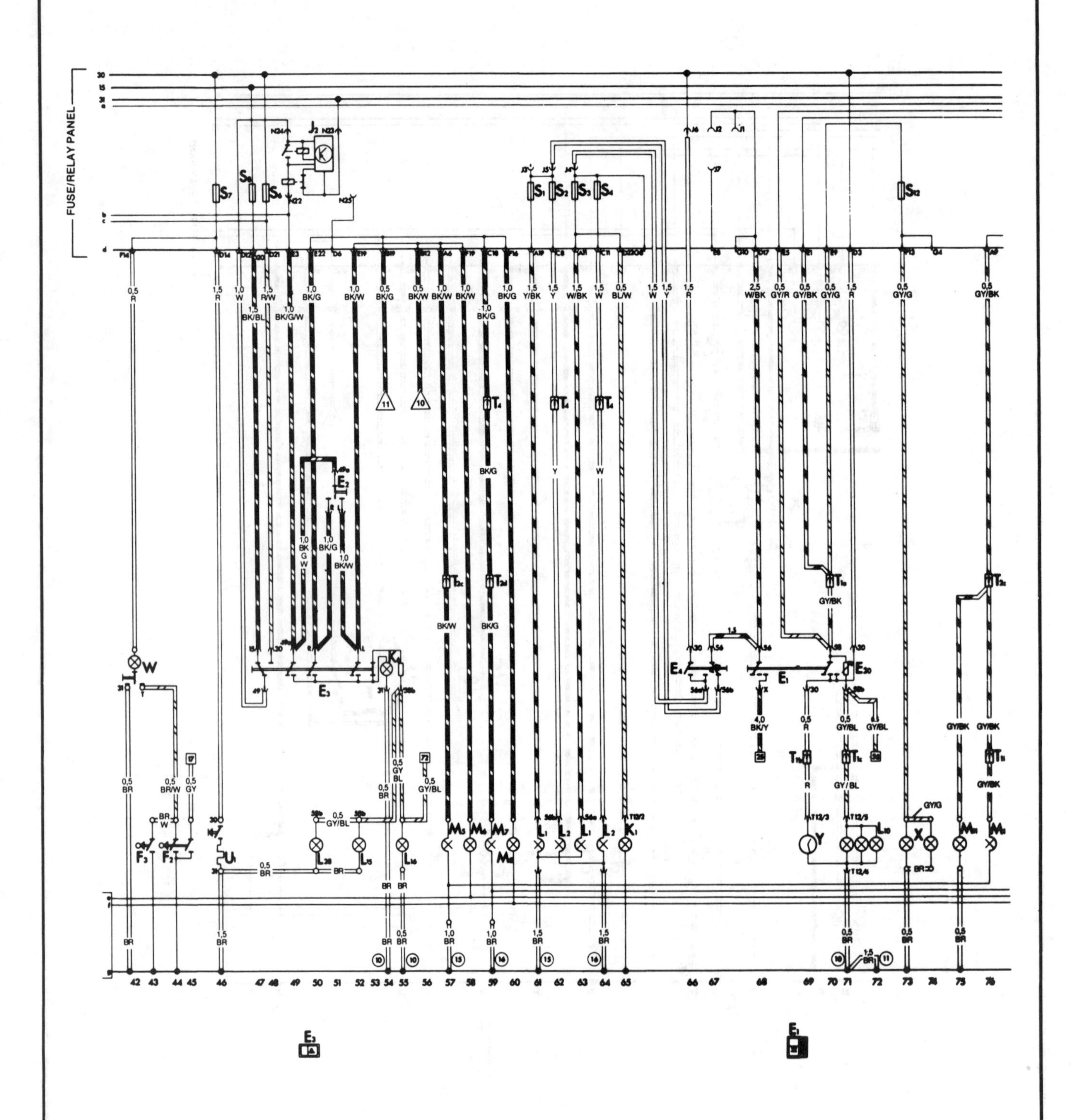

Fig. 7.43C Wiring diagram for 1977 Rabbit (continued) (see page 163 for key and page 175 for colour codes)

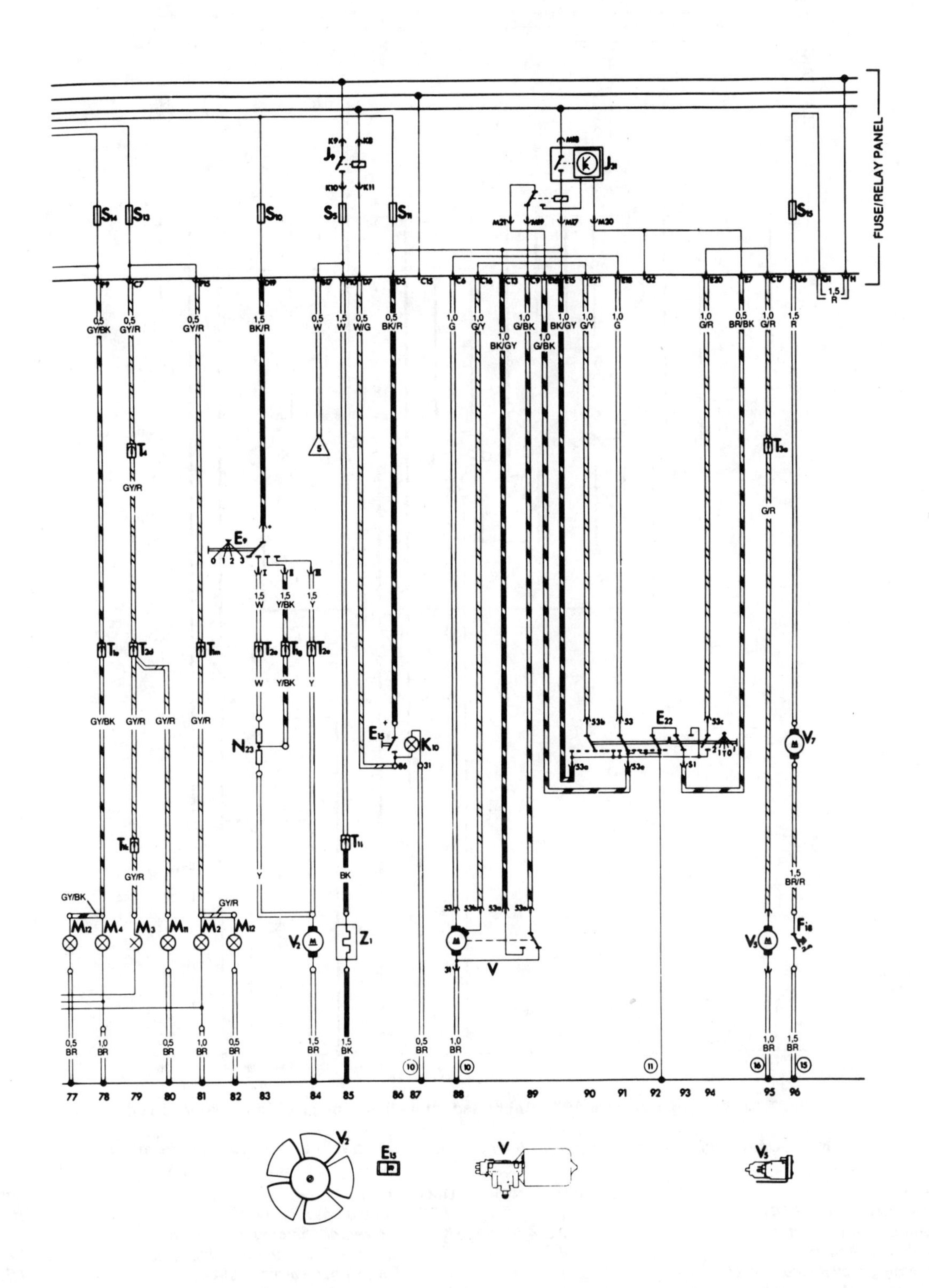

Fig. 7.43D Wiring diagram for 1977 Rabbit (continued) (see page 163 for key and page 175 for colour codes)

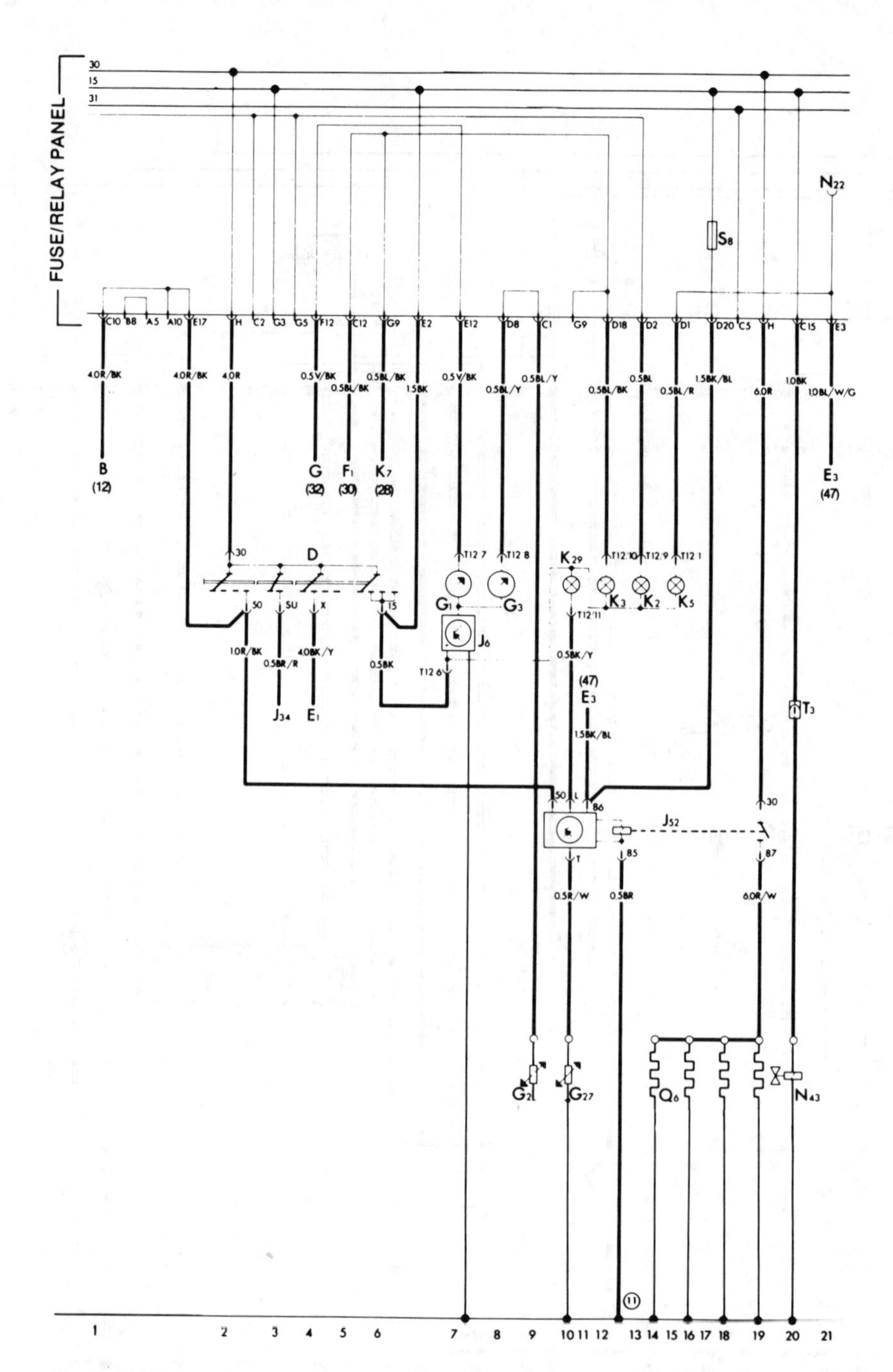

Fig. 7.44 Wiring diagram for 1977 Rabbit (additional) (see page 175 for colour codes)

Note: *Use in conjunction with Fig. 7.43 but note that current track numbers do not correspond*

Description		in current track
B	*to starter, terminal 50*	*1*
D	*ignition/starter switch*	*2,3,4,5,6*
E1	*to light switch, terminal X*	*4*
E3	*to emergency flasher switch*	*11,21*
F1	*to engine oil pressure switch*	*5*
G	*to fuel gauge sending unit*	*4*
G1	*fuel gauge*	*7*
G2	*Coolant temperature sending unit*	*9*
G3	*Coolant temperature gauge*	*8*
G27	*Engine oil temperature sending unit*	*10*
J 6	*Voltage stabilizer*	*7*
J 34	*to relay for safety belt warning system*	*3*

Description		in current track
J 52	*Glow plugs relay (with pre-glow sending unit)*	*10,12,19*
K2	*Alternator charging warning light*	*13*
K3	*Engine oil pressure warning light*	*12*
K5	*Turn signal warning light*	*15*
K7	*to dual circuit brake warning light*	*6*
K29	*Pre-glow warning light*	*10*
N43	*Fuel cut-off valve*	*20*
Q6	*glow plugs*	*14,16,18,19*
S8	*Fuse (in fuse box)*	*17*
T3	*Wire connector, 3 point; behind dashboard*	
T 12	*Wire connector, 12 point; on dashboard cluster*	
(11)	*Ground connector, steering column support*	

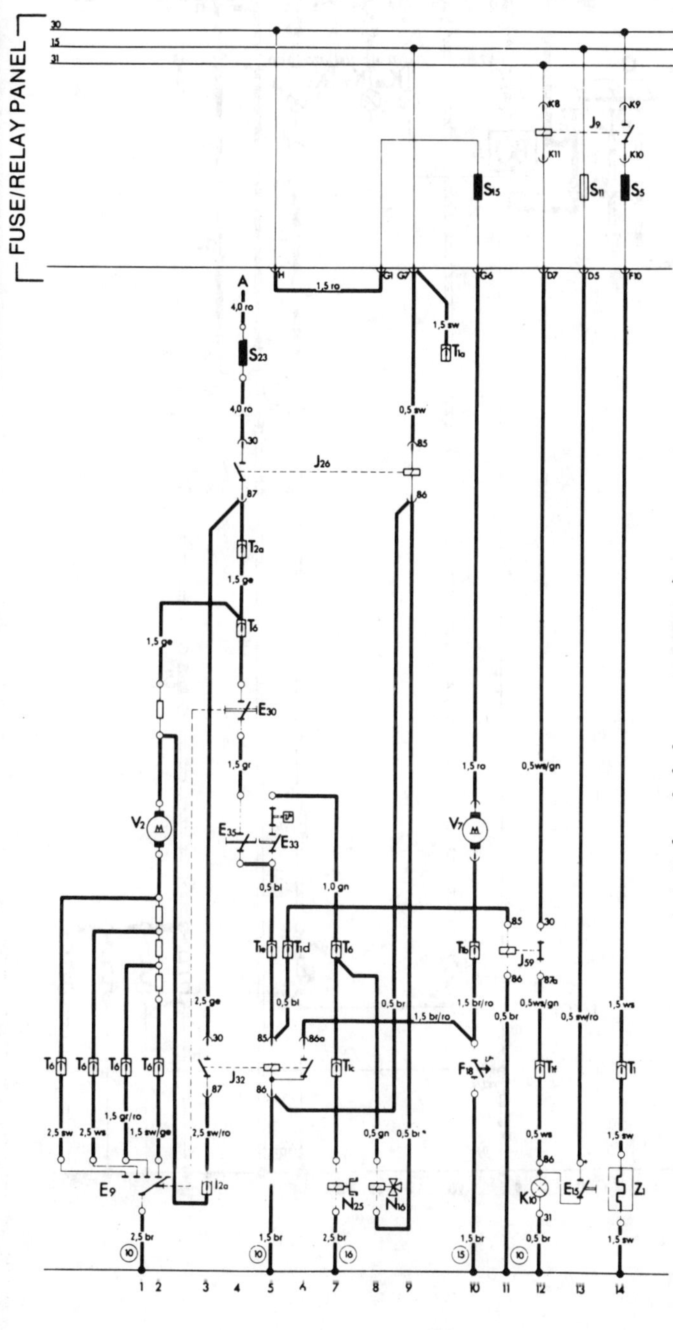

Fig. 7.45 Wiring diagram for air conditioner on 1978 Rabbit (see page 175 for colour codes)

Description		in current track
A	*to battery*	*4*
E9	*A/C speed control*	*1,2*
E15	*defogger switch*	*12,13*
E30	*A/C clutch switch (on speed control)*	*4*
E33	*temp. control*	*5*
E35	*A/C clutch switch (on temp.control)*	*4*
F18	*radiator fan thermo sw.*	*10*
J 9	*defogger relay*	*12-14*
J 26	*A/C on/off relay (on evap. housing)*	*4-9*
J 32	*A/C blower motor/rad. fan relay (on evap. housing*	*3-6*
J59	*defogger cut-out relay*	*11,12*
K10	*defogger warn. light*	*12*
N16	*two-way valve*	*8*
N25	*compressor clutch*	*7*
S5	*fuse (16A)*	*14*
S11	*fuse (8A)*	*13*
S15	*fuse (16A)*	*10*
S23	*A/C fuse (25A – on evap. housing)*	*3*

Wire Connectors

T1	*single, in trunk*	
T1 a	*single, vacant (on fuse/relay panel)*	
T1 b	*single, in engine comp. (left)*	
T1 c	*single, on compressor*	
T1 d	*single, on evaporator*	
T1 e	*single, behind dash (left)*	
T 1f	*single, behind dash (center)*	
T 2a	*double, behind dash (center)*	
T 6	*6 point, behind dash (center)*	
V 2	*A/C blower motor*	*2*
V 7	*radiator cooling fan*	*10*
Z 1	*rear window defogger*	*14*

(10) *grnd. conn., behind dash*

(15) *grnd. conn., in engine comp. (left)*

(16) *grnd. conn., in engine comp. (right)*

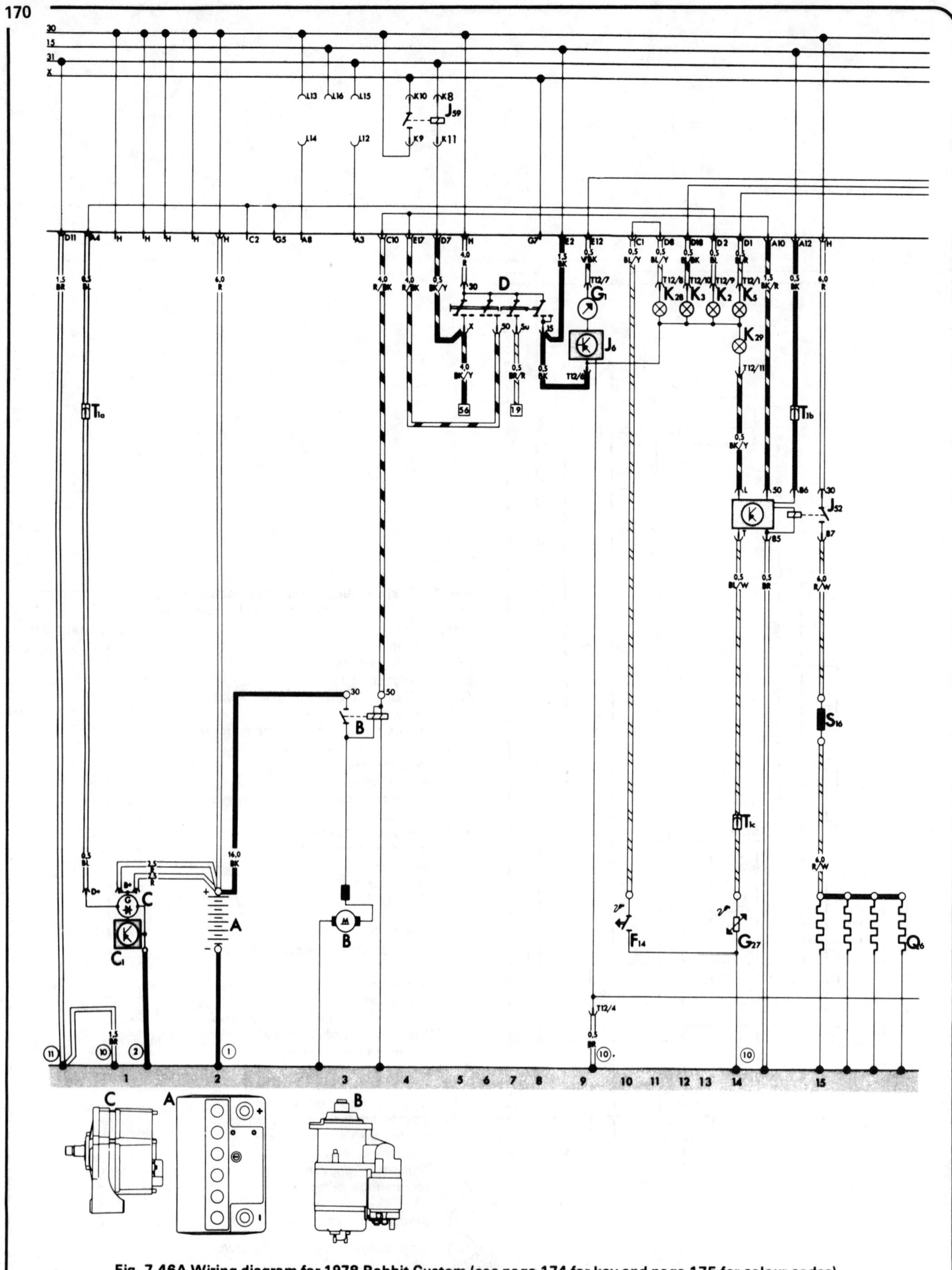

Fig. 7.46A Wiring diagram for 1978 Rabbit Custom (see page 174 for key and page 175 for colour codes)

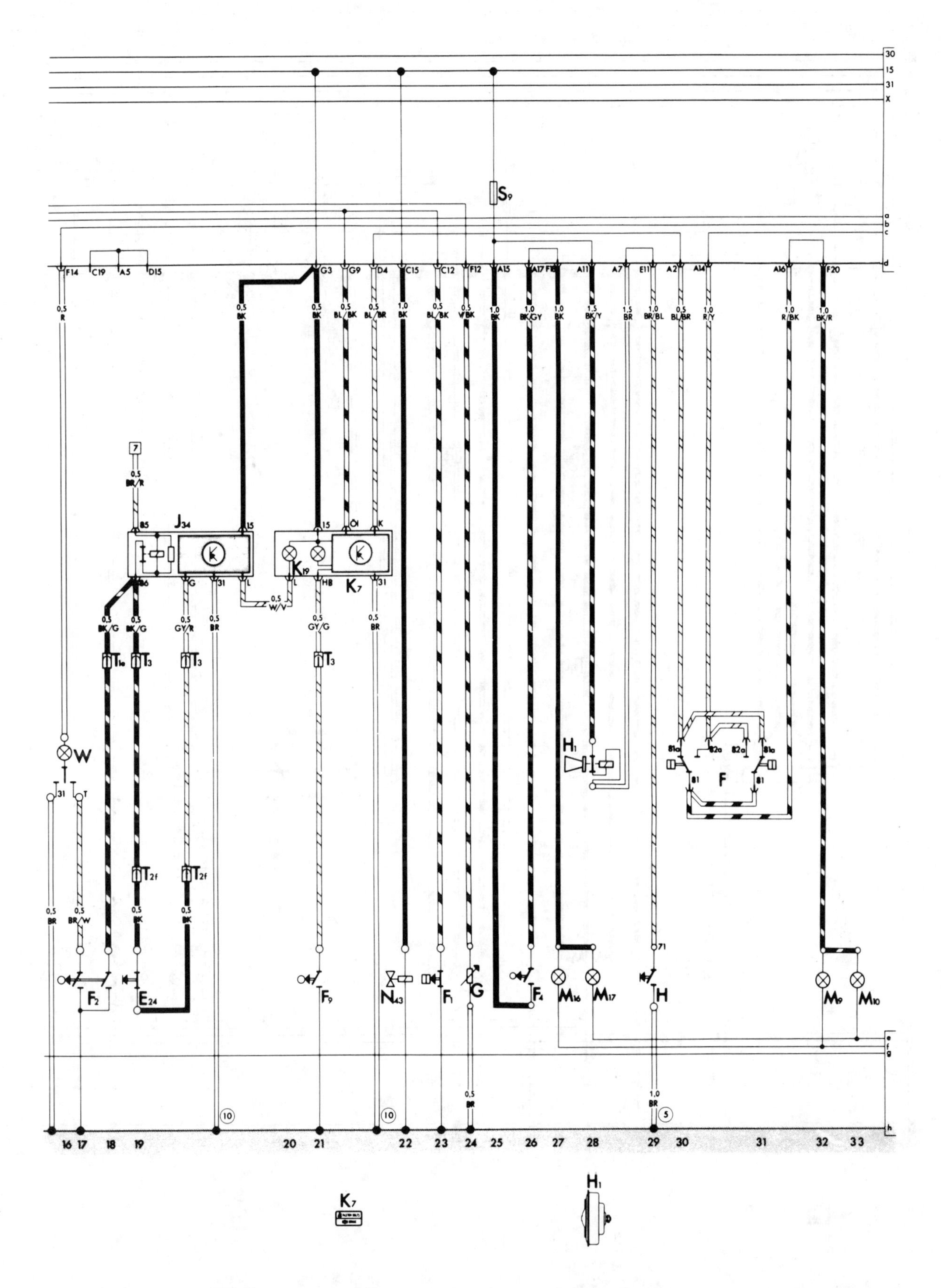

Fig. 7.46B Wiring diagram for 1978 Rabbit Custom (continued) (see page 174 for key and page 175 for colour codes)

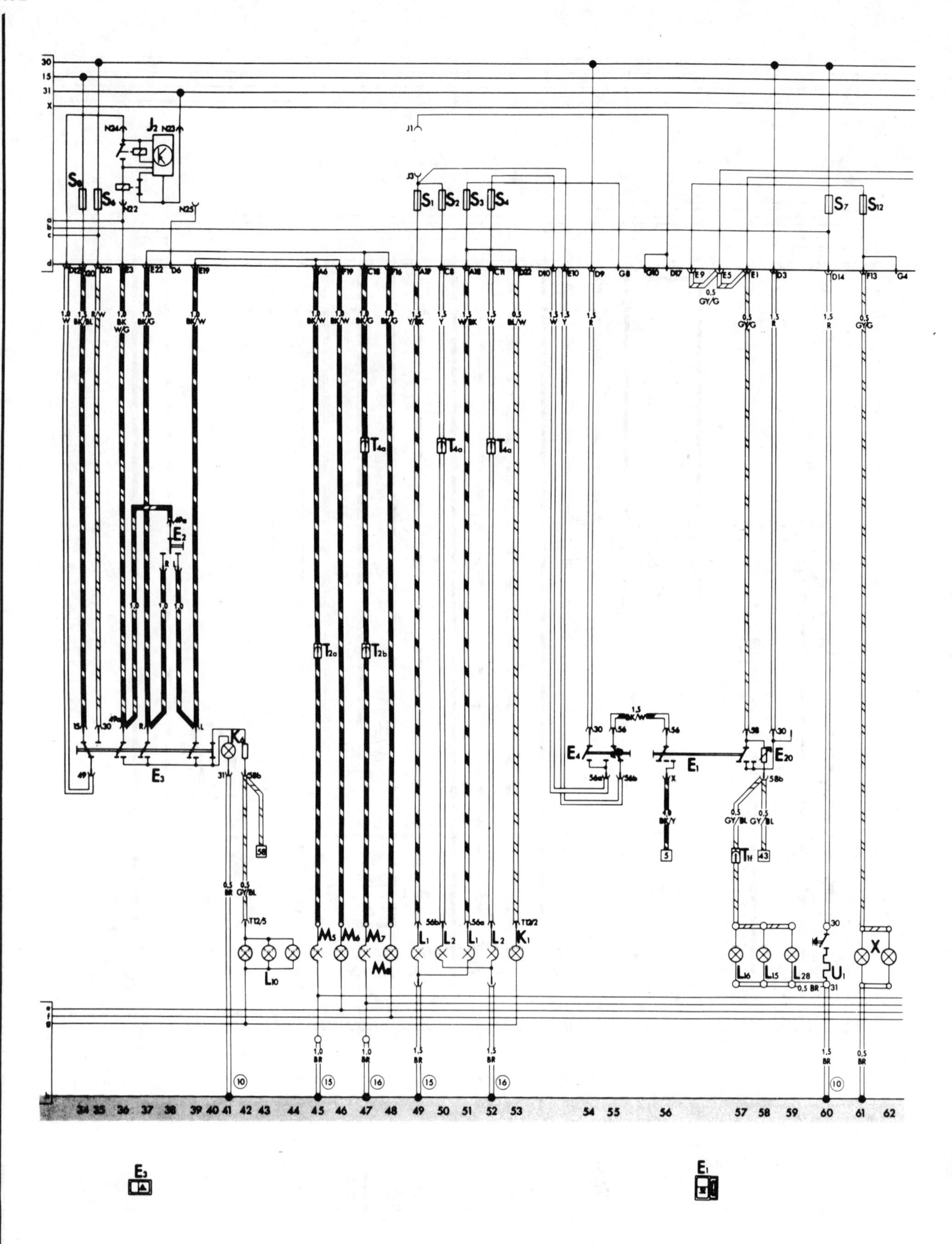

Fig. 7.46C Wiring diagram for 1978 Rabbit Custom (continued) (see page 174 for key and page 175 for colour codes)

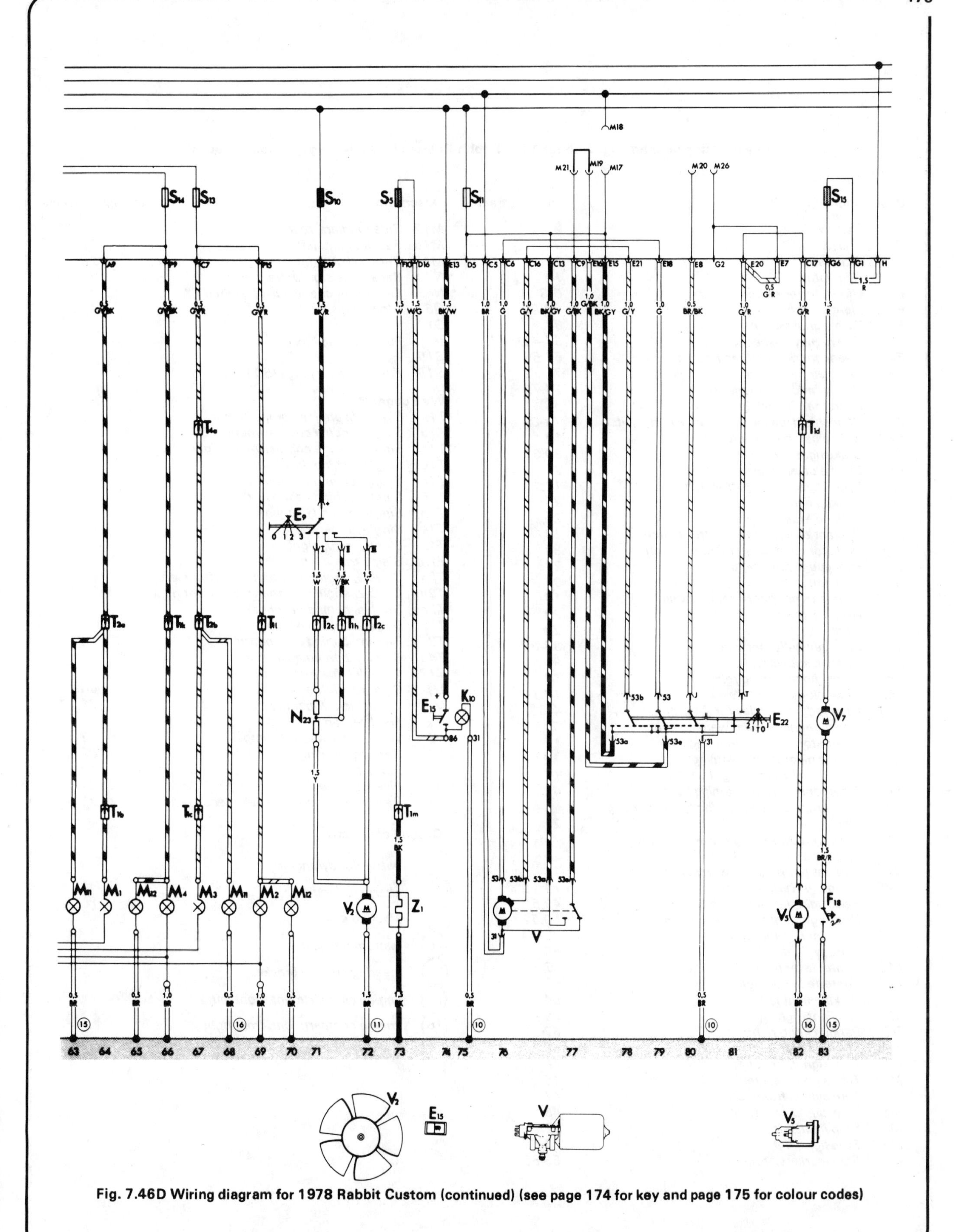

Fig. 7.46D Wiring diagram for 1978 Rabbit Custom (continued) (see page 174 for key and page 175 for colour codes)

Key to wiring diagram Fig. 7.46 for 1978 Rabbit Custom (see next page for colour codes)

Description		in current track
A	Battery	2
B	Starter	3
C	Alternator	1
C1	Regulator	1
D	Glow/starter switch	5-8
E1	Light switch	56-58
E2	Turn signal switch	38
E3	Emergency flasher switch	34-42
E4	Headlight dimmer/flasher switch	54,55
E9	Fresh air fan switch	71,72
E15	Rear window defogger switch	74,75
E20	Instrument panel light switch	58
E22	Windshield washer/wiper intermitt. switch	78-81
E24	Safety belt switch left	19
F	Brake light switch	30,31
F1	Engine oil pressure switch	23
F2	Door/buzzer switch, front left	17,18
F4	Backup light switch	26
F9	Parking brake warning light switch	21
F14	Coolant temperature control switch	10
F18	Radiator cooling fan thermo switch	83
G	Fuel gauge sending unit	24
G1	Fuel gauge	9
G27	Engine temperature sending unit	14
H	Horn button	29
H1	Horn	28
J2	Emergency flasher relay	36-38
J6	Voltage stabilizer	9
J34	Safety belt warning relay	19
J52	Glow plugs relay	14,15
J59	Load reduction relay for X-terminal	4-5
K1	Headlight high beam warning light	53
K2	Alternator charging warning light	13
K3	Engine oil pressure warning light	12
K5	Turn signal warning light	14
K6	Emergency flasher warning light	42
K7	Dual brake system/parking brake warning Light	21
K10	Rear window defogger warning light	75
K19	Safety belt warning light	20
K28	Coolant temperature warning light	11
K29	Pre-glow warning light	14
L1	Head light, left, low/high beam	49,51
L2	Head light, right, low/high beam	50,52
L10	Instrument panel light	42-44
L15	Ashtray light	58
L16	Heater lever light	57
L28	Cigarette lighter light	59
M1	Parking light left	64
M2	Tail light, right	69
M3	Parking light, right	67
M4	Tail light, left	66
M5	Turn signal, front left	45
M6	Turn signal, rear left	46
M7	Turn signal, front right	47
M8	Turn signal, rear right	48
M9	Stop light, left	32
M10	Stop light, right	33
M11	Side markers, front	63,68
M12	Side markers, rear	65,70
M16	Backup light, left	27
M17	Backup light, right	28
N23	Ballast resistor for fresh air fan	71
N43	Electro-magnetic cut-off switch	22
Q6	Glow plugs	15
S1 to S15	Fuses in fuse box	
S16	Fuse for glow plugs (50 amp)	

Wire connectors

T1a	Single, in engine compartment	
T1b	Single, engine compartment, left	
T1c	Single, engine compartment, right	
T1d	Single, behind dashboard	
T1e	Single, behind dashboard	
T1f	Single, behind dashboard	
T1h	Single, behind dashboard	
T1k	Single, trunk, left	
T1l	Single, trunk, right	
T1m	Single trunk, left	
T2a	Double, engine compartment, front left	
T2b	Double, engine compartment, front right	
T2c	Double, behind dashboard	
T2f	Double, behind dashboard	
T3	3-point, behind dashboard	
T4a	4-point, behind dashboard	
T12	12-point, instrument cluster	
U1	Cigarette lighter	60
V	Windshield wiper motor	76,77
V2	Fresh air fan	72
V5	Windshield washer pump	82
V7	Radiator cooling fan	83
W	Interior light	16
X	Licence plate light	61,62
Z1	Rear window defogger element	73

Ground connectors

(1)	battery/body/engine
(2)	alternator/engine
(5)	steering gear
(10)	instrument cluster
(11)	steering column bracket
(15)	engine compartment, front left
(16)	engine compartment, front right

Black	–	BK
Brown	–	BR
Red	–	R
Orange	–	O
Yellow	–	Y
Green	–	G
Blue	–	BL
Violet	–	V
Grey	–	GY
White	–	W

Wiring diagram colour code for Figs. 7.36, 7.43, 7.44 and 7.46

Where a wire has two or three colours, the main colour is given first, followed by the tracer colour(s)

Blue	–	bl
Brown	–	br
Yellow	–	ge
Green	–	gn
Grey	–	gr
Lilac	–	li
Red	–	ro
Black	–	sw
White	–	ws

Wiring diagram colour code for Figs. 7.37 to 7.42 and 7.45

Where a wire has two or three colours, the main colour is given first, followed by the tracer colour(s)

Chapter 8 Front suspension, steering and driveshafts

Refer to Chapter 11 for specifications and information applicable to later models

Contents

Specifications

Front suspension

Type	Independent, MacPherson strut, co-axial coil spring and shock absorber with wishbones
Shock absorbers	Hydraulic, telescopic, double acting
Steering knuckle balljoints – max play	2·5 mm (0·10 in)

Steering gear

Type	Rack and pinion with safety column
Overall ratio	17·37 : 1
Steering wheel turns, lock-to-lock	$3\frac{1}{2}$
Turning circle, between kerbs	10 metres (33 ft)

Steering alignment

Camber – wheels straight ahead	+ 20′ ± 30′
Max permissible difference between sides	1°
Castor	1° 50′ ± 30′ (not adjustable)
Toe angle (vehicle unladen. Negative angle denotes toe-out)	$-15'^{+10'}_{-15'}$

Driveshafts

	LH – solid	RH – tubular
Length	445 mm (17·5 in)	658 mm (25·9 in)

Torque wrench settings

	kgf m	lbf ft
Suspension		
Strut to wing securing nuts	2·0	14
Camber adjusting bolt nut	8·0	58
Strut to steering knuckle bolt	6·0	43
Steering balljoint clamp bolt	3·0	21
Wishbone pivot (front hinge) bolt	6·0	43
Wishbone U-clamp bolt	4·5	32
Driveshaft flange bolt	4·5	32
Hub nut	24·0	173
Shock absorber piston rod nut	8·0	58
Balljoint securing bolts	2·5	18
Steering		
UJ clamp bolt to steering column	3·0	21
UK clamp bolt to steering gear	3·0	21
Steering wheel nut	5·0	36
Tie-rod locknut	3·0	21
Tie-rod balljoint nut	3·0	21
Steering gear securing nuts	3·0	21
Steering column switch housing	1·0	7

1 General description

1 The suspension is simple and easily dismantled once the axleshaft nut is undone. This is tightened to a very high torque, and will present a problem for most owner drivers. The ways of overcoming this are discussed in Section 2, paragraph 5.
2 A straightforward MacPherson strut is housed in a bearing in the body and is located at the bottom by a balljoint attached to a wishbone bracket hinged to the frame. The roadwheel and brake are carried on a steering knuckle which is clamped to the bottom of the suspension strut and houses the wheel bearing. The top bolt of the clamp is eccentric and is used to adjust the camber angle. The steering knuckle is referred to throughout as the wheel bearing housing.
3 The drive to the roadwheel from the differential unit is supplied by a shaft with a constant velocity joint (CVJ) at each end. The inner end is flanged and secured by bolts to the final drive flange. The outer end is splined and passes into the splined inner part of the outer CVJ.
4 The CVJ outer race is an integral part of the short wheel axle which is splined into the hub. The hub is carried in a ball bearing in the wheel bearing housing and the whole wheel hub is held in place by the axle nut.
5 An integral part of the wheel bearing housing is a lever extending to the rear of the suspension strut. To this is fastened, by means of a balljoint, the tie-rod which is connected at the inner end to the steering gear. Thus as the tie-rod moves the suspension strut rotates, and with it the wheel is turned in a vertical plane, so steering the car.
6 The coil spring is mounted on a platform attached to the body of the shock absorber. The upper end of the spring is held by a cap and compressed slightly by a nut on the piston rod of the shock absorber. Thus as the spring is compressed the body of the piston moves up over the piston rod, and as the spring expands it pushes the body of the shock absorber down. In this way a controlled damping force is applied to the suspension.
7 The steering column is held in a tube which is fixed to the body. The column runs in a spacer sleeve at the top and a ball bearing at the bottom. A safety feature of the system is the universal joint (UJ) shaft which connects the bottom of the steering column to the steering gear. This is contained in a rubber cover. Besides catering for a change in direction of the drive, it ensures that should the steering gear be driven backwards in an accident such motion would not be transmitted to the steering column.
8 The lower end of the UJ shaft is clamped to the pinion shaft of the rack and pinion mechanism. The casing of this is bolted to the frame with clamps and rubber bushes. As the pinion turns the rack is moved to the right or left. Attached to each end of the rack shaft, by means of balljoints, are the tie-rods which transmit the steering motion to the front wheels. The tie-rods are easily removable. On new vehicles, the right-hand rod only is adjustable for length. Should a replacement for the left-hand rod be required only the adjustable type is available. This is discussed in Section 12.
9 The steering gear may be removed without disturbing the column or front wheels. Adjustment is simply both for wear and correct track.
10 Toe-in and camber angles are easily adjustable but the measurement of these angles requires special equipment and is best left to the VW agent. The castor angle is fixed. If the geometry is incorrect the tyres will wear rapidly. If such wear is apparent the owner is advised to have the geometry checked forthwith.
11 The text is written about LHD vehicles. Obviously the suspension does not differ for RHD but the steering gear differs, if only in the position of the pinion. Repair techniques are the same, and the photographs are of a RHD vehicle.

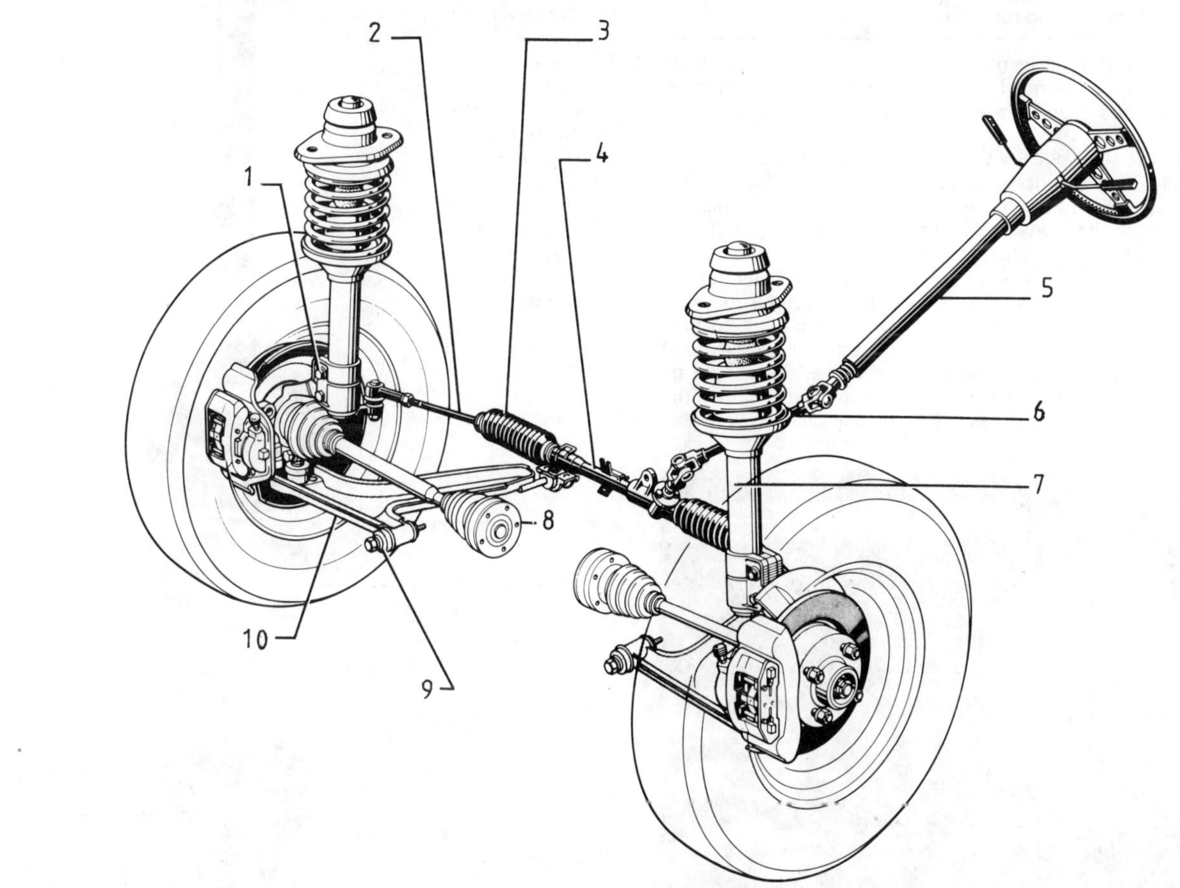

Fig. 8.1 Front suspension and steering layout (Sec 1)

1 Camber adjustment bolt
2 Tie-rod
3 Bellows
4 Steering gear
5 Steering column
6 Lower steering shaft
7 Suspension strut
8 Driveshaft
9 Bonded rubber bush (front)
10 Suspension wishbone

2 Maintenance

1 Apart from careful periodic inspection, maintenance is not necessary. All bearings and joints are prepacked and have no method of replenishment.
2 Check the state of wear of the tyre treads regularly and if there are signs of uneven wear, action, as described later, is necessary right away.
3 Measure the free-play of the steering wheel as soon as you acquire the vehicle. If this measurement subsequently increases, the reason must be located and corrected.
4 The rubber gaiters and covers must be checked for splits and age hardening. They may be renewed as required.
5 The problem of undoing the axle nut may be solved by using a large lorry type spanner but it should only be done while the car is standing on its wheels. The force required could easily pull the vehicle off an axle stand. It **must** be tightened to the correct torque so this will probably mean going to the agent if the hub has been dismantled. At the same time the steering geometry may be checked.
6 We have deliberately omitted any discussion on the setting of the steering geometry. This operation requires special gauges and extensive experience in using them. For those who have that experience and access to the gauges the necessary angles are given in the Specifications. For the remainder our strong recommendation is to get this job done by the VW agent.
7 With the reservations given in paragraphs 5 and 6 there is no reason why the owner cannot dismantle the steering and suspension and renew worns parts as required. The method of doing these jobs is discussed in the rest of the Chapter.

3 Driveshaft – removal and refitting

1 The driveshafts differ in length and construction. The right-hand shaft is longer than the left-hand shaft and is of tubular construction, while the shorter shaft is solid. In all other respects the shafts are the same.
2 Removal of the shaft does not present many problems. While the vehicle is standing on its wheels, undo the axle nut and slacken the roadwheel studs. Jack-up the front wheels and support the vehicle on axle stands. Remove the roadwheel and turn the steering to full lock, left lock for the left shaft and right for the right shaft.
3 Using an 8 mm Allen key remove the socket head bolts holding the inner CVJ to the final drive flange (photo). Be careful to use a proper key for if the socket head is damaged the result will be time consuming to say the least. These bolts are quite tight for the type of spanner used. Ideally a special hexagon 'screwdriver' bit should be used which will fit a torque wrench, but if one is not available then the bolts should be tightened as hard as possible on reassembly.

3.3 Driveshaft inner joint and final drive flange

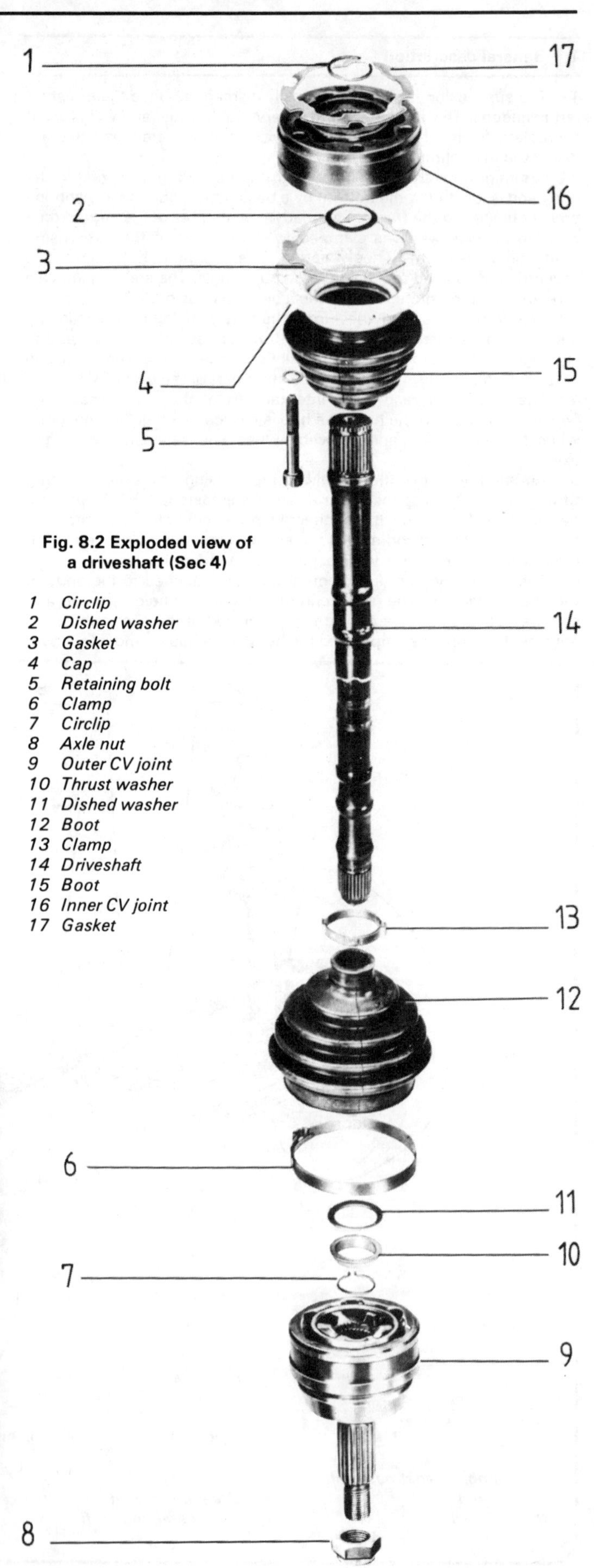

Fig. 8.2 Exploded view of a driveshaft (Sec 4)

1 Circlip
2 Dished washer
3 Gasket
4 Cap
5 Retaining bolt
6 Clamp
7 Circlip
8 Axle nut
9 Outer CV joint
10 Thrust washer
11 Dished washer
12 Boot
13 Clamp
14 Driveshaft
15 Boot
16 Inner CV joint
17 Gasket

3.4 Removing the driveshaft from the final drive flange

4 Once all the bolts are removed the CVJ may be pulled away from the final drive and the shaft removed from the hub (photo).
5 Refitting is the reverse of removal. Clean the splines carefully and smear a little molydbenum based grease on them. The problem of tightening the axle nut is again emphasised, and the nut fitted must be a new one.

4 Driveshaft – dismantling and reassembly

1 Having removed the shaft from the car it may be dismantled for the individual parts to be checked for wear.
2 The rubber boots, clips and thrust washers may be renewed if necessary but the CV joints may only be renewed as complete assemblies. It is not possible to fit new hubs, outer cases, ball cages or balls separately for they are mated to a tolerance on manufacture; it is not possible to buy them either.
3 Start by removing the outer joint (refer to Fig. 8.2). Cut open the 34 mm clip and undo the 88 mm hose clamp. The boot may now be pulled away from the joint. Open the circlip with a pair of circlip pliers and tap the side of the joint. The circlip should spring out of its groove.
4 The outer CV joint may now be removed from the shaft and set aside for examination. The boot, clamp, dished washer and thrust washer may be pulled off the end of the shaft. See Fig. 8.3.
5 If the boot of the inner CV joint is damaged it should be removed and refitted from the outer end and the CV joint on the inner end left in place. If however, the inner joint is suspected then this too must be removed. Use circlip pliers to remove the circlip from the end of the shaft and press off the CV joint. It will be necessary to ease the plastic cap away from the joint before pressing off the CV joint. The CV joint may now be set aside for inspection.
6 The CV joints should be washed clean and lightly oiled. Press the ball hub and cage out of the joint. Pivot the case through 90° to extract it. Examine the balls and ball tracks for scoring and wear. There should be a bright ring where the ball track runs round the joint but no further scoring. If the joint does not seem satisfactory take it to the VW agent for advice and possible renewal.
7 Assembly is the reverse of dismantling. Fit the inner CV joint first. It should be filled with 90 grams of molydbenum disulphide (MOS_2) grease, filling in equal amounts from each side. The grease is obtainable in packets from official agents. Pump the grease in while pressing the CV joint onto the shaft. Pull the joint on until a new circlip may be fitted. The outer diameter of the dished washer should now rest against the CV joint. Refit the plastic cover and the boot.
8 Push a new 34 mm diameter clip onto the shaft, then the boot for the outer CV joint and the boot clamp. Fit the dished washer so that the outer diameter rests against the spacer. Now fit the spacer with the convex side towards the joint.
9 Press a new circlip into the joint and drive the joint onto the shaft using a rubber hammer until the circlip seats in the groove. Pack the joint with 90 grams of MOS_2 grease and then refit the boot.
10 Refit the shaft as described in Section 3.

5 Suspension strut – removal and refitting

1 Refer to Fig. 8.4. There are several ways in which the strut may be removed, depending upon the reason for removal and the tools available. If it is possible to remove the components without disturbing the axle nut or the camber, then so much the better.
2 It is quite simple to remove the whole assembly and then remove the damaged parts and renew them. If only the coil spring or shock absorber are to be removed then turn to paragraph 9.
3 Slacken the wheel nuts and support the front of the car on axle stands. Remove the wheel. Remove the socket-head bolts connecting the inner CVJ to the flange and encase the CVJ in a polythene bag. Remove the front brake caliper and hang it up with a piece of wire taking care not to strain the hydraulic hose, refer to Chapter 6.
4 Remove the clamp bolt (photo) from the wheel bearing housing which holds the steering balljoint in place and separate the wishbone from the wheel bearing housing. Undo the tie-rod securing nut and using a puller break the tie-rod balljoint (photo).
5 Remove the two nuts holding the top of the suspension strut to the bodyframe inside the engine compartment (photo). Lower the strut from the frame and take it away complete with driveshaft and disc.
6 The coil spring must be held in compression while the top nut on the rod is removed. There is a special tool VW340 and VW340/5 to do this, see Fig. 8.5, but there are various types of spring compressor

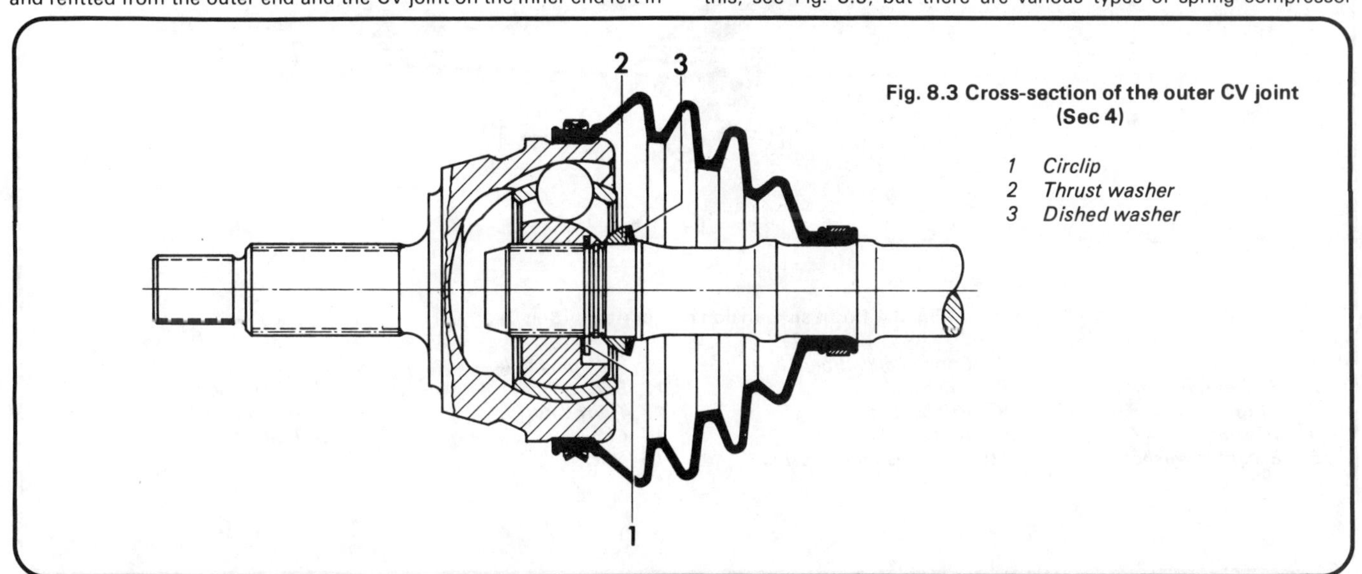

Fig. 8.3 Cross-section of the outer CV joint (Sec 4)

1 *Circlip*
2 *Thrust washer*
3 *Dished washer*

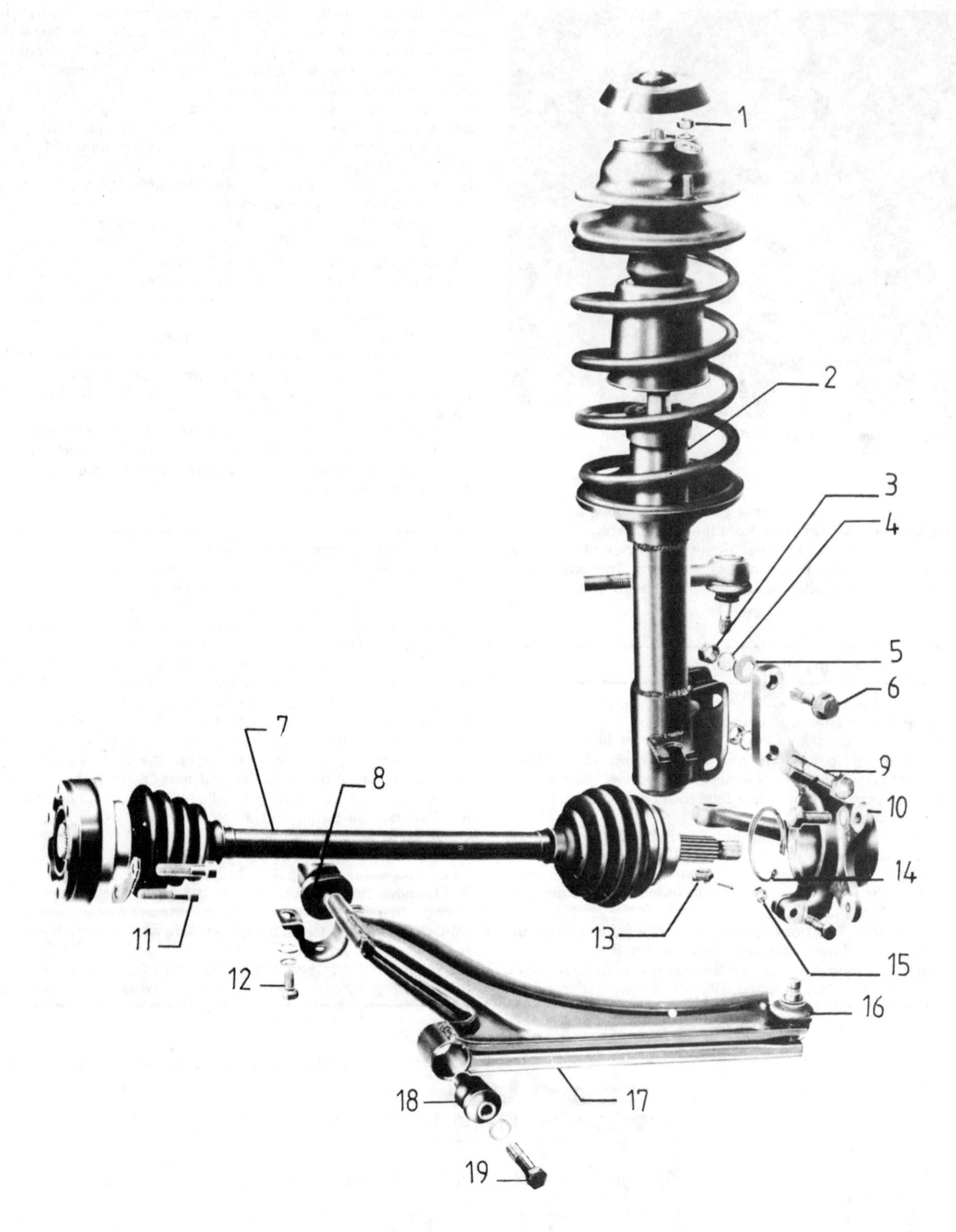

Fig. 8.4 Front suspension components (Sec 5)

1 *Nut*
2 *Suspension strut*
3 *Nut*
4 *Washer*
5 *Eccentric washer*
6 *Camber adjustment bolt*
7 *Driveshaft*
8 *Rubber bush*
9 *Bolt*
10 *Wheel bearing housing*
11 *Retaining bolt*
12 *Bolt*
13 *Tie-rod nut*
14 *Circlip*
15 *Nut*
16 *Ball joint*
17 *Wishbone*
18 *Rubber bush*
19 *Bolt*

5.4A Steering balljoint and clamp bolt location

5.4B Tie-rod end and retaining nut location (right-hand side) and driveshaft outer bellows

5.5 Suspension strut upper mounting

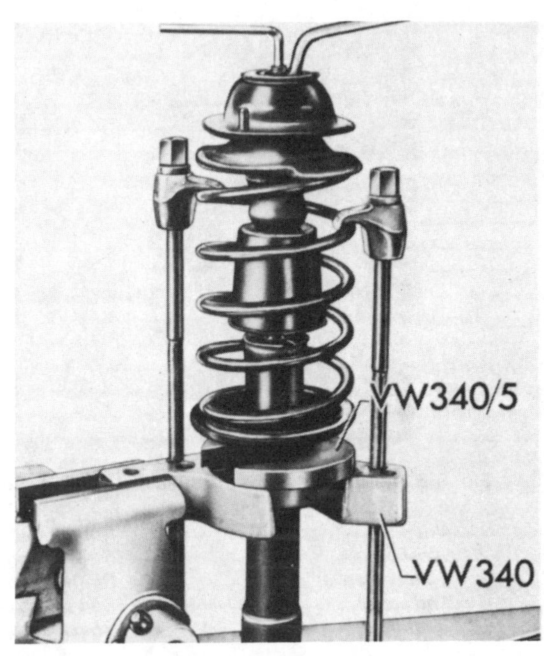

Fig. 8.5 Compressing the front spring with the special tool (Sec 5)

tools which will do the job. The nuts on the compressor tool should be tightened a little at a time on each side until the tension on the piston rod is relieved. The piston rod must be held tight while undoing the nut. There are flats formed on the end of the piston rod which can be held with a spanner. Fig. 8.6 shows details of the suspension strut.

7 Remove the coil spring and test the shock absorbers as in Section 7.

8 If the shock absorber is to be removed then the camber angle setting will be disturbed. To be able to reassemble the strut with the correct camber angle proceed as follows. Locate the top bolt holding the shock absorber clamp to the wheel bearing housing. This is an eccentric bolt (photo). Mark the relative position of the bolt head to the clamp by two centre-punch marks. Undo the camber adjusting bolt and the lower bolt and remove them. Open the clamp with a lever and extract the shock absorber.

9 It is possible to extract the strut without undoing the driveshaft or wishbone. Remove the bolts from the clamp as in paragraph 8, disconnect the top of the strut from the body as in paragraph 5, prise the clamp apart and push the wheel bearing housing off the strut. Lower the strut and remove it.

10 Assembly is the reverse of removal. Refit the shock absorber and coil spring to the wheel bearing housing. The spring must be compressed to start the securing nut. Tighten the nut to the specified torque. Adjust the camber eccentric correctly. Push the top of the piston rod through the wing into the engine compartment and secure it correctly.

11 Refit the steering balljoint and wishbone to the wheel bearing housing, refit the tie-rod balljoint and then refit the driveshaft to the final drive.

12 Refit the brake caliper, refer to Chapter 6. Refit the roadwheel and lower the car to the ground.

13 If the camber adjusting eccentric bolt has been disturbed, it is advisable to have the camber angle checked at the earliest opportunity to ensure that the bolt has been fitted correctly.

6 Coil springs – inspection and renewal

1 There is not a lot to go wrong but even so the spring may require renewal. If it is broken read Chapter 9, Section 8, paragraphs 2 and 3 and examine the whole suspension very carefully.

2 If a knocking noise is heard from the front suspension when travelling over rough ground then the end coils are bottoming on the next coils. Read Chapter 9, Section 8, paragraph 4 but this time fit the damping sleeve to the top **and** bottom coils.

3 There are six tolerance groups for the coil spring, each group denoted by paint spots. If a spring is faulty make sure the new one is from the same tolerance group. Consult the VW storeman if in doubt about the marking.

4 The method of removal and refitting is given in Section 5.

7 Shock absorbers – inspection and renewal

1 A defective shock absorber will upset the steering and braking so it should be attended to right away. Apart from steering and braking difficulties it usually complains audibly as it operates.

2 A quick and reliable test is done by pushing the car down on the front corner and releasing the pressure suddenly. The car should run to its correct level without oscillation. Any oscillation indicates that the shock absorber is empty.

3 The shock absorber, removed from the suspension strut, should be held by the cylinder body in a vice. Push the piston rod right down and then pull it right out. There should be a firm, even resistance to movement throughout the stroke in both directions. If there is not then the unit is defective. If you are not sure take it to the VW agent and ask for it to be tested against a new one.

4 There is no repair, only renewal. It is not necessary to renew both shock absorbers if only one is faulty but it **must** be replaced by one of the same part number.

5 The method of removal and refitting is given in Section 5.

8 Wishbones – inspection, removal and refitting

1 The balljoint is attached to the wishbone by three 7 mm rivets. Renewal of the balljoint is by drilling out the rivets with a 6 mm drill and attaching the new balljoint with three 7 mm bolts and nuts, tightened to the specified torque.

2 Apart from fitting a new balljoint the pivot bushes may be replaced

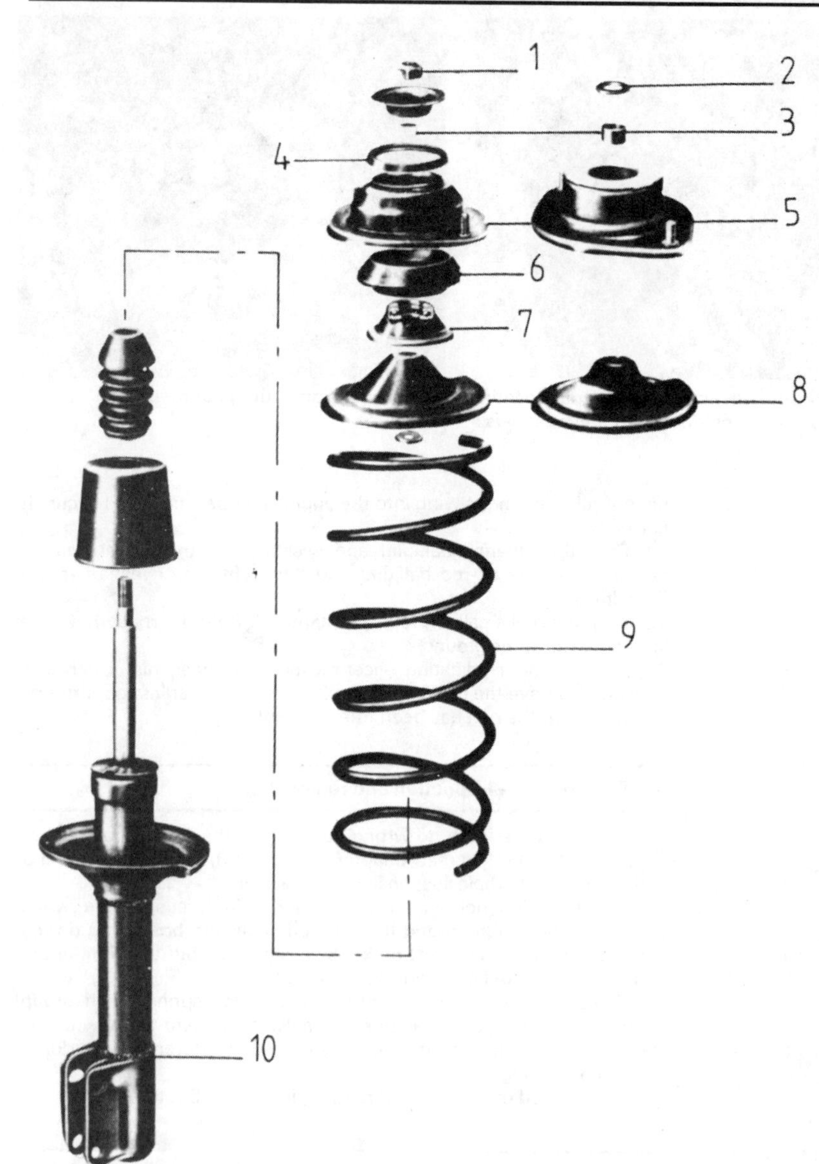

Fig. 8.6 Exploded view of the suspension strut (Sec 5)

1	*Nut*	*6*	*Rubber damper*
2	*Washer*	*7*	*Strut bearing*
3	*Spacer sleeve*	*8*	*Spring seat*
4	*End collar*	*9*	*Coil spring*
5	*Bearing*	*10*	*Shock absorber*

5.8 Wheel bearing housing upper retaining bolt, showing eccentric for adjusting camber angle

if worn. To do this properly the wishbone must be removed from the vehicle. This is discussed below.

3 Accident damage to the wishbone may result in the camber angle being altered and rapid tyre wear. Check the underside for kinks and wrinkles and check that the front and profile sides are straight, using a straight-edge or an engineer's square.

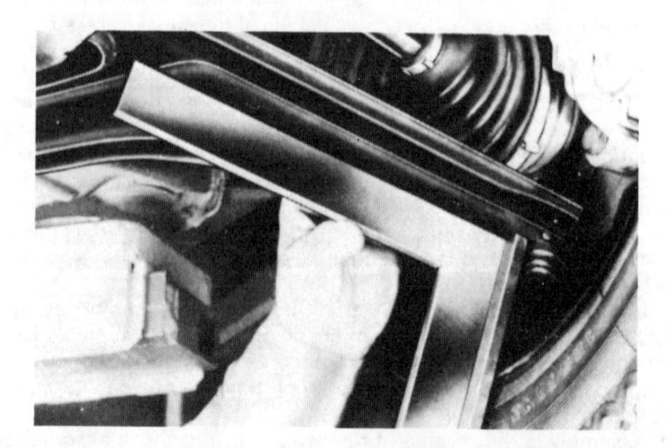

Fig. 8.7 Checking the wishbone for accident damage (Sec 8)

4 Although the wishbone may be removed with the roadwheel in-situ it is easier and safer to support the vehicle on axle stands and remove the front wheel. Support the wheel hub with a jack. Undo the clamp bolt holding the steering balljoint to the wheel bearing housing and lower the wishbone as far as possible. Remove the nuts holding the U-clip and rear pivot, hold the wishbone level and undo and extract the front hinge bolt (photo). Remove the wishbone from the vehicle.

5 The rubber bush at the rear may be pressed off and after the pivot has been cleaned up a new one tapped into place with a hammer.

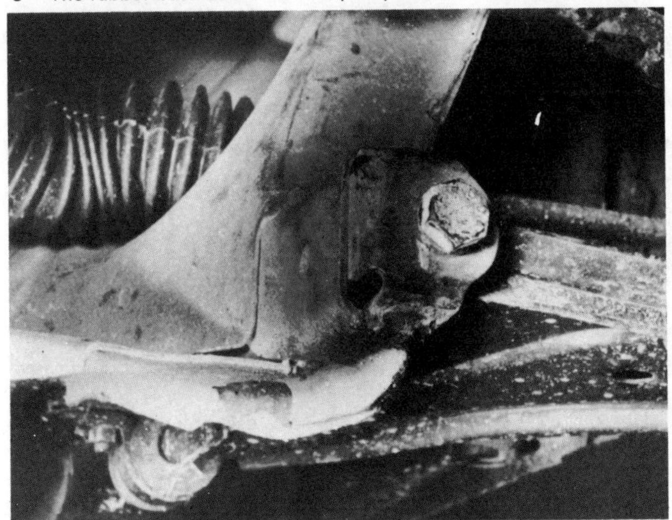

8.4 Suspension lower wishbone front pivot bolt

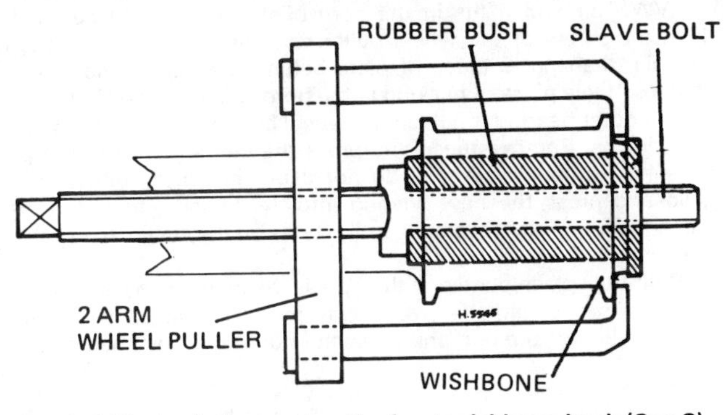

Fig. 8.8 Method of removing the front wishbone bush (Sec 8)

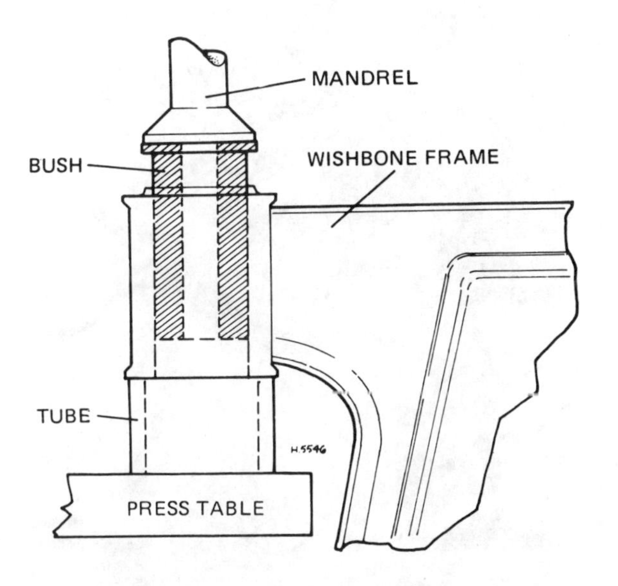

Fig. 8.9 Method of fitting the front wishbone bush (Sec. 8)

6 The front bush is a little more difficult. The simple way to press out the bush is to use a two arm puller. Fit a 10 mm bolt through the bush from the inside surface and push on the bolt with the puller centre bolt as shown in the diagram (Fig. 8.8) so that the bush is driven out from the wishbone. Smear the bore with a little brake grease and push a new bush in from the outside. When fully home, the bush protrudes past the bore of the wishbone so that the wishbone must be supported with a piece of tube large enough to allow the bush to slide inside it. A diagram shows this arrangement (Fig. 8.9).

7 When renewing the balljoint it is possible to drill out the rivets without removing the wishbone from the car, but it is possible that the drill may run out and enlarge the hole in the wishbone as well as removing the rivets. Therefore it is recommended that the wishbone is removed and the drilling carried out under a pillar drill with the wishbone clamped in position on the drilling machine table. Fit the new balljoint as described in paragraph 1.

8 When refitting the wishbone to the vehicle great care must be taken not to destroy the thread of the nut which holds the front hinge bolt in position. The bolt must be tightened to the specified torque. If this torque is not attainable then a new nut must be fitted and as the nut is welded **inside** the front crossmember its renewal is a major task only possible in a fully equipped workshop by professionals who have the right equipment. It involves opening up of the front crossmember – you have been warned! VW recommend the use of a new bolt the thread of which should be coated with D6 locking fluid (ask the VW storeman) and stress that lockwasher part no 411 399 267 must be fitted under the head of the bolt.

9 Before refitting the wishbone try the new bolt in the nut. Make sure it screws in easily. Remove it and position the wishbone in the correct place. Fit the U-clip for the back hinge but do not tighten it. Next try the front bolt and ease the wishbone until the bolt engages the thread in the nut easily. Keep the wishbone in that position, remove the bolt, fit the washer and coat the thread with D6 and install the bolt. Tighten to 1 kgf m (7 lbf ft) and then adjust the back bearing again. Check that the wishbone moves easily about the hinge bolts and then tighten the front one to the correct torque. Then tighten the rear hinge clip securing bolt. Check that the wishbone moves easily about the hinge bolts with no tight spots. If it does not, readjust the rear clip.

10 Finally refit the steering balljoint to the wheel bearing housing. Remove the jack from under the housing, refit the wheel and lower the vehicle to the ground.

9 Steering balljoints – inspection and renewal

1 A simple way to measure the vertical play of the balljoint is to fit a vernier caliper between the bottom of the wishbone and a convenient surface on the wheel bearing housing or brake caliper while the car is on its wheels (Fig. 8.10). Note the measurement. Jack-up the car so that the wheel is clear of the ground and recheck the measurement. Force the wheel down while measuring this distance. If the two measurements differ by more than 2.5 mm (0.10 in) then the joint

Fig. 8.10 Checking the steering balljoint for wear (Sec 9)

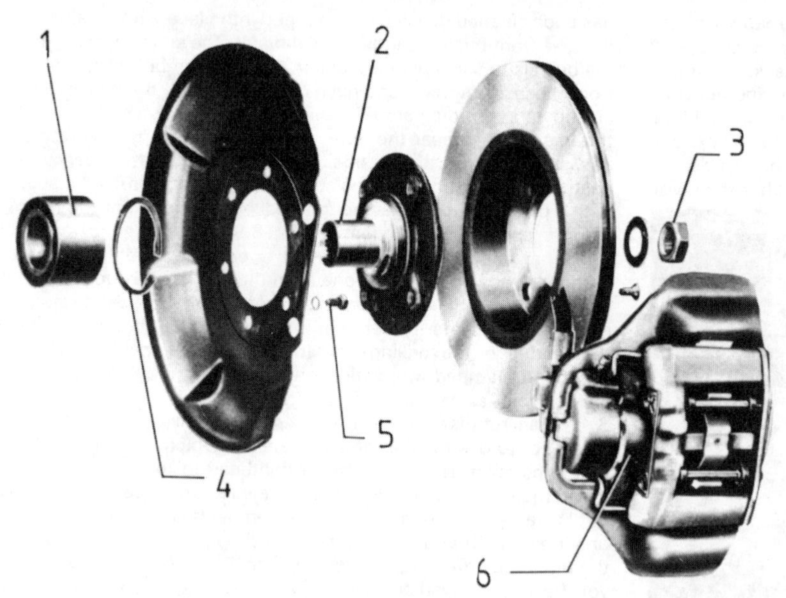

Fig. 8.11 Front hub and wheel bearing components (Sec 11)

1 *Wheel bearing*
2 *Hub*
3 *Axle nut*
4 *Circlip*
5 *Bolt*
6 *Brake caliper*

requires renewal. However it will be best to have this rechecked by the agent with special tools before proceeding further.

2 The method of removal and refitting is described in Section 8, paragraphs 1 and 7.

10 Wheel bearing housing – removal and refitting

1 With the vehicle standing on all four wheels undo the axle nut and slacken the wheel studs. Jack-up the front wheels and support the vehicle on axle stands. Remove the front wheel.

2 Remove the nut from the tie-rod balljoint and break the joint using a wheel puller.

3 Undo the clamp bolt and remove the steering balljoint from the wheel bearing housing. Push the wishbone down out of the way.

4 Remove the brake caliper, refer to Chapter 6, and tie it up out of the way with a piece of wire or cord. Do not allow it to hang by the brake hose.

5 Undo the crosshead screw securing the brake disc to the hub and remove the disc.

6 At this point there is a choice in procedure. Either remove the strut complete by undoing the nuts securing it to the frame in the engine compartment or remove the wheel bearing housing from the strut. The second method involves marking the camber adjustment bolt, so that it may be reassembled with the same camber adjustment, removing it and the lower bolt and then drawing the housing off the bottom of the strut and at the same time pulling it off the driveshaft splines. We found it easier to remove the wheel bearing housing from the strut. Whichever way you do it the bearing housing may now be removed from the car.

7 Assembly is the reverse of removal. If you have disturbed the camber angle set it as near as you can and have it checked when you go to the VW agent to have a **new** axle nut fitted to the correct torque.

11 Wheel bearing (front) – removal and refitting

1 The removal of a wheel bearing is a lengthy job so it is as well to be sure that it is actually necessary before doing the job. A defective bearing will make a rumbling noise under load. Jack-up the suspect wheel and test the rim rock. Spin the wheel and listen for noise indicating wear. A good bearing will make little or no noise. If you are not sure consult someone with experience because a defective race could cause a lot of trouble and, if it collapses, a nasty accident.

2 Having decided that it is necessary, remove the wheel bearing housing as described in Section 10. The next job is to press the hub out of the bearing race and for this job a press is necessary with the correct sized mandrels. **Do not** try to hammer it out. Using a feeler gauge measure the clearance between the hub and the wheel bearing housing to give you an idea of how much clearance to leave when refitting the hub in the housing. Once the hub is removed the splash plate may be removed. The task will not present much difficulty if the wheel bearing housing has been separated from the strut but if it is still attached to the strut suitable provision must be made to support the strut during pressing operations.

3 The inner bearing race will come away on the hub and must be pulled off with a wheel puller. Hold the flange of the hub in soft jaws in a vice so that there is room for the inner race to be removed. Ideally a plug VW 431 which fits in the bore of the hub should be used on which the screw of the wheel puller can push. If such a plug is not available, then put a piece of plate behind the hub in the vice and, using a suitable piece of packing in the bore, push against that.

4 The outer bearing is held in the wheel bearing housing bore by two large circlips. Remove these and press the bearing out using a press and suitable mandrel. Again, do not use a hammer. Refit the outer circlip and press the new bearing into the housing from the inside pushing on the outer race only until it rests on the circlip. Now refit the inner circlip.

5 Refit the splash plate and then press the bearing housing on to the hub. This may sound the wrong way round but actually it is much easier to support the hub and press the housing on by pushing on the

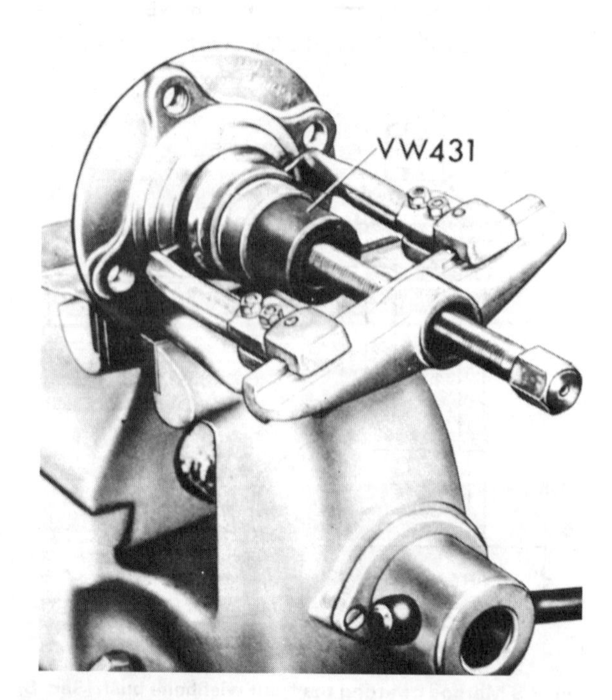

Fig. 8.12 Removing the wheel bearing inner race (Sec 11)

inner bearing race. **On no account** push on the outer race. If you measured the clearance between the hub and the housing before removing the hub then you know how much room to leave.

6 A word of caution. You will have gathered that the old race has been destroyed and a new one has to be bought. Unless you have a press and experience at this job we recommend you to ask the VW agent from whom you buy the race to fit it. It is very easy to damage the race and possibly the bore of the hub if you do not have the correct tools. This will mean a new hub and another race.

7 When the hub is in position reassemble the wheel bearing housing, strut and wishbone. Refit the brake caliper, refer to Chapter 6. If the camber angle has been disturbed, have it checked when you go to have the new axle nut tightened to the specified torque.

12 Tie-rods – removal and refitting

1 Refer to Fig. 8.15. On new vehicles the left-hand tie-rod is made in one piece and is not adjustable. The right-hand tie-rod consists of two pieces screwed together so that its length may be altered to adjust toe-in.

2 When a new tie-rod is required on the left-hand side the solid type is not supplied, an adjustable one is provided which must be adjusted to the same length as the original solid rod.

3 When setting the tie-rods in position the steering rack must protrude equally from each side of the casing, see Fig. 8.13. Note that the right-hand datum is the back (inside) of the flange.

4 When the rack is correctly set the tie-rods should be screwed on to the dimensions given in paragraph 8 below (Fig. 8.14). Once the rods are set to the correct dimension the locknuts should be tightened and the rubber cover fitted in position.

5 The outer ends of the tie-rods are attached to the steering arms by balljoints. These joints are pre-packed with lubricant. If the rubber seal perishes and the balljoint becomes worn a new tie-rod end must be fitted (on original left-hand side rods the complete rod must be renewed).

6 To renew a tie-rod end, mark the position of the locknut and then slacken the locknut. Remove the split-pin and castellated nut attaching the balljoint to the steering arm. Use a balljoint separator to disconnect the balljoint from the steering arm. Screw the new tie-rod end onto the same position on the tie-rod and tighten the locknut. Fit the balljoint to the steering arm, then tighten the securing nut to the specified torque and fit the split-pin.

7 Toe-in must always be adjusted on the right-hand tie-rod. Whenever the tie-rods or tie-rod ends have been disturbed, have the wheel alignment checked at your local VW garage.

8 Tie-rod dimensions are as follows:

Tie-rod length – LHS	*379 mm (14.92 in)*
Adjustable tie-rod length – RHS (Fig. 8.16)	*379 mm (14.92 in)*
Protrusion of rack (dimension [a] in Fig. 8.13)	*Equal*
Dimension (a) in Fig. 8.14 – LHS	*69 mm (2.716 in)*
Dimension (b) in Fig. 8.14 – RHS	*69 mm (2.716 in)*

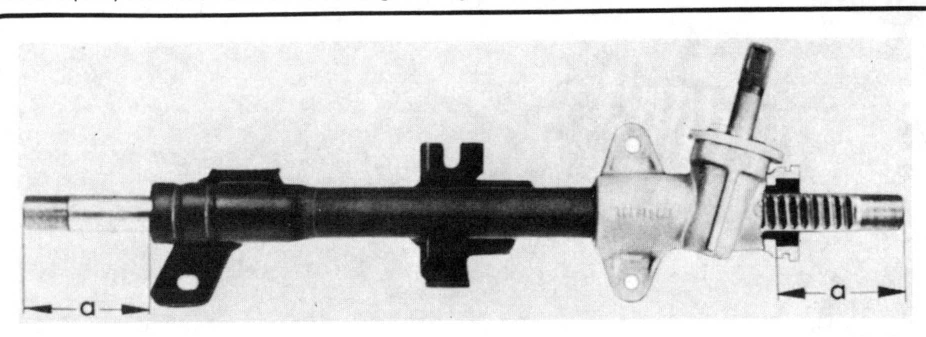

Fig. 8.13 Steering rack in central position – a = a (Sec 12)

Fig. 8.14 Tie-rod setting dimensions (Sec 12) *For a and b see text*

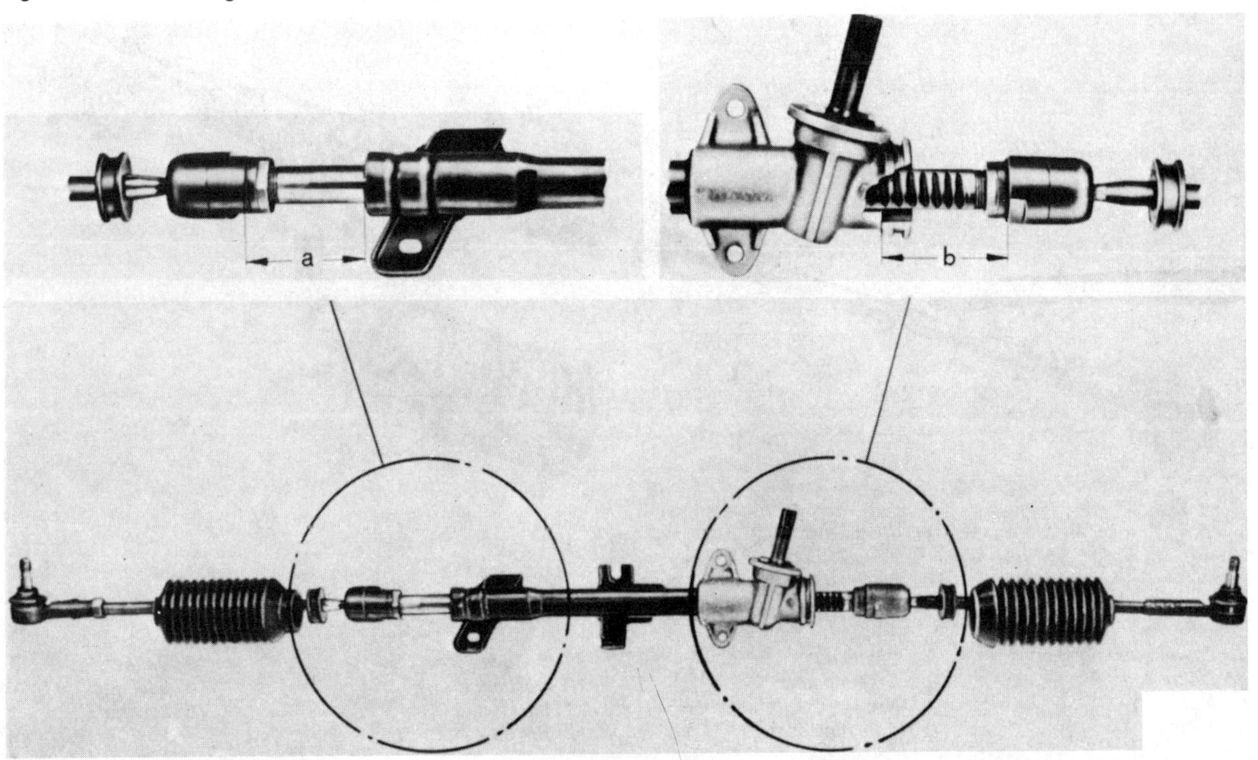

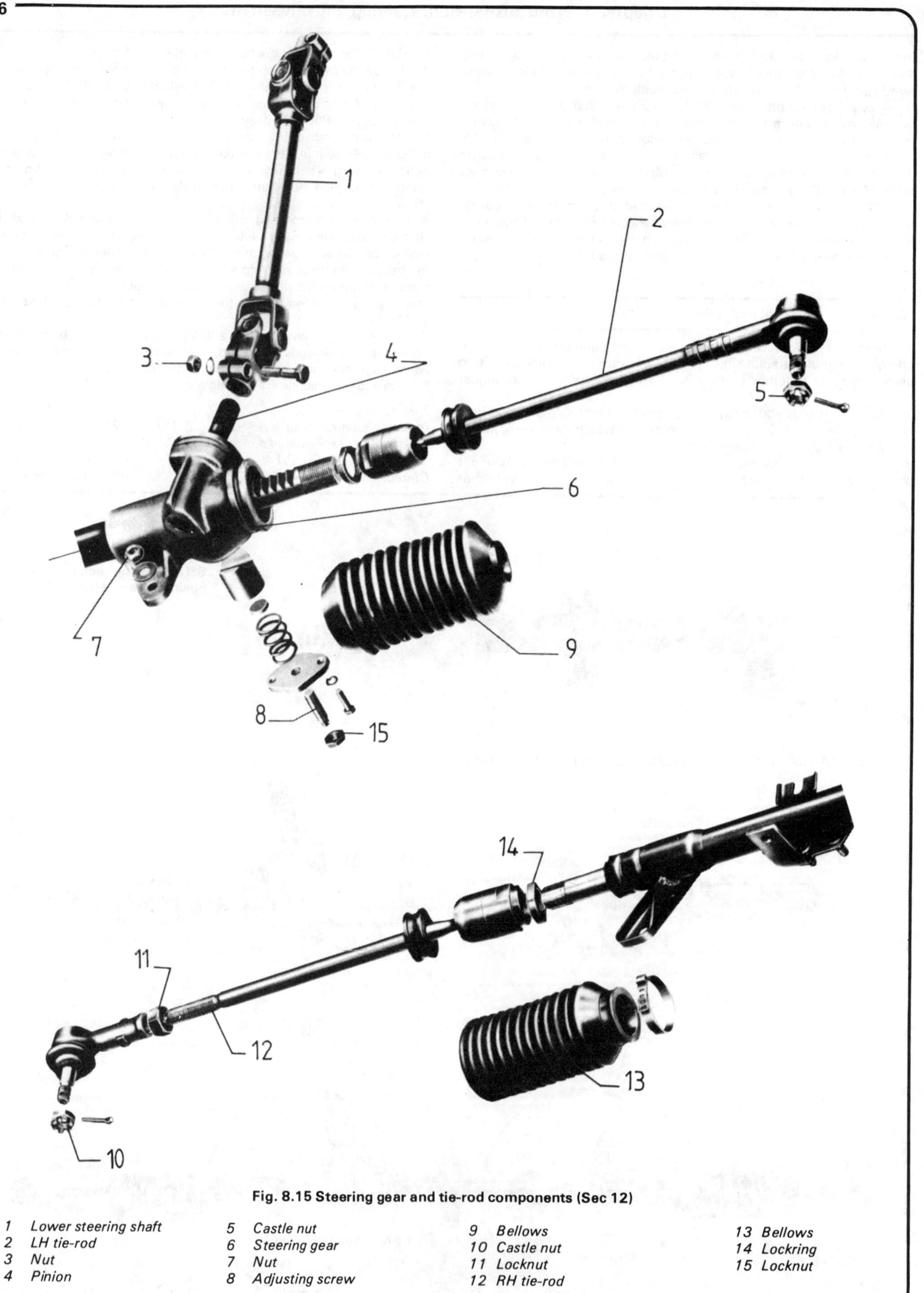

Fig. 8.15 Steering gear and tie-rod components (Sec 12)

1 Lower steering shaft
2 LH tie-rod
3 Nut
4 Pinion
5 Castle nut
6 Steering gear
7 Nut
8 Adjusting screw
9 Bellows
10 Castle nut
11 Locknut
12 RH tie-rod
13 Bellows
14 Lockring
15 Locknut

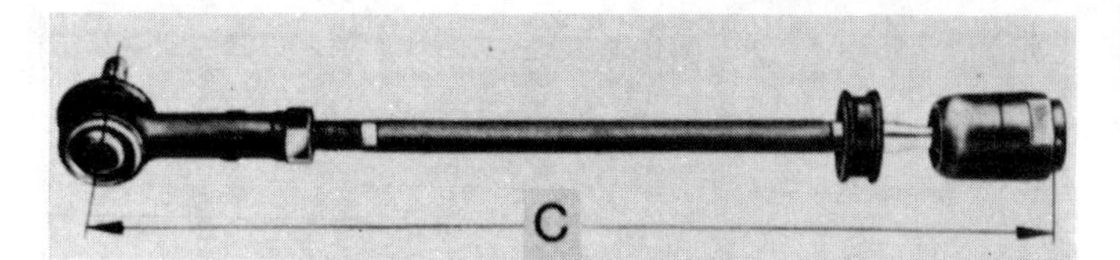

Fig. 8.16 Basic tie-rod adjustment dimension C (Sec 12)

13 Steering gear – removal and refitting

1 Refer to Fig. 8.15. The steering gear is located in the engine compartment behind the engine. Push back the rubber boot from the universal shaft joint and remove the clamp bolt from the lower universal joint (photo). Push the universal joint up the pinion shaft splines as far as possible. It will not come off the pinion shaft at this stage.

2 Working underneath the car, disconnect the bracket which attaches the gearchange linkage to the steering gear.

3 Refer to Section 12 and remove the tie-rods. There is no need to disconnect the outer joints. Slacken the locknuts and unscrew the balljoints from each end of the rack.

4 Undo the nuts securing the U-clamps which attach the steering gear to the body (Fig. 8.17), then remove the clamps and at the same time lift away the steering gear, which will now come away from the

13.1 Steering column to steering gear universal joint bellows

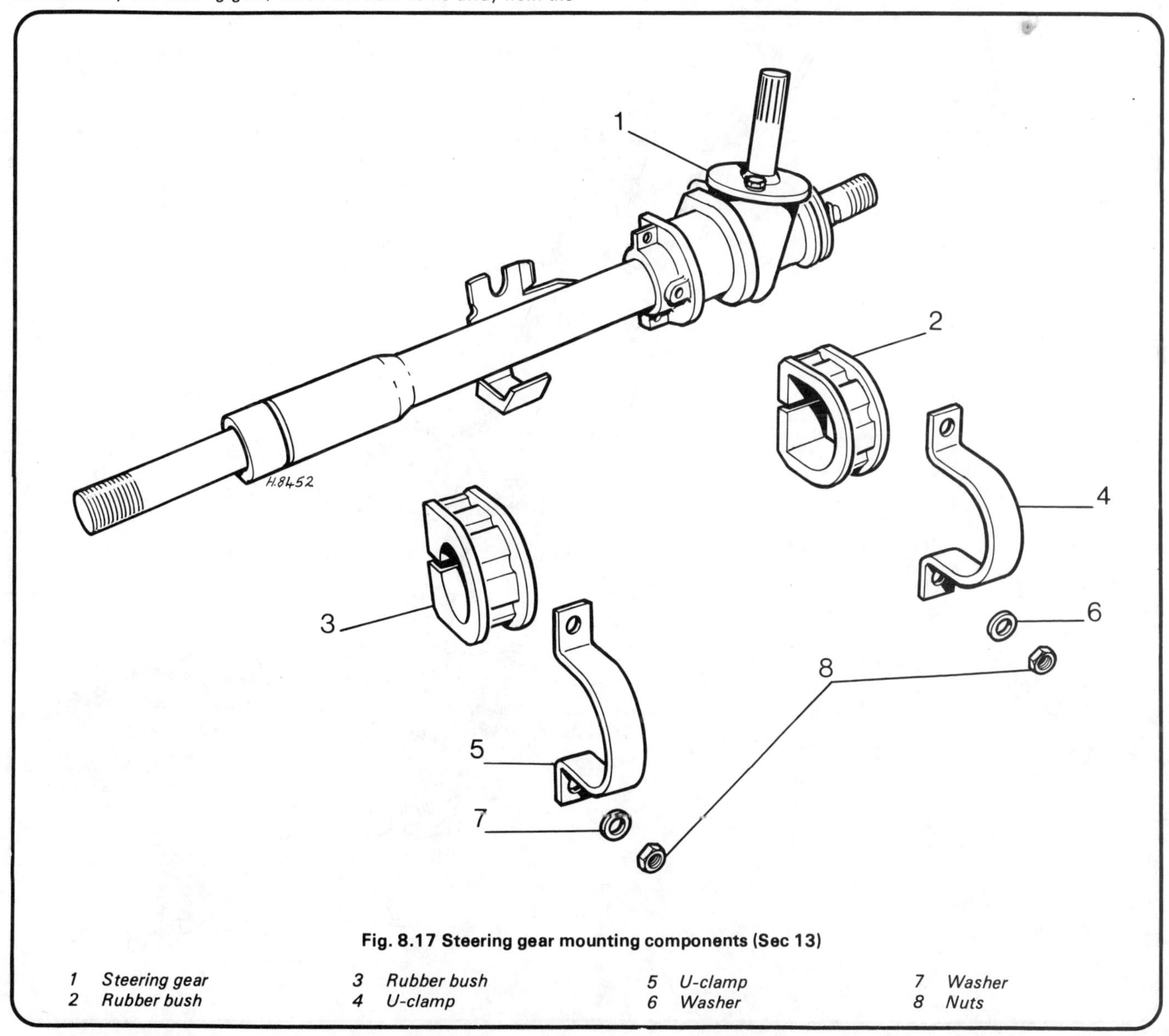

Fig. 8.17 Steering gear mounting components (Sec 13)

1 *Steering gear*
2 *Rubber bush*
3 *Rubber bush*
4 *U-clamp*
5 *U-clamp*
6 *Washer*
7 *Washer*
8 *Nuts*

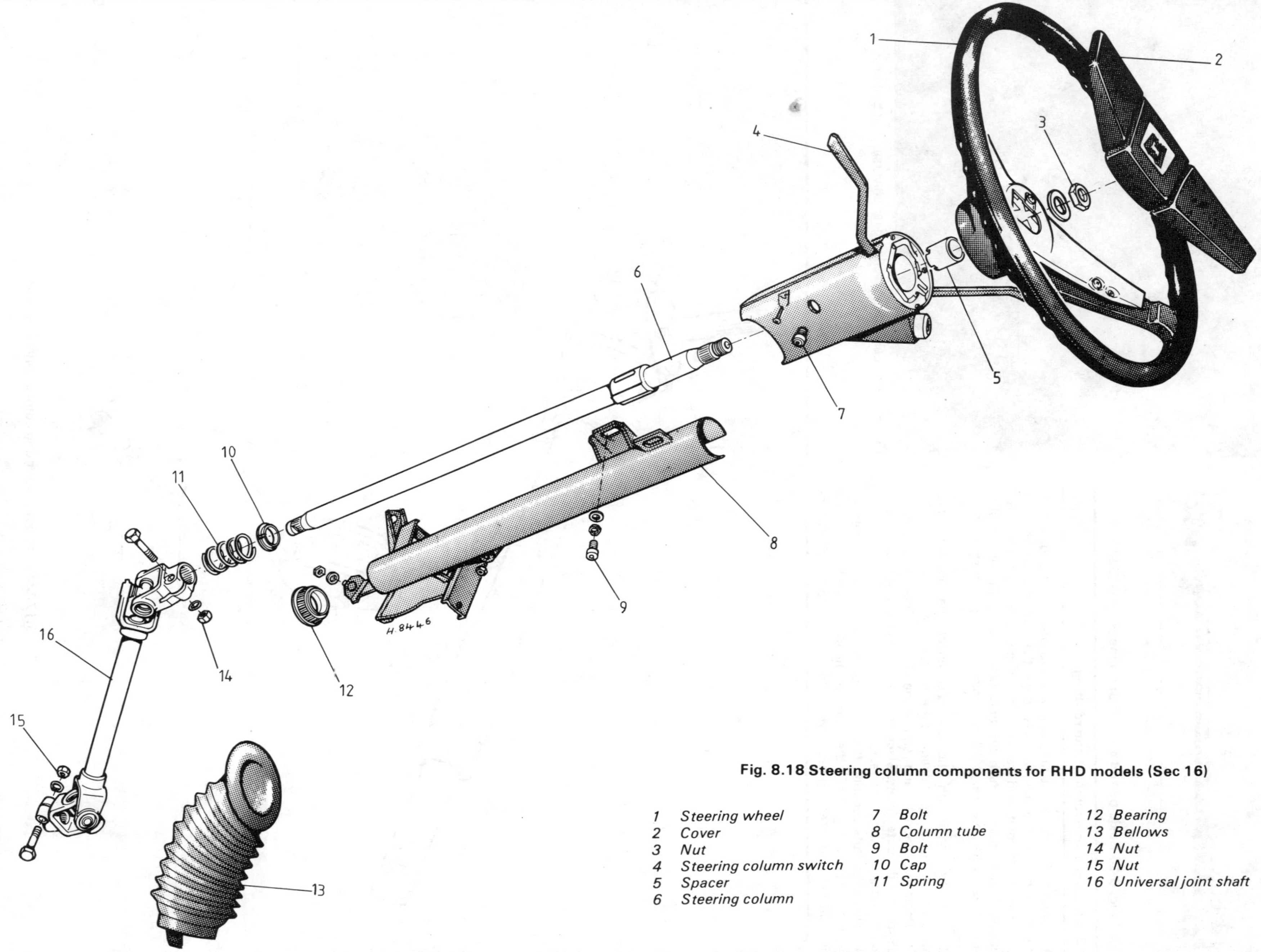

Fig. 8.18 Steering column components for RHD models (Sec 16)

1	*Steering wheel*	7	*Bolt*	12	*Bearing*
2	*Cover*	8	*Column tube*	13	*Bellows*
3	*Nut*	9	*Bolt*	14	*Nut*
4	*Steering column switch*	10	*Cap*	15	*Nut*
5	*Spacer*	11	*Spring*	16	*Universal joint shaft*
6	*Steering column*				

universal joint shaft. The steering gear can now be removed through the right-hand wheel arch. It may be necessary to remove the roadwheel to pull the gear clear.
5 Refitting is the reverse of the removal procedure. Ensure that the rubber mounting bushes on the gear casing are in good condition. When fitting the universal joint shaft take care that the flat on the pinion shaft lines up with the clamp bolt hole. Fit the clamp bolt loosely, then tighten the U-clamp to the body. Tighten all the nuts to the specified torque.
6 Refit the tie-rods, refer to Section 12. Adjust the gearchange linkage, refer to Chapter 5. Have the front wheel alignment checked and adjusted as necessary.

14 Steering gear – adjustment

1 The steering gear is not repairable: if it is damaged or worn beyond adjustment, a complete new assembly must be fitted.
2 The only adjustment is the pressure on the engagement between the rack and pinion. This is adjusted by a screw which is screwed in to increase the pressure and out to reduce the pressure.
3 The adjustment screw is located behind the gear assembly and it is difficult to gain access to adjust the steering.
4 If rattling noises are heard coming from the steering gear when the vehicle goes over rough surfaces then the pinion is loose in the rack and must be adjusted. To do this first jack-up the front of the car and set it on axle stands. Undo the adjuster locknut and back off the adjuster a little. Now tighten it until it touches the thrust washer, you will note more resistance at this point. Hold the adjuster in this position and tighten the locknut.
5 A creaking sound from the rack and pinion denotes lack of lubrication. Move the steering to full lock and pull back the rubber rack cover on the extended side, wipe the grease away and spray on Molykote 321 R. Allow this to dry for ten minutes and then apply a thin film of lithium based grease and refit the rubber cover.
6 When correctly adjusted there should be a natural self-centring force when cornering. If this is absent and the steering is stiff then the pinion is binding on the rack and must be adjusted.

15 Steering wheel – removal and refitting

1 Disconnect the battery earth lead to avoid accidental short circuits. The horn pad pulls off, after a struggle, to expose the steering column nut. Disconnect the lead, undo the nut and pull the wheel off the splines (photos).
2 When refitting the wheel be careful to install the wheel in the central position.
3 Do not forget to reconnect the battery earth lead.

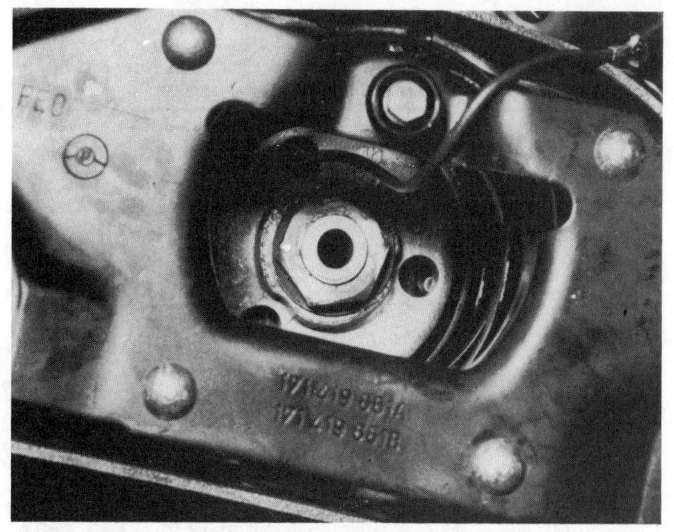
15.1A The steering wheel retaining nut

15.1B Steering wheel horn lead connection

16 Steering column (RHD models) – removal and refitting

1 Disconnect the battery earth strap and remove the steering wheel (Section 15).
2 Refer to Fig. 8.18. Remove the steering column switches as described in Chapter 7 and then prise out the spacer sleeve.
3 At the bottom of the column tube there is a cross bracket which supports the pedals. Remove the pedal connections (for the clutch refer to Chapter 4 and for the brake pedal refer to Chapter 6).
4 Remove the nuts securing the bottom bracket to the body. Remove the screws securing the upper column bracket to the dashboard.
5 Undo the nut and remove the bolt clamping the universal joint to the steering column. Remove the steering column and tube from the car.
6 Remove the column shaft from the tube. At the bottom of the tube there is a spring and spring seat: take care not to lose these. If the steering column bearing has to be renewed it can be pushed out and a new bearing pressed in.
7 Assembly is the reverse of removal. Fit the spring seat and spring, insert the column and reconnect the UJ shaft at the bottom. Refit the nuts and bolts holding the steering column to the body and reconnect the brake pedal and clutch pedal. Refit the column switch and the multipin plugs and now press in the spacer sleeve until the distance from the top of the column to the top of the spacer sleeve is 41.5 mm (1.63 in). This will set the gap between the steering wheel and the column switch at 2 to 4 mm (0.08 to 0.16 in). Fig. 8.19 shows the measurement. Now refit the steering wheel and the covers. Check the clutch and brake pedal movement.

17 Steering column (LHD models) – removal and refitting

1 Refer to Fig. 8.23 and it will be seen that the steering column lower attachment is different from that on RHD models. Otherwise the removal and refitting procedure is the same as described in Section 16.
2 Instead of the column being fixed rigidly to the pedal bracket, which is bolted to the frame, the column is attached to the bracket by a leaf spring. The bracket is reshaped so that the column may be inserted from the left-hand side and then the leaf spring fitted across the open end of the bracket.
3 In the event of a collision the shaft will be pushed sideways, removing the spring. In this event the complete steering gear should be dismantled and thoroughly inspected for damage before using the car again.
4 When removing the steering column, the leaf spring is removed by using a screwdriver to press down on the spring to release it from the retaining clip, see Fig. 8.20, and then withdrawing it from the slot in

the bracket.
5 To refit the leaf spring, insert one end of the spring in the bracket slot (arrow A in Fig. 8.21) and then press down on the other end of the spring until it locates in the retaining clip.

6 Note that the mounting ring shown in Fig. 8.23 has been modified and a split bush fitted instead. To remove the bush, pry it open and slide it from the column tube. When fitting a new bush, ensure that the lug enages with the slot in the column tube (see Fig. 8.22).

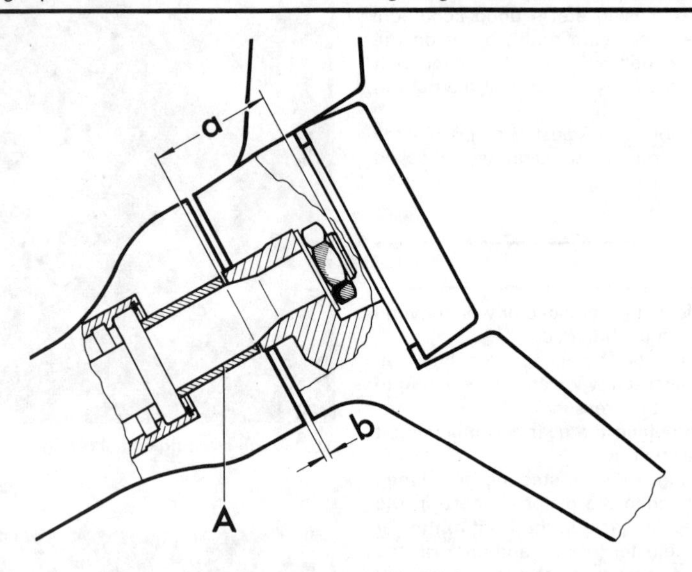

Fig. 8.19 Steering column spacer sleeve A adjustment dimensions (Sec 16)

a = 41.5 mm (1.63 in)
b = 2.0 to 4.0 mm (0.08 to 0.16 in)

Fig. 8.20 Removing the steering column retaining leaf spring on LHD models (Sec 17)

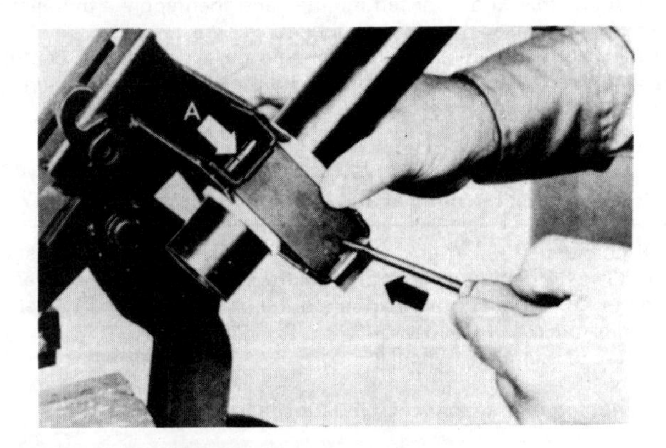

Fig. 8.21 Refitting the steering column retaining leaf spring on LHD models (Sec 17)

A *Spring end in bracket slot*

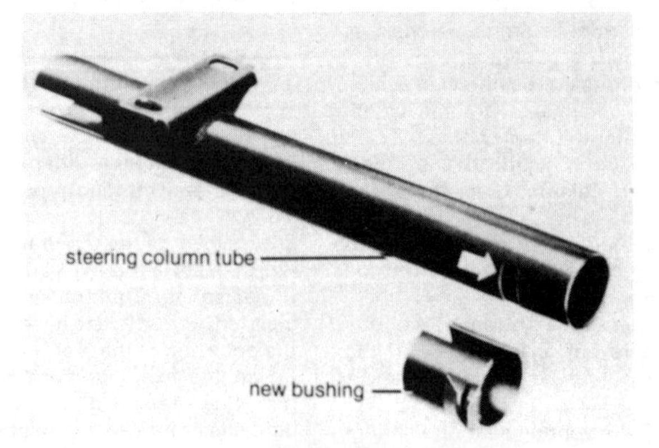

Fig. 8.22 Modified mounting ring location slot – arrowed (Sec 17)

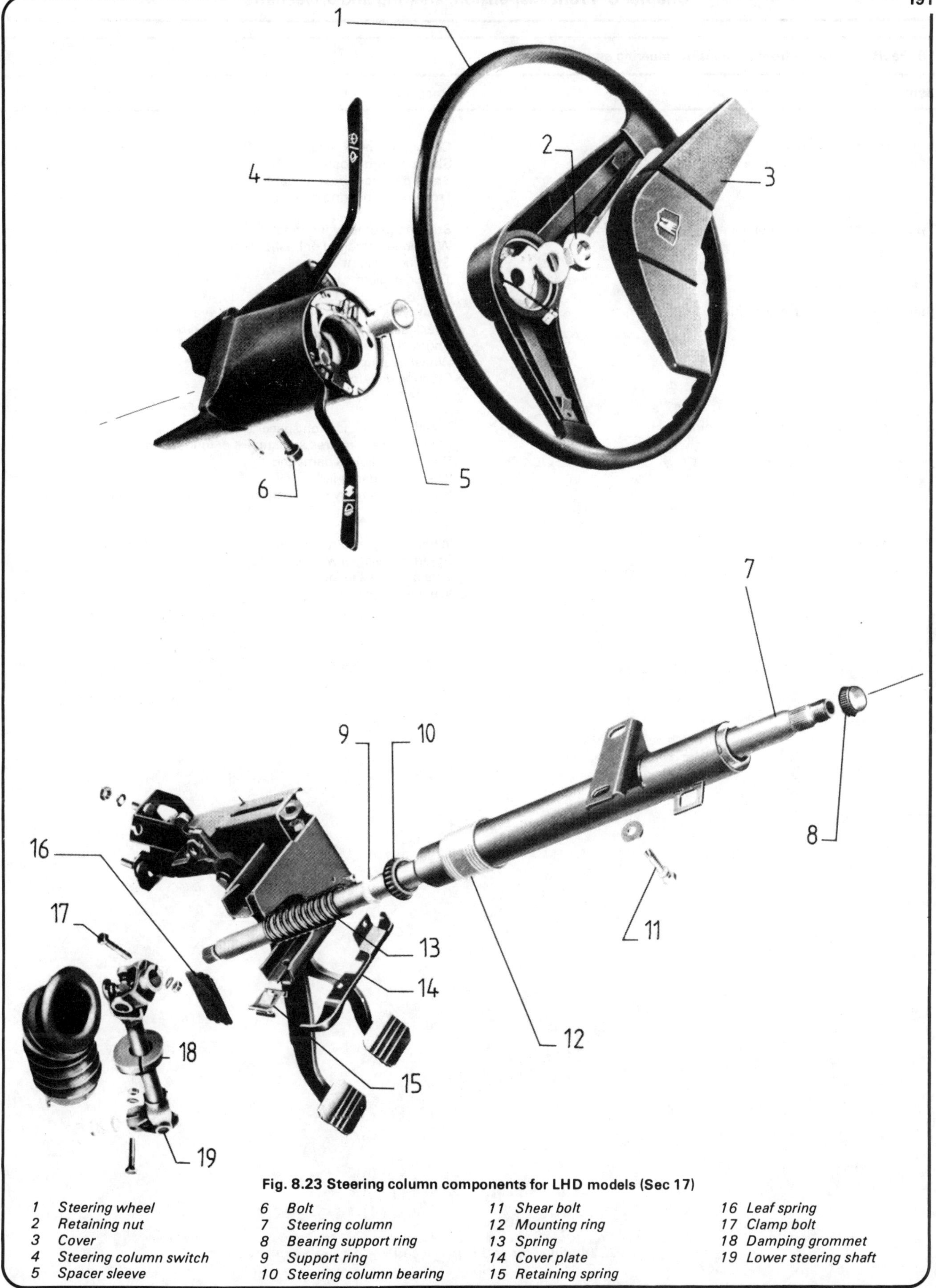

Fig. 8.23 Steering column components for LHD models (Sec 17)

1 *Steering wheel*
2 *Retaining nut*
3 *Cover*
4 *Steering column switch*
5 *Spacer sleeve*
6 *Bolt*
7 *Steering column*
8 *Bearing support ring*
9 *Support ring*
10 *Steering column bearing*
11 *Shear bolt*
12 *Mounting ring*
13 *Spring*
14 *Cover plate*
15 *Retaining spring*
16 *Leaf spring*
17 *Clamp bolt*
18 *Damping grommet*
19 *Lower steering shaft*

18 Fault diagnosis – front suspension, steering and driveshafts

Symptom	Reason/s
Noise from front suspension	Coil springs bottoming Shock absorber defective Brake pads require renewal Front wheel bearing defective
Tyres wearing excessively on inner or outer edges of the tread	Steering geometry incorrect Wishbone damaged or bushes worn Steering balljoint worn Tie-rod balljoints worn
Steering wanders and is unstable	Front tyres soft Check all items from fault 2, (excessive tyre wear) Shock absorbers defective Wheel bearing defective Steering gear loose UJ shaft bolts not tight
Steering stiff and does not centre correctly	Steering wheel hard on spacer sleeve Steering gear requires adjustment or lubrication Steering balljoints damaged Wishbones damaged Steering geometry incorrect (this will be a first sign before tyre wear occurs)
Wheel wobble or 'shimmy'	Wheels or tyres out of balance Steering linkages worn or loose Tyre pressures incorrect Wear in suspension

Chapter 9 Rear axle and rear suspension

Refer to Chapter 11 for specifications and information applicable to later models

Contents

Specifications

Type

Torsion axle beam with trailing arms. MacPherson type struts with coil spring and co-axial shock absorber. Wheels carried on stub axles bolted to trailing arms

Coil springs

Initial fitting: three grades with 1, 2 or 3 paint marks. New springs available only with two green paint marks

Shock absorbers

Hydraulic, telescopic, double-acting

Geometry

Adjustment	All angles fixed – non-adjustable
Camber	$-1° 15' \pm 5'$
Maximum difference between sides	40'
Wheel alignment (total toe-in)	$+20' \pm 30'$
Maximum deviation from driving direction	30'
Track	1358 mm (53·4 in)

Hub bearings

Taper roller, adjustable

Torque wrench settings

	kgf m	lbf ft
Axle beam to mounting bracket	6·0	43
Mounting bracket to body	4·5	32
Shock absorber to trailing arm	4·5	32
Shock absorber to body	3·5	25
Shock absorber piston rod nut	2·0	14
Stub axle to trailing arm	6·0	43

1 General description

1 A fabricated axle beam is located across the vehicle body held on either side in mounting brackets which are bolted to the underside of the body. The axle is held in the mounting by a hinge bolt which passes through a bonded rubber bush. The cross beam is a T-section.

2 Welded to the beam and supported by gusset plates are trailing arms of circular section. These terminate in fabricated brackets which serve as lower anchors for the shock absorbers and as mounting pads for the stub axles and rear brake backplates.

3 The stub axle carries the rear hub, brake drum and wheel on two taper roller bearings. The movement of the wheel is thus a radial one in the vertical plane. The weight of the body and its contents is supported by a suspension strut on each side which is secured to the body at its upper end and the axle at the lower end. The strut consists of a coil spring and shock absorber mounted co-axially. The shock absorbers are of hydraulic telescopic construction. Approximately half way up the outer casing of the shock absorber is welded a spring platform which houses the lower end of the spring (photo). The upper end is housed in spring cup which is kept in place by a slotted nut fitted to the piston rod of the shock absorber. A rubber buffer is fitted on the piston rod to stop the spring being compressed until its coils are solid.

4 The suspension strut is assembled complete away from the vehicle and then fitted, first to the body by a nut removable from inside the vehicle body and at the lower end by a hinge bolt to the bracket at the end of the trailing arm.

5 The different types of spring fitted are discussed in Section 8.

6 The camber angle and toe-in are fixed. Details are given in the Specifications for checking, should distortion be suspected.

7 The suspension struts may be removed without dismantling the axle, and once the brake hoses and the handbrake cables are disconnected the whole assembly may be removed from the car by removing six nuts.

8 If the axle beam or the stub axles are distorted or otherwise damaged they must be renewed. Repair by straightening or welding is not permitted.

2 Rear wheel bearings – adjustment

1 The rear wheel bearings are normal taper bearings. To adjust

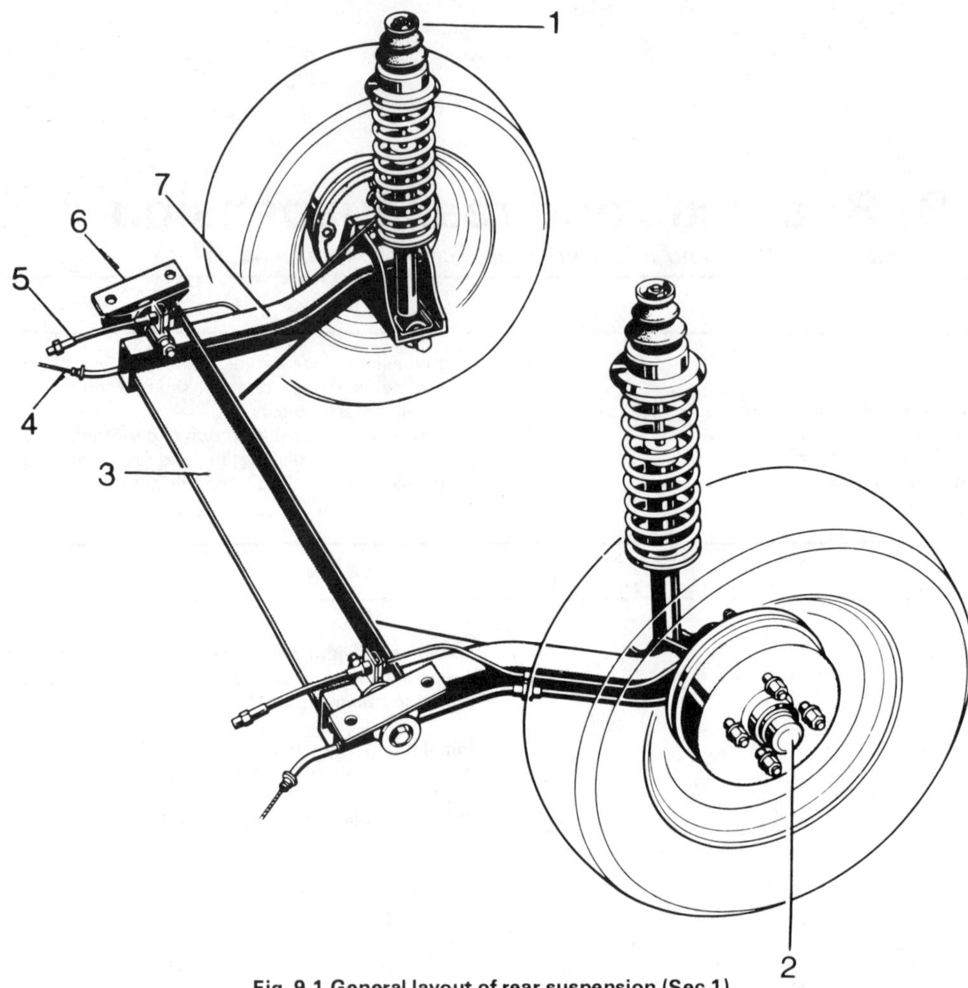

Fig. 9.1 General layout of rear suspension (Sec 1)

1 Suspension strut
2 Wheel bearing hub
3 Axle beam
4 Handbrake cable
5 Hydraulic brake pipe
6 Mounting
7 Trailing arm

1.3 Rear suspension strut and coil spring

2.1 Checking the rear wheel bearing for wear. Rocking movement indicates the need for adjustment

2.2 Rear wheel bearing nut locking ring and retaining split-pin

them, first raise the rear of the car and support it on axle stands or other suitable supports (photo).

2 Remove the dust cap with a suitable lever, remove the split-pin and withdraw the nut locking ring (photo).

3 Slacken off the hub nut and then tighten it while turning the wheel to settle the bearings.

4 Now slacken off the hub nut again, then tighten it until the thrust washer can be moved (rotated) slightly with a screwdriver (or similar tool) as shown in Fig. 9.2. Do not twist or lever the screwdriver to obtain movement.

5 Refit the nut locking ring and a new split-pin.

6 Fill the dust cap with multi-purpose grease and tap it into position on the hub.

7 Remove the axle stands and lower the car to the ground.

Fig. 9.2 Checking the rear wheel bearing adjustment (Sec 2)

3.2 Removing the rear wheel bearing adjusting nut

3 Rear wheel bearings – removal and refitting

1 Slacken the wheel nuts, jack-up the rear wheel, support the vehicle on an axle stand and remove the wheel. Refer to Fig. 9.3.

2 Remove the hub grease cap, remove the split-pin and locking ring. Remove the adjusting nut (photo). Slacken the brake adjusters (Chapter 6), and remove the brake drum. The thrust washer and the outer bearing complete will come away with the drum, together with the oil seal and the outer race of the inner bearing. The inner race of the inner bearing will remain on the stub axle.

3 Remove the oil seal and tap the bearings out of the brake drum with a brass drift (photo). Clean the drum and stub axle and oil them lightly. Wash the bearings in clean paraffin and then swill out all residue with more clean paraffin. Dry them carefully with a non-fluffy rag.

4 Examine the tracks of the races for scoring or signs of overheating. If these are present then the complete bearing should be renewed. Inspect the roller bearings carefully, there should be no flats or burrs. Lubricate the race with a light oil and spin the inner race in the outer race. Any roughness indicates undue wear and the bearing should be renewed. Once a bearing begins to wear the wear accelerates rapidly and may be the cause of damage to the stub axle or drum so if you are in doubt get an expert opinion.

5 Examine the oil seal. Ideally it should be renewed; if there is any damage or sign of hardening renew it anyway. Failure of an oil seal will mean contamination of the brake shoes and the minimum penalty will be new shoes on **both** rear wheels.

6 If all the items are correct and the stub axle bearing surfaces and brake drum bore are in good order then proceed to assemble the hub.

7 Pack the taper bearings with a lithium based grease, and smear the stub axle with a light film of grease. Fit the inner race, the inner bearing and the oil seal into the brake drum. Press the seal into place,

Fig. 9.3 Exploded view of the rear wheel bearings (Sec 3)

1 Dust cap
2 Split-pin
3 Nut locking ring
4 Nut
5 Thrust washer
6 Outer bearing
7 Brake drum and hub
8 Inner bearing
9 Oil seal

3.3 Rear wheel hub oil seal location

tap it home with a rubber hammer taking care that it is squarely seated.

8 Fit the drum to the stub axle then the outer race and finally the thrust washer. Fit the nut and adjust the bearing as in Section 2. Fit the nut locking ring and a new split-pin. Refit the dust cap and the road-wheel and test the bearing by spinning the wheel. Lower the vehicle to the ground.

4 Stub axles – removal, inspection and refitting

1 Apart from scoring on the bearing surfaces or ovality due to excessive bearing wear the only other defect to check on the the stub axle is distortion due to bending. This could happen if the vehicle has been in an accident or has been driven heavily laden at excessive speed over rough ground.

2 The effect of a bent stub axle will be excessive tyre wear due to the wrong camber angle and possibly incorrect toe-in. The angles are small with fine limits and, in our opinion, not measurable without the proper equipment. If irregular wear occurs the owner is advised to have these angles checked right away. Before deciding that the stub axle is bent read Section 5.

3 Alternatively the stub axle may be removed, bolted to a lathe face plate and checked for alignment with a dial gauge. The allowable run out is 0.25 mm (0.010 in).

4 To remove the stub axle proceed as in Section 3 and remove the

Fig. 9.4 Rear wheel stub axle components (Sec 4)

1 *Stub axle*
2 *Axle beam*
3 *Backplate*
4 *Spring washer*
5 *Retaining bolt*

brake drum and bearings. The brake backplate and stub axle are held to the axle beam by four bolts. Unfortunately these are only accessible after the brake shoes have been removed (see Chapter 6). Remove the shoes and then the four bolts. The backplate and wheel cylinder may be moved out of the way to extract the stub axle.

5 Refitting is the reverse of removal. Make sure the joint faces of the stub axle and axle beam are clean and free from burrs before assembly.

5 Axle beam – visual check for distortion

1 If irregular wear of the rear tyres is observed the reason will probably be incorrect camber angle. Before dismantling the stub axle check that the gusset plate of the axle beam is not distorted. If this fault is present the axle beam will be out of alignment. This will require a new axle beam. The removal and refitting of the complete rear axle is discussed in Section 10. Before deciding to renew the axle beam we suggest that the camber angle should be checked by the VW agent and your diagnosis confirmed.

6 Suspension strut – removal and refitting

1 Open the tailgate and lift the parcel shelf. On each side of the body on top of the wheel arches are plastic caps which cover the rear suspension strut upper mounting. Remove the plastic cap (photo).

2 Slacken the wheel nuts, raise the rear of the car and support the body on axle stands. The best place for this is the recommended jacking point, just forward of the wheel, which is a narrow flat surface with a hole in it. Remove the roadwheel.

3 Remove the nut from the bolt securing the shock absorber to the trailing arm and tap out the bolt (photo). It may be necessary to take the weight of the hub when removing this bolt. If the bolt fouls the brake drum then the brake backplate must be removed, refer to Section 4.

4 Once the lower anchorage is clear remove the nut securing the upper mounting of the shock absorber piston inside the body, and lift off the special washer (photo). This operation needs two people: one to undo the nut and one to hold the strut. Remove the strut from the car.

5 Refitting is the reverse of the removal procedure. Tighten the upper and lower mounting nuts to the specified torque. On models from July 1978 ensure that the protective tube, see Fig. 9.6, is

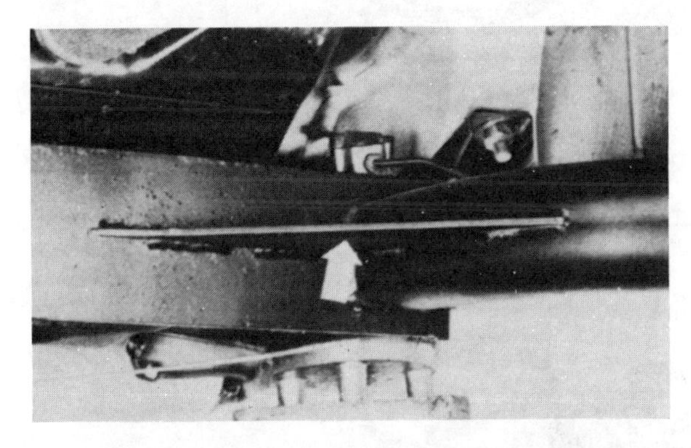

Fig. 9.5 Rear axle gusset plate – arrowed (Sec 5)

Fig. 9.6 Suspension strut components fitted to later models (Sec 6)

1 *Plastic cap*
2 *Plate*
3 *Upper bush*
4 *Lower bush*
5 *Spacer sleeve*
6 *Alternative hexagon nut*
7 *Spring cap*
8 *Protective tube*
9 *Coil spring*

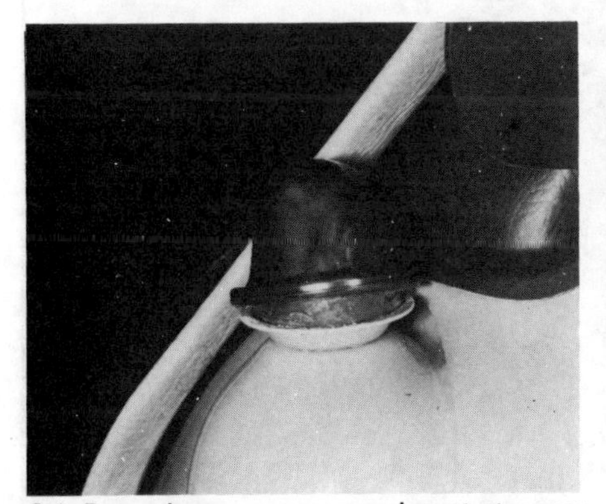

6.1 Removing a rear suspension strut upper mounting cap

6.3 Rear suspension strut lower mounting

6.4 Rear suspension strut upper mounting

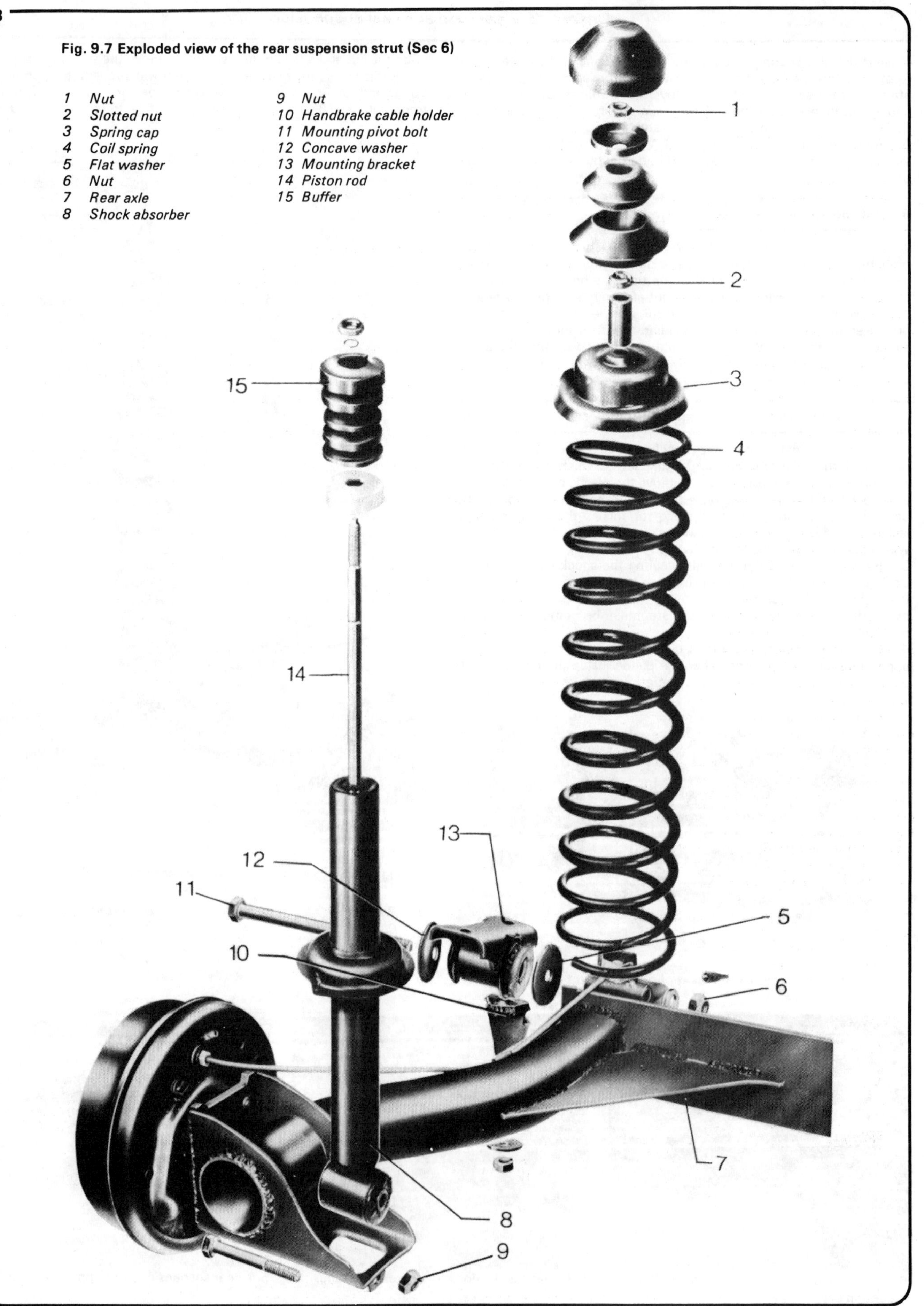

Fig. 9.7 Exploded view of the rear suspension strut (Sec 6)

correctly fitted to prevent the entry of dirt and water which would result in premature failure of the shock absorber.

7 Suspension strut – dismantling and reassembly

1 Refer to Fig. 9.7. The spring is held in compression by a special slotted nut (an ordinary nut may be found instead on some models). It is not necessary to clamp the spring with a spring compressing tool as described for the front springs (Chapter 8). Mount the strut assembly in a vice as shown in Fig. 9.8.
2 If the special slotted nut is fitted, VW tool No. 50-200 will be required to unscrew the nut while holding the shock absorber piston rod with a suitable spanner, to prevent the rod from turning.
3 Remove the slotted nut and separate the spring from the shock absorber.
4 To assemble the strut fit the protective cap and the rubber buffer on the shock absorber piston rod, followed by the circlip and washer.
5 Fit the spring, spring retainer and spacer sleeve. Fit the retaining nut on the piston rod and tighten it to the specified torque. The spring may have to be compressed a little to get the nut started.
6 The remaining parts of the upper mounting are fitted when the strut assembly is fitted on the car.

8 Coil springs – inspection

1 Three faults may occur; either the spring is broken, which is very rare, or more often a rattling noise coming from the region of the back axle when the vehicle is driven over rough surfaces. The latter fault may be due to the end coil of the spring at the top coming into contact with the next coil. Finally the spring may just have got tired and distorted. In any case the suspension strut must be removed and dismantled as in Sections 6 and 7.
2 If the spring is broken then a careful examination of the fracture is indicated. Should the surface of the fracture show a crystalline form plus a smaller portion of clean metal then the spring was probably faulty when assembled. It would be worthwhile pointing this out to the agent. Anyway renewal is indicated.
3 If the metal shows a clean break with no flaws, then a careful examination of the rest of the suspension is indicated for the force required for such a fracture will be of very large proportions indeed.
4 If the top coil has been knocking on the next coil there will be evidence in the form of marking. In this case a piece of damping sleeve may be fitted (Part No 321 511 123) over the top two coils. This is plastic tube. There is enough for top and bottom so cut it in half. The spring must be cleaned and the upper part of the coil spring and the inside of the plastic tube coated with petroleum jelly. Slide one half of the tube over the upper coils of the spring. If necessary, fit the other half of the tube on the bottom coils of the spring.
5 Once the plastic tube is in position the spring may be put back on the strut. The coils will still bottom but the noise will stop, for a while at least.
6 If the spring has distorted, then it has just got tired and it must be renewed. The cause could be one of several: incorrect material, wrong heat treatment, or even the wrong spring. These occasions are mercifully rare and distortion is usually brought on by continuous overloading.
7 When the springs are fitted they are selected for the purpose. Three grades are used, denoted by one, two or three paint stripes. When renewal is required VW state that only the spring with two stripes is available. They further say that it is not necessary to renew both springs if only one is faulty. So fit a spring with two green paint stripes.

9 Shock absorbers – testing

1 A defective shock absorber usually makes a rumbling noise as the car goes over rough surfaces. A quick test to confirm this suspicion is to press the rear of the car down on the suspected side and release it sharply. The car should rise and settle at once, if it oscillates at all the shock absorber is not working correctly. A small seepage of fluid round the piston seal is normal and need cause no worry, but a large leakage indicates that the shock absorber must be changed. There is no repair possible.
2 Remove and dismantle the suspension strut as in Sections 6 and 7. Hold the lower portion of the shock absorber in a vice with the piston rod in a vertical position. Pull the rod right up and press it down again. The force required in each direction should be even and equal. Repeat this test several times. If the stroke is uneven or jerky then the unit must be renewed.
3 Examine the lower anchorage. The bush must be a good fit on the locating bolt, and the rubber mounting in good condition. If either of these are faulty enquire whether a new bush and mounting are available from the VW store. Should this be so get a new locating bolt at the same time. Press the bush out of the eye with the mounting and press both of them into the eye. Lubricate the eye with a little rubber grease or soap before pressing the mounting home and be careful to enter it squarely. This must be done on a flypress or in a large vice, do not attempt to hammer it in. Alternatively an arrangement of a draw bolt with a nut and two large washers may be used.

10 Rear axle – removal and refitting

1 If the rear axle beam is distorted the entire rear suspension must be removed. It is not possible to remove the pivot bolts from the mountings as the bolts foul the body.
2 Support the vehicle body on axle stands (Section 6) and remove the wheels. Remove the brake drums and dismantle the brake shoes to free the handbrake cables (Chapter 6). Unclip the handbrake outer cables from the trailing arms (Fig. 9.9).

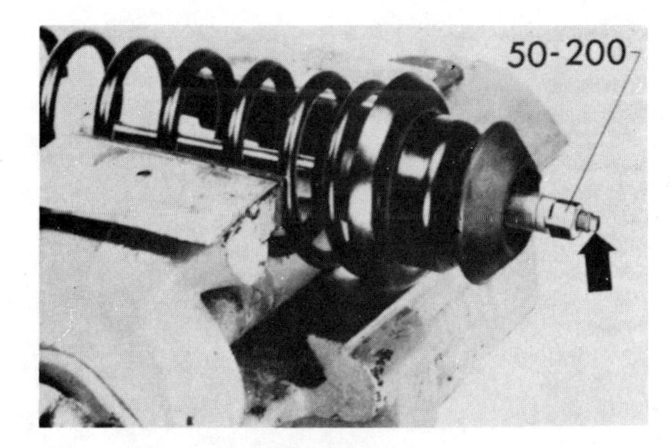

Fig. 9.8 Rear suspension strut mounted in a vice, showing holding flats (arrowed) and special tool (Sec 7)

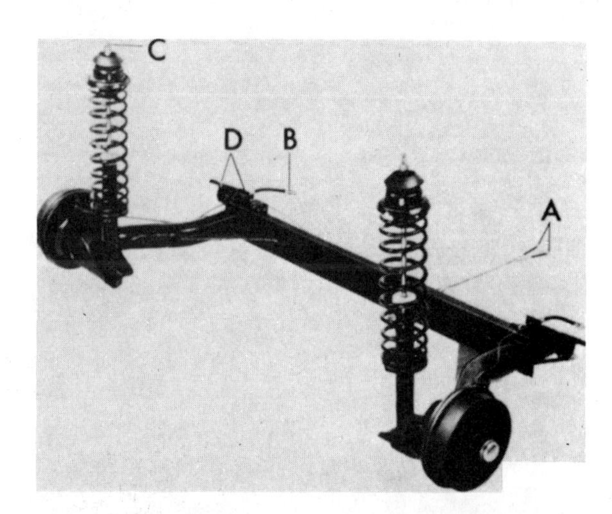

Fig. 9.9 Rear axle assembly removal (Sec 10)

A Handbrake cables
B Brake hydraulic hose
C Suspension strut upper mounting
D Pivot bolt

3 Disconnect the hydraulic brake lines and plug the hoses to prevent dirt or grease entering the system.
4 It may be necessary to remove parts of the exhaust system of the Rabbit before proceeding further.
5 Each mounting is held by two studs and nuts. These must now be slackened off. Be careful not to shear off the studs. If you do, refer to Section 11.
6 From inside the body remove the caps from the suspension struts and slacken off the nuts holding the shock absorber pistons to the body.
7 Support the axle and remove all the retaining nuts. It will be best to have three people to do this job. The axle beam and suspension struts may now be lowered to the ground and removed.
8 Separate the mountings by removing the hinge bolts. Note how the large washer is situated between the axle beam and the mounting. The dished washer is fitted with the bolt.
9 When assembling, lubricate the dished washer to prevent squeaks.
10 Refitting is the reverse of the removal procedure. Remember to adjust the handbrake and bleed the hydraulic system as described in Chapter 6.
11 When the car is back on its wheels, check that the hinge bolt nuts are tightened to the specified torque.

11 Rear axle mountings – removal and refitting

1 If you have the misfortune to shear one of the studs then the stud must be drilled out and the resultant hole tapped for a 10 mm thread.

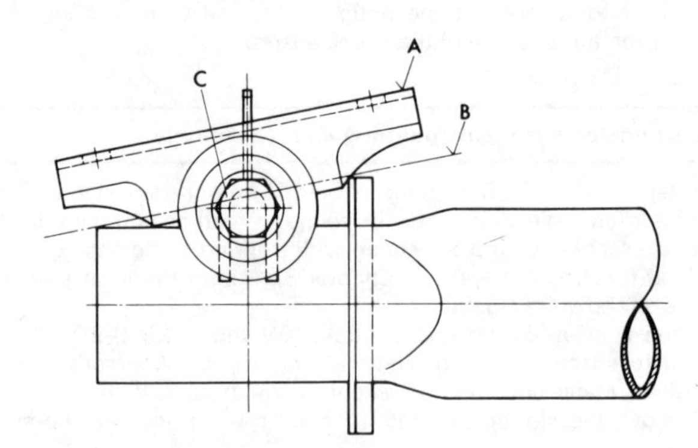

Fig. 9.10 Rear axle pivot bush tightening diagram (Sec 11)

Raise axle until line B is approximately parallel to upper face of mounting bracket A, then tighten nut C

Use an 8 mm drill and be careful to drill exactly in the centre of the stud. We advise caution. Unless you have experience in this type of work get a trained mechanic to do it for you. If you drill in the wrong place, the result will be very expensive. After tapping the hole fit a 10 mm bolt x 40, tensile class 10.9.
2 When refitting the mounting align it as shown in Fig. 9.10.

12 Fault diagnosis – rear axle and rear suspension

Symptom	Reason/s
Excessive or uneven tyre wear	Stub axle bent, check camber Axle beam distorted Wheel bearings defective Shock absorber defective Brake binding
Rumbling noise from rear axle	Shock absorber defective Wheel bearing defective
Chattering or knocking from rear axle	Top coil spring knocking on lower coil Lower mount of shock absorber worn Broken spring
Squeak from rear axle	Disc washer on hinge bolt rubbing on the bonded rubber bush Guide sleeve of handbrake cable requires lubrication

Chapter 10 Bodywork and fittings

Refer to Chapter 11 for specifications and information applicable to later models

Contents

1 General description

The body is of all-steel unit construction with impact-absorbing front and rear crumple zones which take the brunt of any accident, leaving the passenger compartment with minimum distortion. The front crumple zones take the form of two corrugated box sections in the scuttle and firewall.

There are two-door and four-door models. All models have a large tailgate hinged at the top, which is propped open by a gas-filled telescopic strut.

The front wings are bolted on to the body and can easily be renewed in the event of damage.

The type of bumpers fitted varies. In the USA and certain other territories, shock absorbing bumpers are mandatory. Where this is not a legal requirement normal bolt-on bumpers are fitted.

All models have individual fully reclining front bucket seats that can be adjusted to any position.

2 Maintenance – bodywork and underframe

1 The general condition of a vehicle's bodywork is the one thing that significantly affects its value. Maintenance is easy but needs to be regular. Neglect, particularly after minor damage, can lead quickly to further deterioration and costly repair bills. It is important also to keep watch on those parts of the car not immediately visible, for instance the underside and inside all the wheel arches.

2 The basic maintenance routine for the bodywork is washing – preferably with a lot of water, from a hose. This will remove all the loose solids which may have stuck to the car. It is important to flush these off in such a way as to prevent grit from scratching the finish. The wheel arches and underbody need washing in the same way to remove any accumulated mud which will retain moisture and tend to encourage rust. Paradoxically enough, the best time to clean the underbody and wheel arches is in wet weather when the mud is thoroughly wet and soft. In very wet weather the underbody is usually cleaned of large accumulations automatically and this is a good time for inspection.

3 Periodically it is a good idea to have the whole of the underside of the vehicle steam cleaned, so that a thorough inspection can be carried out to see what minor repairs and renovations are necessary. Steam cleaning is available at commercial vehicle garages but if not, there are one or two excellent grease solvents available which can be brush applied. The dirt can then be hosed off.

4 After washing paintwork, wipe it with a chamois leather to give an unspotted clear finish. A coat of clear protective wax polish will give added protection against chemical pollutants in the air. If the paintwork sheen has dulled or oxidised, this requires a little more effort, but is usually caused because regular washing has been neglected. Always check that drain holes are completely clear so that water can drain out (photo). Brightwork should be treated the same way as paintwork. Windscreens and windows can be kept clear of the smeary film which often appears if a little ammonia is added to the water. If they are scratched, a good rub with a proprietary metal polish will often clear them. Do not use any form of wax or chromium polish on glass.

3 Maintenance – upholstery and carpets

1 Mats and carpets should be brushed or vacuum cleaned regularly to keep them free of grit. If they are badly stained remove them from the car for scrubbing or sponging and make quite sure they are dry before replacement. Seats and interior trim panels can be kept clean by a wipe over with a damp cloth. If they do become stained (which can be more apparent on light coloured upholstery) use a little liquid

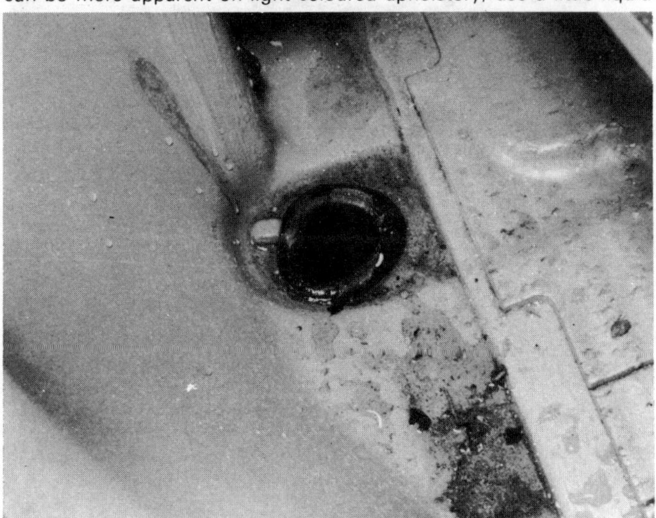

2.4 Bulkhead rubber grommet type water drain

This photo sequence illustrates the repair of a dent and damaged paintwork. The procedure for the repair of a hole is similar. Refer to the text for more complete instructions

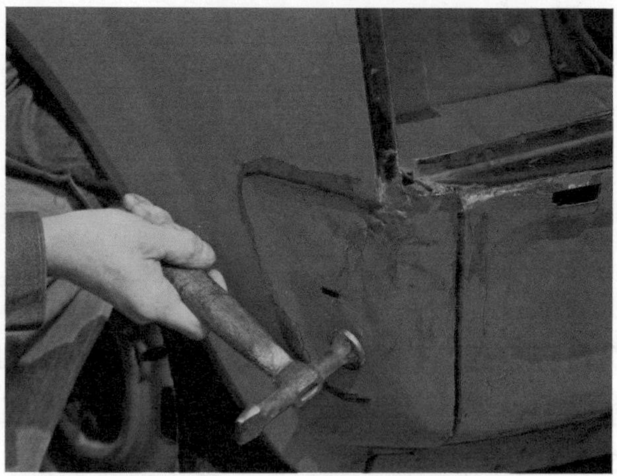
After removing any adjacent body trim, hammer the dent out. The damaged area should then be made slightly concave

Use coarse sandpaper or a sanding disc on a drill motor to remove all paint from the damaged area. Feather the sanded area into the edges of the surrounding paint, using progressively finer grades of sandpaper

The damaged area should be treated with rust remover prior to application of the body filler. In the case of a rust hole, all rusted sheet metal should be cut away

Carefully follow manufacturer's instructions when mixing the body filler so as to have the longest possible working time during application. Rust holes should be covered with fiberglass screen held in place with dabs of body filler prior to repair

Apply the filler with a flexible applicator in thin layers at 20 minute intervals. Use an applicator such as a wood spatula for confined areas. The filler should protrude slightly above the surrounding area

Shape the filler with a surform-type plane. Then, use water and progressively finer grades of sandpaper and a sanding block to wet-sand the area until it is smooth. Feather the edges of the repair area into the surrounding paint.

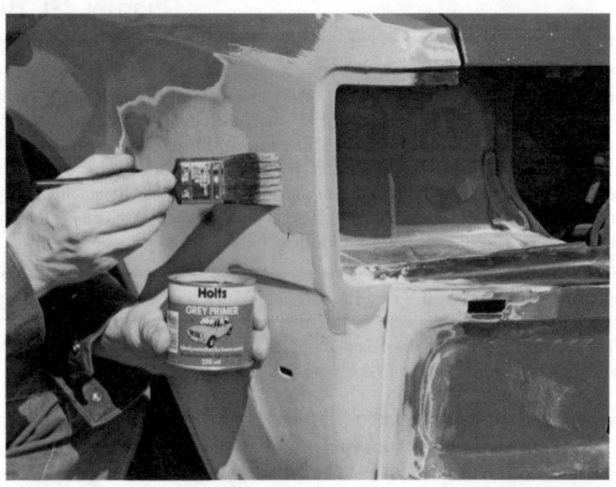

Use spray or brush applied primer to cover the entire repair area so that slight imperfections in the surface will be filled in. Prime at least one inch into the area surrounding the repair. Be careful of over-spray when using spray-type primer

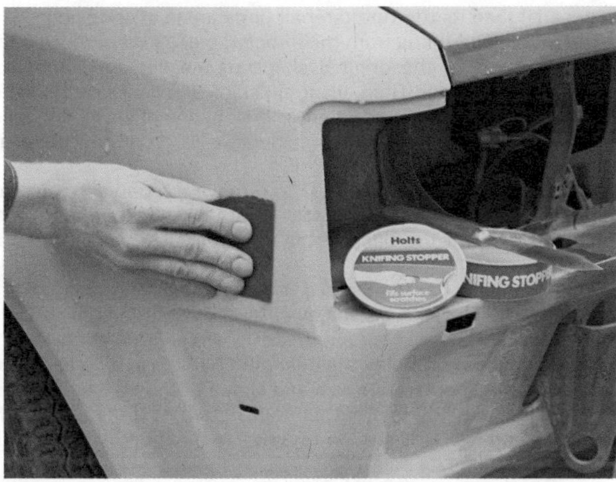

Wet-sand the primer with fine (approximately 400 grade) sandpaper until the area is smooth to the touch and blended into the surrounding paint. Use filler paste on minor imperfections

After the filler paste has dried, use rubbing compound to ensure that the surface of the primer is smooth. Prior to painting, the surface should be wiped down with a tack rag or lint-free cloth soaked in lacquer thinner

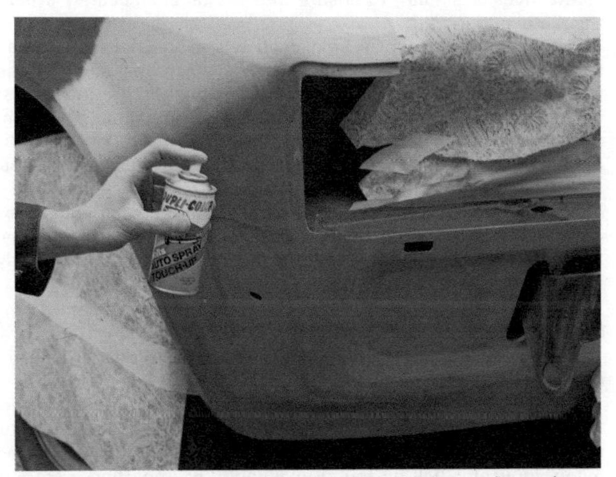

Choose a dry, warm, breeze-free area in which to paint and make sure that adjacent areas are protected from over-spray. Shake the spray paint can thoroughly and apply the top coat to the repair area, building it up by applying several coats, working from the center

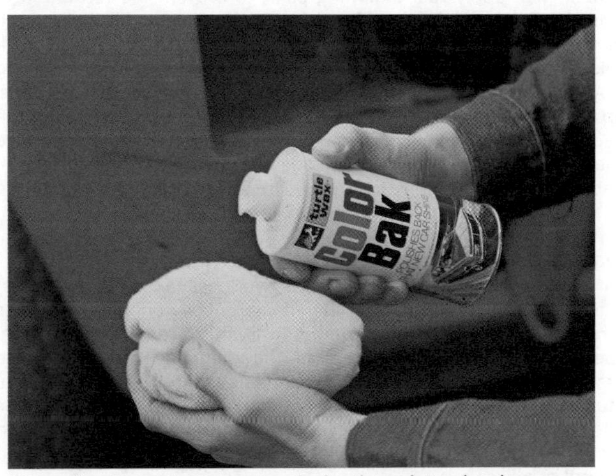

After allowing at least two weeks for the paint to harden, use fine rubbing compound to blend the area into the original paint. Wax can now be applied

detergent and a soft nailbrush to scour the grime out of the grain of the material. Do not forget to keep the head lining clean in the same way as the upholstery. When using liquid cleaners inside the car do not overwet the surfaces being cleaned. Excessive damp could get into the seams and padded interior causing stains, offensive odours or even rot. If the inside of the car gets wet accidentally it is worthwhile taking some trouble to dry it out properly, particularly where carpets are involved. **Do not** leave heaters inside for this purpose.

4 Minor body damage – repair

The photographic sequences on pages 202 and 203 illustrate the operations detailed in the following sub-sections.

Repair of minor scratches in the car's bodywork

If the scratch is very superficial, and does not penetrate to the metal of the bodywork, repair is very simple. Lightly rub the area of the scratch with a paintwork renovator, or a very fine cutting paste, to remove loose paint from the scratch and to clear the surrounding bodywork of wax polish. Rinse the area with clean water.

Apply touch-up paint to the scratch using a thin paint brush; continue to apply thin layers of paint until the surface of the paint in the scratch is level with the surrounding paintwork. Allow the new paint at least two weeks to harden: then blend it into the surrounding paintwork by rubbing the paintwork in the scratch area with a paintwork renovator or a very fine cutting paste. Finally, apply wax polish.

If the car is painted with a two-coat metallic finish an entirely different technique is required. The materials may be obtained from the official agent. Two types of repair are possible, the 80°C drying method and the Air-drying method. A 'wet-on-wet' procedure for the topcoat and clear varnish is used. The repair can be done satisfactorily only if the specified top coat and varnish are used with the specially developed synthetic thinner. After filling with Filler L145, if required, sand down with 400-500 wet and dry paper. Apply the first top coat using synthetic resin metallic paint LKL of spraying viscosity 15-17 seconds (DIN cup 4 mm). Let the paint flash off for 25 minutes, then apply the second layer of Air-drying L100 clear varnish with hardener L101 mixed in proportion 8:1. This becomes unusable after six hours. The repair is dust dry after 30 minutes but requires up to 8 days for complete drying. As can be seen it is a complicated process and you are advised to go to the official agent for advice if you have not done the job before. If you have other than a metallic finish then proceed as follows.

Where the scratch has penetrated right through to the metal of the bodywork, causing the metal to rust, a different repair technique is required. Remove any loose rust from the bottom of the scratch with a penknife, then apply rust inhibiting paint to prevent the formation of rust in the future. Using a rubber or nylon applicator fill the scratch with bodystopper paste. If required, this paste can be mixed with cellulose thinners to provide a very thin paste which is ideal for filling narrow scratches. Before the stopper-paste in the scratch hardens, wrap a piece of smooth cotton rag around the top of a finger. Dip the finger in cellulose thinners and then quickly sweep it across the surface of the stopper paste in the scratch; this will ensure that the surface of the stopper paste is slightly hollowed. The scratch can now be painted over as described earlier in this Section.

Repair of dents in the bodywork

When deep denting of the car's bodywork has taken place, the first task is to pull the dent out, until the affected bodywork almost attains its original shape. There is little point in trying to restore the original shape completely, as the metal in the damaged area will have stretched on impact and cannot be reshaped fully to its original contour. It is better to bring the level of the dent up to a point which is about (3 mm) $\frac{1}{8}$ in below the level of the surrounding bodywork. In cases where the dent is very shallow anyway, it is not worth trying to pull it out at all.

If the underside of the dent is accessible, it can be hammered out gently from behind, using a mallet with a wooden or plastic head. Whilst doing this, hold a suitable block of wood firmly against the impact from the hammer blows and thus prevent a large area of the bodywork from being 'belled-out'.

Should the dent be in a section of the bodywork which has double skin or some other factor making it inaccessible from behind, a different technique is called for. Drill several small holes through the metal inside the area – particularly in the deeper section. Then screw long self-tapping screws into the holes just sufficiently for them to gain a good purchase in the metal. Now the dent can be pulled out by pulling on the protruding heads of the screws with a pair of pliers.

The next stage of the repair is the removal of the paint from the damaged area, and from an inch or so of the surrounding 'sound' bodywork. This is accomplished most easily by using a wire brush or abrasive pad on a power drill, although it can be done just as effectively by hand using sheets of abrasive paper. To complete the preparation for filling, score the surface of the bare metal with a screwdriver or the tang of a file, or alternatively, drill small holes in the affected area. This will provide a really good key for the filler paste.

To complete the repair see the Section on filling and respraying.

Repair of rust holes or gashes in the bodywork

Remove all paint from the affected area and from an inch or so of the surrounding 'sound' bodywork, using an abrasive pad or a wire brush on a power drill. If these are not available a few sheets of abrasive paper will do the job just as effectively. With the paint removed you will be able to gauge the severity of the corrosion and therefore decide whether to renew the whole panel (if this is possible) or to repair the affected area. New body panels are not as expensive as most people think and it is often quicker and more satisfactory to fit a new panel than to attempt to repair large areas of corrosion.

Remove all fittings from the affected areas except those which will act as a guide to the original shape of the damaged bodywork (eg headlamp shells etc). Then, using tin snips or a hacksaw blade, remove all loose metal and any other metal badly affected by corrosion. Hammer the edges of the hole inwards in order to create a slight depression for the filler paste.

Wire brush the affected area to remove the powdery rust from the surface of the remaining metal. Paint the affected area with rust inhibiting paint; if the back of the rusted area is accessible treat this also.

Before filling can take place it will be necessary to block the hole in some way. This can be achieved by the use of one of the following materials: Zinc gauze, Aluminium tape or Polyurethane foam.

Zinc gauze is probably the best material to use for a large hole. Cut a piece to the approximate size and shape of the hole to be filled, then position it in the hole so that its edges are below the level of the surrounding bodywork. It can be retained in position by several blobs of filler paste around its periphery.

Aluminium tape should be used for small or very narrow holes. Pull a piece off the roll and trim it to the approximate size and shape required, then pull off the backing paper (if used) and stick the tape over the hole; it can be overlapped if the thickness of one piece is insufficient. Burnish down the edges of the tape with the handle of a screwdriver or similar, to ensure that the tape is securely attached to the metal underneath.

Polyurethane foam is best used where the hole is situated in a section of bodywork of complex shape, backed by a small box section (eg where the sill panel meets the rear wheel arch – most cars). The usual mixing procedure for this foam is as follows: put equal amounts of fluid from each of the two cans provided in the kit, into one container. Stir until the mixture begins to thicken, then quickly pour this mixture into the hole, and hold a piece of cardboard over the larger apertures. Almost immediately the polyurethane will begin to expand, gushing out of any small holes left unblocked. When the foam hardens it can be cut back to just below the level of the surrounding bodywork with a hacksaw blade.

Bodywork repairs – filling and respraying

Before using this Section, see the Sections on dent, deep scratch, rust holes and gash repairs.

Many types of bodyfiller are available, but generally speaking those proprietary kits which contain a tin of filler paste and a tube of resin hardener are best for this type of repair. A wide, flexible plastic or nylon applicator will be found invaluable for imparting a smooth and well contoured finish to the surface of the filler.

Mix up a little filler on a clean piece of card or board – use the hardener sparingly (follow the maker's instructions on the packet) otherwise the filler will set very rapidly.

Using the applicator, apply the filler paste to the prepared area: draw the applicator across the surface of the filler to achieve the correct contour and to level the filler surface. As soon as a contour that approximates the correct one is achieved, stop working the paste – if you carry on too long the paste will become sticky and begin to 'pick up' on the applicator. Continue to add thin layers of filler paste at twenty-minute intervals until the level of the filler is just proud of the surrounding bodywork.

Once the filler has hardened, excess can be removed using a metal plane or file. From then on, progressively finer grades of abrasive paper should be used, starting with a 40 grade production paper and finishing with 400 grade wet-and-dry paper. Always wrap the abrasive paper around a flat rubber, cork, or wooden block – otherwise the surface of the filler will not be completely flat. During the smoothing of the filler surface the wet-and-dry paper should be periodically rinsed in water. This will ensure that a very smooth finish is imparted to the filler at the final stage.

At this stage the 'dent' should be surrounded by a ring of bare metal, which in turn should be encircled by the finely 'feathered' edge of the good paintwork. Rinse the repair area with clean water, until all of the dust produced by the rubbing-down operation has gone.

Spray the whole repair area with a light coat of grey primer – this will show up any imperfections in the surface of the filler. Repair these imperfections with fresh filler paste or bodystopper, and once more smooth the surface with abrasive paper. If bodystopper is used, it can be mixed with cellulose thinners to form a really thin paste which is ideal for filling small holes. Repeat this spray and repair procedure until you are satisfied that the surface of the filler, and the feathered edge of the paintwork are perfect. Clean the repair area with clean water and allow to dry fully.

The repair area is now ready for final spraying. Paint spraying must be carried out in a warm, dry, windless and dust free atmosphere. This condition can be created artificially if you have access to a large indoor working area, but if you are forced to work in the open, you will have to pick your day very carefully. If you are working indoors, dousing the floor in the work area with water will help settle the dust which would otherwise be in the atmosphere. If the repair area is confined to one body panel, mask off the surrounding panels; this will help to minimise the effects of a slight mis-match in paint colours. Bodywork fittings (eg chrome strips, door handles etc) will also need to be masked off. Use genuine masking tape and several thicknesses of newspaper for the masking operations.

Before commencing to spray, agitate the aerosol can thoroughly, then spray a test area (an old tin, or similar) until the technique is mastered. Cover the repair area with a thick coat of primer; the thickness should be built up using several thin layers of paint rather than one thick one. Using 400 grade wet-and-dry paper, rub down the surface of the primer until it is really smooth. While doing this, the work area should be thoroughly doused with water, and the wet-and-dry paper periodically rinsed in water. Allow to dry before spraying on more paint.

Spray on the top coat, again building up the thickness by using several thin layers of paint. Start spraying in the centre of the repair area and then using a circular motion, work outwards until the whole repair area and about 2 inches of the surrounding original paintwork is covered. Remove all masking material 10 to 15 minutes after spraying on the final coat of paint.

Allow the new paint at least two weeks to harden, then, using a paintwork renovator or a very fine cutting paste, blend the edges of the paint into the existing paintwork. Finally, apply wax polish.

5 Major body damage – repair

Where serious damage has occurred or large areas need renewal due to neglect it means certainly that completely new sections or panels will need welding in and this is best left to professionals. If the damage is due to impact it will also be necessary to check the alignment of the body structure. In such instances the services of an agent with specialist checking jigs are essential. If a body is left misaligned it is first of all dangerous as the car will not handle properly – and secondly, uneven stresses will be imposed on the steering, engine and transmission, causing abnormal wear or complete failure. Tyre wear will also be excessive.

6 Front wings – removal and refitting

1 A badly damaged front wing may be removed complete and a new one fitted. The wing is secured with 10 fixing screws. Refer to Fig. 10.2 in which the locations of the screws are shown. Those marked 'A1' are under the wing and not easily accessible.
2 It will be necessary to remove the side bumper before removing the wing. The screws fastening the wing are fitted very tightly and may not come out without considerable force. Do not use an impact screwdriver or you will distort the frame. It may be necessary to grind off the heads and drill out the shanks.
3 Once all the screws are out, the wing may be levered away pulling it out of the guides. If it does not come out easily it may be necessary to warm the line of the joint with a blow lamp to melt the adhesive underseal. Be careful how you do this, for apart from the fire risk to the car the adhesive is also inflammable.
4 When removing, lever the wing away from the wheel housing and the door pillar, work it to-and-fro, pulling it forwards to the front of the car a little.
5 Clean up the frame and paint with inhibitor if any rust is present. Use a good sealing tape along the line of the bolts before installing the wing, and once the wing is securely in place treat the underside with underseal compound. Refer to Section 4 for respraying techniques.

7 Radiator grille – removal and refitting

The grille is held in position by eight screws. There are four along the top., visible when the bonnet is opened. There are two more by each headlamp. Fig. 10.1 refers. Note that some later models have plastic clips instead of screws (photo).

8 Tailgate, strut and lock – removal and refitting

1 Disconnect the battery earth strap.
2 Open the tailgate and remove the straps which support the rear shelf. Disconnect the gas-filled support strut from the tailgate.
3 Ease the headlining back to uncover the hinge securing bolts. Take care not to tear the lining. Now slacken the hinge bolts. This will

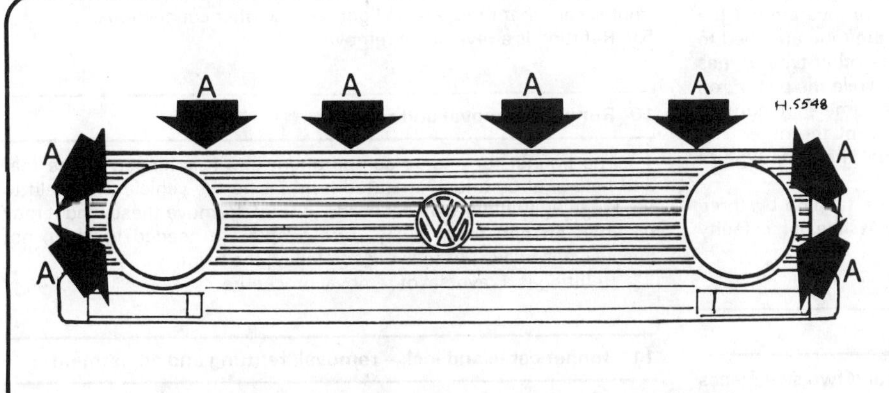

Fig. 10.1 Radiator grille fixing screw locations (A) (Sec 7)

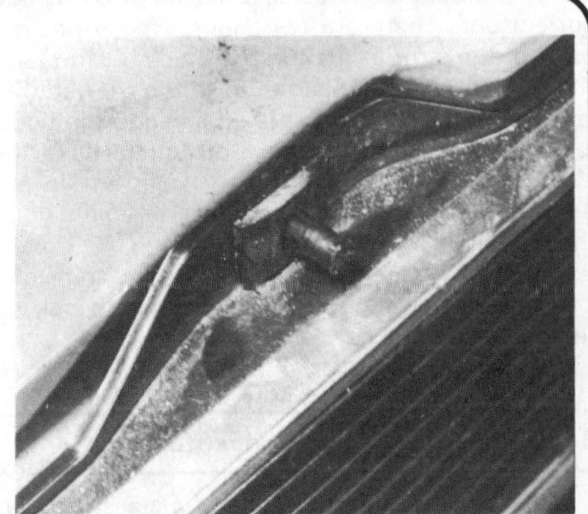
7.1 Radiator grille retaining clip

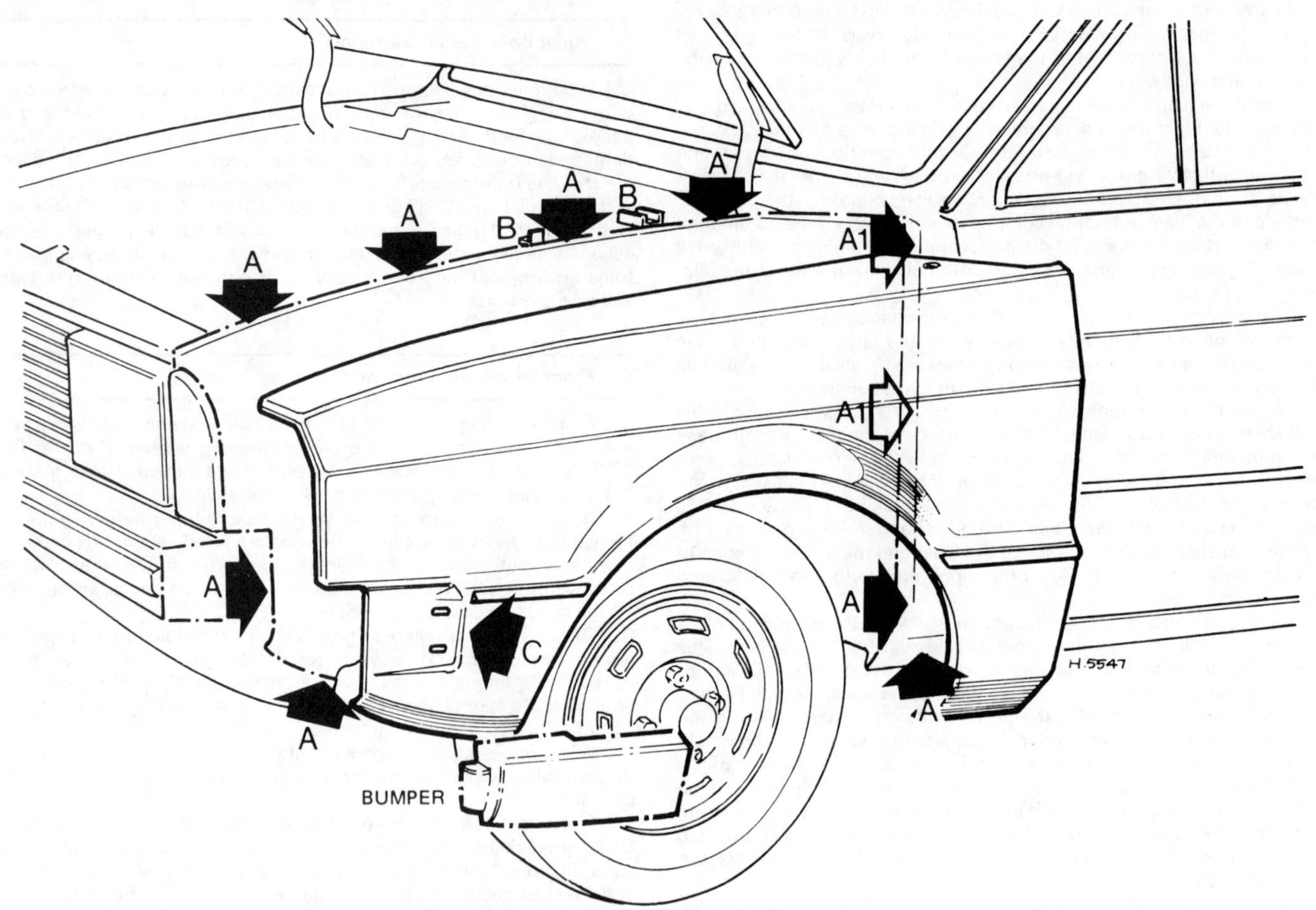

Fig. 10.2 Front wing fixing screw locations (Sec 6)

Note: Wing is shown in continuous line, car frame in broken line

A1 Two screws beneath wing
A Eight screws along top, front and rear of wing
B Guides
C Bumper fixing location

require an impact screwdriver with a special bit.

4 Disconnect the electrical wiring to the heated rear window at the connector.

5 With the help of an assistant remove the hinge bolts and lift away the tailgate.

6 When refitting the tailgate, insert the hinge bolts and tighten them hand-tight. Close the tailgate and check that there is an even gap all round and that the lock works correctly. If it does not, adjust the hinge position until it does. Tighten the hinge bolts fully and refit the headlining in position over them.

7 Refit the tailgate support strut.

8 If the support strut is to be renewed, it is important to get the correct one. There are two kinds, type A has one groove around the body of the strut at the top. This type of strut has the tube attached to the tailgate and the piston rod to the car body. The other type, B, has two grooves and the tube is attached to the body while the piston rod is attached to the tailgate. Both struts have the same part number (171 827 550A) but if you try to fit one in place of the other, the tailgate will not function properly. Make sure you get the same type as the one taken off.

9 The lock is secured to the bottom edge of the tailgate by three crosshead screws (photo). No repair to the lock is possible. If it is faulty remove it and fit a new one.

9 Bumpers – removal and refitting

1 The front bumper consists of a crossmember and two side pieces on early Golf models, and the sections are fitted to each other and the wings by studs and nuts. To remove the bumper, first disconnect the bumper lamp wires and unscrew the stud nuts. Withdraw the side pieces then unbolt the centre section from the underframe.

2 Later Golf models are fitted with a similar centre section to which is attached a single plastic covering. To remove this type, use a screwdriver to prise the side covering from each wing, disconnect the bumper lamp wires and headlamp washer tubing where fitted, and unbolt the centre section from the underframe. The plastic covering is obtainable as a separate item but a press will be required to attach it to the centre section (Fig. 10.6).

3 The front bumper fitted to Rabbit models is removed in a similar manner but bolts holding the recuperators to the underframe must be removed first.

4 Removal of the rear bumper is similar to the procedure for the front except that there are no lighting or washer connections.

5 Refitting is a reversal of removal.

10 Bonnet – removal and refitting

1 The bonnet, or engine compartment cover, is hinged at the rear and held shut by a lock operated from inside the vehicle. It is held to the hinge by two bolts on each side (photo). Remove these, and it may be lifted off and taken away. Two people are needed to lift it, not because it is heavy but to avoid scratching the paint.

2 Refitting is a reversal of removal procedure.

11 Bonnet cable and lock – removal, refitting and adjustment

1 Working inside the car, unscrew the two crosshead retaining screws and release the bonnet lock handle from the left-hand side panel.

Fig. 10.3 Type B tailgate support strut showing the two identification rings (arrowed)(Sec 8)

2 Disconnect the cable at the lock end after removing trhe radiator grille.

3 Bend the clamping plate in the operating handle outwards, and remove the handle from the cable.

4 Remove the cable from the car.

5 Refitting is a reversal of removal, but make sure that the cable is routed free of strain without any sharp curves. To adjust the cable, pull it through the clamp on the lock as far as possible then, with the lever fully released, tighten the clamp screw and bend the excess cable over.

8.9 The tailgate lock

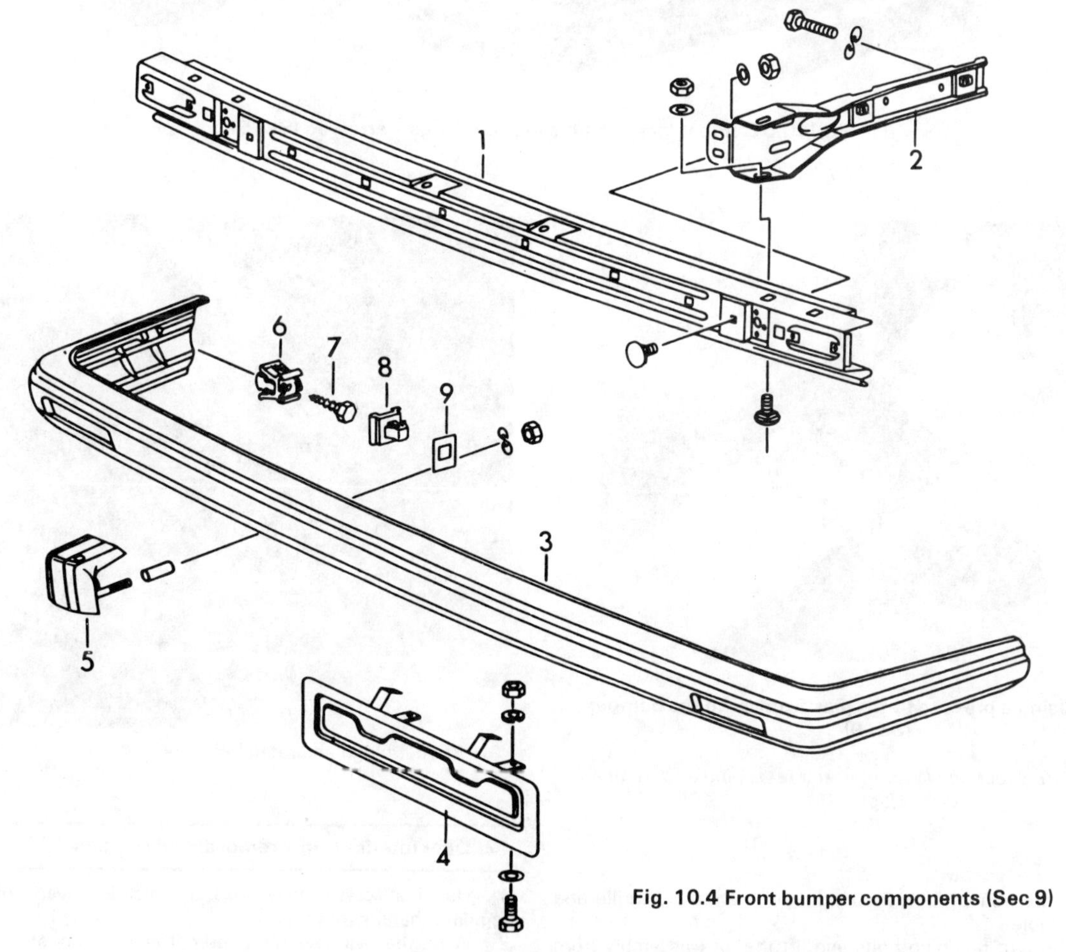

Fig. 10.4 Front bumper components (Sec 9)

1 *Centre section*
2 *Bracket*
3 *Plastic cover*
4 *Number plate backing*
5 *Overrider*
6 *Clamp*
7 *Bolt*
8 *Clip*
9 *Seal*

Fig. 10.5 Bumper location hole in front wings – arrowed (Sec 9)

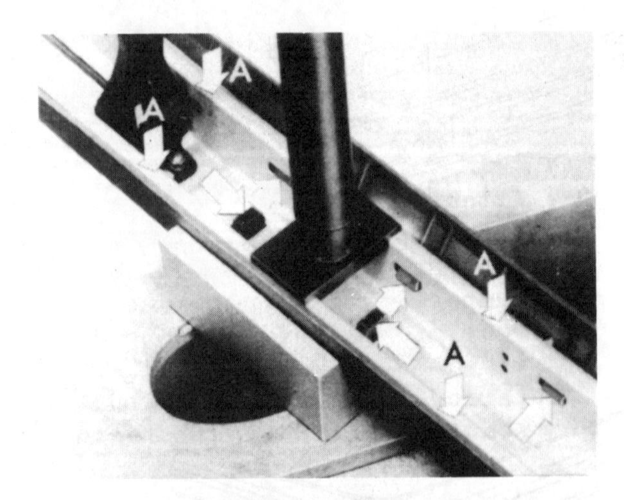

Fig. 10.6 Using a press to fit the plastic cover to the bumper (Sec 9)

Arrows A indicate direction of pressure, and remaining arrows location lugs

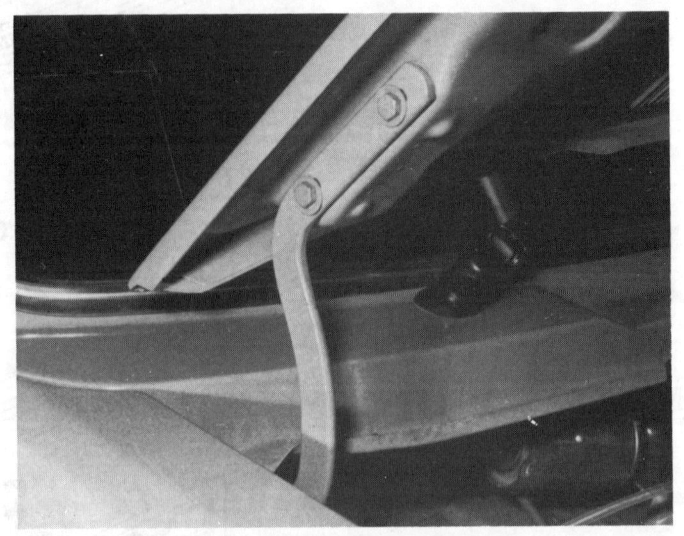

10.1 Bonnet hinge and retaining bolts

6 To remove the bonnet lock, first withdraw the radiator grille and disconnect the cable.
7 Unscrew the two retaining bolts and lift the lock assembly from the crossmember.
8 Refitting is a reversal of removal but it will be necessary to adjust the cable as described in paragraph 5.

12 Door interior trim – removal and refitting

1 Using a screwdriver, prise the plastic cover from the window regulator handle (photo).
2 With the window shut, note the position of the handle, then remove the retaining screw and slide the handle from the shaft (photo).
3 Unscrew and remove the two crosshead retaining screws and withdraw the door pull (photo).

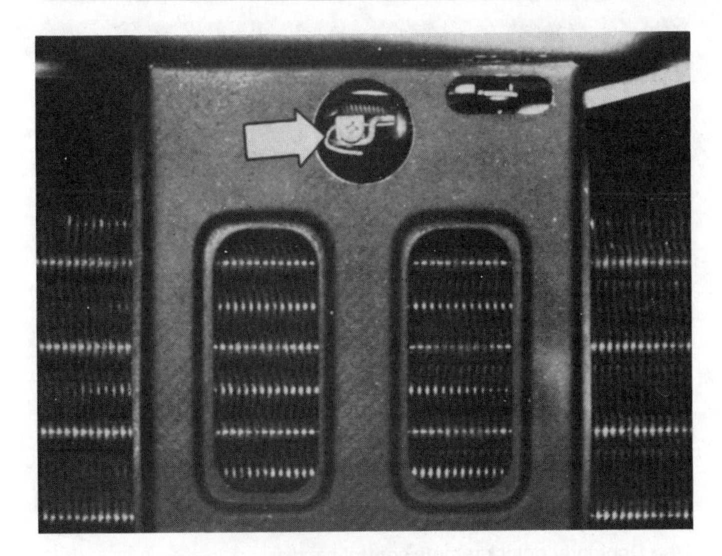

Fig. 10.7 Bonnet lock cable adjusting screw – arrowed (Sec 11)

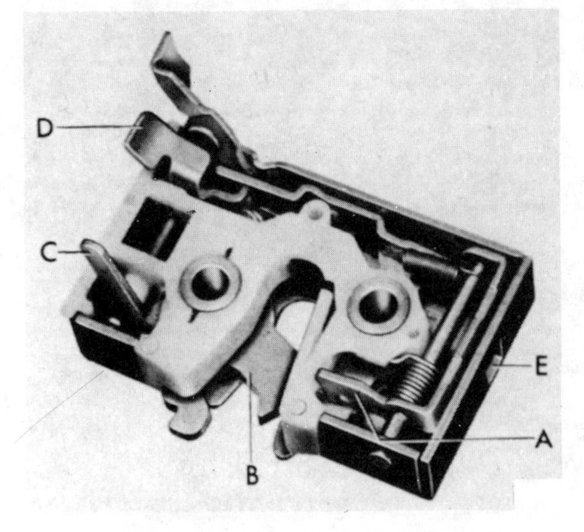

Fig. 10.8 Door lock components (Sec 13)

A Remote control lever
B Latch
C Locking lever
D Locking lever
E Access hole

4 Prise the recessed cover from the door interior handle assembly. Unscrew the retaining screw and withdraw the handle cover (photos).
5 Where fitted, prise off the cover and remove the mirror adjusting knob and escutcheon.
6 Using a wide blade screwdriver inserted between the trim and the door, prise the retaining clips out of the holes. Lever as near to the clips as possible to avoid tearing them from the trim panel (photo).
7 Refitting is a reversal of removal, but make sure that the plastic sheeting is correctly positioned before mounting the trim on the door.

13 Door lock – removal and refitting

Note: *It is not necessary to remove the door interior trim in order to remove the door lock.*
1 Open the door and set the lock in the locked position, either by moving the interior knob or by turning the exterior key.
2 Using a hexagon key, unscrew and remove the two retaining screws, then withdraw the lock approximately 12 mm (0·5 in) to

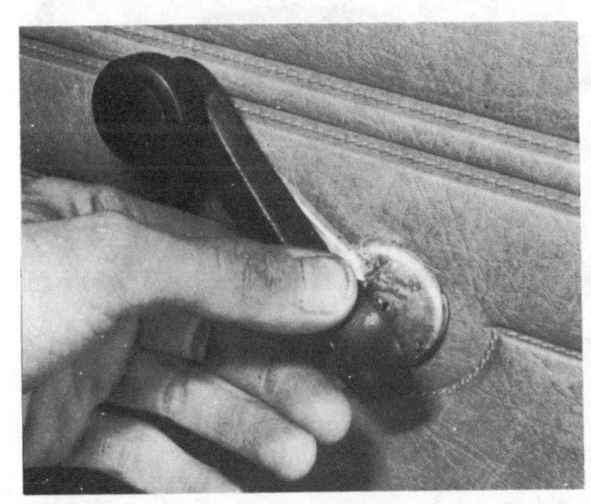

12.1 Removing the window regulator handle cover

12.2 Removing the window regulator handle

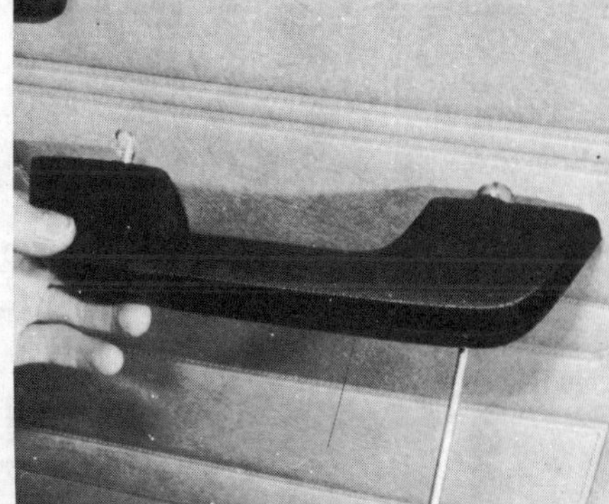

12.3 Removing the door pull/armrest

12.4A Removing the door interior handle recessed cover

12.4B Door interior handle cover retaining screw

12.6 Removing the door interior trim panel

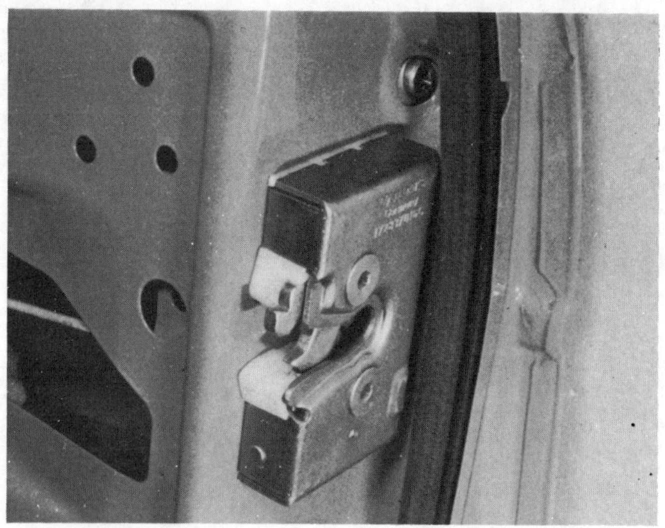

13.2 The door lock assembly

14.3 Door interior lock remote control handle

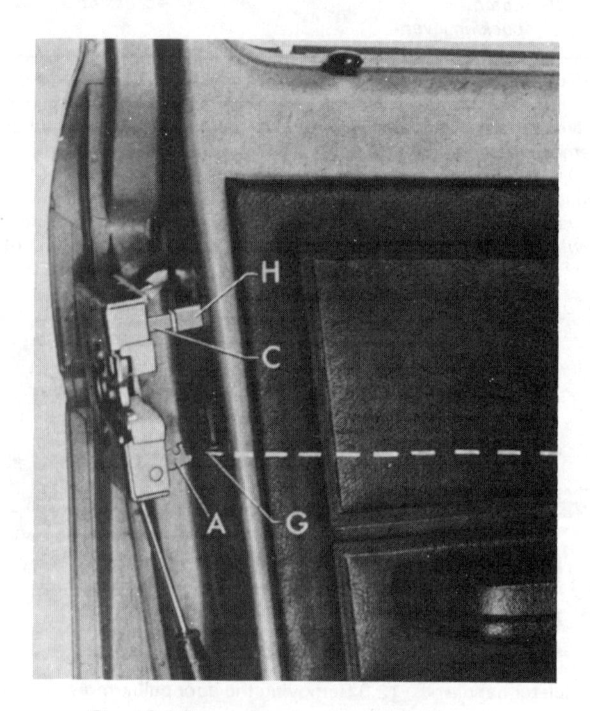

Fig. 10.9 Removing the door lock (Sec 13)

A Lever
C Locking lever
G Remote control rod
H Sleeve

expose the operating lever (photo) (Figs. 10.8 and 10.9).

3 Retain the lower lock lever in the extended position by inserting a screwdriver through the hole in the bottom of the lock.

4 Unhook the lower lever from the remote control rod and pull the upper lever from the sleeve. Remove the lock from the door.

5 Refitting is a reversal of removal, but remember to set the lock in the locked position first.

14 Door lock remote control – removal and refitting

1 Remove the door lock as described in Section 13.

2 Remove the door interior trim as described in Section 12.

3 Unbolt the remote control handle from the door and remove it (photo).

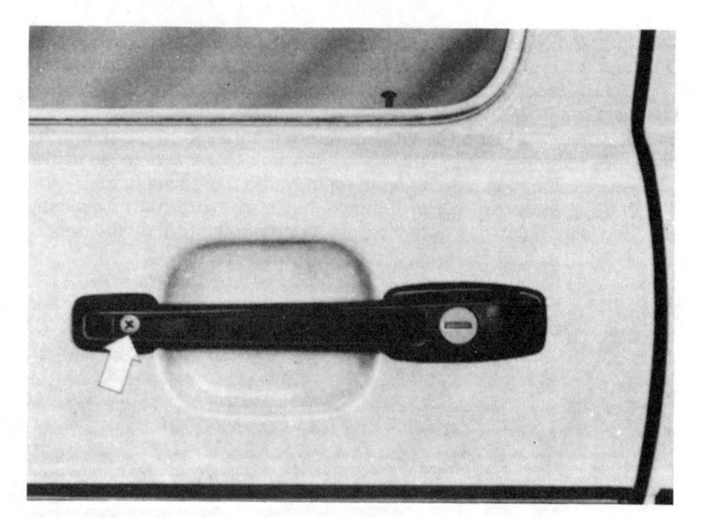

Fig. 10.10 Door exterior handle outer retaining screw – arrowed (Sec 15)

4 Refitting is a reversal of removal, but make sure that any sound deadening foam is positioned correctly on the control rod.

15 Door exterior handle – removal and refitting

1 Open the door and remove the retaining screw located above the door lock.

2 Prise the plastic insert from the handle and remove the second crosshead retaining screw. Note that some early models may not have this screw.

3 If the driver's side handle is being removed, make sure that it is unlocked.

4 Withdraw the handle from the door.

5 Refitting is a reversal of removal.

16 Window regulator – removal and refitting

1 Remove the interior trim as described in Section 12, after fully opening the window, then pull back the plastic sheeting (photo).

2 Remove the bolts securing the winding gear to the door, and the bolts securing the lifting plate to the window channel (photos).

3 Unclip the regulator tube from the door and remove it through the aperture.

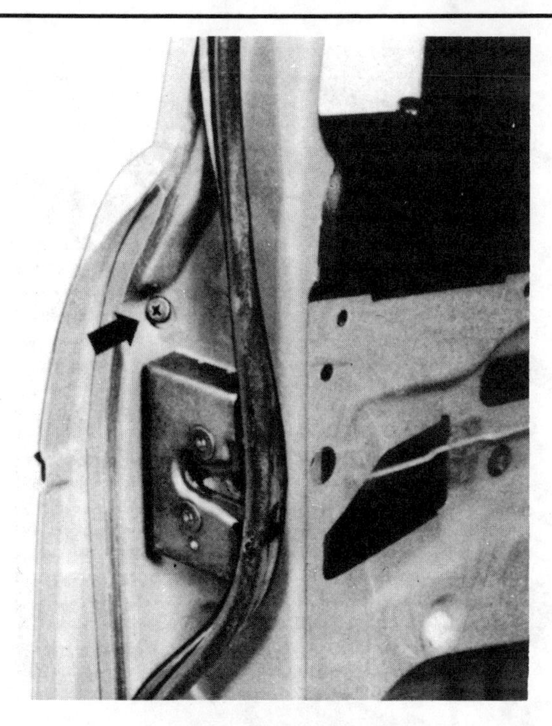

Fig. 10.11 Door exterior handle side retaining screw – arrowed (Sec 15)

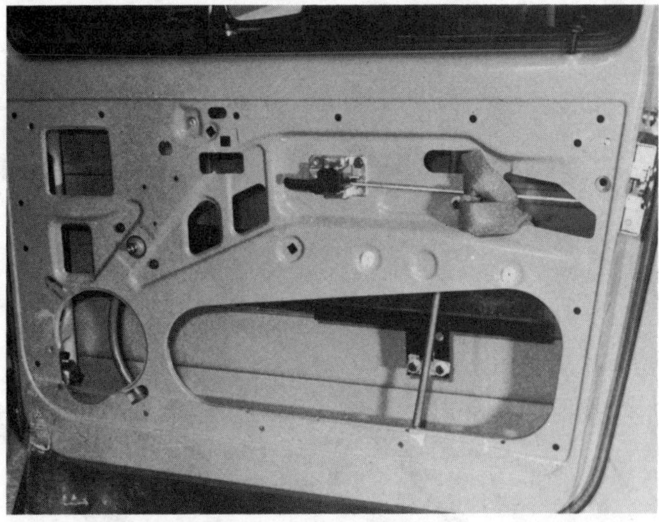

16.1 View of door inner panel with plastic sheeting removed

4 Refitting is a reversal of removal, but ensure that the cable is adequately lubricated with grease, and if necessary adjust the position of the regulator tube so that the window moves smoothly.

17 Door window glass – removal and refitting

1 Remove the window regulator as described in Section 16.
2 Remove the chrome moulding from the window aperture.
3 Remove the front window channel by unscrewing the upper retaining screw and drilling out the lower rivet (Fig. 10.12).
4 Unclip the window seals from the door.
5 Withdraw the window from the door.
6 If necessary, the triangular vent window can also be removed from the door at this stage.
7 Refitting is a reversal of removal, but a rivet inserting tool will be required.

16.2A Window regulator gear

18 Doors (front) – removal and refitting

1 The hinges on the Golf/Rabbit are secured to the pillar with shallow socket headed bolts which are very difficult to undo unless the special tools are available. They are also in awkward position. If the socket heads are damaged and the hexagon hole is converted into a round one then you are in serious trouble, as it will be necessary to drill the bolt and extract it with a special bolt extractor. For this reason we recommend that door hinges should be undone by professionals with the correct equipment. However, if you can undo these bolts the door is easily removed.
2 Take the pin out of the check strap, slacken the hinge bolts, have someone hold the door while you remove the bolts, and lift the door away (photo). Check the weather strip and fit a new one, if necessary.
3 When refitting the door first remove the striker bolt from the frame. You cannot adjust the hinge while this is in place. Fit the door to the pillar and tighten the hinge bolts enough to hold the door in place. Close the door and check that the gap all round the edge is symmetrical. Adjust the hinge until it is correct then tighten the hinge bolts fully. Refit the striker bolt.
4 Note that on late models, self-locking M8 x 18 hinge bolts are fitted, and those must always be renewed when the door is removed. They should be tightened to 5 kgf m (37 lbf ft).

16.2B Window regulator lifting plate and window channel

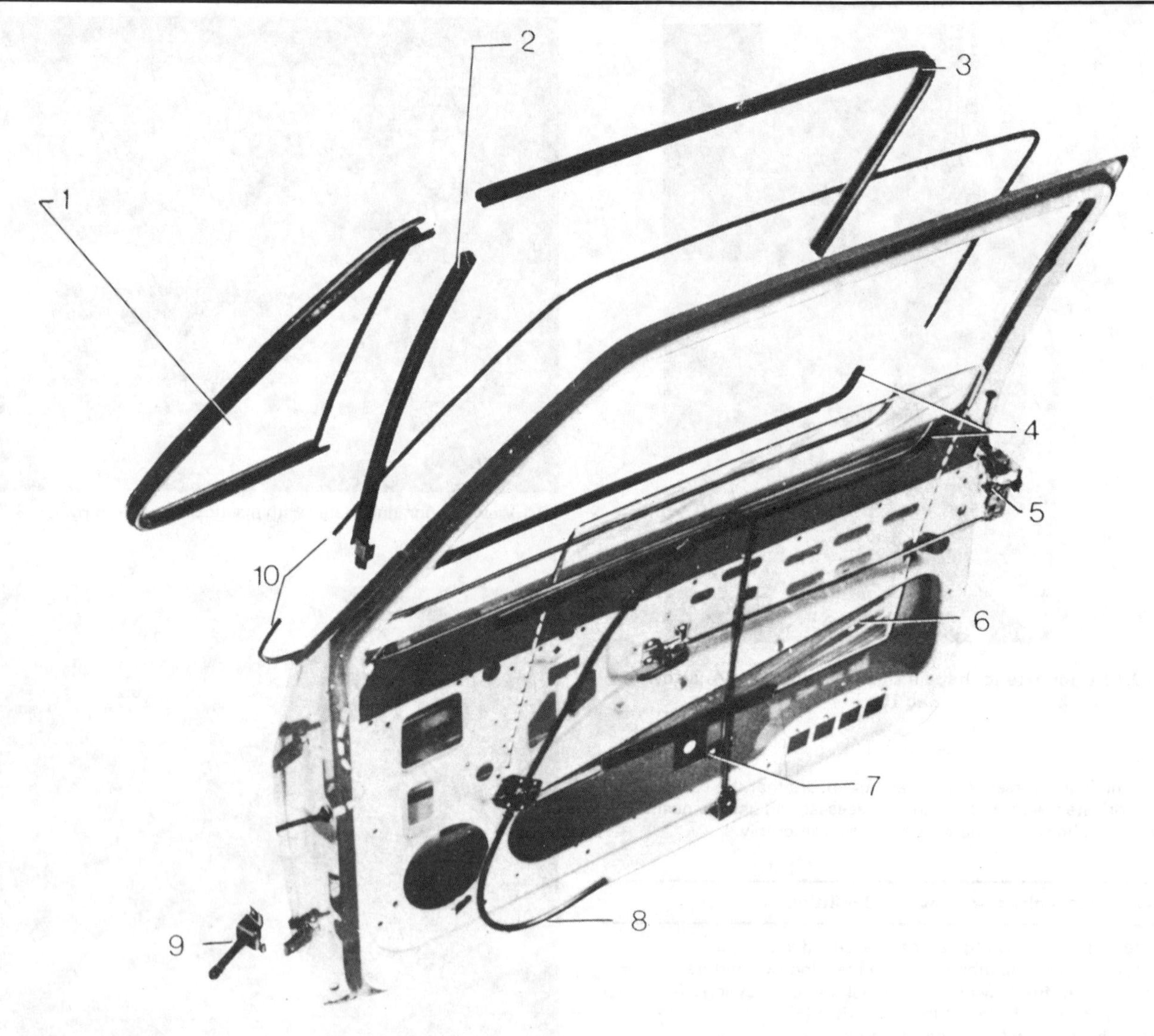

Fig. 10.12 Front door window components (Sec 17)

1 Vent window
2 Front window channel
3 Guide
4 Seals
5 Lock
6 Glass
7 Lifting plate
8 Regulator
9 Check strap
10 Trim moulding

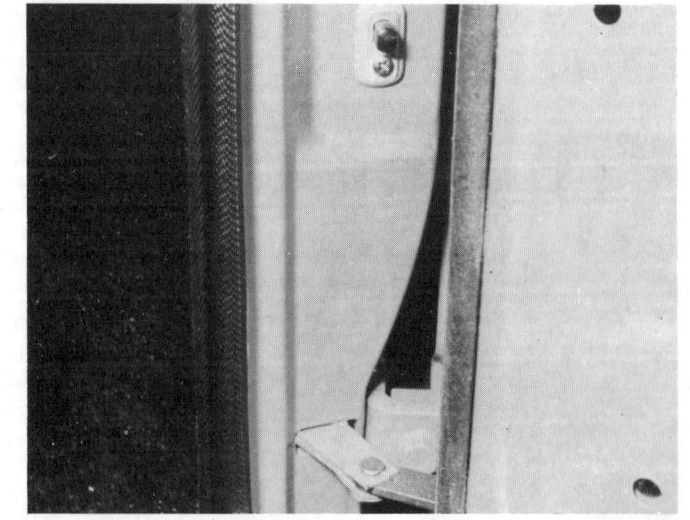

18.2 The door check strap

19 Remote controlled exterior mirror – removal and refitting

1 Remove the door interior trim panel as described in Section 12.
2 Unscrew the countersunk screws securing the mirror to the door exterior and lift it out together with the reinforcement plate.
3 From the door interior panel, unscrew the three retaining bolts, and withdraw the remote control lever and bracket together with the anti-rattle foam.
4 Refitting is a reversal of removal.

20 Front seats – removal and refitting

Removal of the front seat from the car is done in the following manner. At the back of the runners are small caps, lever these off the runners. Now look under the seat. There is an extra clip which must be removed and then the seat can be pushed towards the rear. When refitting fit the spring clips in the upper slide, install the seat from the rear, guiding the adjusting lever into the latch bolt mountings. Refit the covers on the runners.

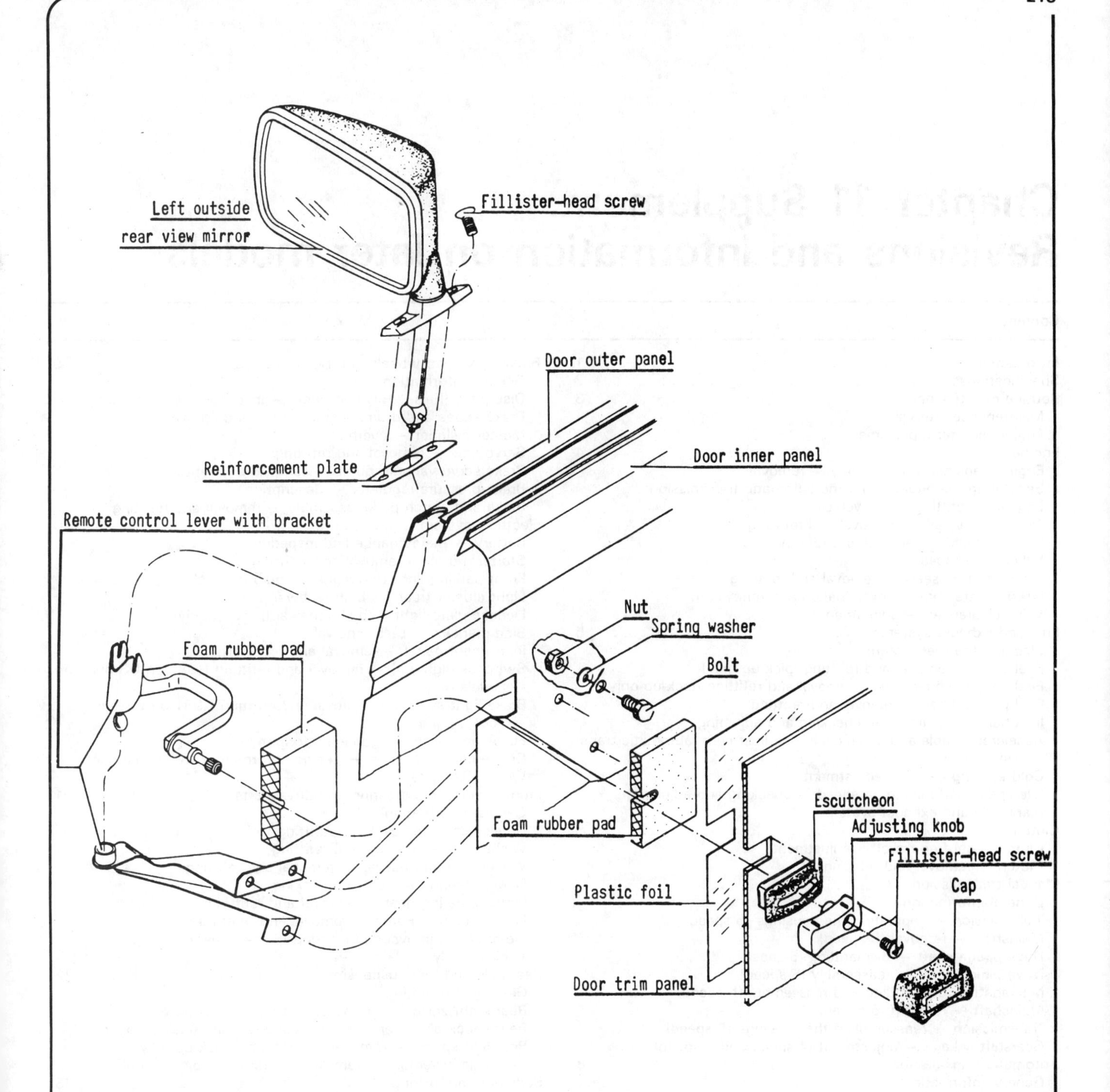

Fig. 10.13 Remotely controlled exterior mirror components (Sec 19)

Chapter 11 Supplement: Revisions and information on later models

Contents

1 Introduction

This Supplement contains specifications and changes which apply to diesel-powered VW Rabbit, Jetta and pick-up models produced between 1980 and 1984. Also included is information related to previous models which was not available at the time of original publication of this manual.

Where no differences, or very minor differences, exist between 1980 through 1984 models and previous models, no information is given. In those instances, the original material included in Chapters 1 through 10 should be used.

2 Specifications

Note: *The specifications listed here include only those items which differ from those listed in Chapters 1 through 10. For information not listed here, refer to the appropriate Chapter.*

Recommended lubricants and fluids

Engine oil quantity	
With filter change	
1977 through 1980 models	(3.5 liters)
1980 through 1984 models	4.7 qts (4.5 liters)
Without filter change	
1977 through 1980 models	3.2 qts (3.0 liters)
1981 through 1984 models	4.2 qts (4.0 liters)
Automatic transmission fluid type	Dexron II
Power steering fluid type	VW 239 902, Texaco 1833 PS 4634 or equivalent

Engine

Connecting rod journal diameter (1981 on)		
Standard	47.76 to 47.78 mm (1.8803 to 1.8811 in)	
1st undersize	47.51 to 47.53 mm (1.8705 to 1.8712 in)	
2nd undersize	47.26 to 47.28 mm (1.8606 to 1.8614 in)	
3rd undersize	47.01 to 47.03 mm (1.8508 to 1.8516 in)	
Torque wrench settings	**Ft-lbs**	**Nm**
Crankshaft sprocket bolt		
Engine code CK	59	80
Engine code CR	111	150
Engine code JK*	148	200
Cylinder head bolts	Tighten only according to procedure in text	

*Lubricate bolt threads before installation

Fuel and exhaust systems

Injection pump timing	
1977 through 1980 (pump without yellow dot)	
Checking	0.83 to 0.93 mm
Adjusting	0.86 ± 0.02 mm
1977 through 1980 (pump with yellow dot)	
Checking	1.10 to 1.20 mm
Adjusting	1.15 ± 0.02 mm
1981 and 1982 (standard setting)	
Checking	0.83 to 0.93 mm
Adjusting	0.86 ± 0.02 mm
1981 and 1982 (improved performance setting)	
Checking	0.93 to 1.03 mm
Adjusting	0.98 0.02 mm
1983 and 1984	
Checking	0.90 to 1.00 mm
Adjusting	0.95 ± 0.02 mm
Idle speed	
1977 through 1982	800 to 850 rpm
1983 and 1984	810 to 850 rpm
Maximum rpm	
1977 through 1980	5500 to 5600 rpm
1981 through 1984	5300 to 5400 rpm
Fuel tank capacity	
Rabbit	
1977 through 1979	45 liters (11.8 US gal)
1980 through 1982	38 liters (10 US gal)
1983 on	41.5 liters (11 US gal)
Jetta	40 liters (10.5 US gal)
Pick-up	56 liters (15 US gal)

Clutch

Clutch pedal free travel	
1977 and 1978	15 mm (5/8 in)
1979 on	15 to 25 mm (5/8 to 1 in)

Automatic transmission

Transmission fluid type	Dexron ATF	
Torque wrench settings	**Ft-lbs**	**Nm**
Pan-to-case bolt	15	20
Strainer-to-transmission (screw)	27 (in-lb)	3
Driveshaft-to-flange (socket-head bolt)	33	45
Engine-to-transmission bolt/nut	41	55
2nd gear band adjusting screw locknut	15	20
Transmission protection plate bolts	18	25
Starter mounting bolts	22	30
Torque converter cover plate bolts	11	15
Torque converter-to-drive plate bolts	22	30

Braking system, wheels and tyres

Brake drums (pick-up models only)		
Inside diameter	7.874 in (200 mm)	
Inside diameter wear limit	7.913 in (201 mm)	
Maximum inside diameter after resurfacing	7.894 in (200.50 mm)	
Maximum out-of-round	0.002 in (0.05 mm)	
Torque wrench settings		
Master cylinder-to-brake servo nuts	**Ft-lbs**	**Nm**
1977 and 1978	9.5	13
1979 on	15	20

Electrical system

Torque wrench settings	**Ft-lbs**	**Nm**
Starter mounting bolts (automatic transmission only)	18	25

Front suspension, steering and driveshafts

Front wheel caster angle (pick-up only)	1°20′ ± 30′	
Torque wrench settings	**Ft-lbs**	**Nm**
Power steering gear U-joint clamp bolt nut	22	30
Power steering gear mounting nuts	22	30
Tie-rod-to-steering arm nut	22	30
Power steering pump mounting nuts	11	15
Power steering pump pulley nuts	13	18

Rear axle and rear suspension

Rear wheel camber (pick-up only)	0° ± 1′	
Total toe-in (pick-up only)	0° ± 1′	
Torque wrench settings		
Stabilizer bar retainer nuts	**Ft-lbs**	**Nm**
(Jetta only)	22	30
Rear shock absorber mounting bolt nuts (pick-up only)	30	40
Rear spring U-bolt nuts (pick-up only)	30	40
Shackle bolt nuts (pick-up only)	44	60
Rear spring front mount nut (pick-up only)	70	95

3 Routine maintenance

Maintenance intervals

Every 10000 miles (16000 km) or 8 months, whichever comes first

Engine
- Check cooling system (through 1982 models)
- Check engine idle speed and adjust if necessary
- Drain water from the fuel filter

Every 15000 miles (24000 km) or yearly, whichever comes first

Engine
- Check cooling system (1983 and 1984 models)
- Check cylinder compression (not required on 1983 and 1984 models

Electrical system
- Check specific gravity of battery electrolyte and perform open-circuit and load voltage tests (not required on 1983 and 1984 models)

Transmission
- Check automatic transmission fluid level
- Check automatic transmission kickdown operation
- Check CV joint boots for holes and loose clamps

Steering
- Check power steering fluid level

Brakes
- Check brake pedal adjustment

General
- Check windshield wipers and washer system (not required on 1983 and 1984 models)

Every 30000 miles (48000 km) or 2 years, whichever comes first

Transmission
- Replace automatic transmission fluid

Brakes
- Inspect brake pressure regulator

Every 50000 miles (80000 km)

Note: *Even though no replacement interval is specified by VW, it is highly recommended that the camshaft drivebelt be replaced with a new one at this mileage interval.*

Engine oil filter replacement

Note: *Correct oil filter installation depends on the type of oil filter flange on the engine. From 1977 through early 1982, the flange has a ridge on the filter sealing surface (the surface in contact with the filter seal). Midway through the 1982 model year, a new type of flange was introduced. This new flange has a machined sealing surface that is approximately 1/2-inch (13 mm) wide. When replacing the oil filter, be sure to follow the procedure for your particular model year or oil leakage and subsequent engine damage could occur.*

Early models (flange with ridge)

1 Wipe the surface of the flange clean, then apply a thin coat of clean engine oil to the seal on the new filter (do not use grease).

2 Install the filter and tighten it by hand until the rubber seal is in firm contact with the flange all the way around. Next, use a cap-type filter wrench to tighten the filter exactly 3/4-turn.

3 After you have refilled the crankcase with the correct quantity and type of oil, start the engine and allow it to idle until the oil warning

light is out. Run the engine at various speeds for three to five minutes, then shut it off.
4 Using a torque wrench and the cap-type filter wrench, tighten the filter to 18 ft-lbs (25 Nm).
5 Check the oil level, then run the engine again and check for leaks around the filter (retighten if necessary).

Later models (smooth flange)

6 Follow the procedure outlined above for early models, but note that the filter should be initially tightened according to the instructions supplied by the filter manufacturer and it is not necessary to retighten it with a torque wrench.

4 Engine

Engine/transmission assembly – removal

1 Use the procedure in Chapter 1, but include the following steps if your vehicle is equipped with an automatic transmission:

a) Place the selector lever in Park.
b) Detach the transmission fluid cooler lines from the radiator and plug them.
c) Separate the selector cable and housing from the lever on the transaxle and from the bracket.
d) Detach the accelerator cable from the transmission lever and the housing from the bracket.
e) Remove the starter cover plate, then disconnect all wires from the solenoid (label them to simplify installation).
f) If you are planning to separate the engine and transaxle, remove the transaxle protection plate. Remove the cover from the engine end of the bellhousing, then detach the torque converter from the drive plate by taking out the bolts.

2 Also, regardless of transmission type, note that on 1980 and later Canadian models and 1981 and later US models, the header pipe spring clips must be detached from the exhaust manifold with a special tool. Try to obtain the tool before beginning engine removal to avoid delays once the procedure is started.
3 The radiators on 1982 and later US-built vehicles have a slightly different mounting design and are held in place by pins.
The removal procedure will have to be modified somewhat on these vehicles.

Separating the engine from the automatic transmission

4 To separate the engine from a transaxle with an automatic transmission, the engine and transaxle must both be securely supported.
5 Remove the bolts that attach the transaxle to the engine.
6 Carefully separate the transaxle from the engine (make sure the drive plate detaches cleanly from the torque converter without pulling the converter off its support).
7 Once the transaxle is separated from the engine, bolt a metal bar across the mouth of the bellhousing to keep the torque converter from falling out.

Engine – refitting in the vehicle

8 Use the procedure in Chapter 1, but note the following if your vehicle is equipped with an automatic transmission:

a) Before you attach the transaxle to the engine, make sure that the torque converter has not slipped off the support inside the bellhousing (remove the bar that was installed to keep it in place).
b) Tighten the transaxle-to-engine bolts to 59 ft-lb (80 Nm).
c) Tighten the drive plate-to-torque converter bolts to 22 ft-lb (30 Nm).
d) On 1982 and 1983 vehicles, secure all of the starter wires under the plastic tie on the solenoid.
e) Adjust the selector cable as described later in this Chapter.

9 For vehicles equipped with a manual transmission, use the procedure in Chapter 1, but note that the starter wiring on 1982 and 1983 vehicles must be routed with the battery cable outside the plastic tie.

Toothed drivebelt – removal and refitting

10 The procedure is correct as described in Chapter 1, but note that if the crankshaft sprocket bolt becomes loose, or is removed, on engines with code letters CK, a new bolt (PN N901 120 01) and a new washer with a collar (PN 068 105 299) should be installed and tightened to

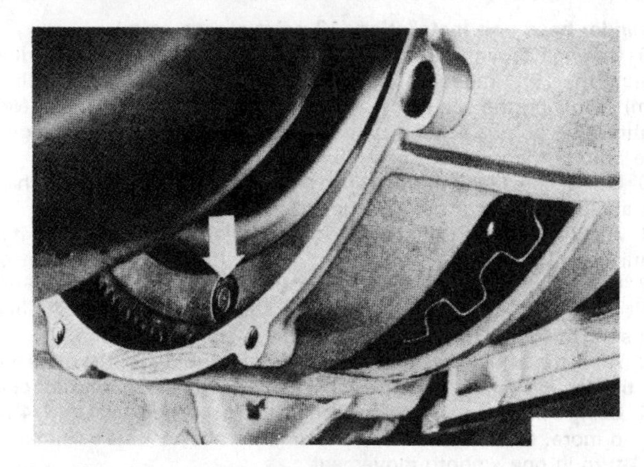

Fig. 13.1 One of three bolts (arrow) that must be removed to separate the torque converter from the drive plate (turn the crankshaft – in a clockwise direction only – with a wrench on the pulley bolt to bring the torque converter bolts into view) (Sec 4)

111 ft-lb (150 Nm). If the new parts are unavailable and the original bolt and washer must be used, be sure to apply thread-locking compound to the bolt threads before installation.
Tighten the original bolt to 59 ft-lb (80 Nm).

Cylinder head – removal and refitting

11 During 1981, a new type of cylinder head bolt was gradually introduced. The new bolt is 12 mm in diameter and 115 mm (5-7/8 inches) long and has a 12 point recess in the head. The old-type bolt was 11 mm in diameter and 98 mm (3-27/32) inches long with a hex-type recess in the head. **Caution:** *The new-type bolts cannot be used in engines that did not originally have them and engine blocks, cylinder heads, head gaskets and head bolt washers used with new-type bolts should never be interchanged with the equivalent parts from engines with the old-type head bolts. The new-type bolts are designed to stretch a predetermined amount when they are tightened and should never be reused (always replace them with new ones). Failure to observe these precautions can result in severe engine damage.*
12 On 1981 and later models, the piston projection above the engine block was changed, necessitating a change in the available head gasket thicknesses. Use the following chart to determine the correct head gasket to use on these models:

Piston projection mm (inches)	Gasket thickness mm (inches)	Identification notches	Part no.
0.67 to 0.82 (0.026 to 0.032)	1.40 (0.055)	1	068 103 383 L
0.83 to 0.92 (0.033 to 0.036)	1.50 (0.059)	2	068 103 383 M
0.93 to 1.02 (0.037 to 0.040)	1.60 (0.063)	3	068 103 383 N

Cylinder head bolt installation (hex-socket bolts)

13 On engines with hexagon-socket head bolts (originally used until mid-1981), tighten the bolts initially to 37 ft-lb (50 Nm), following the sequence shown in Fig. 1.12 in Chapter 1. Next, follow the sequence a second time, tightening the bolts to 52 ft-lb (70 Nm). Finally, tighten the bolts a third time to 66 ft-lb (90 Nm).
14 After the engine is installed in the vehicle, start it up and let it run until the oil temperature is at least 122°F (50°C), then shut it off.
15 Remove the cylinder head cover, if necessary, then retighten each bolt to 66 ft-lb (90 Nm) (do not loosen the bolts first). Be sure to follow the sequence in the illustration.
16 After the vehicle has been driven for about 1000 miles, the bolts must be tightened again (it can be done with the engine either hot or cold).
17 Remove the cylinder head cover, loosen each of the bolts about 30°, then tighten them to 66 ft-lb (90 Nm) in the correct sequence.
18 Check and adjust the valve clearances, then install the cylinder head cover.

Cylinder head bolt installation (12-point recess bolts)

19 On engines with 12-point recess head bolts (introduced gradually during the 1981 model year), tighten the bolts initially to 30 ft-lb (40 Nm) following the sequence shown in Fig. 1.12 in Chapter 1. Next, tighten the bolts a second time to 44 ft-lb (60 Nm). Finally, tighten the bolts a third time to 55 ft-lb (75 Nm).

20 Using a breaker bar rather than a torque wrench, tighten each bolt an additional 1/2-turn in the correct sequence.

21 After the engine is installed in the vehicle, start it up and let it run until the oil temperature is at least 122°F (50°C), then shut it off.

22 Remove the cylinder head cover, if necessary, then tighten each bolt exactly 1/4-turn with a breaker bar (do not loosen the bolts first). Be sure to follow the sequence shown in the illustration.

23 After the vehicle has been driven about 1000 miles, the bolts must be tightened again (it can be done with the engine either hot or cold).

24 Using a breaker bar, tighten each bolt, in sequence, exactly 1/4-turn – no more. Once a bolt starts moving, do not hesitate – make the 1/4-turn in one smooth movement.

25 Check and adjust the valve clearances, then install the cylinder head cover.

Valves – removal

26 This procedure is correct as described in Chapter 1, but note that later model valves have three keeper (collet) grooves rather than one. The corresponding keepers (collets) have three ridges and are tinted a copper color for identification (early one-ridge keepers are a steel gray color). Do not try to interchange early and late model keepers unless the valves are replaced as well.

Crankshaft oil seals – removal and refitting

27 To replace the front crankshaft oil seal on 1983 and later models (with the engine in the vehicle), different special tools than the ones listed in Chapter 1 are required. Tool number 2085 should be used to remove the old seal and tool number 3083 should be used to install the new one.

Intermediate shaft – examination and renovation

28 Note that a new intermediate shaft oil seal (PN 056 103 085 B), which does not have an arrow indicating the counterclockwise rotation of the shaft, has been made available. The new seal can be identified by its brown inner ring and black outer ring.

Valve clearances – adjustment

29 The special tool required to depress the cam followers when removing and installing the valve adjustment discs was changed for 1983 and 1984 models. When working on 1977 through 1982 engines, tool number VW 546 is required to depress the followers and tool number US 4476 is used to withdraw the disc. On 1983 and 1984 engines, tool number 2078 is needed to depress the followers (tool number US 4476 is used to withdraw the discs on these models as well).

5 Fuel and exhaust systems

Bleeding the fuel system

1 Later model vehicles do not have a priming pump and the fuel system must be bled using a slightly different method.

2 Carefully loosen the fuel line connections at all four injectors, then crank the engine (without using the glow plugs) until fuel emerges from all of the loosened connections.

3 Tighten the fuel line connections, then start the engine and allow it to run until you are sure that no air that could cause stalling remains in the system.

Fuel tank – removal and refitting (pick-up only)

4 This procedure is essentially the same as the one outlined in Chapter 3, but note that the fuel filler neck connecting hose must be removed. Also, it is not necessary to disconnect the brake line fittings, lower the axle or move the exhaust system components on pick-up models. When installing the tank, position rubber insulating strips between the tank and mounting straps, then tighten the nuts to the specified torque.

Fuel gauge sender unit – removal and refitting (pick-up only)

5 Follow the procedure in Chapter 3, but note that instead of removing the rear seat, the fuel tank must be removed from the vehicle to gain access to the sender unit.

Fuel pickup filter – removal and refitting

6 All models have a fuel pickup filter inside the fuel tank, which does not normally have to be checked or replaced. If it becomes clogged, remove the fuel gauge sending unit, then, on all models except pick-ups, use a wire hook inserted into the loop on top of the filter to pull the filter off the pickup tube inside the tank and withdraw the filter through the sending unit opening. Use needle-nose pliers to insert the filter into the tank and slip it onto the end of the pickup tube.

7 On pick-up models, the pickup tube is part of the fuel gauge sending unit. When the sending unit is withdrawn from the tank, the filter can be slipped off, a new one installed and the sending unit inserted back into the tank.

Injection pump timing – checking and adjusting

8 Follow the procedure in Chapter 3, but note the new specifications at the beginning of this Supplement. Also, a 1981 or 1982 vehicle in 3rd gear (4-speed transmission) should accelerate from 35 to 55 mph in no more than 16.1 seconds. A 1981 or 1982 vehicle in 4th gear (5-speed transmission) should accelerate from 35 to 55 mph in no more than 18.6 seconds. If the acceleration time is excessive, use the 'improved performance' setting when adjusting the injection pump timing.

Accelerator cable and pedal cable – adjustment (automatic transmission only)

9 Vehicles equipped with an automatic transmission have two cables to adjust. One, called the pedal cable, connects to the accelerator pedal and a lever on the transmission. The second cable, called the accelerator cable, is connected to the lever on the transmission and the control lever on the fuel injection pump. Refer to the accompanying illustration to identify the components involved in the following steps.

10 Place the selector lever in Park and set the parking brake, then raise the hood and remove the plastic cover from the fuel injection pump.

11 Make sure the idle speed and maximum rpm are as specified, then raise the vehicle and support it securely on jackstands.

12 Working beneath the vehicle, loosen the pedal cable adjusting nut and detach the cable end from the transmission lever.

13 Working at the injection pump, carefully slide the metal clip off the accelerator cable end, then detach the spring housing from the ball stud on the speed control lever.

14 Check the ball stud travel between the control lever's idle and maximum rpm positions. It should be 1-1/4 ± 1/32 in (32 ± 1 mm). If necessary, loosen the ball stud retaining nut and reposition the stud on the lever to obtain the correct travel.

15 Reconnect the accelerator cable spring housing to the ball on the speed control lever, then separate the rubber boot from the spring housing and loosen both accelerator cable adjusting nuts.

16 Have an assistant hold the lever on the transmission against the pedal released stop. With the transmission lever in this position, hold the injection pump speed control lever against the idle speed adjusting screw.

17 Without compressing the kickdown engagement spring, and keeping both levers in position, move the accelerator cable housing as far as possible in a direction away from the pump, then tighten the adjusting nuts.

18 While watching the cable at the pump, have an assistant move the operating lever on the transmission until the speed control lever on the pump contacts the maximum rpm adjusting screw (the kickdown engagement spring should not be compressed).

19 If necessary, correct the adjustment of the accelerator cable. Have an assistant push the transmission operating lever in the direction of the arrow in the accompanying illustration which indicates that it is 'against stop during kickdown'. The kickdown engagement spring should now be compressed and the kickdown detent spring should be partially compressed.

20 While an assistant beneath the vehicle keeps the transmission lever against the kickdown stop, depress the accelerator pedal all the way and keep it against the stop on the floorboard. Have the assistant reconnect the pedal cable to the transmission lever and turn the pedal cable adjusting nut until all slack is removed from the cable.

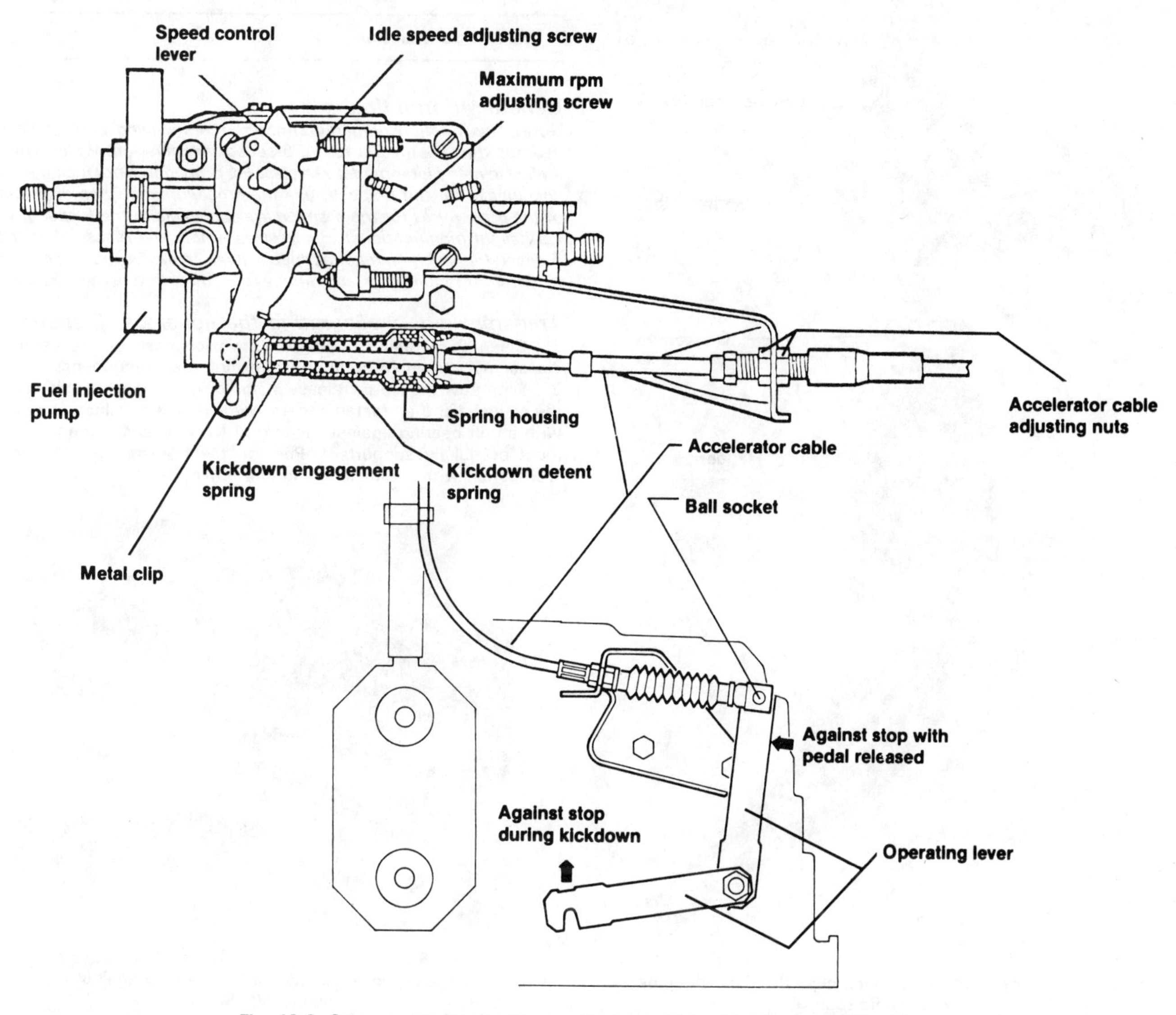

Fig. 13.2 Components involved in accelerator cable/pedal adjustment (Sec 5)

Fig. 13.3 Pedal cable adjusting nut (1) and point where cable should be detached from the transmission lever (2) (Sec 5)

21 Make sure that when the accelerator pedal is depressed, the injection pump speed control lever is moved into contact with the maximum rpm adjusting screw (the kickdown spring should not be compressed).

22 When the accelerator pedal is depressed to the stop, the kickdown engagement spring should be compressed and the operating lever on the transmission should be against the kickdown stop.

23 Make any necessary corrections to the adjustments, then reattach the rubber boot to the spring housing and reinstall the plastic cover on the injection pump.

Cold starting cable – adjustment

24 Follow the procedure in Chapter 3, but be sure to pull the cold starting knob out of the dashboard a distance of 5/64-inch (2 mm) before proceeding with Paragraph 4 in the procedure.

Idle speed and maximum speed – checking and adjusting

25 Follow the procedure in Chapter 3, but note the new specifications included at the beginning of this Supplement.

Manifolds and exhaust system

26 Beginning with 1983 models, a positive alignment system is built into the exhaust system components. The system consists of a dimple in one component and a corresponding slot in the mating compo-

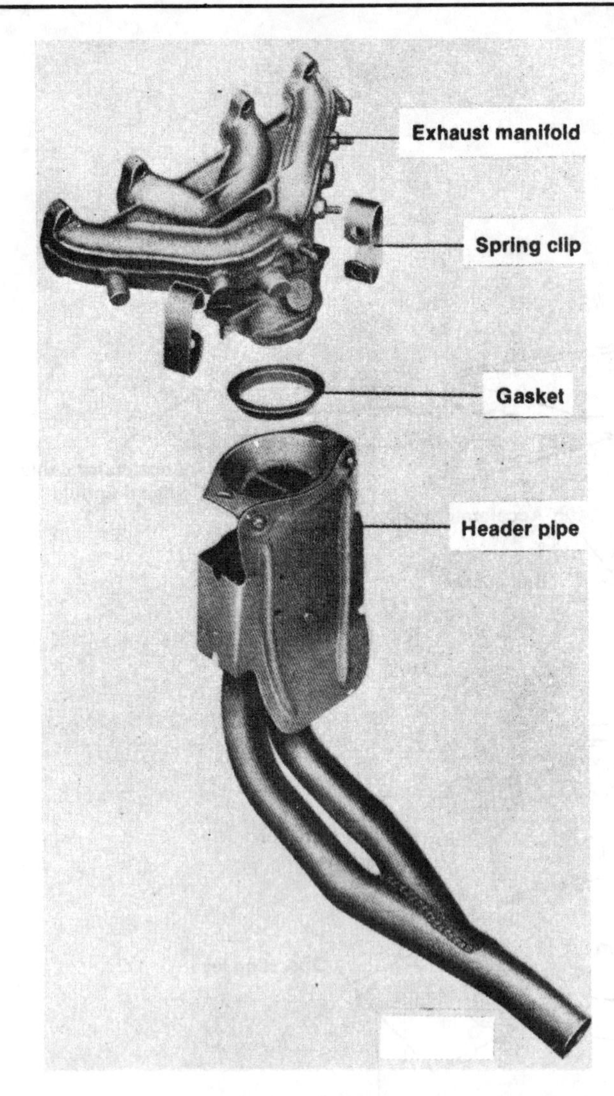

Fig. 13.4 Later models have spring clips that hold the pipe to the manifold (Sec 5)

nent. If all components are installed correctly (with the slots and dimples mated), the clamps can be tightened without being concerned that exhaust system components will come into contact with other components, causing rattles and noises.

27 Canadian models from 1980 on and US models from 1981 on have the header pipe attached to the exhaust manifold by two spring steel clips. To detach this type of pipe, first remove the heat shield from the header pipe (if installed), then insert the pins of the special tool into the holes of one of the spring clips. Turn the knob of the special tool until it reaches the stop, then, while pushing up on the header pipe, disengage the spring clip from the recesses. Repeat the operation on the second clip. Installation is the reverse of removal. Be sure to use a new gasket and make sure the spring clips are correctly engaged in the recesses.

6 Clutch

Clutch pedal free travel — adjustment

1 The procedure in Chapter 4 is correct for later models, but note the new specifications included at the beginning of this Supplement.

Clutch — removal and refitting

2 Beginning with engine number CK 130 178, a larger clutch assembly was used in these vehicles. When replacing components, make sure that you obtain the correct replacement parts for your vehicle.

7 Manual transmission

General information

Note: *This Section primarily contains the servicing and repair procedures that are unique to the 5-speed transmission, which was installed in many later model vehicles. For procedures not included here specifically for the 5-speed, use the procedures in Chapter 5 (many of the procedures for the 4-speed are applicable to the 5-speed, with very slight modifications). Since disassembly and reassembly of the 5-speed transmission requires special tools, it may be more economical to obtain a rebuilt or used transmission than to overhaul yours.*

Transmission — separating the housings (5-speed)

1 Clean the outside of the transmission case before beginning disassembly to avoid getting dirt into internal components.

2 Withdraw the clutch release pushrod from the bellhousing end of the mainshaft, then fasten a metal bar across the bellhousing mouth with a bolt bearing against the end of the mainshaft (the mainshaft must be solidly supported). Position the transmission on a clean workbench with the bellhousing end down.

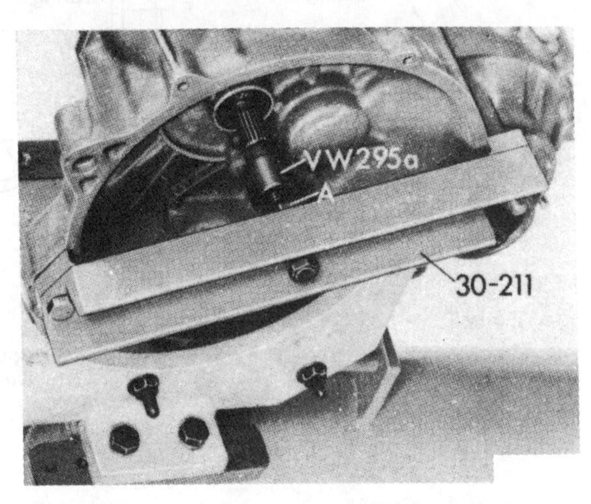

Fig. 13.5 A metal bar must be installed across the bellhousing mouth with a bolt bearing against the mainshaft (Sec 7)

3 Remove the entire speedometer drive gear assembly by withdrawing it through the top of the case. Remove the back-up light switch.

4 Remove the selector shaft boot, then clean the exposed end of the shaft so that when the shaft is removed, dirt on the shaft will not damage the oil seal or transmission case.

5 Make sure the transmission is in Neutral (move the selector shaft if necessary).

6 Loosen the locknut for the main adjuster and the locknut for the 5th gear adjuster, then remove the adjusters from the case.

7 Remove the selector shaft cover, then take out the selector control spring. Remove the selector shaft by pressing it out through the cover opening.

8 Remove the bolts that fasten the 5th gear/clutch release bearing housing to the transmission case, then separate the housing from the transmission.

9 Pry the plastic dust cap out of the center of the left drive flange, then remove the circlip and dished washer. Using the special puller recommended by VW, remove the left drive flange.

10 Push down on the shift fork indicated in the accompanying illustration to engage 5th and Reverse gears and keep the mainshaft from turning. Using a 12 mm multi-point driver, remove the 17 mm bolt from the end of the mainshaft (where applicable, remove the separate dished washer as well).

11 Return the transmission gears to Neutral. On transmissions from number 01 030 on, pry up the locking plate, then turn the 5th gear shift link tube to unscrew it from the shift fork (special tool number 3059 must be used for this procedure). **Caution:** *Do not pull off the selector rod or the shift fork assembly inside the case will fall apart.*

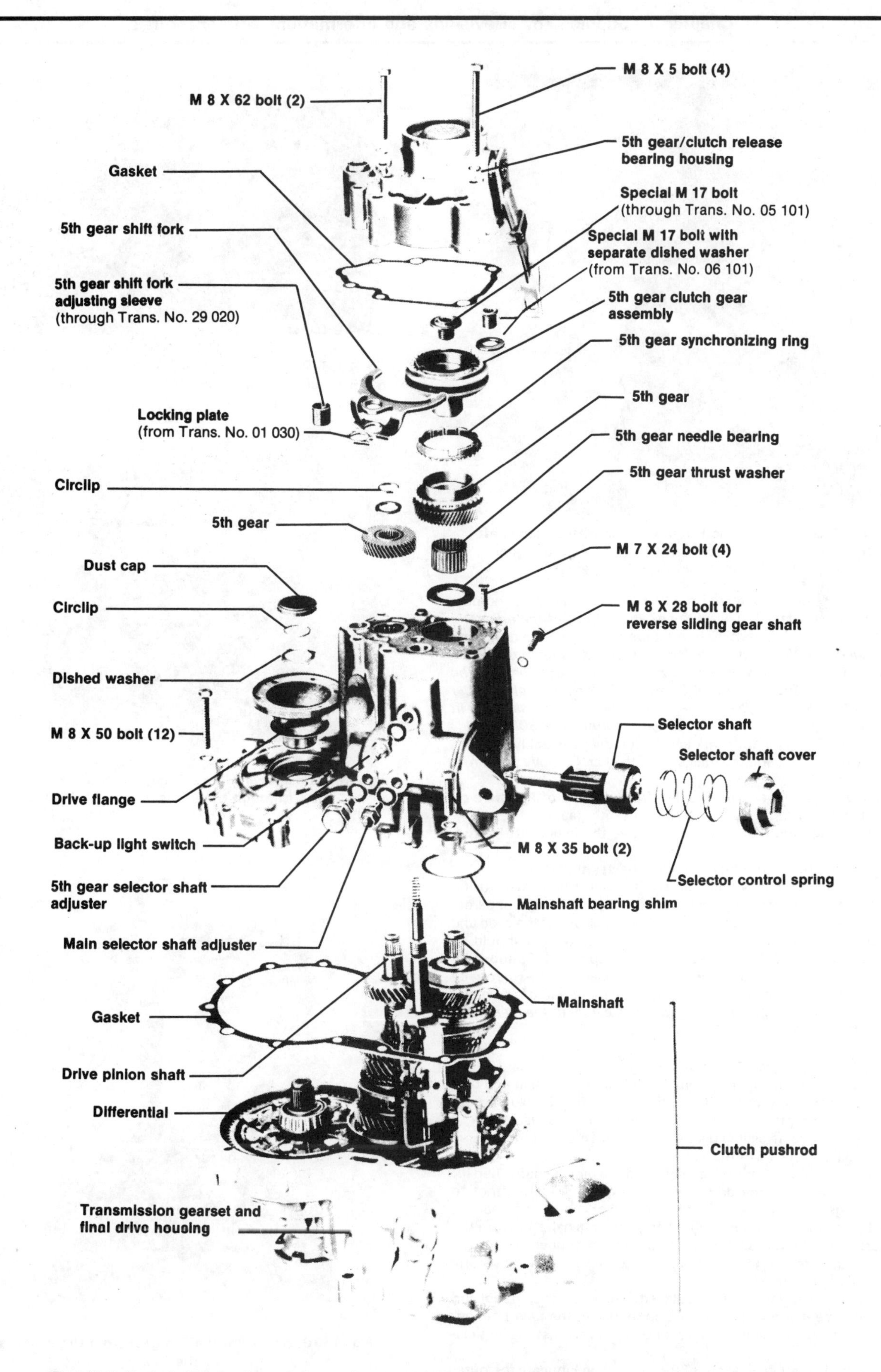

Fig. 13.6 Exploded view of components that must be removed to separate the transmission cases (Sec 7)

Fig. 13.7 Removing the bolt from the end of the mainshaft (push down on the shift fork [arrow] to engage the gears and keep the shaft from turning) (Sec 7)

12 On transmissions through number 29 020, use special tool number 3038 to unscrew the 5th gear shift link tube. **Note:** *If the adjusting sleeve interferes with loosening of the tube, either grip the tube with pliers between the 5th gear shift fork and the case and apply additional force with the pliers or very carefully grind or drill through the peened area of the locking sleeve. On transmissions through number 08 128, install the later-type 5th gear shift link during reassembly (it has 5 mm wide slots for use with special tool number 3059).*
13 Lift off the 5th gear shift fork and clutch gear assembly, the 5th gear synchronizer ring and the mainshaft 5th gear. Remove the thrust washer.
14 Remove the circlip and the washer from the end of the drive pinion shaft, then, using a puller, remove the 5th gear.
15 Using a 5 mm multi-point driver, remove the mainshaft bearing clamping bolts.
16 Remove the bolts that hold the case halves together.
17 Remove the reverse sliding gear shaft support bolt, then, using a puller that will exert force against the end of the mainshaft, separate the case halves. The mainshaft bearing should pull out of the case and remain on the shaft. The original mainshaft bearing shim should be stored in a safe place after the case has been separated. **Caution:** *Do not attempt to pry the case halves apart by inserting tools between them; doing so will damage the castings. The input end of the mainshaft must be solidly supported as mentioned earlier or the mainshaft bearings may be damaged.*

Mainshaft — removal (5-speed)

18 Withdraw the shift fork rod from the hole in the housing and turn the shift fork assembly to disengage the forks from the clutch gear assemblies, then remove the assembly from the transmission.
19 Pull out the reverse sliding gear and shaft, then unbolt the reverse sliding gear shift fork from the case.
20 Remove the circlip from the end of the drive pinion shaft, then lift out the mainshaft and the drive pinion 4th gear. Be careful not to damage the gear teeth, the mainshaft oil seal or the needle bearing.
21 The ball bearing must be removed from the mainshaft before the transmission can be reassembled. Be sure that the inner race of the bearing is supported as the shaft is pressed out and be very careful not to drop the mainshaft.
22 If the needle bearing must be replaced, use an expansion tool and puller to remove it from the case. When installing the new bearing, make sure the thickest side of the bearing cage is against the installation tool.
23 Pry the mainshaft oil seal out of the case, then lubricate the outer edge of the new seal with grease and drive it in very carefully until it is completely seated.

Fig. 13.8 Pry up on the locking plate to disengage it from the 5th gear shift link tube (trans. no. 01 030 on) (Sec 7)

Fig. 13.9 Unscrewing the 5th gear shift link tube with tool no. 3038 (through transmission no. 29 020) (Sec 7)

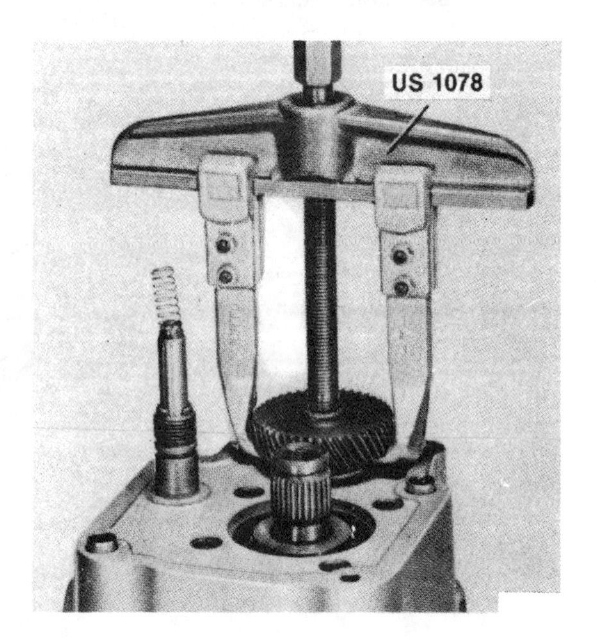

Fig. 13.10 Removing the 5th gear from the pinion shaft (Sec 7)

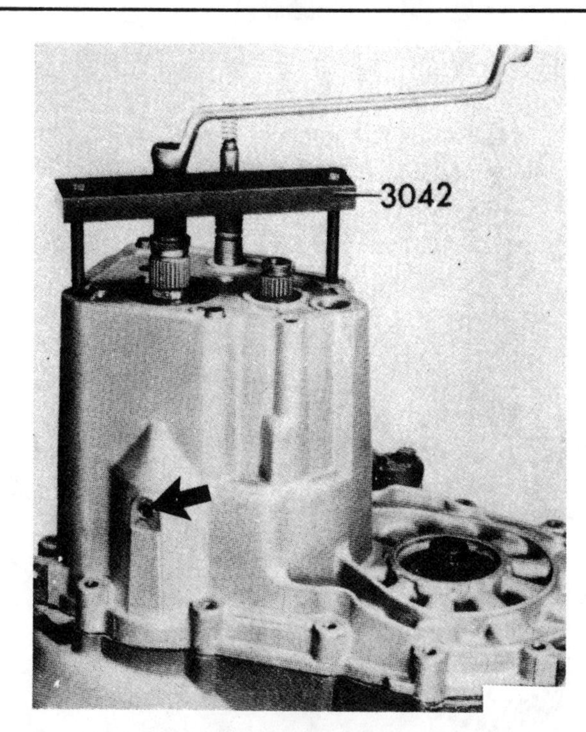

Fig. 13.11 Remove the reverse sliding gear shaft support bolt (arrow), then separate the cases with a tool that will push on the mainshaft (Sec 7)

Fig. 13.12 Removing the shift fork assembly (turn it in the direction of the arrow to disengage the forks from the clutch gear assemblies) (Sec 7)

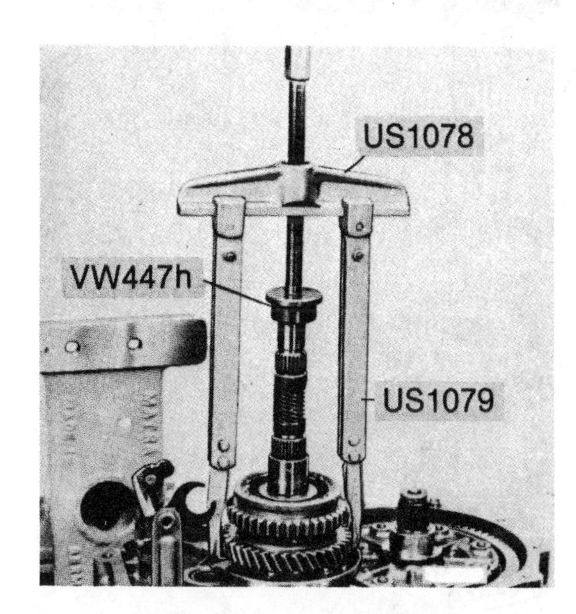

Fig. 13.13 A puller must be used to remove the 1st/2nd clutch gear assembly (Sec 7)

Fig. 13.14 Checking the synchronizer ring-to-gear clearance ('a' must be as specified) (Sec 7)

Drive pinion shaft — dismantling (5-speed)

24 Remove the 3rd gear circlip, then lift off 3rd gear, 2nd gear, the 2nd gear needle bearing and the 2nd gear synchronizing ring. If 3rd gear cannot be lifted off, use a puller to remove it.

25 Using a puller, remove 1st gear, the 1st gear synchronizing ring, the 1st/2nd gear clutch gear assembly and the 2nd gear needle bearing inner race as a unit. Remove the 1st gear needle bearing and thrust washer.

26 If necessary, remove the spring rings and separate the hub from the synchronizer sleeve.

27 If the drive pinion shaft needle bearing in the case must be replaced, drive it out toward the inside of the case. Install the new bearing with the thickest side of the bearing cage against the installation tool.

28 If the drive pinion shaft, the bearings or the differential require removal for servicing or replacement, take the transmission to an authorized dealer to complete the repairs. If the drive pinion shaft is removed, the bearing preload must be carefully adjusted or noise and rapid wear of the components will occur.

Drive pinion shaft — reassembly (5-speed)

29 Inspect all parts and replace any that are worn or damaged. Keep in mind that there have been changes in gear designs and tooth ratios so make sure any new parts match the originals exactly. Also, if a gear is replaced on the drive pinion shaft, the matching gear on the mainshaft must also be replaced (they are available only as matched pairs).

30 Hand press the synchronizing rings into the gears and measure the clearance (a in the accompanying illustration) with a feeler gauge. The wear limit is 0.020 in (0.50 mm). With new parts, the clearance should be 0.043 to 0.067 in (1.10 to 1.70 mm).

31 Assemble the 1st/2nd gear clutch gear assembly. Insert the hub into the operating sleeve, then insert the keys and install the spring rings 120° apart on opposite sides of the clutch gear assembly (one key will have a spring ring hooked into each of its ends and each of the other two keys will have a spring ring hooked into only one end).

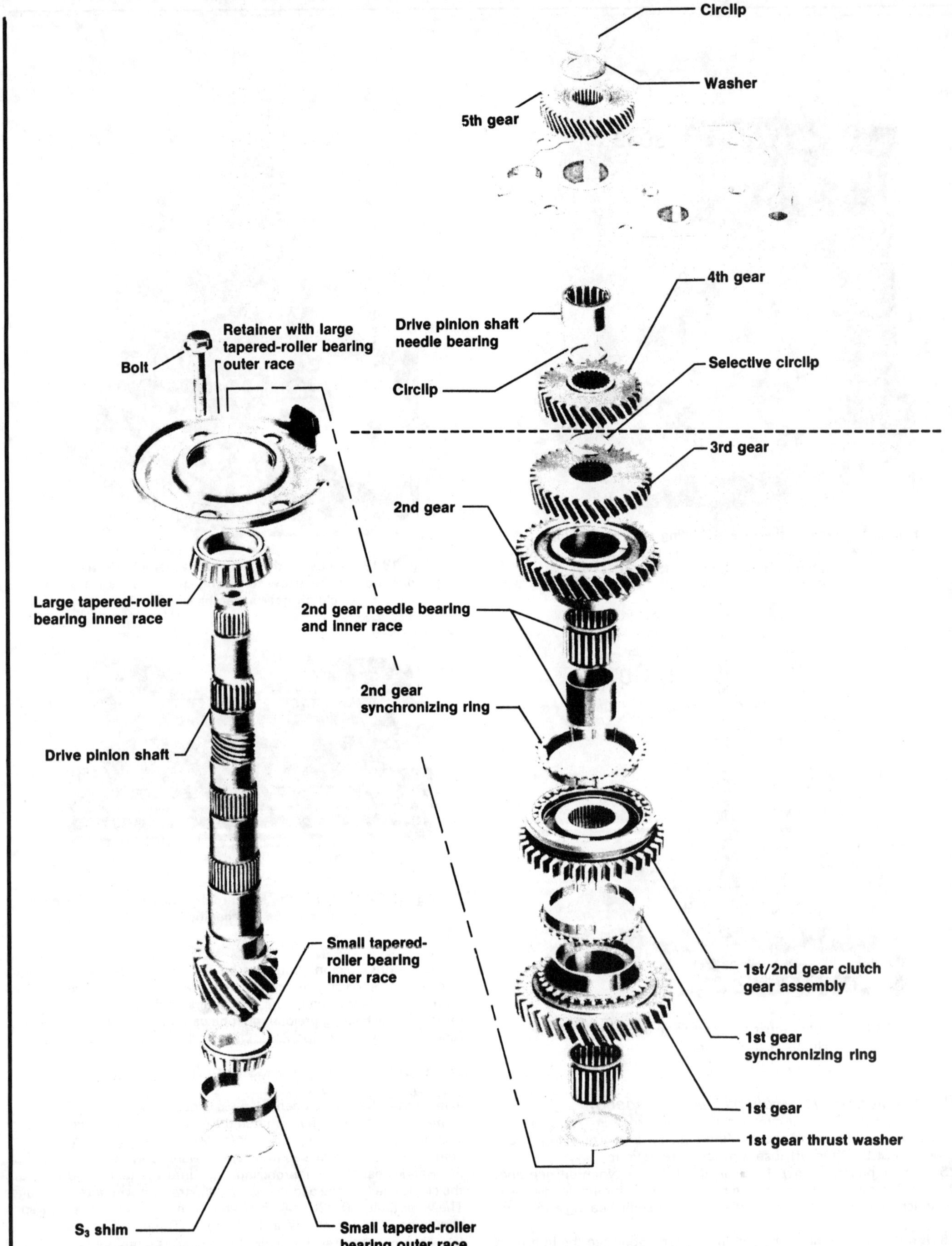

Circlip
Washer
5th gear
4th gear
Drive pinion shaft needle bearing
Retainer with large tapered-roller bearing outer race
Bolt
Circlip
Selective circlip
3rd gear
2nd gear
Large tapered-roller bearing inner race
2nd gear needle bearing and inner race
2nd gear synchronizing ring
Drive pinion shaft
Small tapered-roller bearing inner race
1st/2nd gear clutch gear assembly
1st gear synchronizing ring
1st gear
1st gear thrust washer
S_3 shim
Small tapered-roller bearing outer race

Fig. 13.15 Drive pinion shaft components – exploded view (Sec 7)

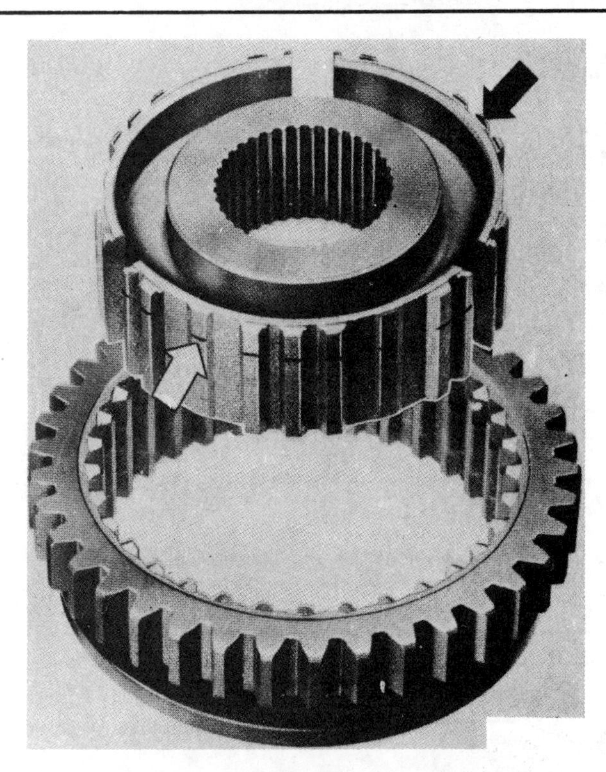

Fig. 13.16 Install the 1st/2nd gear clutch gear assembly with the grooved teeth (white arrow) or the groove on the face (black arrow) facing 1st gear (Sec 7)

32 Install the 1st gear thrust washer, the 1st gear needle bearing, 1st gear and the 1st gear synchronizing ring on the drive pinion shaft. **Note:** *The 1st gear synchronizing ring is different from all others and has three teeth missing from its circumference.*

33 Heat the 1st/2nd gear clutch gear assembly to 250°F (120°C) in an oven, then press it onto the shaft. **Caution:** *To prevent damage, you must align the 1st gear synchronizing ring with the keys of the clutch gear assembly before it is pressed all the way onto the shaft. The shift fork groove in the sleeve must be toward the press tool — facing away from the previously installed 1st gear.*

34 *Drive on the 2nd gear needle bearing inner race (it must be completely seated against the synchronizer hub of the clutch gear assembly).*

35 Install the 2nd gear needle bearing, the 2nd gear synchronizing ring and 2nd gear.

36 Heat the 3rd gear to 212°F (100°C) in an oven, then install it on the shaft with the side that has the collar facing 2nd gear.

37 Install a new circlip that will limit the end play of 3rd gear to 0.000 to 0.008 in (0.00 to 0.20 mm). The closer the end play is to zero the better. Circlips are available in thicknesses from 2.50 mm to 3.00 mm in 0.10 mm increments.

38 Do not install 4th gear or the circlip until after the mainshaft is in place. Always use a new 4th gear circlip.

Mainshaft — dismantling and reassembly (5-speed)

39 Follow the procedure for the 4-speed transmission outlined in Chapter 5, but use the exploded view drawing included here to make sure everything is reassembled correctly. Note that the 5-speed mainshaft has an additional clutch gear assembly to check. Also, when checking the synchronizer ring-to-gear clearance, note that the clearance with new parts should be 0.051 to 0.075 in (1.30 to 1.90 mm) for 4th and 5th gear and 0.045 to 0.069 in (1.15 to 1.75 mm) for 3rd gear.

Mainshaft — refitting (5-speed)

40 Being careful not to damage the needle bearing, the mainshaft oil seal or the gear teeth, install the mainshaft so that its teeth mesh with those of the drive pinion shaft. If you have changed the position of the support bar bolt, turn the bolt as required until the mainshaft is supported with 0.040 (1 mm) of clearance between 2nd gear on the drive pinion shaft and 3rd gear on the mainshaft.

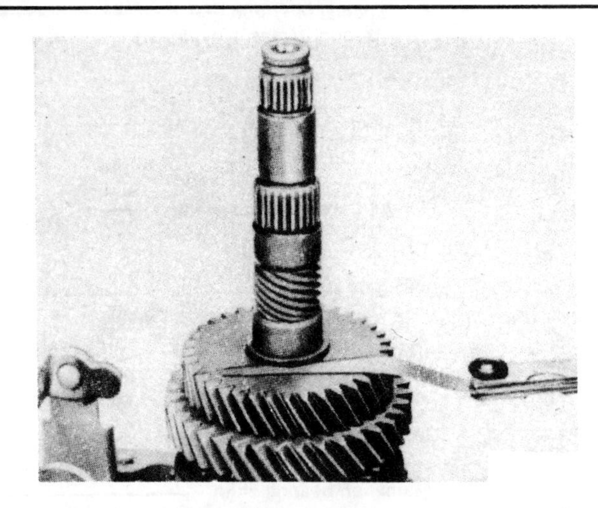

Fig. 13.17 Checking 3rd gear end play with a feeler gauge (Sec 7)

41 Install 4th gear on the drive pinion shaft, then install the washer and press the circlip down over the drive pinion shaft until it is seated in the groove.

42 Place the mainshaft ball bearing shim (the original one) in the recess inside the case, then press the ball bearing into place with the wide shoulder on the inner race facing 4th gear.

43 Install the clamping plate and tighten the bolts to 11 ft-lb (15 Nm). Insert one of the two shift fork rod springs into the housing.

44 Engage the 1st/2nd gear shift fork in the groove on the 1st/2nd clutch gear assembly, then lift the selector rod slightly and swing the shift fork assembly around the pinion shaft, guiding the 3rd/4th shift fork into the groove of the 3rd/4th clutch gear assembly. Engage the reverse shift fork with the reverse shift link before inserting the shift fork rod into the hole in the case. Install the remaining spring on the shift fork rod.

45 Inspect the reverse sliding gear and shaft and the sliding gear stop on the drive pinion shaft bearing retainer. Replace any worn or damaged parts with new ones.

46 Install the reverse sliding gear and shaft. Loosely install the support bolt in the reverse sliding gear shaft and position the bolt so that it is equidistant from the two bolt holes in the flange of the case.

Transmission — reassembling the housings (5-speed)

47 Position a new gasket on the final drive half of the case (do not use sealant on the gasket). Position the gear carrier/shift housing portion of the case on the gear train, making sure the mainshaft is aligned with the ball bearing and the drive pinion shaft is aligned with the needle bearing. The dowels should align one half of the case with the other.

48 Make sure the mainshaft is supported by the bolt in the metal bar, then carefully drive the bearing onto the mainshaft. Be sure to apply force only to the bearing inner race. The case halves will move together as this is done.

49 Install the support bolt in the reverse sliding gear shaft and tighten it to 22 ft-lb (30 Nm). Install the case bolts and tighten them in a diagonal pattern to 18 ft-lb (25 Nm). Work up to the final torque in three steps.

50 Check the left drive flange oil seal. If it is worn, hard, cracked or otherwise damaged, pry it out carefully with a large screwdriver. Pack the open side of the new seal with multi-purpose grease, then install it in the case. Apply pressure evenly around the outer circumference of the seal.

51 Inspect the drive flange. Replace it if there is a groove worn into the surface that contacts the oil seal. Install the drive flange with special tool number VW 391, which screws into the threads in the center of the differential sidegear shaft.

52 Make sure the drive flange is completely seated, then install the dished washer with the convex side away from the drive flange. Use circlip pliers to install a new circlip, but do not attempt to seat the circlip in the groove.

53 Drive the circlip in, against the dished washer, until it snaps into the groove, then install a new dust cap.

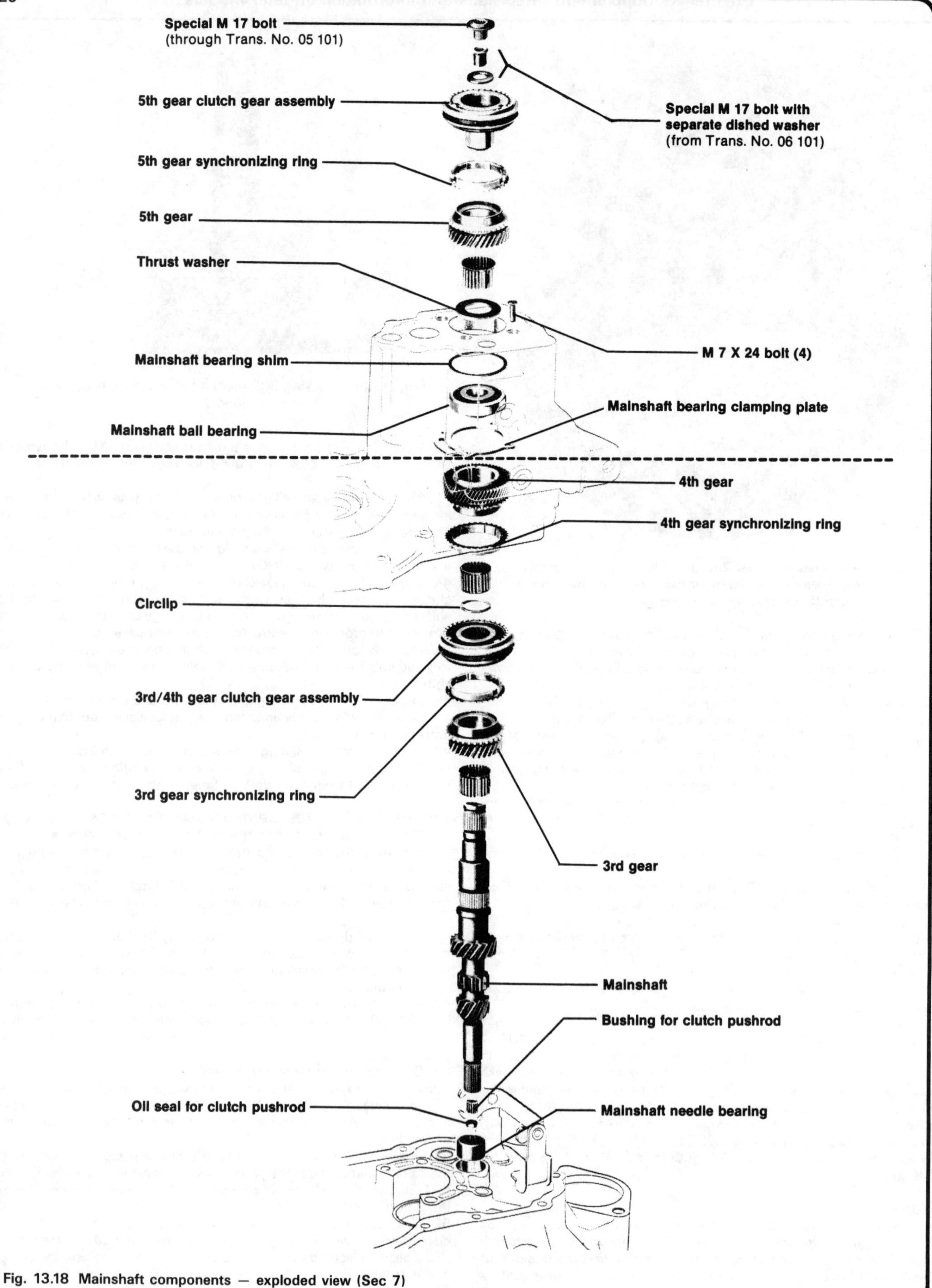

Fig. 13.18 Mainshaft components – exploded view (Sec 7)

Fig. 13.19 Correct alignment of reverse sliding gear shaft (distance 'x' should be equal) (Sec 7)

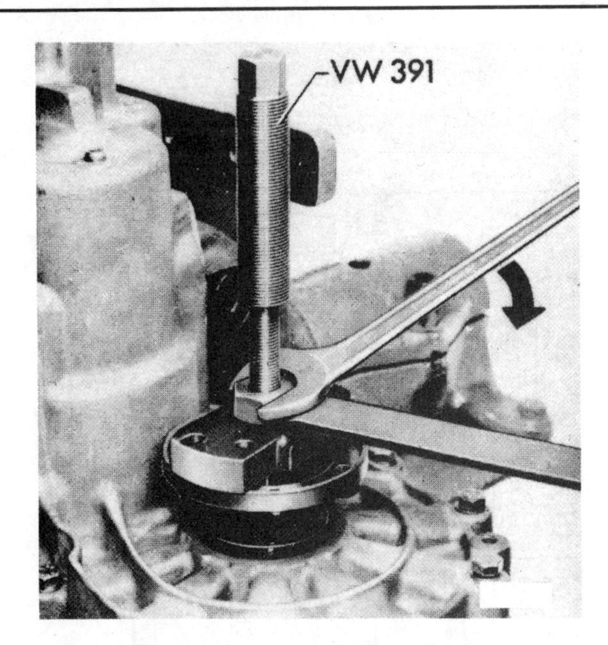

Fig. 13.20 Installing the left drive flange with the special tool (Sec 7)

Fig. 13.21 The 5th gear shift link tube projection (x) must be as specified (Sec 7)

Fig. 13.22 Selector shaft (A) in 5th gear position (note shift fork being lifted up to remove play so operating sleeve-to-5th gear synchronizer teeth clearance can be checked) (Sec 7)

54 Install the speedometer drivegear assembly, being careful to mesh the drivegear with the helical teeth on the drive pinion shaft.
55 Heat the drive pinion shaft 5th gear to 212 °F (100 °C) in an oven, then install it on the shaft with the grooved side of the gear visible. Use gloves or rags to avoid burning your hands.
56 Install the washer and circlip on the shaft.
57 Loosely assemble the 5th gear clutch gear assembly, the 5th gear synchronizer ring, 5th gear for the mainshaft, the 5th gear needle bearing, the 5th gear thrust washer and the 5th gear shift fork, then install them as an assembly on the mainshaft. **Note:** *As you install the 5th gear components, screw the tube of the 5th gear selector link into the 5th gear shift fork. Tubes with a groove use an adjusting sleeve; tubes with a shoulder use a locking plate. When turning the tube with special tool number 3059, be very careful not to accidentally pull the shift fork rod up by tilting the wrench. If it is pulled up, the shift forks will fall apart inside the transmission and the cases will have to be disassembled again.*
58 Adjust the tube position on the shift link until it projects 0.200 in (5 mm) above the shift fork.
59 Engage both 5th and Reverse gears so the mainshaft cannot rotate, then install a new 17 mm bolt in the end of the mainshaft (use Locktite on the threads). Tighten the bolt to 111 ft-lb (150 Nm).
60 Return the transmission gears to Neutral, then install the selector shaft. Adjust the main selector shaft by referring to Section 14 in Chapter 5. Adjust the 5th gear selector shaft by removing the plastic cap and loosening the locknut. Tighten the adjusting sleeve until the detent plunger just starts to move out. From this position, loosen the sleeve 1/3-turn, then, while holding the sleeve stationary, tighten the locknut to 15 ft-lb (20 Nm).
61 Install the back-up light switch. Install the selector shaft lever on the shaft, then engage 5th gear by pulling the shaft out all the way and turning it counterclockwise.

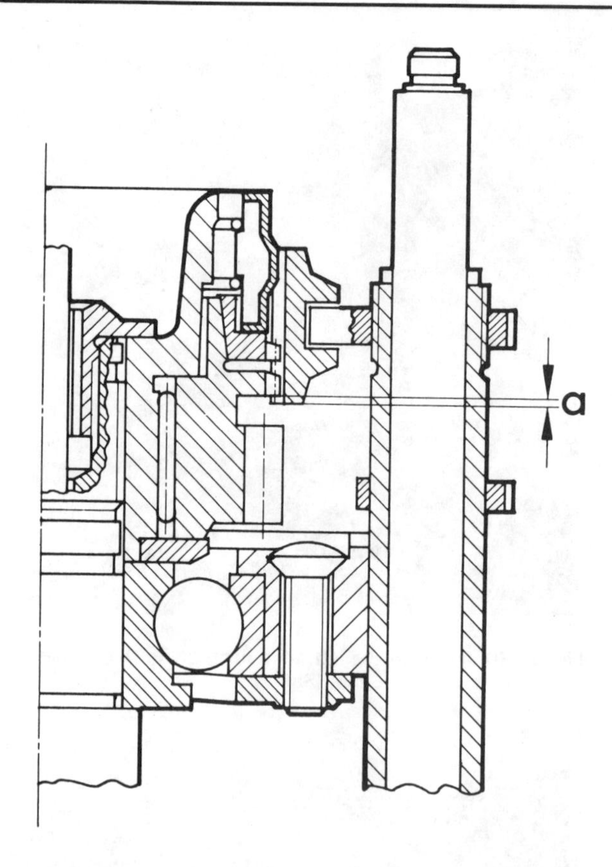

Fig. 13.23 The operating sleeve-to-5th gear synchronizer teeth clearance (a) must be as specified in the text (Sec 7)

Fig. 13.24 Shift link tube being secured by peening adjusting sleeve ('a' = 3/4-inch) (Sec 7)

Fig. 13.25 Shift link tube being secured by driving on locking plate with special tool (Sec 7)

62 With 5th gear engaged, lift up on the shift fork so there is no play between it and the operating sleeve of the clutch gear assembly. The distance from the lower edge of the operating sleeve to the edge of the 5th gear synchronizer teeth should be 0.039 in (1 mm). If not, readjust the position of the 5th gear shift link tube in the shift fork.
63 When the adjustment is correct, secure it as follows. **Note:** *Be careful not to deform the shift link tube in the process. The selector rod should spring back automatically when depressed and released.*
64 On transmissions through number 29 020, peen the adjusting sleeve with a blunt cold chisel at a point 3/4-inch (19 mm) from the edge of the selector link tube.
65 From transmission number 01 030, support the shift fork with tools 15/32-inch (12 mm) thick (open end wrenches should work), then drive on the locking plate with tool number 3097.
66 Position a new gasket (without sealant), then install the 5th gear/clutch release bearing housing. Tighten the bolts in a diagonal pattern to 18 ft-lb (25 Nm). Work up to the final torque in three steps.
67 Remove the bar from the mouth of the bellhousing, then lubricate the clutch pushrod with multi-purpose grease and insert it into the mainshaft.

Gearshift linkage — adjustment (4-speed and 5-speed)

68 The procedure is correct as described in Chapter 5, but note that starting with VIN 17 C 079822, the Rabbit and pick-up have a new-type clamp bolt and a self-locking nut installed. When making the adjustment, tighten the nut to 11 ft-lb (15 Nm). Tighten the old-style nut to 15 ft-lb (20 Nm). If the shift rod adjustment cannot be maintained on an older vehicle, use the new-type clamp bolt and nut and tighten them to the new specified torque.

8 Automatic transmission

General information

Due to the complexity of the clutches and the hydraulic control system, and because of the special tools and expertise required to perform an automatic transmission overhaul, it should not be undertaken by the home mechanic. Therefore, the procedures included here are limited to general trouble diagnosis, routine maintenance and adjustment and transmission removal and installation.

If the transmission requires major repair work, it should be left to a dealer service department or a reputable automotive or transmission repair shop. You can, however, remove and install the transmission yourself, and save the expense, even if the repair work is done by a transmission specialist.

Problem diagnosis

Automatic transmission malfunctions may be caused by four general conditions: poor engine performance, improper adjustments, hydraulic malfunctions and mechanical malfunctions. Diagnosis of these problems should begin with a check of the easily repaired items — fluid level and condition and cable/band adjustments. Next, perform a road test to determine if the problem has been corrected. If the problem persists after the preliminary checks and corrections are completed, take the vehicle to a dealer service department or a reputable automotive or transmission repair shop.

Transmission band — adjustment

1 The second gear band can be adjusted with the transmission either in or out of the vehicle. In any case, the axis of the planetary gear system must be in a horizontal plane during adjustment — as it would be when the transmission is in the vehicle with the vehicle on level ground. If this precaution is not observed, especially with the transmission removed, the band may jam during adjustment. This would require at least partial disassembly of the transmission to realign the band.

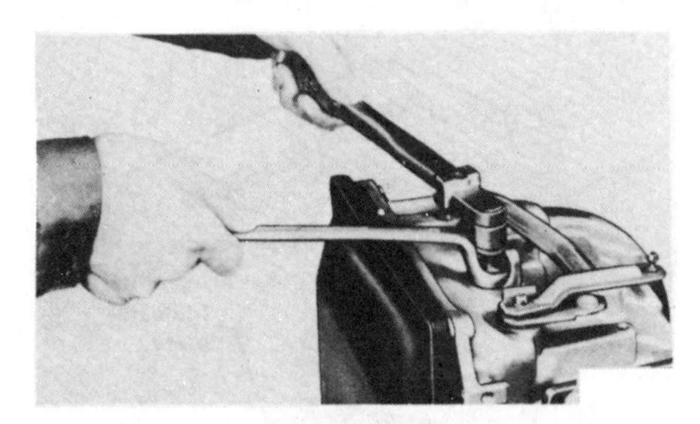

Fig. 13.26 Adjusting the automatic transmission 2nd gear band (Sec 8)

2 If the transmission has not been removed, raise the vehicle and support it securely on jackstands for access underneath.
3 Loosen the adjusting screw locknut, then tighten the adjusting screw to about 7.5 ft-lb (89 in-lb/10 Nm).
4 Loosen the adjusting screw, then retighten it to 3.5 ft-lb (44 in-lb/5 Nm).
5 From this setting, back the screw off exactly 2-1/2 turns. Hold the screw and tighten the locknut to 15 ft-lb (20 Nm).

Changing the automatic transmission fluid

6 Before beginning this procedure, obtain a new transmission pan gasket and, if available, a new strainer. Raise the vehicle and support it securely on jackstands, then remove the transaxle protection plate. Position a drain pan underneath the transmission.
7 Remove the rear transmission pan bolts and loosen the front bolts. Carefully break the gasket seal, then pull the pan down at the rear and allow as much of the fluid as possible to drain into the container. Remove the front bolts and lower the pan, then dump out the remaining fluid.
8 Wash the pan in solvent, then dry it with compressed air. Do not use rags to wipe the pan — lint from the rags may jam the transmission valves.
9 Remove the strainer from the bottom of the transmission and inspect it carefully. If it is dirty or clogged, clean it with solvent and dry it with compressed air. If available, install a new strainer. Tighten the mounting screws to 27 in-lb (3 Nm).
10 Position a new gasket on the transmission flange (do not use sealant on the gasket), then install the pan and bolts.
11 Tighten the bolts in a criss-cross pattern to 15 ft-lb (20 Nm).
Caution: *Do not overtighten the bolts or the pan will be deformed and leaks will result.*
12 Install the transmission protection plate, then lower the vehicle and open the hood. Refill the transmission with 3.2 quarts (3 liters) of Dexron-type ATF, then drive the vehicle to warm up the transmission fluid. With the fluid warm, the engine idling in Neutral and the parking brake set, check the fluid level with the dipstick. Do not fill above the top mark on the dipstick.
13 Check for leaks around the transmission fluid pan.

Selector lever — removal and refitting

14 Disconnect the cable from the negative battery terminal. Loosen the set screw that holds the knob to the top of the selector lever, then lift off the knob assembly.
15 Carefully pry up the indicator plate and remove it. Remove the screws and detach the console.
16 Remove the four bolts that hold the selector lever assembly in place. Pry off the E-clip, then detach the cable and housing from the selector lever and mount.
17 Disconnect the wires from the indicator light and contact plate. If further disassembly is required, use the accompanying illustration as a guide.
18 Installation is the reverse of removal. Before installing the console, adjust the cable as described below. Also, make sure the back-up lights work properly and that the starter operates only with the selector lever in Park or Neutral. If necessary, reposition the contact plate so the switches operate correctly.

Fig. 13.27 Selector lever cable clamping pin and nut (1) and transmission lever (2) locations (Sec 8)

Selector lever cable — removal and refitting

19 If the cable is not broken, position the selector lever in Park. Refer to the selector lever removal and refitting procedure and detach the cable from the lever and floorboard (Paragraphs 14 through 16 above).
20 Raise the vehicle and support it securely on jackstands. Remove the transmission protection plate, then loosen the nut for the cable clamping pin so the cable end can be withdrawn from the transmission lever.
21 Loosen the housing nut so the housing can be separated from the bracket.
22 Detach the cable from the lever and the bracket, then pull them out as a unit.
23 Installation is the reverse of removal. Tighten the cable clamping pin nut to 71 in-lb (8 Nm) and adjust the cable as described below.

Selector lever cable — adjustment

24 The cable must be adjusted whenever it is replaced or if the correct letter or number on the indicator plate is not illuminated with the selector lever in a particular position. If you have not replaced the cable, you must remove the protection plate under the transmission before beginning this procedure.
25 Place the selector lever in Park. Try pushing the vehicle back-and-forth to make sure the parking pawl has engaged inside the transmission.
26 Loosen the nut on the pin that holds the cable end to the transmission lever. Make sure the operating lever on the transmission is fully engaged in the Park position by pushing it to the left, then tighten the nut on the cable clamping pin to 71 in-lb (8 Nm).
27 Start the engine with the selector lever in Neutral, then, with the parking brake set, keep the engine running at a steady 1000 to 1200 rpm throughout the following tests.
28 Move the selector lever to Reverse. The engine speed should drop (indicating that the transmission has engaged reverse gear). Move the lever to Park. The engine speed should now increase (indicating that reverse gear has disengaged).
29 Without depressing the button on the knob, try to engage the lever in Reverse (the engine speed must not drop, as this would indicate that reverse gear has reengaged).
30 Depress the button on the knob, then repeat Paragraph 28.
31 Move the selector lever to Neutral. The engine speed should in crease, indicating that reverse gear has disengaged.
32 Move the selector lever to Drive. The engine speed should decrease, indicating that the transmission has engaged 1st gear.
33 Readjust the cable as necessary, then install the protection plate under the transmission.

Automatic transmission — removal and refitting

Removal

34 Disconnect the cables from the battery (negative first, then positive).

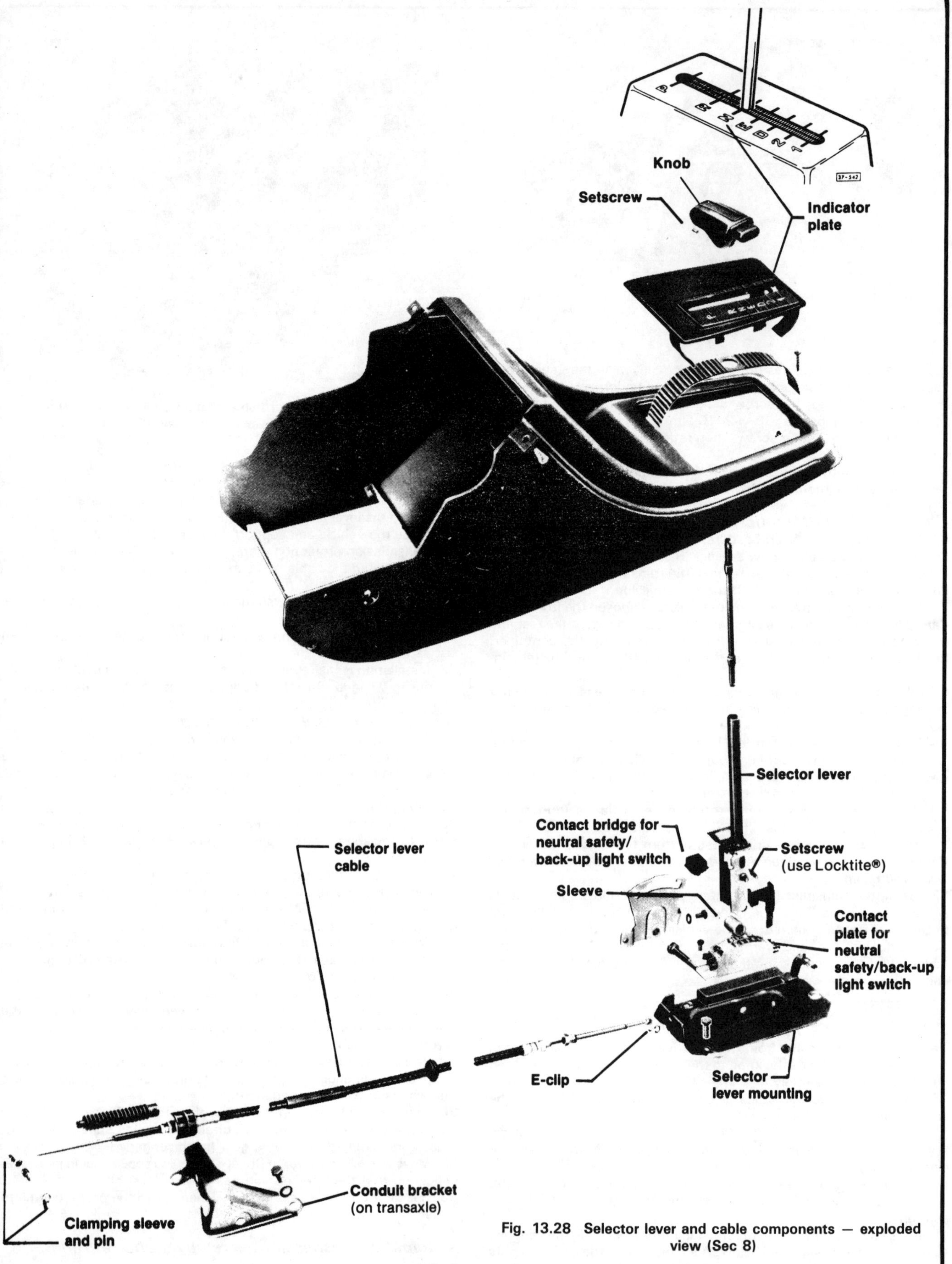

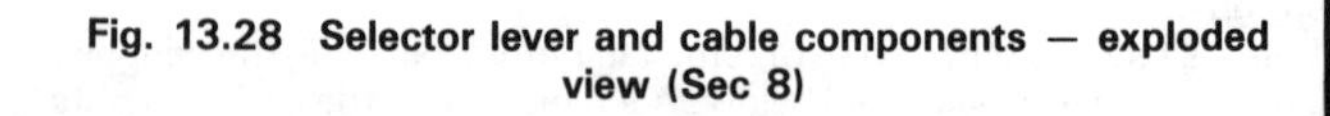

Fig. 13.28 Selector lever and cable components — exploded view (Sec 8)

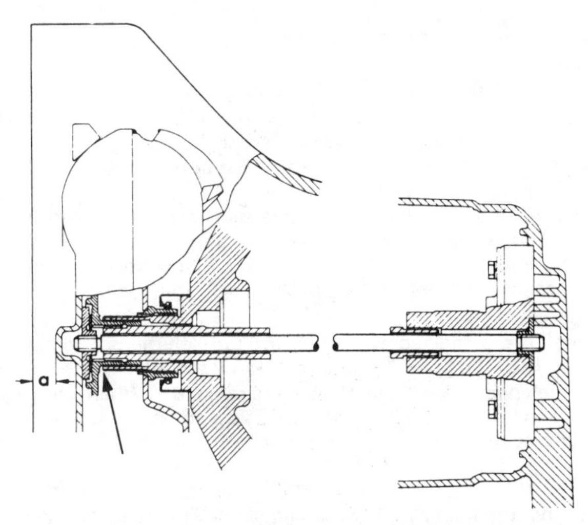

Fig. 13.29 When the torque converter is correctly positioned on the support, distance 'a' will be 1-13/16 inch (30 mm) (Sec 8)

35 Detach the speedometer cable from the driven gear assembly.
36 Install a device to support the engine from above (a standard engine hoist would work fine), then remove the two upper engine-to-transmission mounting bolts.
37 Unscrew the bolts that hold the transmission protection plate and torque converter cover plate in place, then remove them.
38 Support the transmission with a jack, then raise it slightly. Separate the transmission side mount from the body.
39 Remove the rear mount after removing the nuts that attach it to the transmission and body.
40 Detach the driveshafts from the drive flanges by taking out the bolts. Support the loose ends of the driveshafts with pieces of stiff wire – do not allow them to hang unsupported.
41 Disconnect the wires from the starter (label them to avoid confusion during installation), then remove the bolts and separate the starter from the engine.
42 Working through the openings in the bellhousing, remove the torque converter-to-drive plate bolts.
43 Position the selector lever in Park, then detach the cable from the lever on the transmission.
44 Remove the accelerator cable/throttle cable bracket from the transmission, then disconnect the cable ends from the lever on the transmission. Do not detach the cable housings from the bracket.
45 Remove the nuts, then separate the side mount from the transmission.
46 Detach the front mount from the transmission only.
47 Make sure the engine is firmly supported, then remove the remaining engine-to-transmission bolts. Raise the engine and transmission slightly so the left-hand driveshaft can be moved up and out of the way.
48 Carefully pull the transmission off the dowels that align it with the engine (make sure the torque converter separates from the drive plate, not the transmission).
49 Lower the transmission and raise the vehicle (and engine), then move the transmission out toward the front of the vehicle.
50 Fasten a metal bar or wire across the mouth of the bellhousing to keep the torque converter from falling out.

Refitting

51 To avoid damage, make certain that the torque converter has not slipped off the support inside the bellhousing. Position the transmission under the vehicle, then raise it with the jack until it can be pushed onto the dowels on the engine. Loosely install two bolts to hold it in place on the engine.
52 Carefully raise the engine and transmission until the left-hand driveshaft can be slipped into the recess in the drive flange, then lower the engine to its normal position.
53 Install all of the engine-to-transmission mounting bolts and nuts and tighten them to 41 ft-lb (55 Nm).
54 Install the side mount but leave the bolt loose.
55 Loosely install the front mount, then attach the cables to the transmission lever and install the bracket.
56 Connect the selector lever cable to the lever on the transmission (make sure the selector lever is in Park and that the transmission lever is correctly engaged).
57 Install the torque converter-to-drive plate bolts and tighten them to 22 ft-lb (30 Nm). If necessary, turn the crankshaft by hand (clockwise only) to bring the bolt holes into view.
58 Install the torque converter cover plate. Install the starter and tighten the bolts to 22 ft-lb (30 Nm), then hook-up the wires.
59 Attach the driveshafts to the drive flanges and tighten the bolts to 33 ft-lb (45 Nm).
60 Loosely install the rear mount, then reconnect the speedometer cable.
61 Remove the jack and install the transmission protection plate.
62 Reconnect the battery cables (positive first, then negative), then remove the engine support.
63 Refer to Chapter 1 (Section 47, Paragraph 3) and align the engine mounts as described there before tightening all of the mount fasteners.
64 Check the selector cable adjustment.

9 Braking system, wheels and tires

General information

Very few changes have been made to the brakes on later model vehicles covered by this manual. However, note that all vehicles from 1980 on are equipped with power-assisted brakes and that different front brake calipers (manufactured by Kelsey-Hayes) were installed on all 1981 and later Rabbits manufactured in the US and on some earlier models. Also, slightly different master cylinders and brake servos were widely used on later model vehicles (they cannot be interchanged).

Disc pads (Kelsey-Hayes caliper) – inspection and renewal

1 Some of the vehicles covered by this manual are equipped with a brake pad wear indicator. When the extension on the disc contacts the pad lip, the driver should feel the brake pedal pulsate. If your vehicle is not equipped with the wear indicator, replace the pads when the friction material has worn to 0.080 inch (2.00 mm).
2 To check the pads on Kelsey-Hayes calipers, look at the rear of the caliper, behind the wheel and tire. Measure the distance from the pad backing plate to the disc. If called for, replace the pads as discussed below.
3 Raise the front of the vehicle and support it securely on jackstands, then remove the front wheels. Carefully pry off both the upper and lower anti-rattle springs.
4 Unscrew and remove the two guide pins. Pull the caliper off the caliper frame and suspend it from a piece of stiff wire. **Caution:** *Do not allow the caliper to hang by the brake hose. If the piston begins to move out of the caliper, wrap a piece of soft wire or a rubber band around the piston to keep it in place.*
5 Separate the brake pads from the caliper frame.
6 Check the anti-rattle springs, the guide pins, the guide pin bushings and the guide pin sleeves in the bushings. Replace any parts that are damaged, worn or corroded. Note that the upper and lower anti-rattle springs and guide pins are not interchangeable.
7 Clean the pad seats and sliding surfaces in the caliper frame. Check the rubber dust seal on the piston. If it is cracked, hardened or swollen, remove the caliper and replace it (Chapter 6). Check the brake disc for wear as well.
8 Install the new pads in the caliper frame. Make sure the pad with the chamfered lining material is on the inside of the caliper.
9 Remove some of the brake fluid from the master cylinder (be very careful not to contaminate the fluid remaining in the reservoir).
10 Carefully push the piston into the caliper bore, then slip the caliper into place over the pads. Be careful not to damage the rubber seal on the piston.
11 Lubricate the sliding surfaces of the guide pins very lightly with silicone grease, then align the holes and install the guide pins. Tighten them to 30 ft-lb (40 Nm). **Note:** *The longer of the two guide pins should be installed in the upper hole.*
12 Install the anti-rattle springs with the looped ends in and the center projections facing the caliper.
13 Repeat the procedure for the remaining caliper.
14 Firmly depress the brake pedal a number of times to seat the pads, then check the brake fluid level in the master cylinder.
15 Install the wheels and test drive the vehicle.

Brake shoes and drums — inspection and renovation

16 Follow the procedure in Chapter 6, but note that the brake drums used on pick-up models are larger (see the Specifications at the front of this Chapter).

Master cylinder — overhaul

17 Master cylinders with slight design differences were used on later model vehicles, but the overhaul procedures are virtually identical, regardless of the type of master cylinder. Use the procedure in Chapter 6, but be sure to lay out the parts in the correct order as they are removed (doing so will ensure that everything goes back together correctly). Also, note that the master cylinder-to-brake servo mounting nut torque is different for later models.

Servo unit — removal and refitting

18 Follow the procedure in Chapter 6, but note that the mounting nut torque has been changed.

Brake servo vacuum pump — removal and refitting

19 The procedure in Chapter 6 is correct as described, but note that many vehicles with air conditioning and pick-ups built after early 1979 do not have an external oil line to disconnect.

Brake pressure regulator — description

20 On 1981 and later US-built Rabbits and pick-ups, the brake pressure regulator was replaced with a brake proportioning valve that is located in the engine compartment, under the master cylinder. The proportioning valve is not load sensitive and cannot be adjusted.

Brake and clutch pedal assembly — removal and refitting

21 On vehicles built after 1980, the brake pedal push rod does not require adjustment. For 1980 models, follow the steps in Chapter 6.

10 Electrical system

Battery — maintenance and inspection

The 'low-maintenance' battery used in some 1982 and later vehicles does not require as much attention as batteries used in earlier models. The 'low-maintenance' battery does have cell caps to enable hydrometer checks of the battery cells, but they are almost flush with the battery's top. The caps are equipped with a small slot at the edge to allow insertion of a screwdriver for removal. In many ways the 'low-maintenance' battery is like a 'maintenance-free' battery in that the electrolyte does not require routine replenishment and normal charging can be done without removing the cell caps.

Starter motor — removal and refitting

Two entirely different types of starter motors are used on the vehicles covered by this manual; one for manual transmissions and one for automatic transmissions. Starters for automatic transmissions have a different installed position and run in the opposite direction of starters for manual transmissions. The following procedure is for starters installed on automatic transmissions (for manual transmission starters, follow the procedure in Chapter 7).

1 Disconnect the cable from the negative battery terminal.

2 Raise the vehicle and support it securely on jackstands. Working beneath the vehicle, remove the starter cover plate (it is held in place by two nuts).

3 Disconnect the wires and battery cable from the solenoid (label the wires to simplify installation).

4 Remove the starter-to-bellhousing bolts and the starter bracket-to-engine bolts, then separate the starter from the engine.

5 Installation is the reverse of removal. Tighten the mounting bolts to 18 ft-lb (25 Nm).

Front parking light/turn signal — bulb renewal

6 On 1981 and later models (except Jetta), the bulbs can be replaced after removing the sockets from the housings. Turn the socket 1/4-turn counterclockwise and pull it out. Depress the bulb and remove it, then press the new bulb into place. Install the socket and turn it 1/4-turn clockwise to lock it in place.

Light cluster (rear) — bulb renewal

Cars only

7 Through 1980 models, the bulbs in the taillight assembly are accessible from inside the luggage compartment. Remove the knurled screw and detach the cover, then depress the spring clip on the inside edge and pull out the bulb holder. The bulbs are removed by depressing and turning them gently.

8 On 1981 and later vehicles, the bulb installation is changed slightly but is still done from inside the luggage compartment. Depress the spring clip, then turn the bulb holder counterclockwise 1/4-turn (except on Jettas) and pull it out.

Pick-ups only

9 Remove the four screws, then separate the lens and housing assembly from the body (leave the wires attached). Detach the bulb holder, then depress and turn the bulb to remove it. When reinstalling the lens and housing assembly, do not overtighten the screws or the lens will crack.

License plate light — bulb renewal (pick-up only)

10 Remove the bulb holder mounting screw, then detach the wires from the bulb assembly. Attach the wires to the new bulb assembly and install it with the screw (do not overtighten the screw or the lens may crack). Repeat the procedure if the remaining license plate light bulb is defective.

Side markers — bulb renewal

11 On 1981 and later models, the bulbs can be replaced after detaching the sockets from the rear of the light housing. The socket is removed by turning it 1/4-turn counterclockwise and locked in place by turning it clockwise.

Instrument panel — removal and refitting

1981 through 1984 Rabbit and pick-up

12 Disconnect the negative cable from the battery, then remove the left-hand speaker grille and speaker.

13 With the headlight switch pulled out all the way, reach through the speaker opening and depress the release button on the bottom of the switch, then pull out the knob and shaft.

14 Pull off the radio knobs and remove the six screws holding the instrument panel bezel in place, then separate the bezel from the dash.

15 Remove the four screws and separate the instrument panel from the dash. Pull the instrument cluster out and disconnect the wiring harness plugs from the switches and printed circuit. Unscrew the nut and separate the speedometer cable from the instruments.

16 Installation is the reverse of removal. Do not overtighten the screws for the instrument panel or the lugs may crack.

1981 through 1984 Jetta

17 Disconnect the negative battery cable from the battery, then remove the steering wheel.

18 Working under the dash, tilt the relay panel cover down and away from the panel, then remove the three screws and detach the relay panel cover and the shelf behind the steering column.

19 Remove the three screws and clips, then separate the lower dash cover from the dash. Remove the two screws that hold the instrument panel insert trim to the instrument cluster, then remove the trim piece.

20 Remove the screw from the top of the instrument cluster ,then tilt the top of the cluster away from the dash. Rea' behind the cluster and depress the speedometer cable tabs, then separate the cable from the instrument cluster.

21 Separate the wiring connector from the cluster and lift it away from the dash.

22 Installation is the reverse of removal. Do not overtighten the screws for the instrument panel or the lugs may crack.

Switches (lighting) — removal and refitting (1981 on US-built models only)

Headlight switch

23 Disconnect the negative cable from the battery, then remove the left-hand speaker grille and speaker.

24 With the headlight switch knob pulled out all the way, reach through the speaker opening and depress the release button on the bottom of the headlight switch, then pull out the knob and shaft.

25 Remove the radio knobs, then remove the six screws and detach the instrument cluster bezel from the dash.

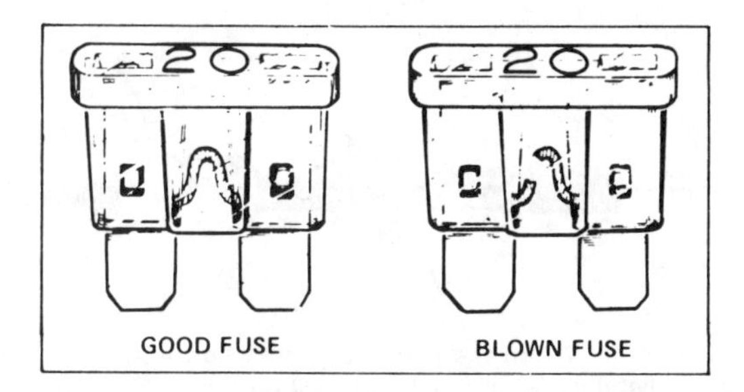

Fig. 13.30 The fuses used in later models look like this (Sec 10)

26 Remove the ring nut, then withdraw the switch from the dashboard assembly and detach the wires.
27 Installation is the reverse of removal. As the knob and shaft are installed, twist it slightly one way or the other until the shaft snaps in and locks.

Emergency flasher/heated rear window switch

28 Follow the procedure in Paragraphs 23 through 25 above.
29 Reaching behind the dash, use a small screwdriver to depress the springs on each side of the switch while pressing the switch out of the dash opening. Detach the wires from the switch.
30 Installation is the reverse of removal.

Brake light switch – removal and refitting (1981 on US-built models only)

31 The brake light switch is attached to the brake pedal assembly. To remove it, first disconnect the wire from the switch, then unscrew the switch from the pedal mount. As this is done note the distance between the switch and the brake pedal arm – the new switch will have to be adjusted to this same distance.
32 Thread the new switch into the mount until the base of the switch is 13/64 to 15/64-inch (5 to 6 mm) from the brake pedal, then reconnect the wire.

Fuses and relays – general

33 The electrical wiring and the fuse/relay box were completely redesigned beginning with 1981 models. Two fusible links are installed between the battery and the main wiring harness and a new fuse/relay box that uses different fuses is incorporated.
34 On 1981 and 1982 models, the fusible links are located near the battery in the left-hand side of the engine compartment (under the windshield washer reservoir). On 1983 and 1984 models, the fusible links are located low in the engine compartment, near the bottom of the left-hand strut tower. A third fusible link, used to supply power to the two-speed radiator fan, is used on automatic transmission-equipped vehicles.
35 If a fusible link is burned out, determine the cause of the overload before replacing the link. The fusible links should never be replaced with ordinary wire and should never be repaired – always replace them with new ones as follows:
36 On 1981 and 1982 models, remove the battery and windshield washer reservoir, then remove the tape from the connections and cut out the faulty fusible link(s).
37 Strip about 1/2-inch (12 mm) from the cut wires, then install the new fusible link with butt connectors and a crimping tool.
Solder the wires to the butt connectors with rosin core solder.
38 Tape all exposed connections, then install the battery and windshield washer reservoir (where applicable).

Cruise control system – general information and adjustment

39 An optional cruise control system is available on 1983 and later models. When the driver selects a speed above 40 mph and activates the system, the computer will maintain the preset speed unless the clutch pedal, brake pedal or accelerator is depressed.
40 If the system fails to operate properly, check/adjust the switches and servo as follows:
41 To adjust the vacuum vent switch or the electrical switch on either the clutch or brake pedal, turn the switch in the mounting bracket until the switch plunger projects 7/32 inch (5.6 mm) with the pedal all the way up.
42 To adjust the servo, remove the cover from the fuel injection pump and check the play in the servo linkage. It should be 0.004 to 0.012-inch (0.10 to 0.30 mm). If necessary, loosen the locknut and turn the threaded rod one way or the other to arrive at the specified play. Be sure to retighten the locknut.

Wiring diagrams

Note that wiring diagrams for 1980 through 1984 models are included at the end of this Supplement. Since space limitations prohibit including every wiring diagram for every model, only a typical sampling is included.

11 Front suspension, steering and driveshafts

General information

The most significant change to the front suspension and steering systems is the addition of power steering to later model vehicles.

Driveshaft – removal and refitting

1 Follow the procedure in Chapter 8, but note that beginning with 1982 models, the driveshaft inboard constant velocity joints are attached to the transaxle drive flanges with Torx-type bolts. These bolts require a special tool for removal and installation – be sure to obtain it before beginning the driveshaft removal procedure.

Wishbones – removal and refitting

2 Follow the procedure in Chapter 8, but note that the front pivot bolt on vehicles equipped with an automatic transmission cannot be removed unless the engine is raised. To do this, remove the front left engine mount, the nut for the rear mount and the engine mount support. After the pivot bolt is reinstalled, lower the engine and reinstall the engine mounts. Tighten the engine-to-mounting support bolts to 41 ft-lb (55 Nm).

Wheel bearing housing – removal and refitting

3 Follow the procedure in Chapter 8, but note that beginning with chassis number 178 3173 584, suspension balljoints with 17 mm studs are used in place of the earlier 15 mm size. A new wheel bearing housing, with a larger hole to accept the 17 mm stud and a 10 mm clamp bolt, was also introduced at the same time. All pick-up models are equipped with the larger-type balljoints.

Power steering system – checking

4 The checks described here will help determine whether power steering system problems are caused by the pump or steering gear. No replacement parts are available for either component, so if they are defective, they must be replaced with new units.
5 Check the power steering fluid level on the reservoir dipstick (the engine does not have to be running). If the level is not between the Full Hot and Full Cold marks, add fluid to raise the level to a point between the marks. Replace the dipstick/cap and tighten it securely.
6 Start the engine and allow it to idle. While an assistant turns the steering wheel from lock-to-lock, check all hoses and connections for leaks. Replace leaking hoses and tighten loose connections.
7 Detach the pinion shaft boot and the right-hand tie-rod boot from the steering gear. If there is a fluid leak at the pinion shaft or rack-to-tie-rod junction, replace the steering gear with a new one.

Power steering gear – removal and refitting

8 Raise the front of the vehicle and support it securely on jackstands. Remove both front wheels, then loosen the left tie-rod end locknut. Mark the position of the tie-rod end to simplify installation.
9 Withdraw the cotter pins, then remove the nuts from the studs in the steering arms.
10 Using a puller, press the tie-rod end studs out of the steering arms. Remove the end from the left tie-rod.
11 Place a drain pan under the power steering pump, then detach the hoses from the pump and allow the fluid to drain into the pan.
12 On vehicles equipped with a manual transmission, separate the gearshift linkage from the power steering gear.
13 Remove the pinion shaft boot, then detach the clamp bolt and nut that hold the lower end of the U-joint shaft to the pinion shaft.
14 Detach the hoses from the steering gear, then remove the rear transaxle mount and the exhaust system header pipe.

Fig. 13.31 When installing a new power steering gear on a vehicle with an automatic transmission, bend the bracket up to approximately a 45° angle (Sec 11)

15 Remove the steering gear clamp nuts, then separate the clamps and steering gear from the vehicle.
16 If the rubber mount bushings are hardened, worn or cracked, replace them with new ones before reinstalling the steering gear.
17 If a new power steering gear is being installed on a vehicle equipped with an automatic transmission, modify the bracket as shown in the accompanying illustration.
18 Attach the steering gear to the vehicle by guiding the pinion shaft into the lower U-joint of the U-joint shaft.
19 Install the clamps and mounting nuts but do not tighten the nuts. With the steering wheel spokes horizontal, make sure the steering gear is centered as shown in the accompanying illustration. Install the U-joint shaft clamp bolt and tighten it to 22 ft-lb (30 Nm). Reinstall the pinion shaft boot.
20 Keep the steering wheel spokes horizontal and the steering gear centered as the mounting nuts are tightened to 22 ft-lb (30 Nm).
21 Reinstall the transaxle rear mount (center it as described in Chapter 1) and the exhaust system header pipe.
22 Reconnect the hoses to the steering gear. On vehicles with a manual transmission, attach the gearshift linkage to the power steering gear.
23 Reconnect the hoses to the pump. Reinstall the tie-rod end on the left tie-rod (thread it on until it is at the marked position).
24 Reconnect the tie-rods to the steering arms and tighten the nuts to 22 ft-lb (30 Nm). Tighten the nuts, if necessary, to uncover the cotter pin holes, then install new cotter pins.
25 Fill the reservoir with power steering fluid and have the front end alignment checked.

Power steering pump — removal and refitting

26 The V-belt that drives the power steering pump also drives the water pump on vehicles equipped with power steering. Since the water pump and power steering pump are both rigidly mounted, a series of shims is built into the power steering pump drive pulley to adjust the belt tension.
27 To remove the pump, first remove the nuts and washers that attach the belt adjusting shims and pulley halves to the pump. Being very

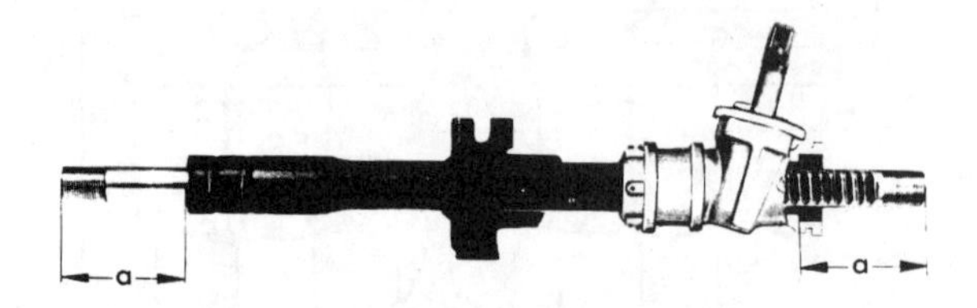

Fig. 13.32 When the steering gear is correctly centered, distance 'a' on each end will be equal (Sec 11)

careful to keep the shims in their original positions, separate the pulley, shims and belt from the pump.
28 Have a container ready to catch the fluid, then loosen the hose clamp that holds the feed hose to the upper connection on the pump. Pull the hose off the connection and allow the fluid to drain into the container.
29 Remove the hoses from the pump, then separate the pump from the engine by removing the mounting nuts.
30 Installation is basically the reverse of removal. Install the inner pulley half on the pump. If you are reinstalling the original V-belt, install the original number of shims between the pulley halves. If you are installing a new belt, install one or two fewer shims between the pulley halves.
31 Slip the belt into place, then install the outer pulley half and the remaining shims. If necessary, turn the pump by hand until one of the studs is facing the crankshaft. Install a nut on that stud finger tight. **Note:** *You can turn the pump pulley by inserting a socket-head driver into the recess in the center of the pump shaft.*
32 Position the two remaining studs (one at a time) facing the crankshaft and install the nuts finger tight. Turn the pump pulley by hand one complete revolution, stopping to tighten the nuts finger tight when they are facing the crankshaft.
33 Repeat the procedure and tighten each nut to 13 ft-lb (18 Nm). Rotate the crankshaft several turns (clockwise only) to seat the belt, then repeat the procedure one more time.
34 Check the belt tension. It is correct if you can depress it 3/8 to 7/16-inch (12 mm) at a point midway between the crankshaft pulley and the power steering pump pulley along the lower run of the belt.
35 If the belt tension is incorrect, disassemble the power steering pump pulley and add or subtract shims as necessary, then repeat the adjustment procedure. To decrease tension, add shims; to increase tension, remove shims from between the pulley halves. **Note:** *Never discard belt adjusting shims. There must always be at least one shim in the storage position. Replace the belt if it cannot be tensioned correctly by removing shims from between the pulley halves.*
36 Be sure to add power steering fluid to the reservoir.

Tie-rods — removal and refitting (power steering-equipped vehicles only)

37 Remove the cotter pin, then loosen and remove the nut from the balljoint stud in the steering arm. Using a puller, separate the stud from the steering arm.
38 Remove the clamp that attaches the boot to the rack tube, then pull the boot down off the tube onto the tie-rod to expose the tie-rod's inner balljoint housing. If necessary, turn the boot inside out on the tie-rod.
39 Grip the flats of the inner balljoint housing with a wrench and unscrew the housing and tie-rod. Considerable effort may be required because the threads are coated with locking compound.
40 Before installing the tie-rod, clean the threads on the rack and inside the inner balljoint housing. If necessary, install the boot on the tie-rod. **Note:** *To install the boot, the end must be removed from the tie-rod. If you are reinstalling the original tie-rod, mark the position of the end so it can be installed in the same relative position. If you are installing a new tie-rod, measure the original and thread the end onto the new one until it is the same length.*
41 Apply thread locking compound to the threads inside the balljoint housing, then attach the tie-rod to the rack. Tighten the inner balljoint housing until the end of the rack bottoms against the inside of the housing.
42 Reposition the boot on the tie-rod and steering rack tube, then install the clamps.

43 Reconnect the tie-rod to the steering arm and tighten the nut to 22 ft-lb (30 Nm). Tighten the nut, if necessary, to expose a cotter pin hole, then install a new cotter pin. Be sure to have the front end alignment checked.

12 Rear axle and rear suspension

General information

The rear suspension on pick-ups is fundamentally different from the rear suspension on cars in that it consists of leaf springs and separate shock absorbers rather than a strut assembly.

Rear stabilizer bar — removal and refitting (Jetta only)

1 To remove the stabilizer bar from the axle beam, drive off the clips that hold together the steel straps that attach the stabilizer bar to the round parts of the axle (use a hammer and punch).
2 Remove the bolts and nuts that hold the U-shaped retainers to the flat part of the axle, then detach the stabilizer bar and rubber bushings.
3 Installation is the reverse of removal. If the rubber bushings are worn or damaged, replace them with new ones. Tighten the retainer bolts and nuts to 22 ft-lb (30 Nm). Clamp the steel straps together and drive on the clips with a hammer.

Rear shock absorber — removal and refitting (pick-up only)

4 Raise the rear of the vehicle and support the axle beam with jackstands.
5 Remove the mounting bolts and separate the shock absorber from the axle and frame.

6 Installation is the reverse of removal. Tighten the mounting bolt nuts to 30 ft-lb (40 Nm).

Rear leaf spring — removal and refitting (pick-up only)

7 Raise the rear of the vehicle and support it with jackstands placed under the frame rails (the axle beam and wheels must be unsupported). Detach the parking brake cable housing from the rear spring front clamp and cut the tie wrap that holds the housing.
8 Position a jack under the axle beam and raise the axle until the shock absorber just begins to compress. Remove the lower shock mounting bolt, then separate the shock from the axle and swing it up and out of the way. Lower the axle to unload it.
9 Remove the nuts, then separate the U-bolts and spring mounting plate from the spring and axle.
10 If you are removing the left-rear spring, remove the three bolts that hold the exhaust flex-pipe to the header pipe. Unhook the exhaust system from the hangers and separate the system from the vehicle.
11 Loosen the upper shackle bolt nut, then remove the lower shackle bolt so the spring can be detached at the rear. Remove the bolt from the front spring mount, then separate the spring from the vehicle.
12 To install the spring, position it on top of the axle with the spring center bolt engaged in the hole in the center of the spring mount.
13 Position the front of the spring in the front mount and install the nut and bolt finger tight. Engage the rear of the spring in the shackle and install the nut and bolt finger tight.
14 Install the U-bolts and spring mounting plate. Tighten the nuts to 30 ft-lb (40 Nm) following a criss-cross pattern.
15 Raise the axle with a jack to load the spring. Reattach the shock absorber to the axle and tighten the bolt/nut to 30 ft-lb (40 Nm). Tighten the shackle bolt nuts to 44 ft-lb (60 Nm) and the front mount bolt to 70 ft-lb (95 Nm). Lower the axle to unload it.
16 Attach the parking brake cable housing to the front clamp and install a new tie wrap in place of the old one.
17 If removed, reinstall the exhaust system. Use a new gasket between the header pipe and flex pipe and tighten the bolts to 18 ft-lb (25 Nm).
18 Lower the vehicle to the ground.

Rear axle assembly — removal and refitting (pick-up only)

19 Raise the rear of the vehicle and support it with jackstands placed under the frame rails (the axle and wheels must be unsupported).
20 Detach the parking brake cables from the lever as described in Chapter 6. Separate the parking brake cable housings from the leaf spring front clamps, then cut the tie wraps that hold the housings and pull the cables out of the tubes leading to the parking brake lever.
21 Unhook the brake pressure regulator coil spring (where applicable). Note the positions of the spring's end hooks so it can be reinstalled correctly.
22 Position a jack under the axle and raise the axle until the shock absorbers just start to compress. Disconnect the brake hoses from the top of the axle beam and plug the lines with clean caps. Remove the brake pressure regulator coil spring bolt from the top of the axle beam, then detach the brake hose mounting plate.
23 Remove the lower shock absorber mounting bolts/nuts and detach the shocks from the axle. Remove the U-bolt nuts, then separate the U-bolts and spring mounting plates from the axle and springs.
24 Carefully lower the axle assembly to the ground and move it out from under the vehicle.
25 Installation is the reverse of removal. When the installation is complete, the brake pressure regulator coil spring must be in its original position to avoid contact with adjacent components.
Also, be sure to bleed the brakes and adjust the parking brake as described in Chapter 6.

13 Bodywork and fittings

The floor plate, which consists of three sub-assemblies that are welded together, is the same on Rabbits and Jettas. Pick-up models, however, are built on a different floor plate and share no rear body panels with sedans and hatchbacks. A sliding steel sunroof was offered as optional equipment on many of the vehicles covered by this manual.

Front wings — removal and refitting

1 Follow the procedure in Chapter 10, but note that on 1980 models the battery ground strap must be disconnected and the wires for the side marker light must be unplugged. On 1981 through 1983 models, the grille must be removed.

Rear bumper — removal and refitting (pick-up only)

2 The rear bumper is composed of two sections, the center support and the brackets. To remove the bumper sections, pry off the impact strip and remove the bolts that attach the bumper section to the center support. Remove the bumper-to-bracket bolts and separate the bumper section from the vehicle.
3 To remove the bumper assembly, first remove the bolts that hold the center support to the bumper brackets, then pry off the impact strip and remove the bolts that attach the bumper section to the bracket.
4 Installation is the reverse of removal. Be sure to center the bumper on the body before tightening the mounting bolts.

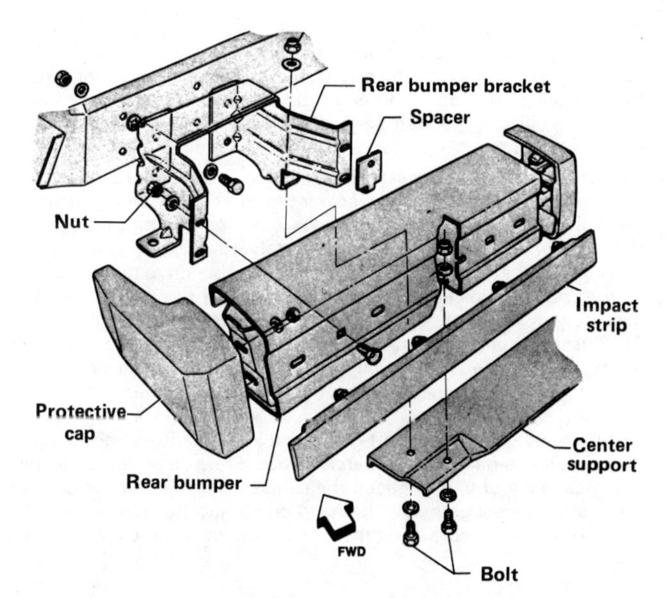

Fig. 13.33 Pick-up rear bumper components — exploded view (Sec 13)

Tailgate handle and latch — removal and refitting (pick-up only)

5 The tailgate handle and each of the two latches can be removed and installed independently of the other two components. The tailgate, however, must be removed before installing a latch.
6 To remove the tailgate handle, work from the inside of the tailgate and remove the three screws that attach the handle to the tailgate.
7 Working from the handle side of the tailgate, unsnap both latch connecting rods from the clips that hold them to the handle (be careful not to let the clips fall into the tailgate). Lift out the handle.
8 To remove the tailgate latch, unsnap the latch connecting rod from the handle (be careful not to let the clip fall into the tailgate). Remove the two screws from the latch and separate it from the tailgate with the rod attached.
9 To install the latch, release the holding straps from each side of the tailgate, then remove the tailgate by lifting it out of the right side hinge and pulling it out of the left side hinge.
10 Stand the tailgate on end with the end that the latch is being installed in upright. Insert the latch and connecting rod into the tailgate, carefully guiding the rod through the hole in the handle opening. Attach the latch to the tailgate with the screws.
11 Snap the latch connecting rod into the handle, then install the tailgate by reversing the removal procedure.

Sunroof — removal and refitting

12 Using the hand crank, open the sunroof half way. Carefully pry the front edge of the sunroof trim panel off with a wooden wedge (work as closely as possible to the clips).
13 Open the sunroof all the way, then slide the trim panel forward as far as possible. Push the trim panel up from below and remove it one side at a time through the top of the vehicle (be careful not to kink it).
14 Close the sunroof almost all the way with the hand crank, then remove the screws and detach the front guides. Close the sunroof (be careful not to go past the closed position), then unhook the leaf springs from the lifters. Swing the springs in toward the center of the vehicle.
15 Remove the sunroof-to-lifter screws, then carefully lift it out through the top of the vehicle.
16 To install the sunroof, begin by inserting it through the opening (rear edge first) and engaging the rear guides in the runners. Slowly push the sunroof to the rear while lowering the front edge into position.
17 Pull it forward to the closed position, then install the screws that hold it to the lifters and hook the springs over the lifters.
18 Install the front guide-to-sunroof screws. Adjust the sunroof as described later in this Section, then open it half way.
Carefully insert the trim panel through the top of the vehicle and guide it into position under the sunroof, then press the clips into the sunroof.

Sunroof — adjustment

19 The cables should be adjusted after removing and installing the sunroof and whenever it fails to open and close evenly. Portions of the following procedure can be used to replace the crank, the drive gear assembly and related parts, if necessary.

Cable adjustment

20 Remove the trim panel as described in Paragraph 12, then push it to the rear as far as possible. Lower the sunroof by turning the crank one-half turn from the completely closed position. Remove the screw, then pull off the hand crank and escutcheon.
21 Loosen the two screws that hold the cable drive gear assembly to the roof approximately six turns each. Pull the cable drive gear assembly down until the gear disengages the cables.
22 Push the sunroof open and shut several times by hand, then shift it until it is square with the opening. If the sunroof is not now flush with the top of the vehicle, adjust the height as described later in this Section.
23 Turn the drive gear shaft clockwise as far as it will go by hand, then turn it one-half turn counterclockwise. Press up on the cable drive gear assembly until it engages the cables, then tighten the screws.
24 Install the escutcheon, the hand crank and the screw.
25 To check the cable adjustment and crank position, open and close the sunroof several times. If necessary, reposition the crank on the shaft so it can be folded into the recess when the sunroof is completely closed.
26 If the sunroof does not open and close evenly on both sides, repeat the adjustment. When the adjustment is complete, pull the trim panel forward and attach it to the sunroof by pressing the clips into place by hand.

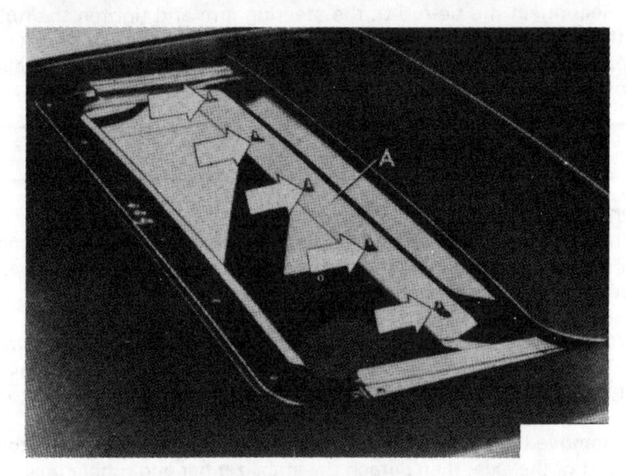

Fig. 13.34 The trim panel is attached to the sunroof with 5 clips (Sec 13)

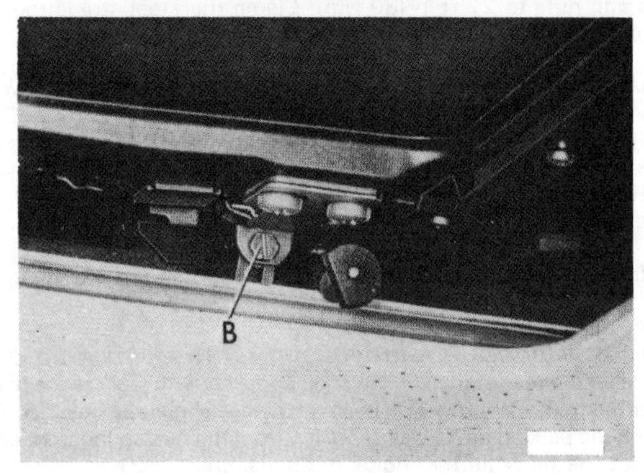

Fig. 13.35 Screw B must be loosened to adjust the height of the rear of the sunroof (Sec 13)

Height adjustment

27 The sunroof height should be adjusted after removal and installation of the sunroof and whenever the top does not lie flush with the top of the vehicle.
28 Remove the trim panel as described in Paragraphs 12 and 13 above, then close the sunroof (be careful not to pass the fully closed position).
29 Loosen the front guide-to-sunroof screws, then turn the front height adjusting screws as required to align the sunroof perfectly with the top of the vehicle. Tighten the guide-to-sunroof screws.
30 Unhook the leaf springs from the lifters, then swing them in toward the center of the vehicle. Loosen the rear height adjustment screws and raise or lower the sunroof by hand until the rear of the sunroof is perfectly aligned with the top of the vehicle, then tighten the screws.
31 Swing the leaf springs back and hook them over the lifters.
Using the hand crank, open and close the sunroof several times to make sure it is flush with the top of the vehicle. Repeat the adjustment if necessary. If the height is satisfactory, but the sunroof does not close evenly on both sides, adjust the cables as described earlier in this Section.
32 When the adjustment is complete, pull the trim panel forward and press the clips into the sunroof by hand.

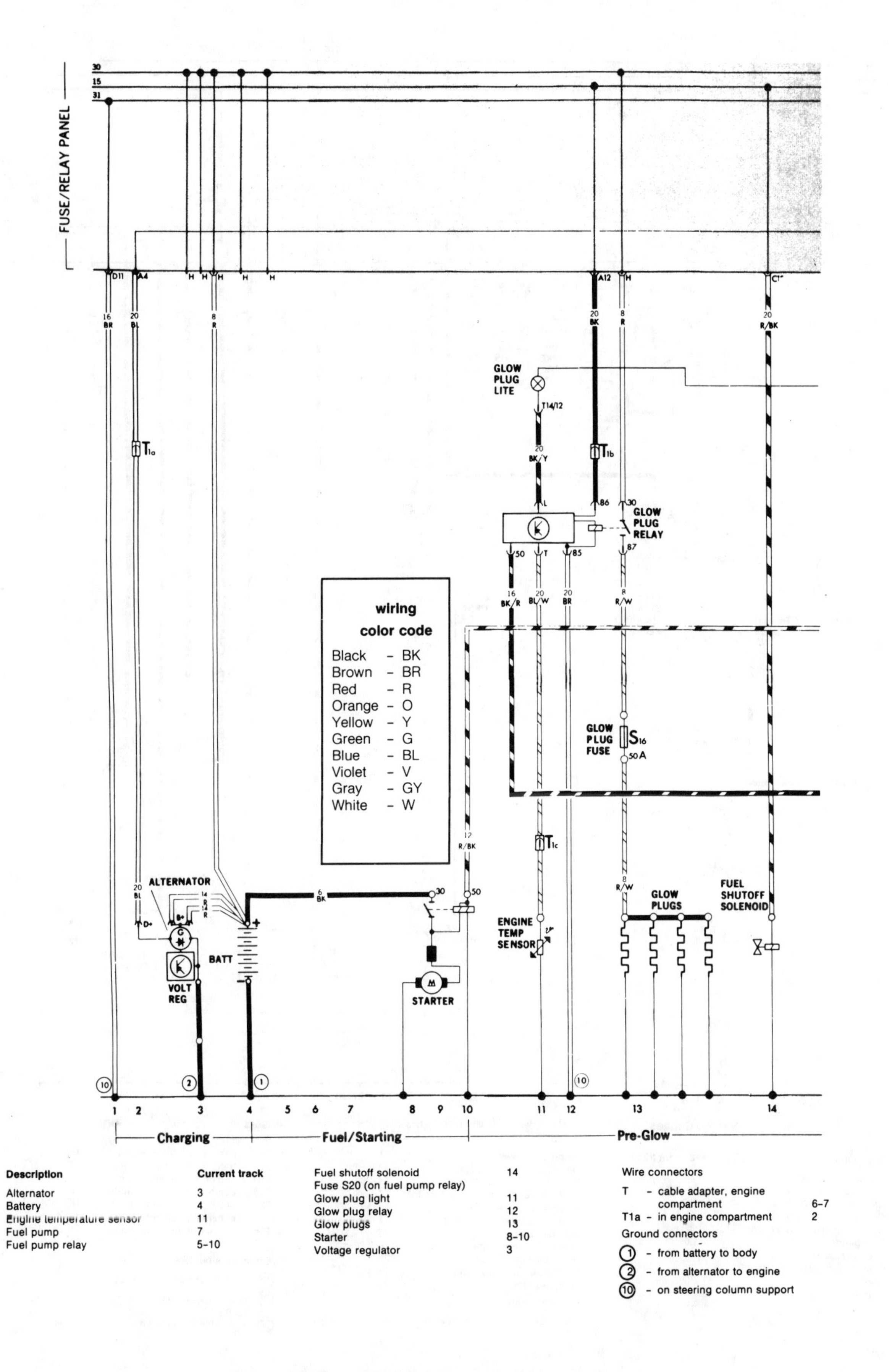

Description	Current track
Alternator	3
Battery	4
Engine temperature sensor	11
Fuel pump	7
Fuel pump relay	5–10
Fuel shutoff solenoid	14
Fuse S20 (on fuel pump relay)	
Glow plug light	11
Glow plug relay	12
Glow plugs	13
Starter	8–10
Voltage regulator	3

Wire connectors

T	- cable adapter, engine compartment	6–7
T1a	- in engine compartment	2

Ground connectors

- (1) - from battery to body
- (2) - from alternator to engine
- (10) - on steering column support

1980 model wiring diagram — typical (1 of 5)

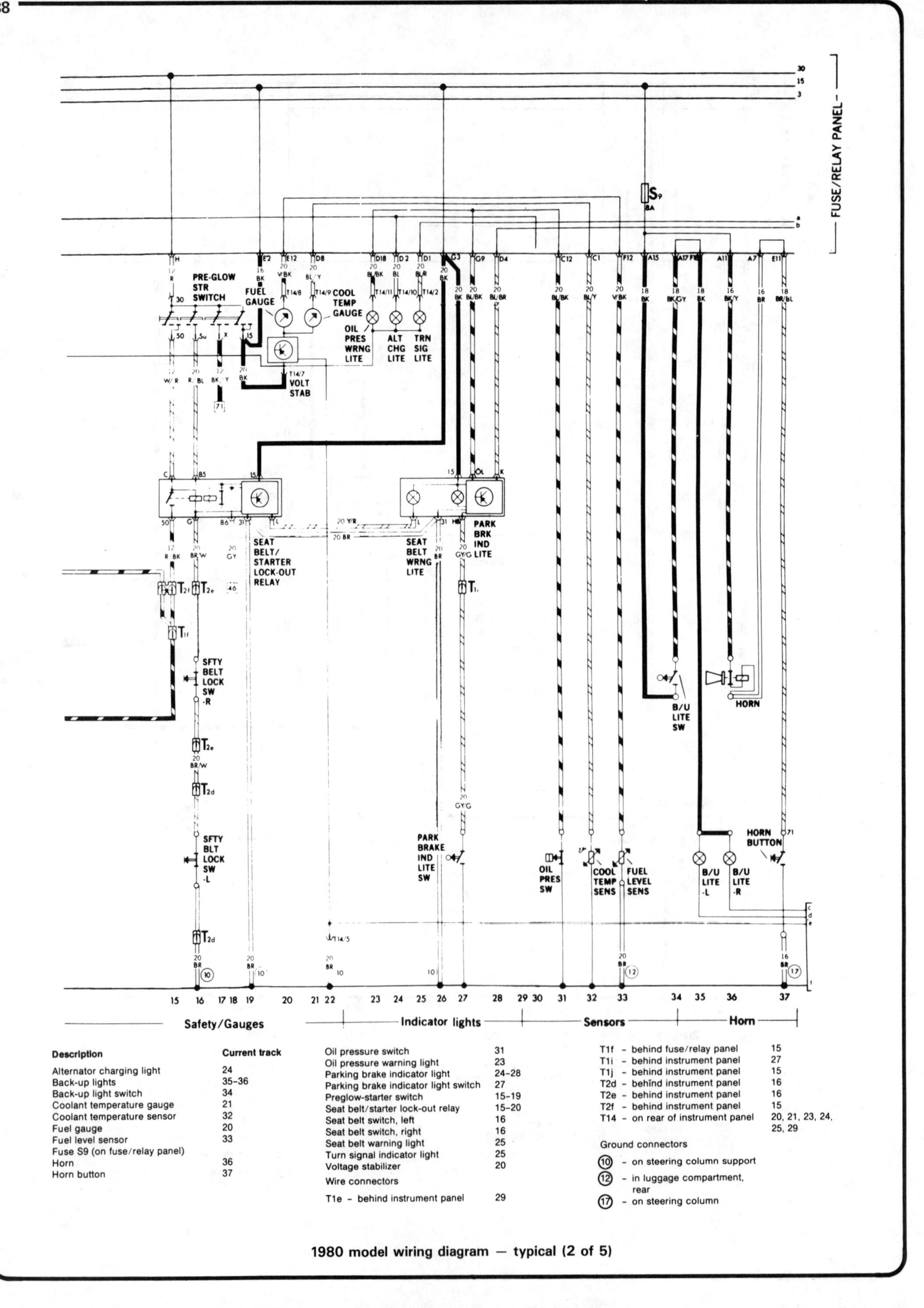

1980 model wiring diagram — typical (2 of 5)

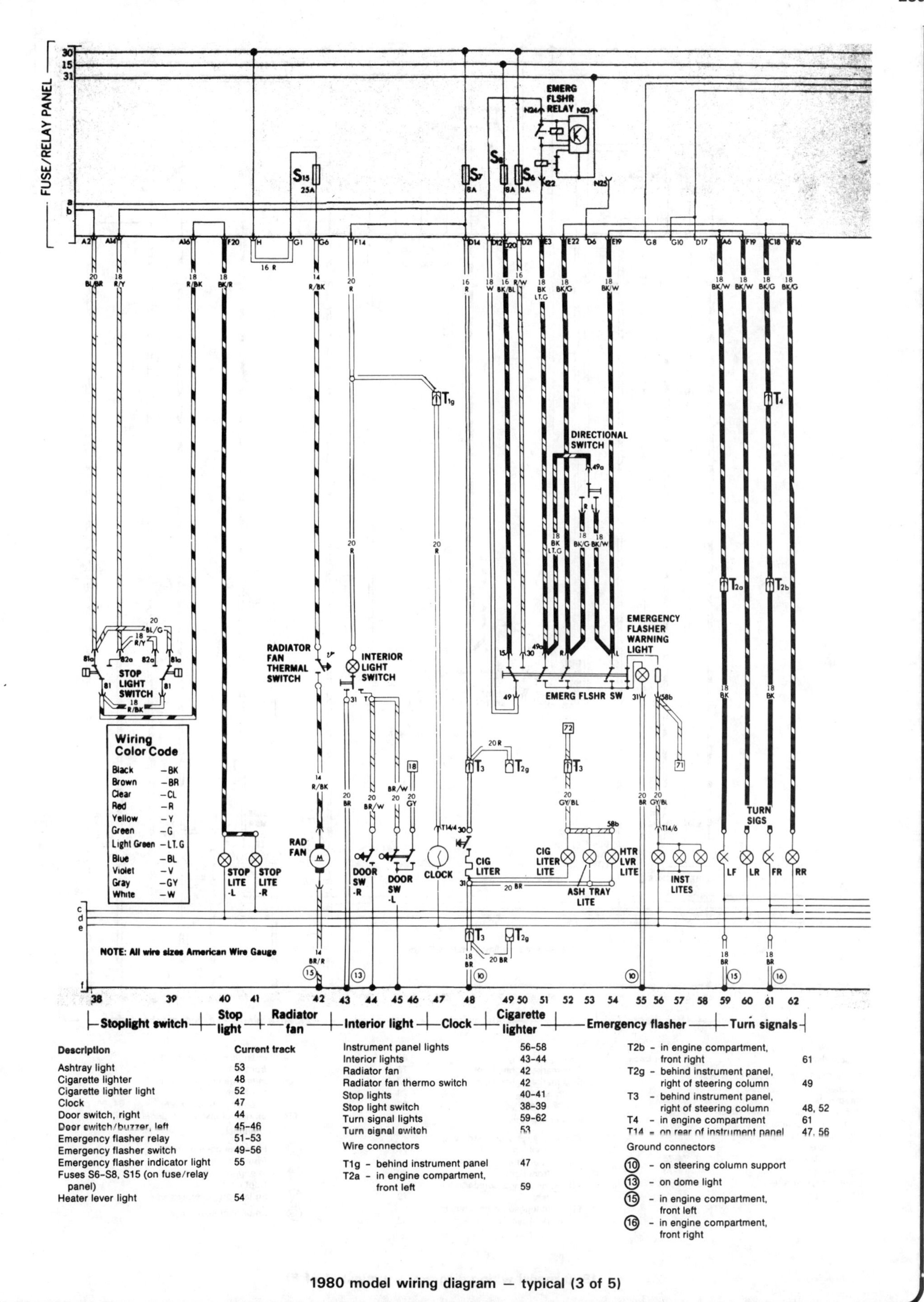

1980 model wiring diagram — typical (3 of 5)

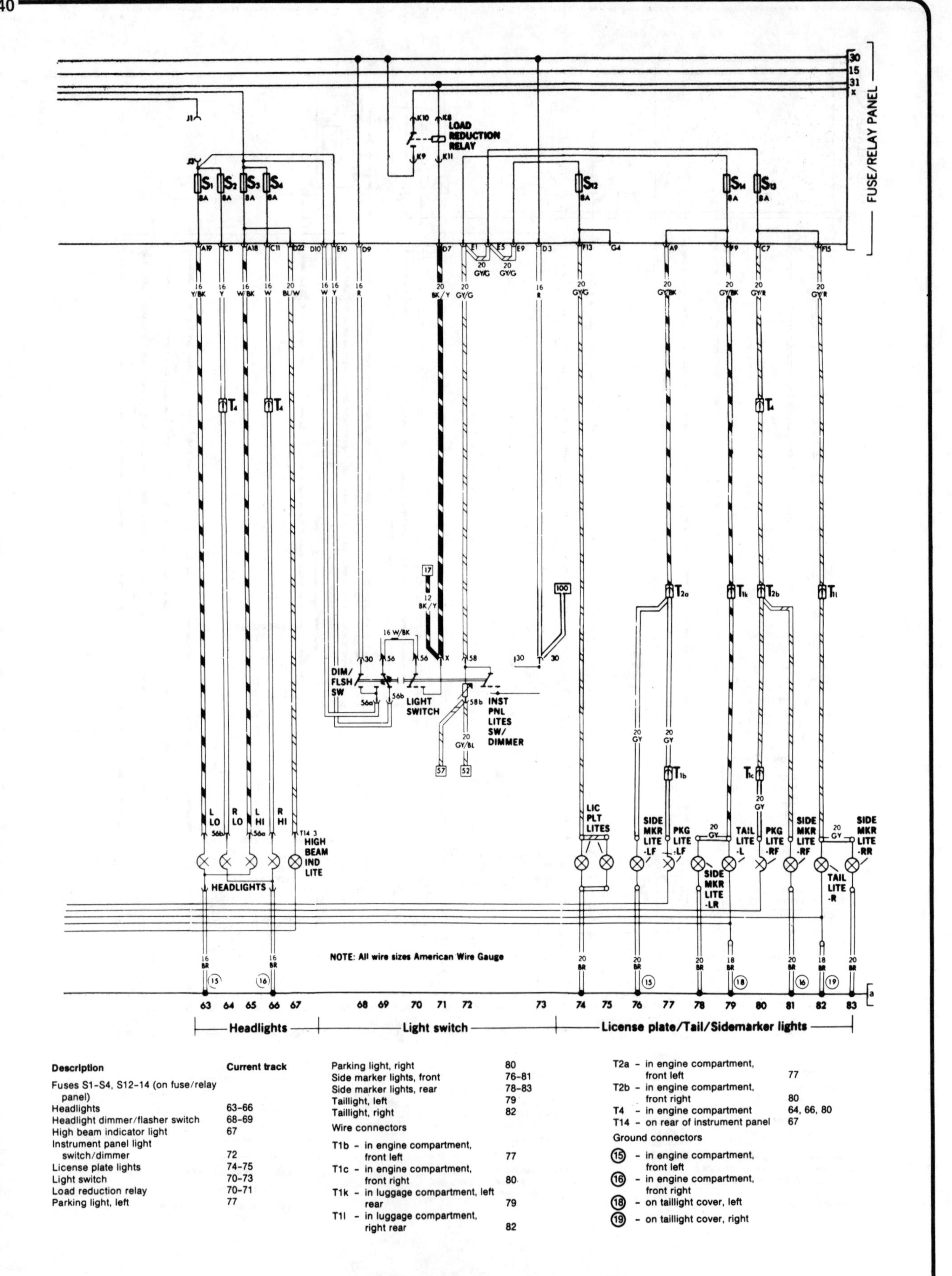

Description	Current track
Fuses S1-S4, S12-14 (on fuse/relay panel)	
Headlights	63-66
Headlight dimmer/flasher switch	68-69
High beam indicator light	67
Instrument panel light switch/dimmer	72
License plate lights	74-75
Light switch	70-73
Load reduction relay	70-71
Parking light, left	77
Parking light, right	80
Side marker lights, front	76-81
Side marker lights, rear	78-83
Taillight, left	79
Taillight, right	82
Wire connectors	
T1b - in engine compartment, front left	77
T1c - in engine compartment, front right	80
T1k - in luggage compartment, left rear	79
T1l - in luggage compartment, right rear	82
T2a - in engine compartment, front left	77
T2b - in engine compartment, front right	80
T4 - in engine compartment	64, 66, 80
T14 - on rear of instrument panel	67
Ground connectors	
(15) - in engine compartment, front left	
(16) - in engine compartment, front right	
(18) - on taillight cover, left	
(19) - on taillight cover, right	

1980 model wiring diagram — typical (4 of 5)

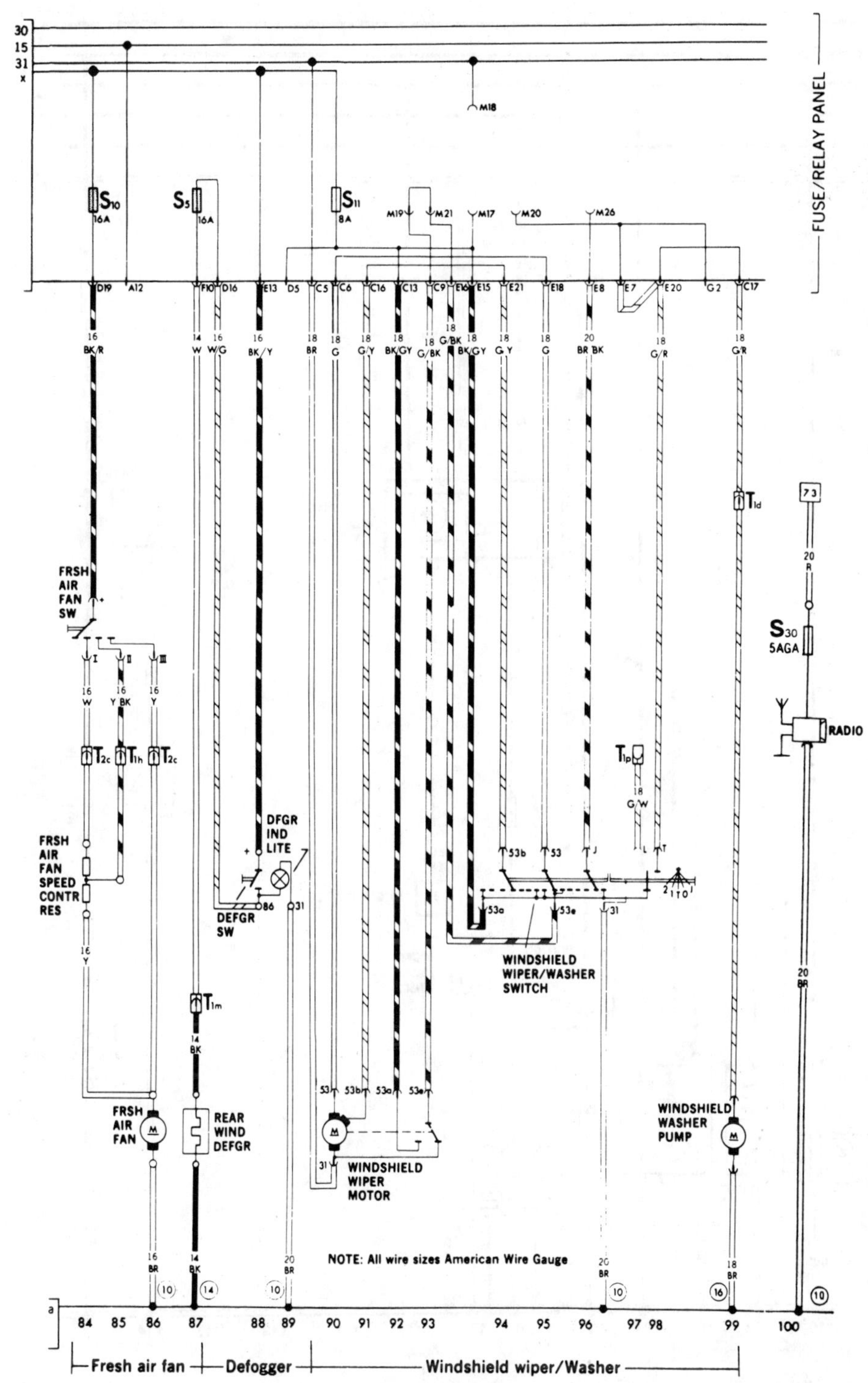

wiring color code

Black	-	BK
Brown	-	BR
Red	-	R
Orange	-	O
Yellow	-	Y
Green	-	G
Blue	-	BL
Violet	-	V
Gray	-	GY
White	-	W

Description	Current track
Fresh air fan	86
Fresh air fan speed control resistors	84
Fresh air fan switch	84-86
Fuses S5, S10, S11 (on fuse/relay panel)	
Radio	100
Rear window defogger	87
Rear window defogger indicator light	89
Rear window defogger switch	88-89
Windshield washer pump	99
Windshield wiper motor	90-93
Windshield wiper switch	94-98
Wire connectors	
T1d - behind instrument panel	99
T1h - behind instrument panel	85
T1m - in luggage compartment, left rear	87
T1p - behind instrument panel	97
T2c - behind instrument panel	84, 86

Ground connectors

- ⑩ - on steering column support
- ⑭ - middle body member, rear
- ⑯ - in engine compartment, front right

1980 model wiring diagram — typical (5 of 5)

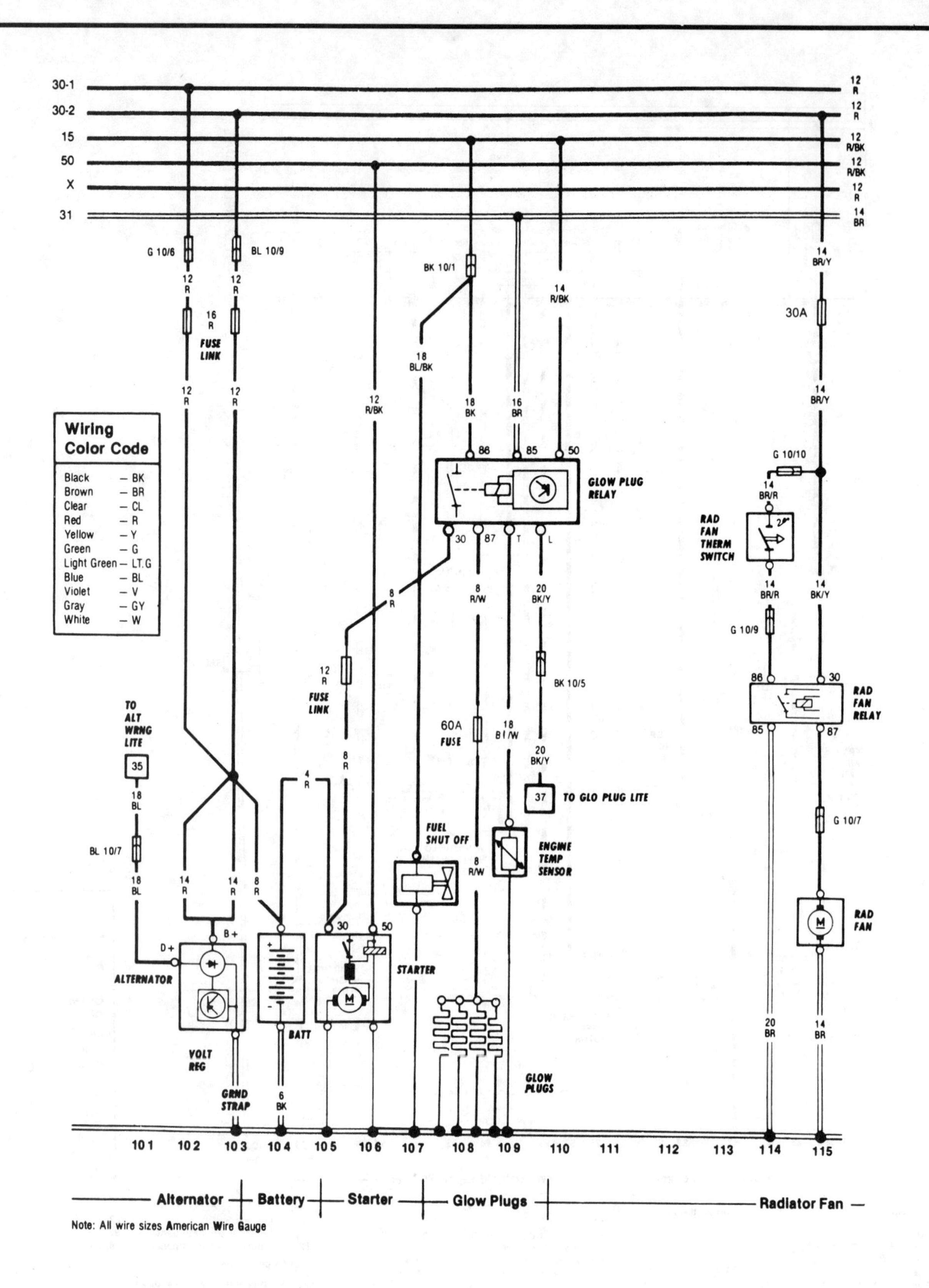

Alternator, battery, starter, glow plug and radiator fan wiring diagram – typical (1981 through 1984 models except Jetta)

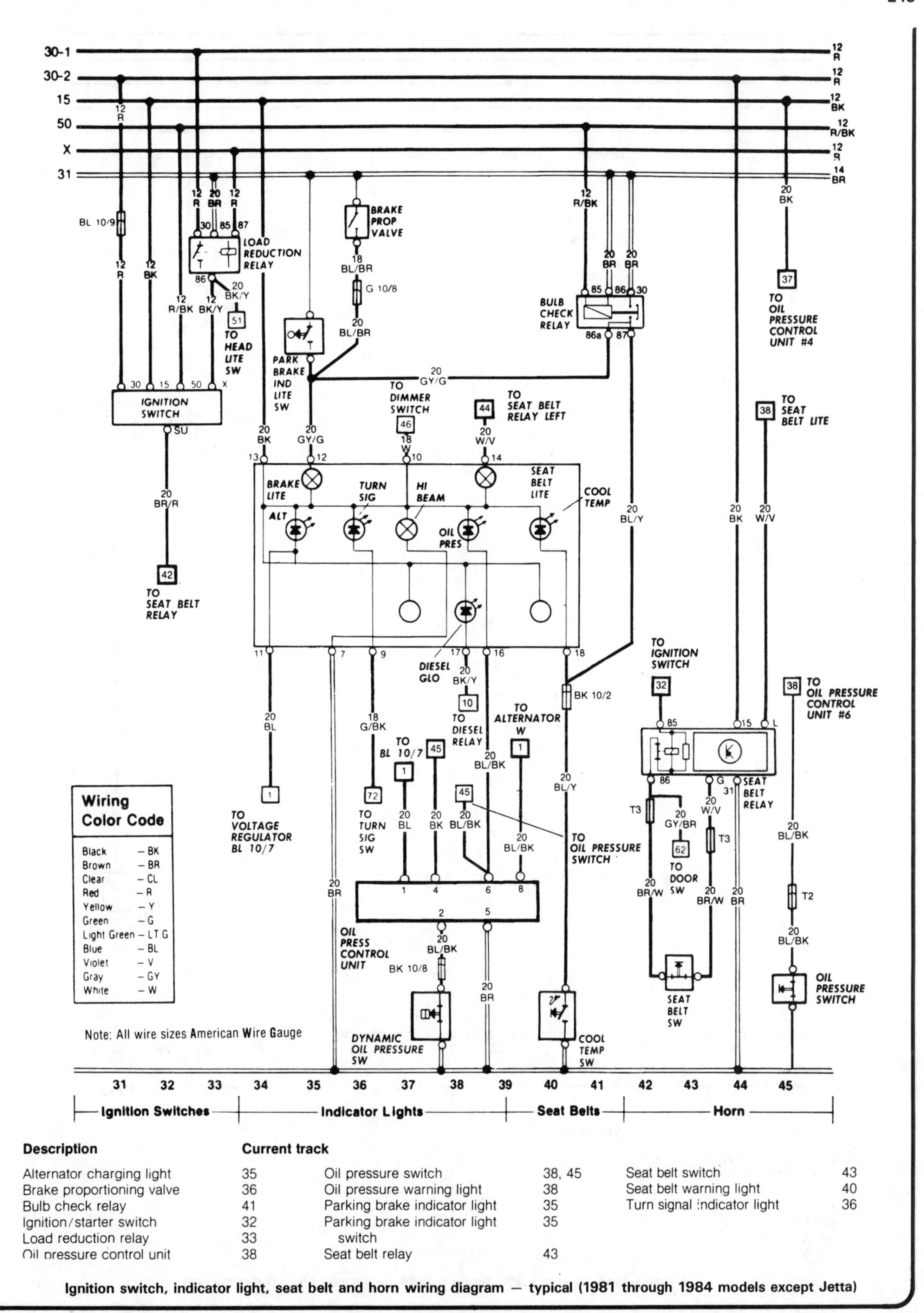

Description	Current track				
Alternator charging light	35	Oil pressure switch	38, 45	Seat belt switch	43
Brake proportioning valve	36	Oil pressure warning light	38	Seat belt warning light	40
Bulb check relay	41	Parking brake indicator light	35	Turn signal indicator light	36
Ignition/starter switch	32	Parking brake indicator light switch	35		
Load reduction relay	33	Seat belt relay	43		
Oil pressure control unit	38				

Ignition switch, indicator light, seat belt and horn wiring diagram — typical (1981 through 1984 models except Jetta)

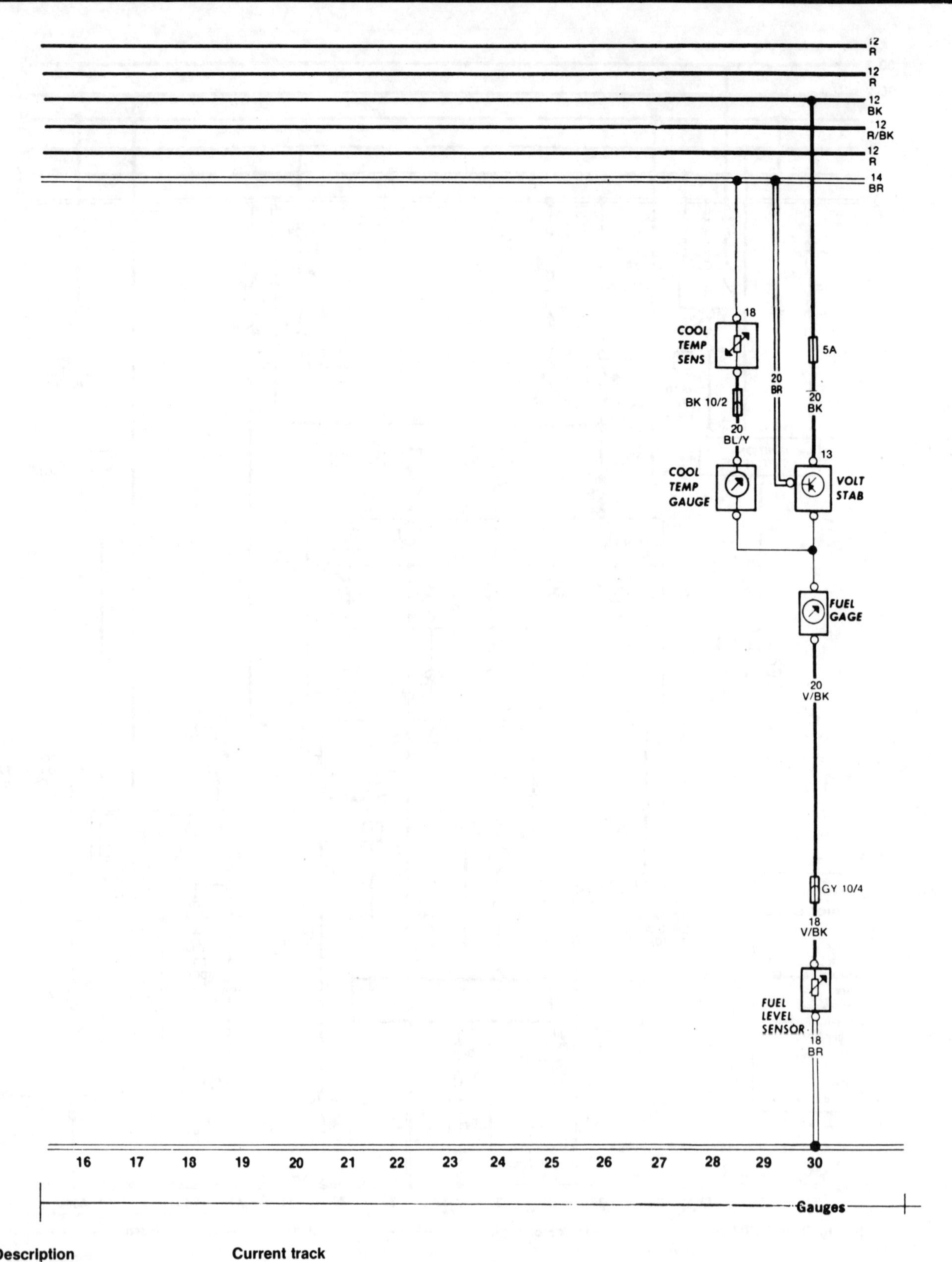

Description	Current track
Coolant temperature gauge	29
Coolant temperature sensor	29
Fuel gauge	30
Fuel level sensor	30
Thermo time switch	29
Voltage stabilizer	30

Gauge wiring diagram (1982 through 1984 Rabbit and pick-up models only)

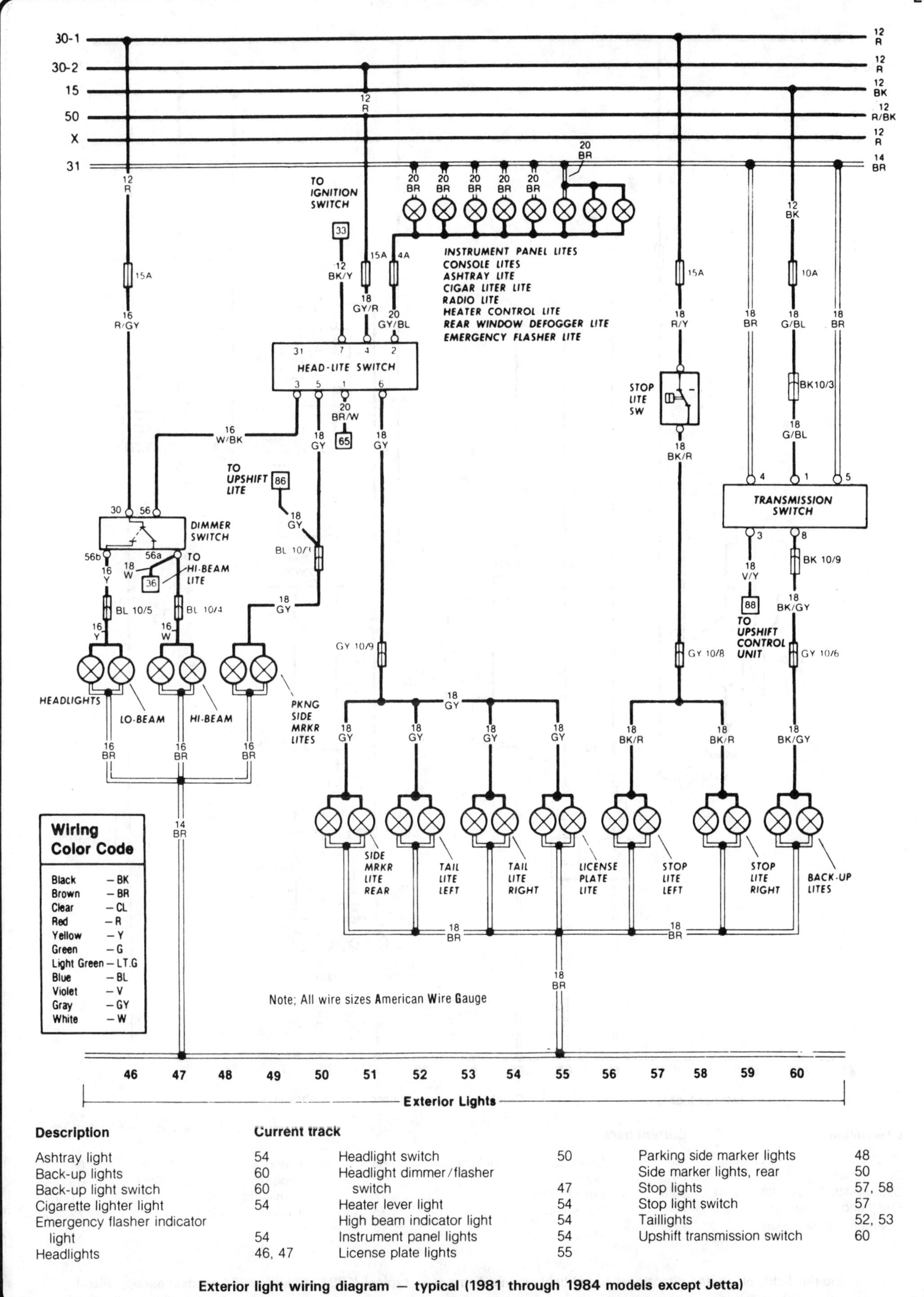

Description	Current track	Description	Current track	Description	Current track
Ashtray light	54	Headlight switch	50	Parking side marker lights	48
Back-up lights	60	Headlight dimmer/flasher switch	47	Side marker lights, rear	50
Back-up light switch	60	Heater lever light	54	Stop lights	57, 58
Cigarette lighter light	54	High beam indicator light	54	Stop light switch	57
Emergency flasher indicator light	54	Instrument panel lights	54	Taillights	52, 53
Headlights	46, 47	License plate lights	55	Upshift transmission switch	60

Exterior light wiring diagram — typical (1981 through 1984 models except Jetta)

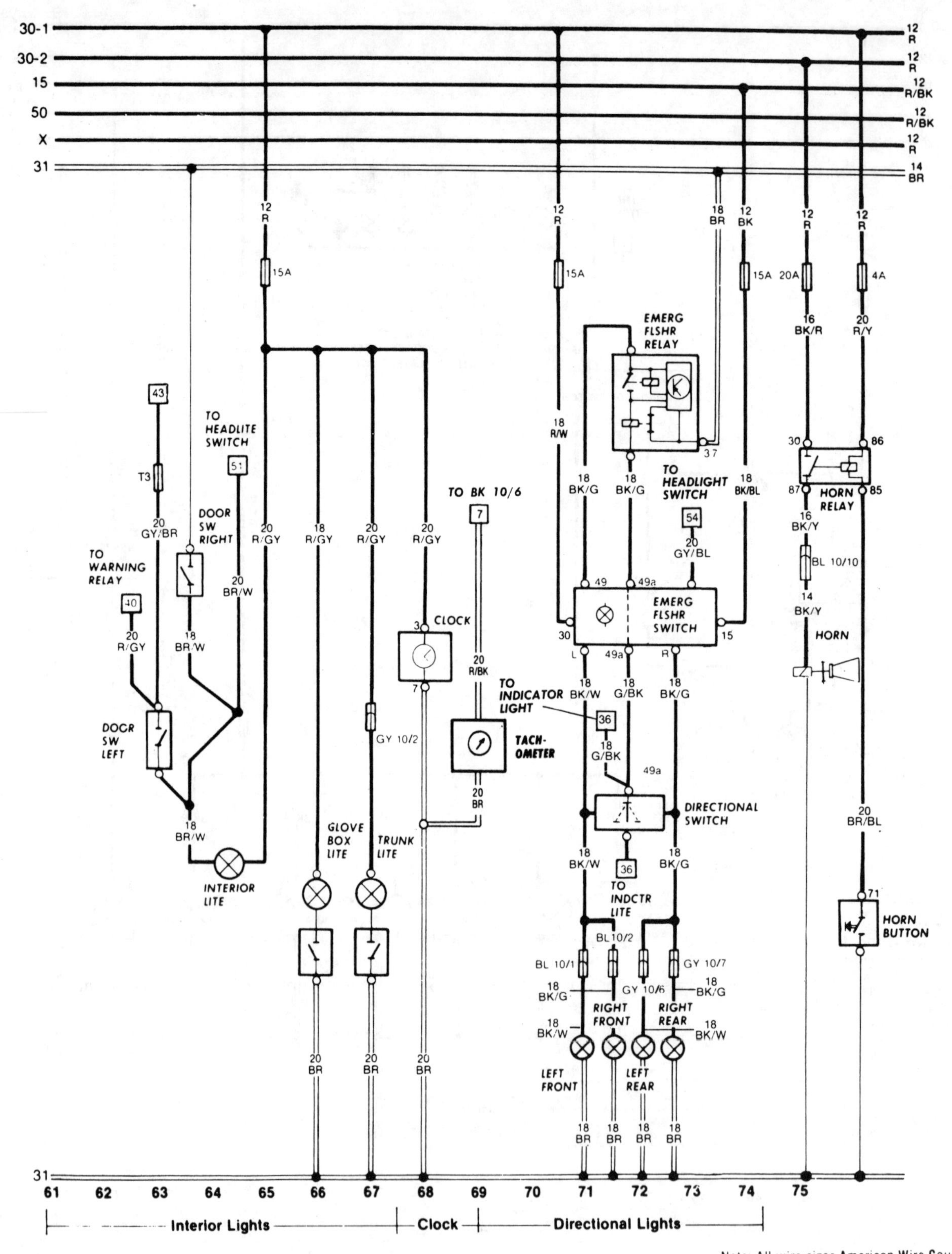

Note: All wire sizes American Wire Gauge

Description	Current track				
Clock	68	Glove compartment light	66	Interior light	64
Door switch, right	64	Horn	75	Luggage compartment light	67
Door switch/buzzer, left	63	Horn button	75	Turn signal lights	71, 73
Emergency flasher relay	72	Horn relay	75	Turn signal switch	72
Emergency flasher switch	72				

Interior light, clock and directional signal light wiring diagram — typical (1981 through 1984 models except Jetta)

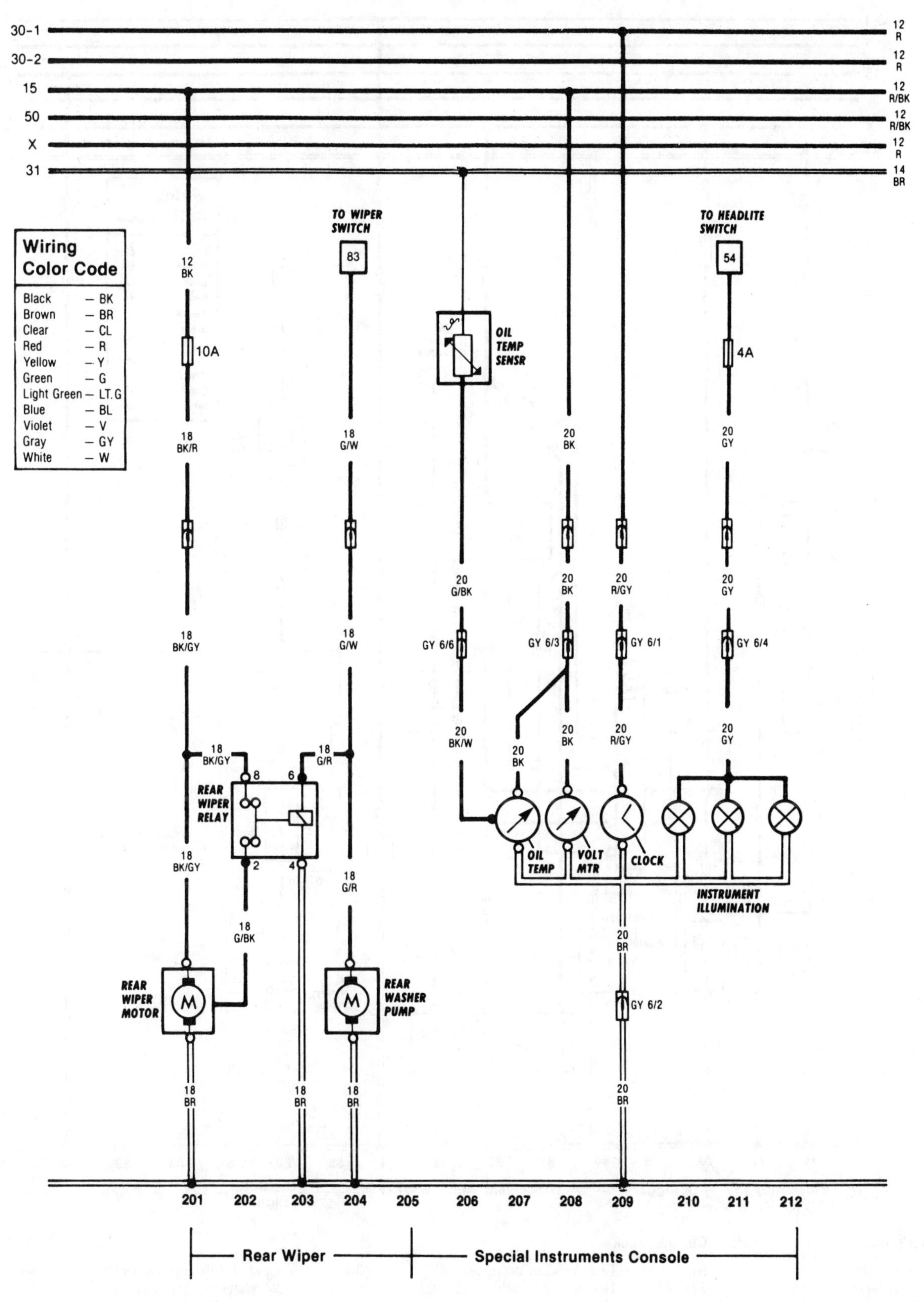

Note: All wire sizes American Wire Gauge

Rear wiper and special instrument console wiring diagram (1981 Rabbit and pick-up models only)

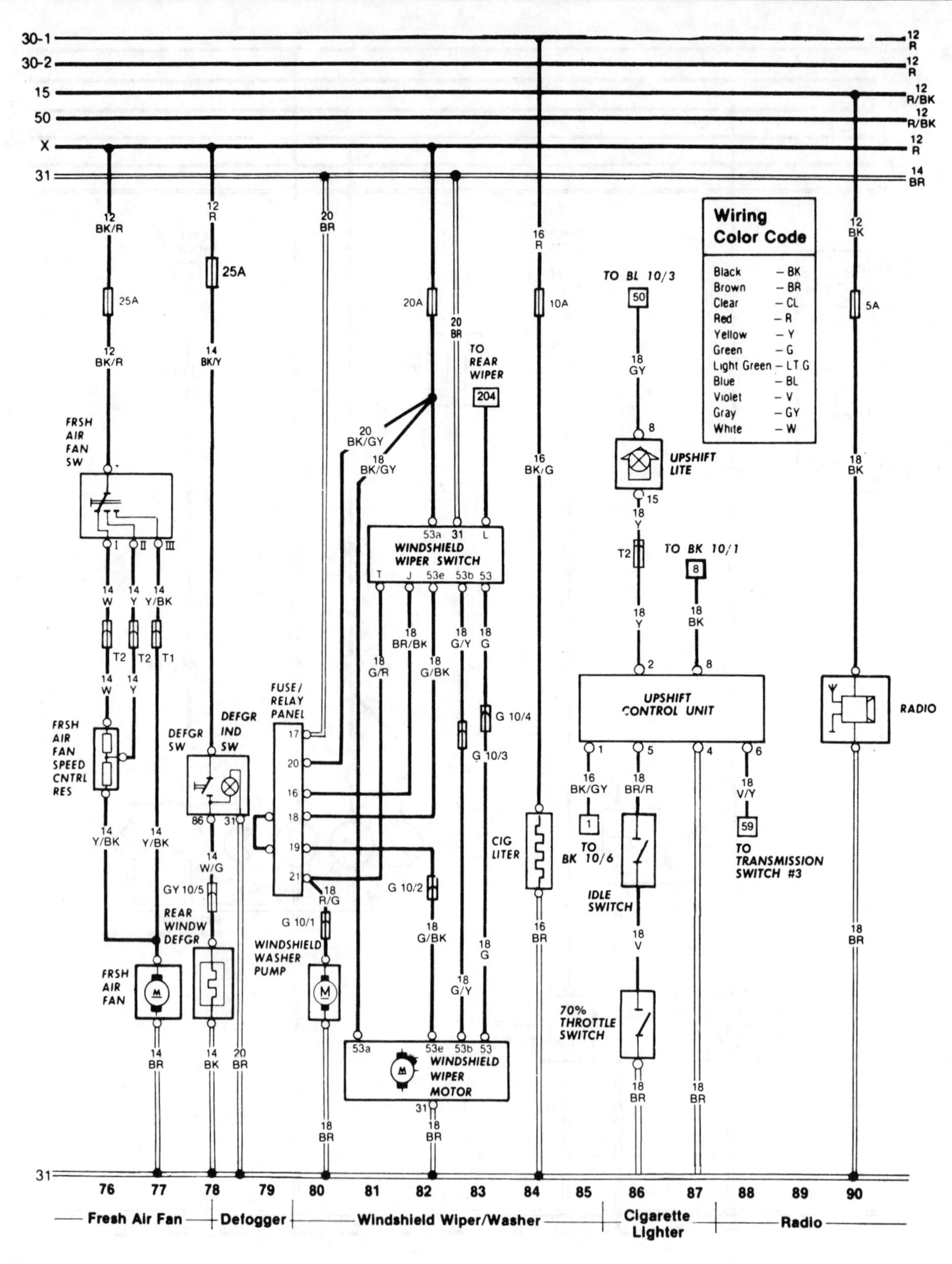

Description	Current track
Cigarette lighter	84
Fresh air fan	77
Fresh air fan speed control resistors	76
Fresh air fan switch	77
Radio	90
Rear window defogger	78
Rear window defogger switch	78
Upshift control unit	87
Upshift idle switch	86
Upshift light	87
Upshift 70% throttle switch	86
Windshield washer pump	80
Windshield wiper motor	82
Windshield wiper switch	82

Fresh air fan, defogger, wiper/washer, cigarette lighter and radio wiring diagram — typical (1981 through 1984 models except Jetta)

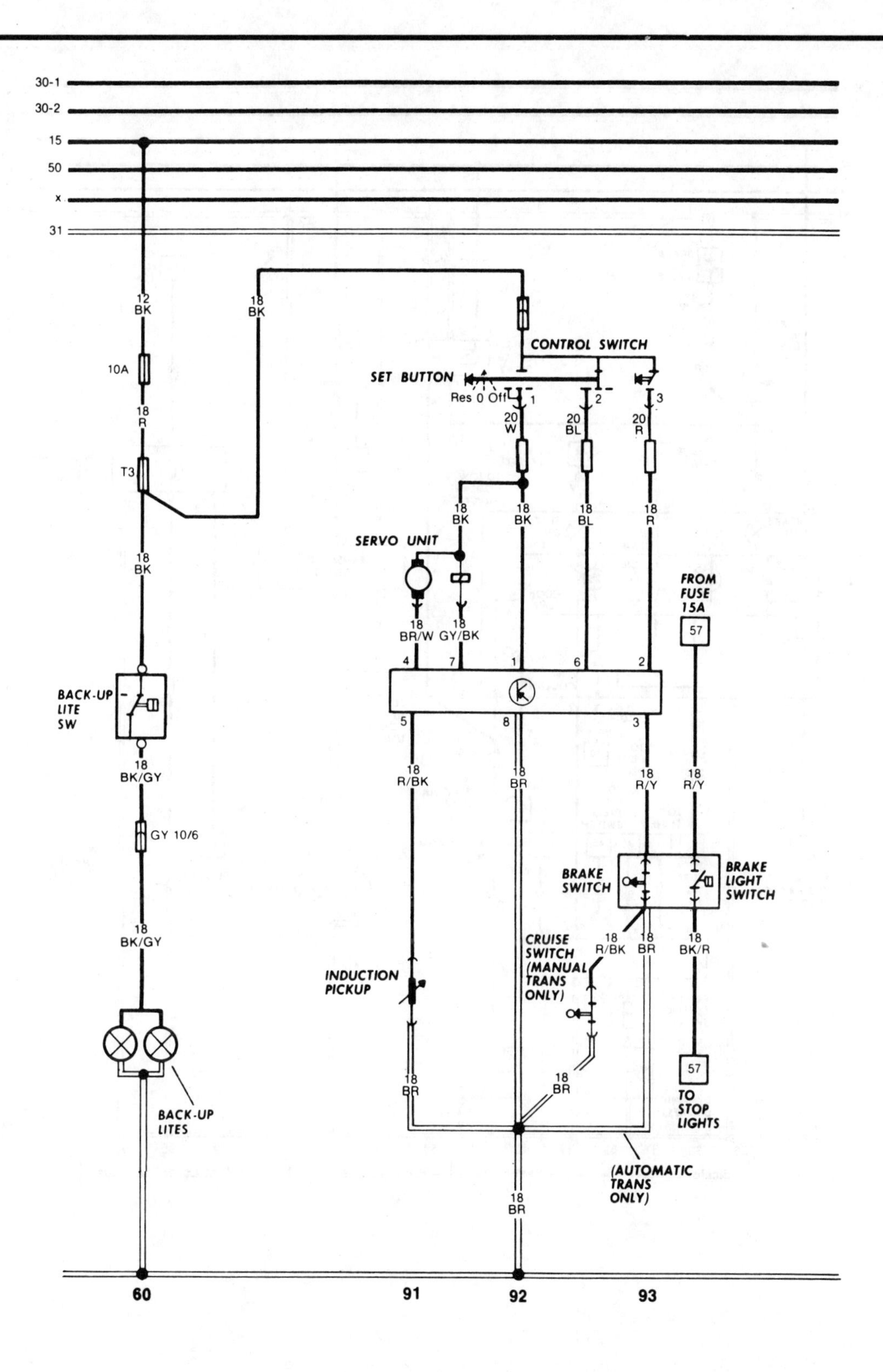

Description	Current track
Back-up lights	60
Back-up light switch	60
Brake light switch	93
Brake switch	93
Control switch	91-93
Cruise control relay	91-93
Cruise control set button	91-93
Cruise control servo	91
Cruise switch (manual transmissions only)	92, 93
Induction pickup	91

Cruise control system wiring diagram (1982 through 1984 Rabbit and pick-up models only)

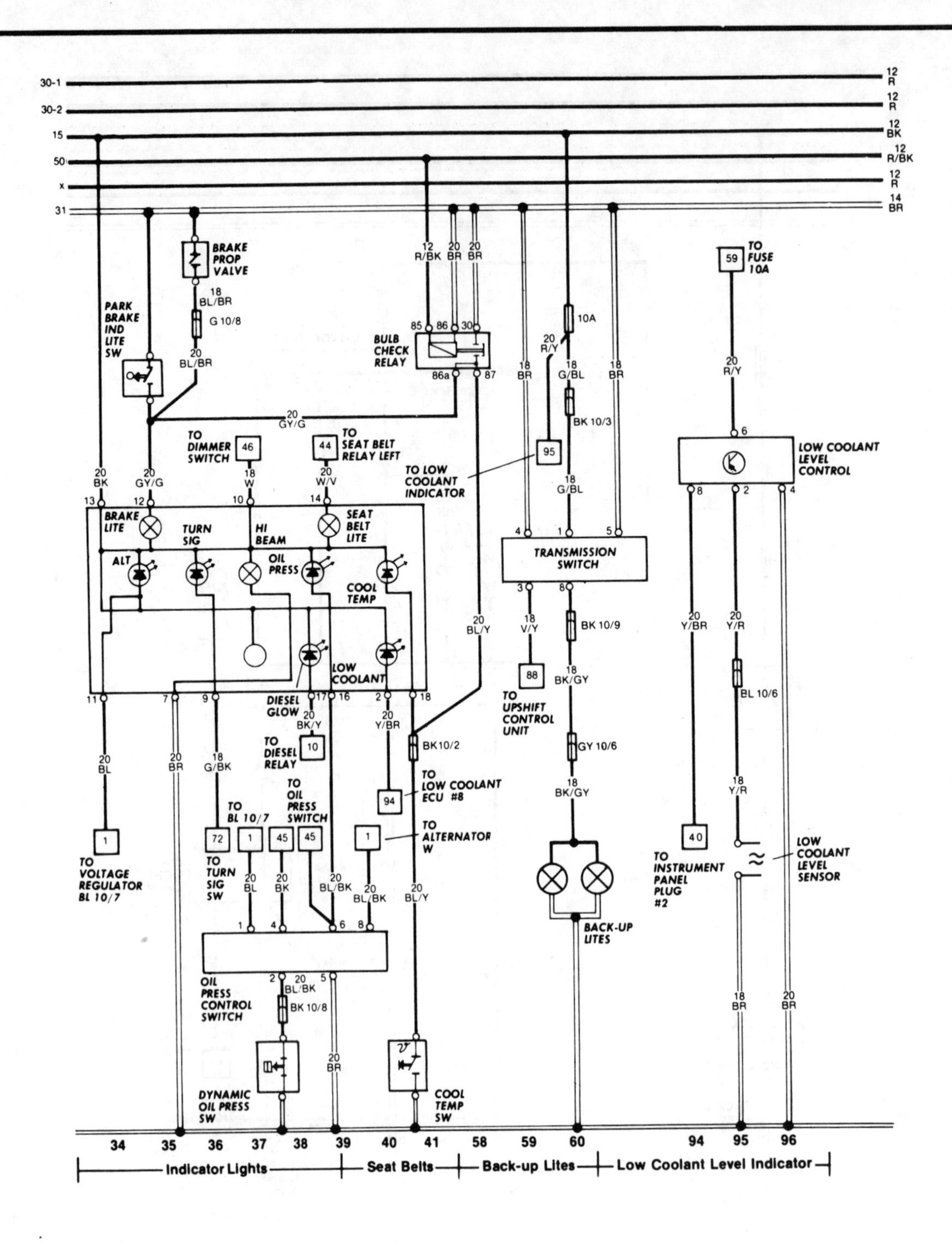

Description	Current track
Alternator warning light	34
Back-up lights	60
Brake proportioning valve	36
Brake warning light	35
Bulb check relay	58
Coolant temperature switch	41
Coolant temperature warning light	41
Glow plug light	38
Oil pressure switch	37, 38
High beam indicator light	37
Low coolant level control relay	94–96
Low coolant level indicator light	40
Low coolant level sensor	95
Oil pressure control switch	36–40
Oil pressure warning light	38
Parking brake indicator light switch	34, 35
Seat belt warning light	39
Turn signal indicator light	36
Transmission switch	58–65

Indicator light, seat belt, back-up light and low coolant level indicator wiring diagram — typical (1981 through 1984 models except Jetta)

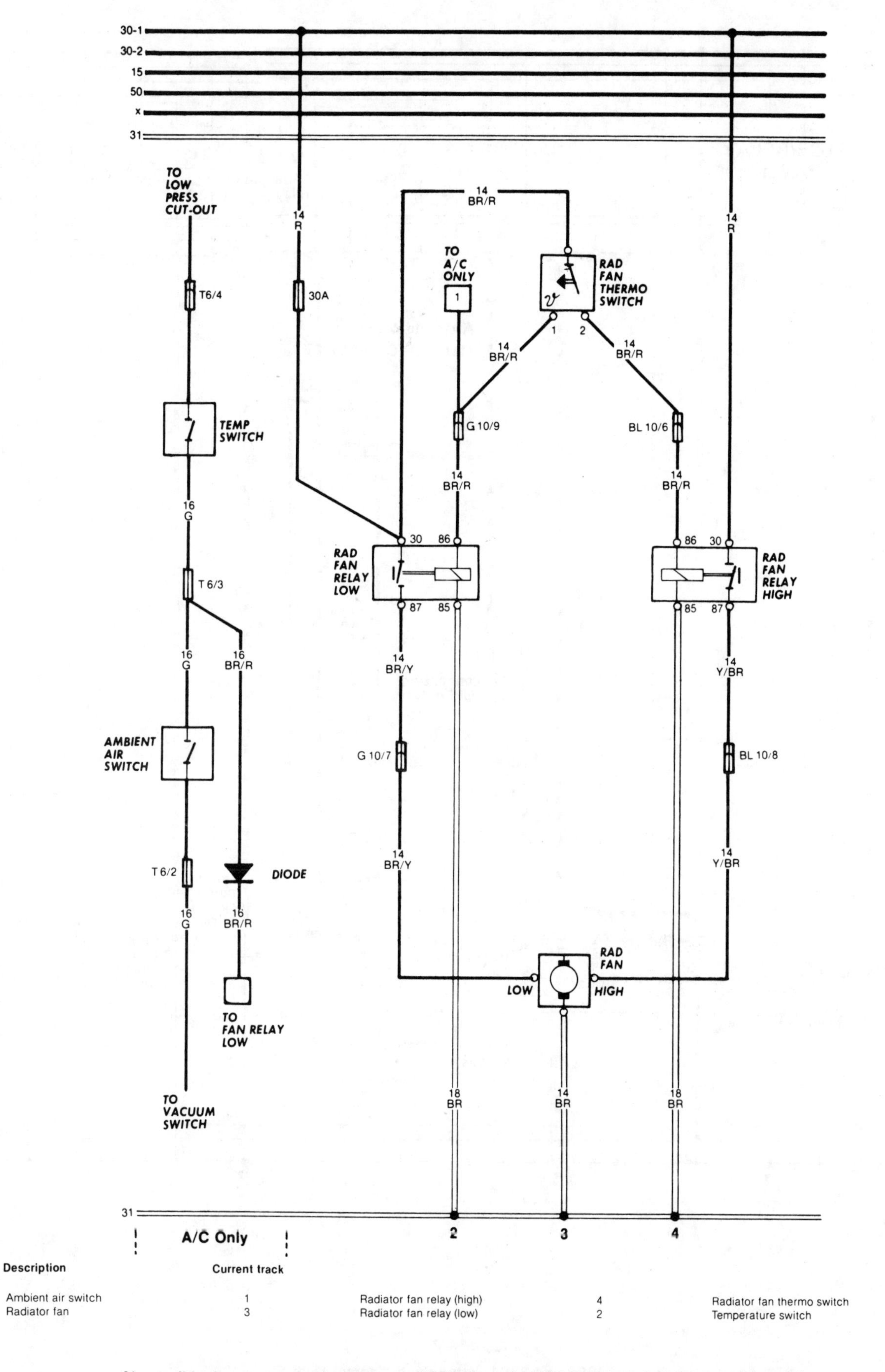

Description	Current track			
Ambient air switch	1	Radiator fan relay (high)	4	Radiator fan thermo switch
Radiator fan	3	Radiator fan relay (low)	2	Temperature switch

Air conditioning system wiring diagram (1982 through 1984 Rabbit and pick-up models only)

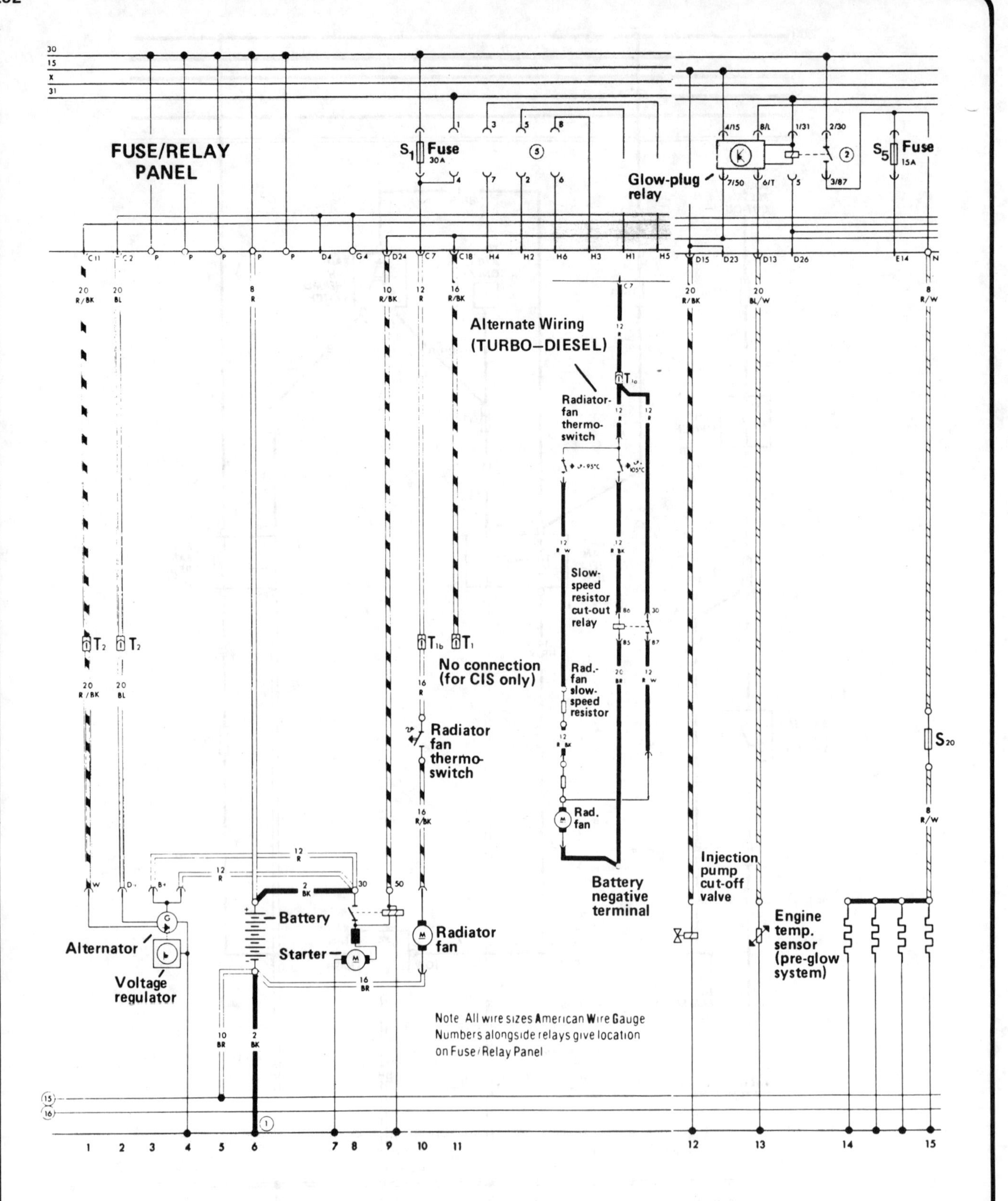

1983 and 1984 Jetta wiring diagram (1 of 9)

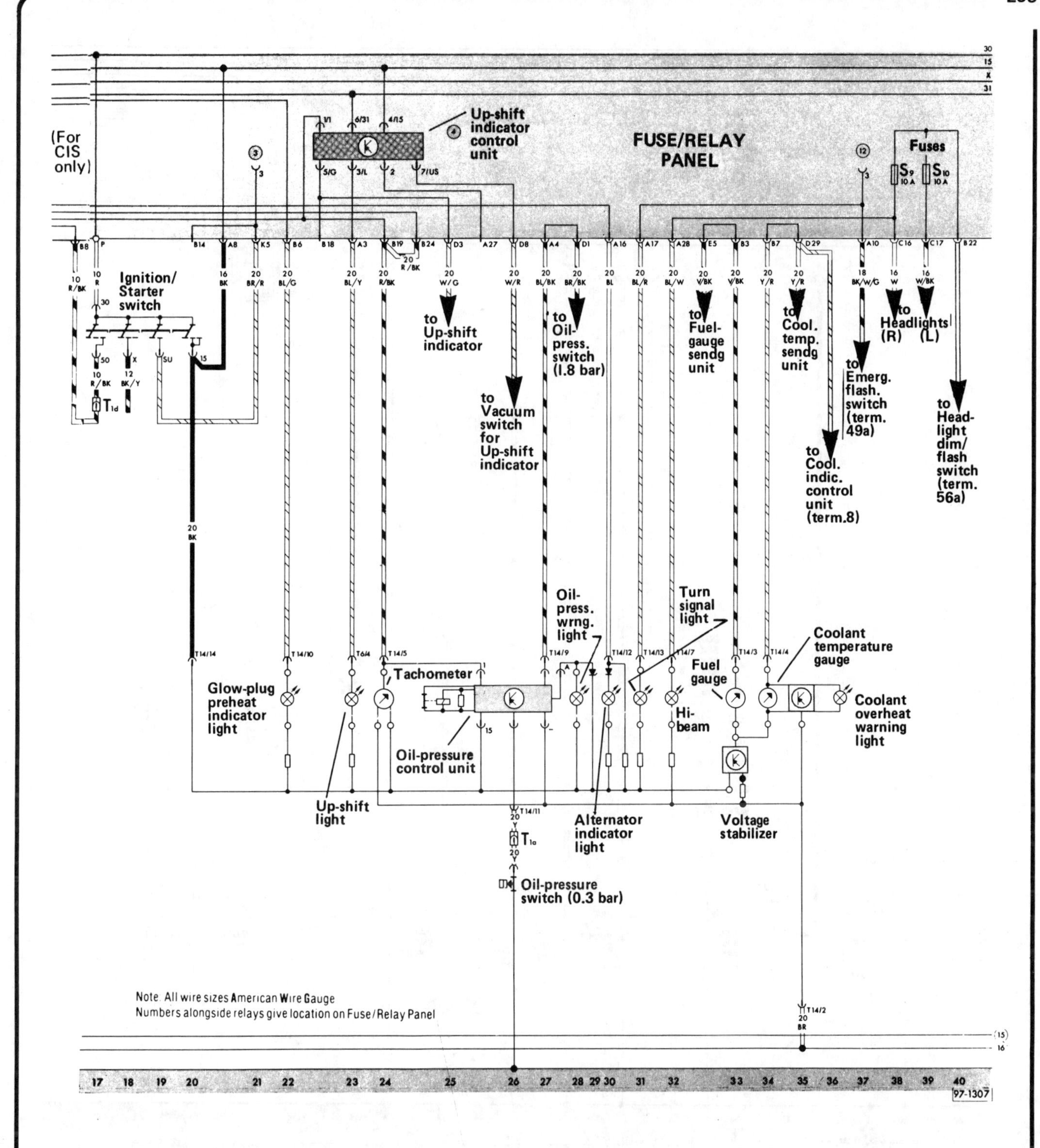

Description	Current Track				
Alternator indicator light	30	High beam	32	Tachometer	24
Coolant overheating warning light	35	Ignition/Starter switch	17–20	Turn signal light	31
Coolant temperature gauge	34	Oil pressure control unit	24 27	Upshift indicator control unit	22–24
Fuel gauge	33	Oil pressure switch (0.3 bar)	26	Upshift light	23
Glow plug preheat indicator light	21	Oil pressure warning light	28	Voltage stabilizer	33

1983 and 1984 Jetta wiring diagram (2 of 9)

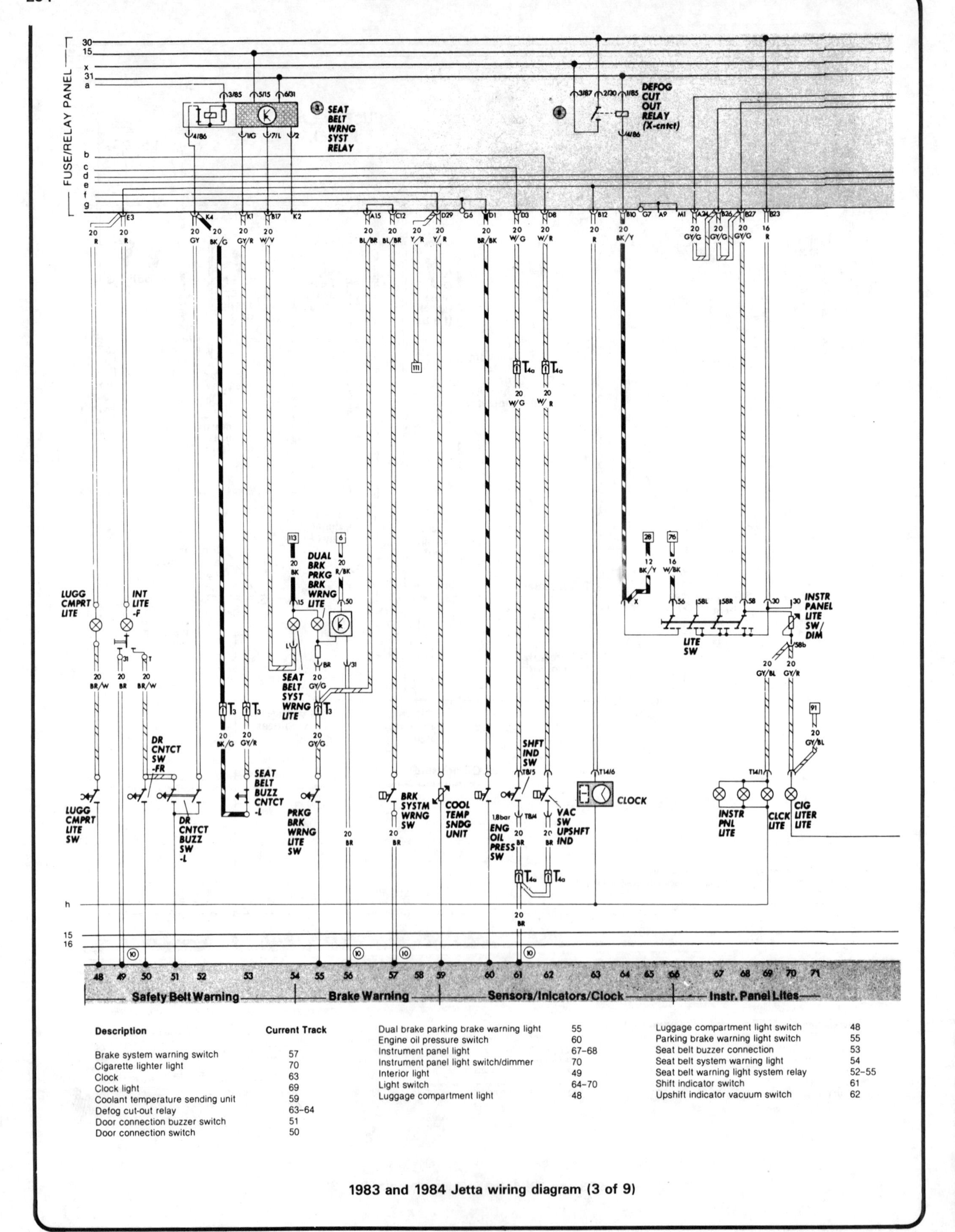

Description	Current Track
Brake system warning switch	57
Cigarette lighter light	70
Clock	63
Clock light	69
Coolant temperature sending unit	59
Defog cut-out relay	63-64
Door connection buzzer switch	51
Door connection switch	50
Dual brake parking brake warning light	55
Engine oil pressure switch	60
Instrument panel light	67-68
Instrument panel light switch/dimmer	70
Interior light	49
Light switch	64-70
Luggage compartment light	48
Luggage compartment light switch	48
Parking brake warning light switch	55
Seat belt buzzer connection	53
Seat belt system warning light	54
Seat belt warning light system relay	52-55
Shift indicator switch	61
Upshift indicator vacuum switch	62

1983 and 1984 Jetta wiring diagram (3 of 9)

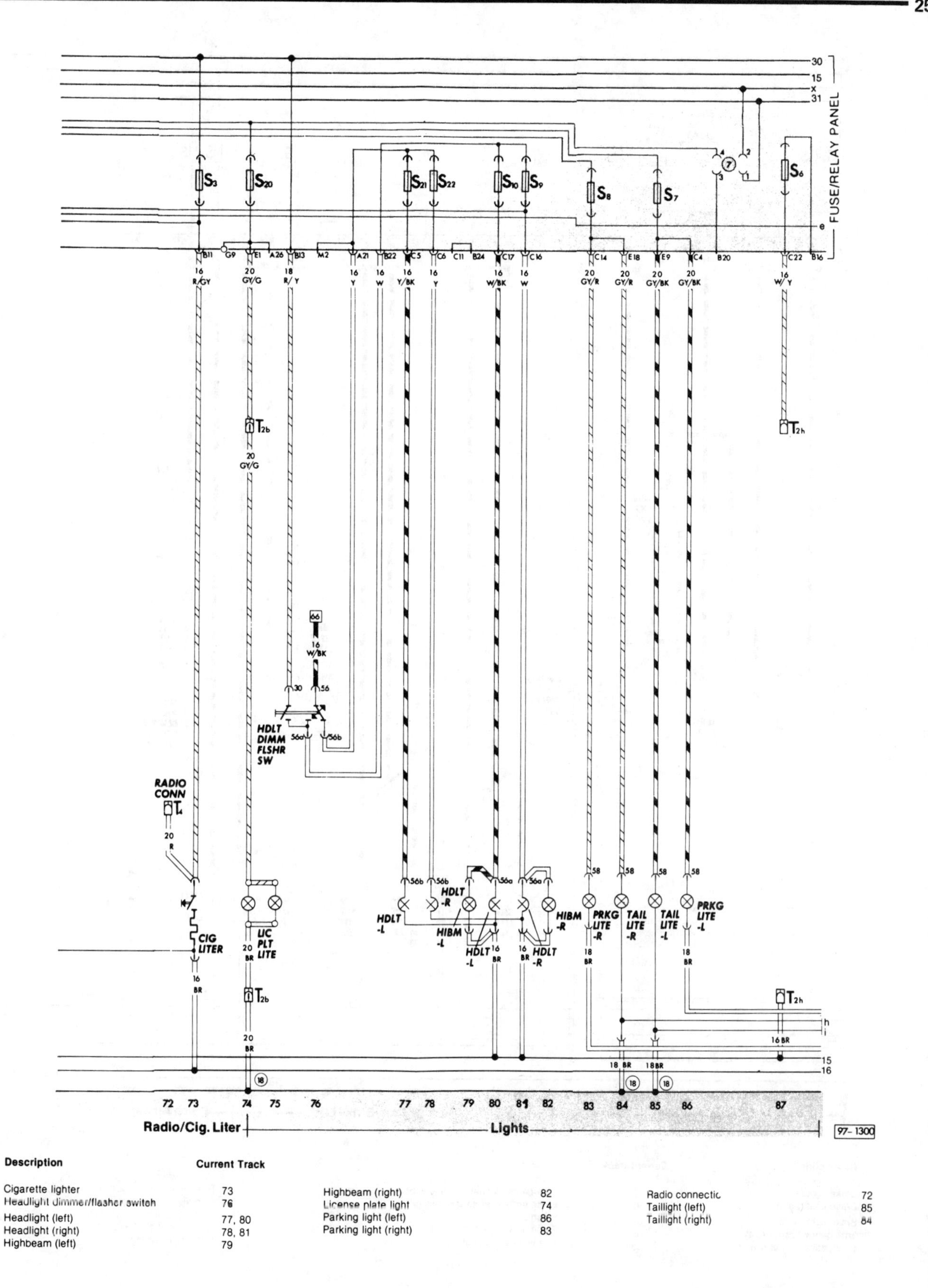

Description	Current Track
Cigarette lighter	73
Headlight dimmer/flasher switch	76
Headlight (left)	77, 80
Headlight (right)	78, 81
Highbeam (left)	79
Highbeam (right)	82
License plate light	74
Parking light (left)	86
Parking light (right)	83
Radio connectic	72
Taillight (left)	85
Taillight (right)	84

1983 and 1984 Jetta wiring diagram (4 of 9)

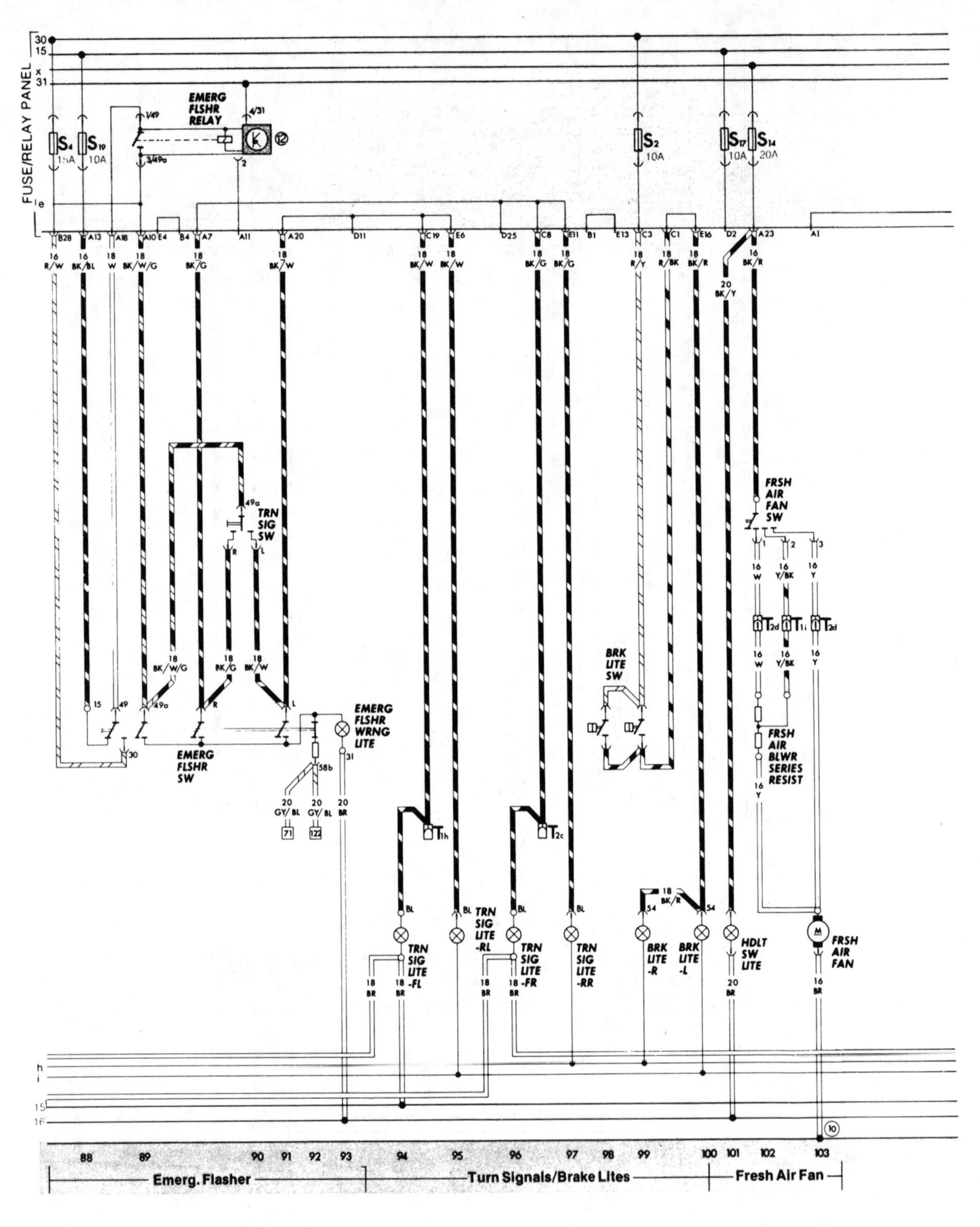

Description	Current track
Brake light (left)	100
Brake light (right)	99
Brake light switch	98–99
Emergency flasher relay	89–91
Emergency flasher switch	88–91
Emergency flasher warning light	93
Fresh air blower series-resistor	102
Fresh air fan	103
Fresh air fan switch	102
Headlight switch light	101
Turn signal light (front/left)	94
Turn signal light (front/right)	96
Turn signal light (rear/left)	95
Turn signal light (rear/right)	97
Turn signal switch	90

1983 and 1984 Jetta wiring diagram (5 of 9)

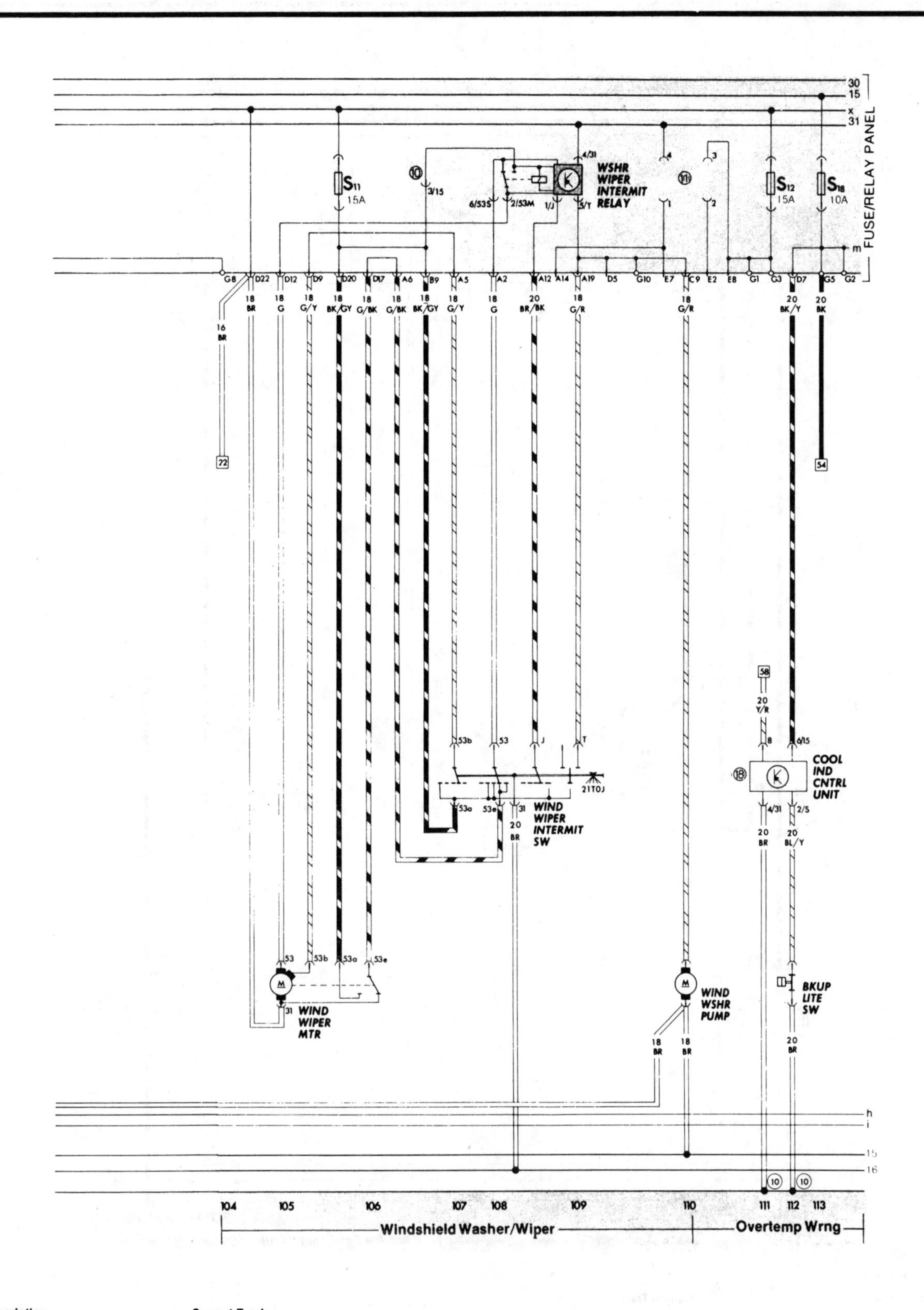

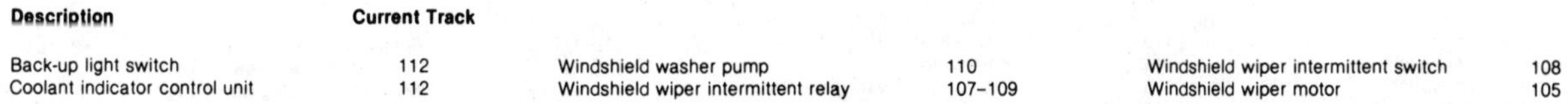

Description	Current Track				
Back-up light switch	112	Windshield washer pump	110	Windshield wiper intermittent switch	108
Coolant indicator control unit	112	Windshield wiper intermittent relay	107–109	Windshield wiper motor	105

1983 and 1984 Jetta wiring diagram (6 of 9)

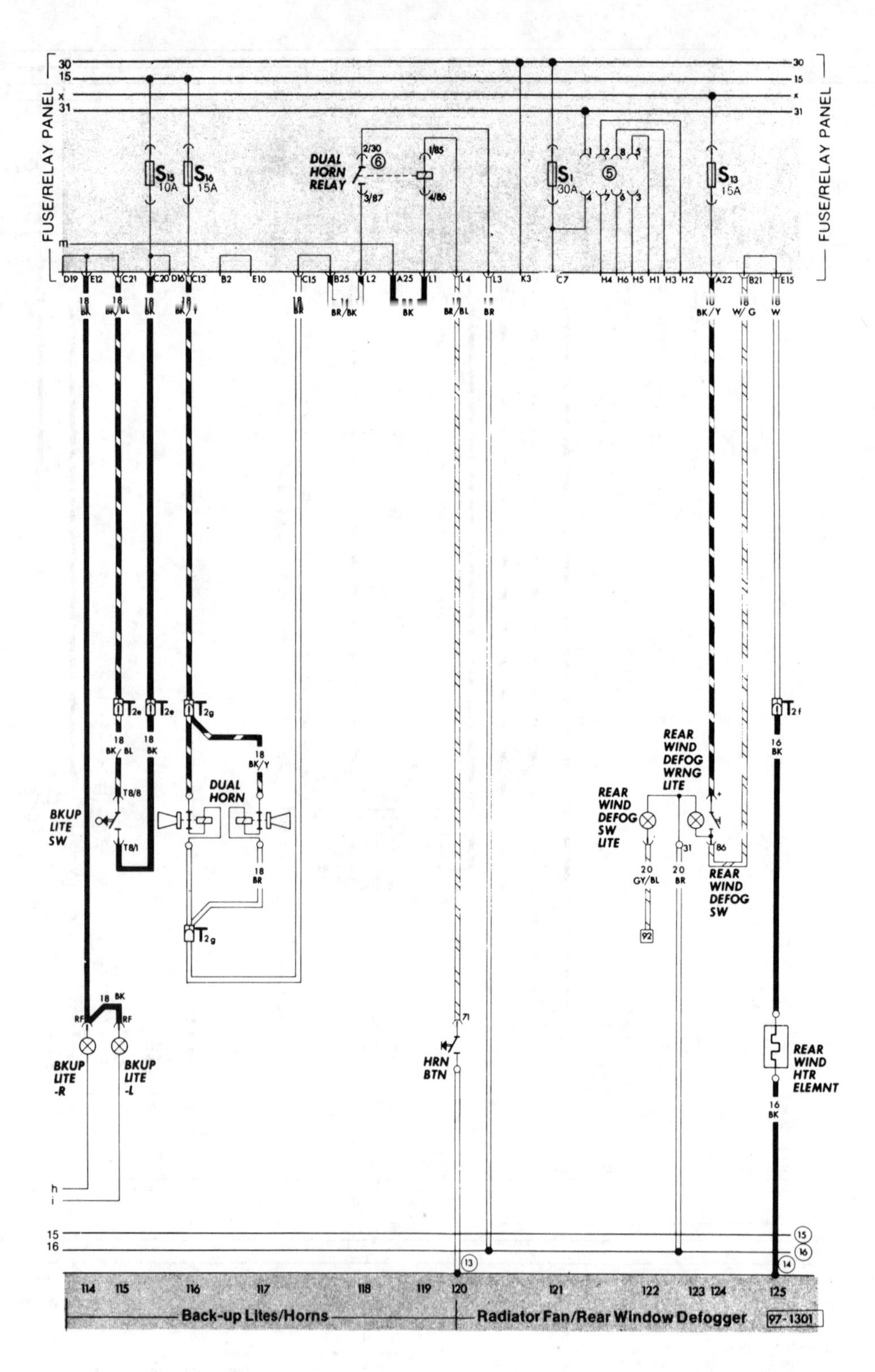

Description	Current Track				
Back up light (left)	115	Dual horns	116–117	Rear window defog switch	124
Back up light (right)	114	Horn button	120	Rear window defog switch light	122
Back up light switch	115	Rear window heating element	125	Rear window defog warning light	123
Dual horn relay	118–119				

1983 and 1984 Jetta wiring diagram (7 of 9)

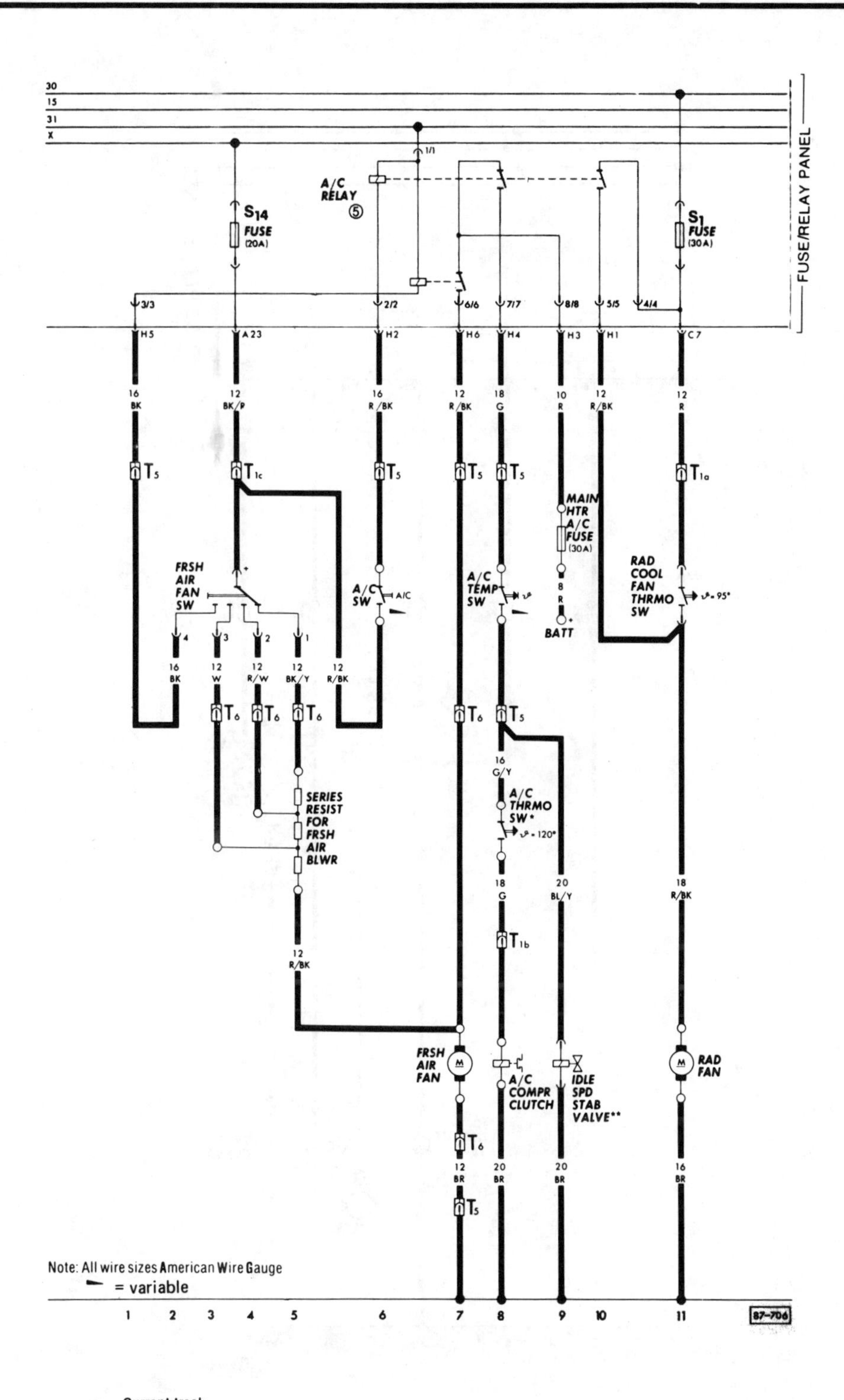

Description	Current track				
Air conditioner compressor clutch	8	Air conditioner thermo switch	8	Radiator cooling fan thermo switch	11
Air conditioner/heater main fuse	9	Battery	9	Radiator fan	11
Air conditioner relay	6–10	Fresh air fan	8	Series resistor for fresh air blower	5
Air conditioner switch	6	Fresh air fan switch	3, 4		
Air conditioner temperature switch	0	Idle speed stabilizer valve	9		

1983 and 1984 Jetta wiring diagram (8 of 9)

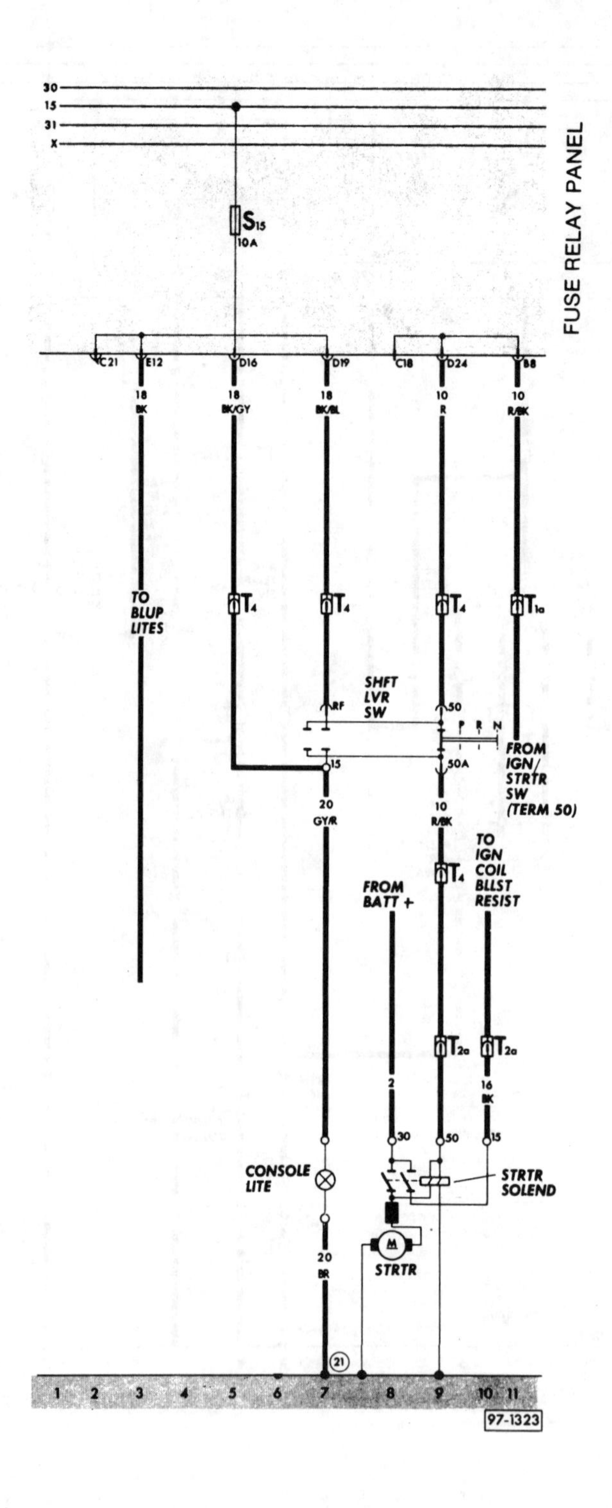

1983 and 1984 Jetta wiring diagram (9 of 9)

Conversion factors

Length (distance)

Unit		Factor		Result		Factor		Result
Inches (in)	X	25.4	=	Millimetres (mm)	X	0.0394	=	Inches (in)
Feet (ft)	X	0.305	=	Metres (m)	X	3.281	=	Feet (ft)
Miles	X	1.609	=	Kilometres (km)	X	0.621	=	Miles

Volume (capacity)

Unit		Factor		Result		Factor		Result
Cubic inches (cu in; in^3)	X	16.387	=	Cubic centimetres (cc; cm^3)	X	0.061	=	Cubic inches (cu in; in^3)
Imperial pints (Imp pt)	X	0.568	=	Litres (l)	X	1.76	=	Imperial pints (Imp pt)
Imperial quarts (Imp qt)	X	1.137	=	Litres (l)	X	0.88	=	Imperial quarts (Imp qt)
Imperial quarts (Imp qt)	X	1.201	=	US quarts (US qt)	X	0.833	=	Imperial quarts (Imp qt)
US quarts (US qt)	X	0.946	=	Litres (l)	X	1.057	=	US quarts (US qt)
Imperial gallons (Imp gal)	X	4.546	=	Litres (l)	X	0.22	=	Imperial gallons (Imp gal)
Imperial gallons (Imp gal)	X	1.201	=	US gallons (US gal)	X	0.833	=	Imperial gallons (Imp gal)
US gallons (US gal)	X	3.785	=	Litres (l)	X	0.264	=	US gallons (US gal)

Mass (weight)

Unit		Factor		Result		Factor		Result
Ounces (oz)	X	28.35	=	Grams (g)	X	0.035	=	Ounces (oz)
Pounds (lb)	X	0.454	=	Kilograms (kg)	X	2.205	=	Pounds (lb)

Force

Unit		Factor		Result		Factor		Result
Ounces-force (ozf; oz)	X	0.278	=	Newtons (N)	X	3.6	=	Ounces-force (ozf; oz)
Pounds-force (lbf; lb)	X	4.448	=	Newtons (N)	X	0.225	=	Pounds-force (lbf; lb)
Newtons (N)	X	0.1	=	Kilograms-force (kgf; kg)	X	9.81	=	Newtons (N)

Pressure

Unit		Factor		Result		Factor		Result
Pounds-force per square inch (psi; lbf/in^2; lb/in^2)	X	0.070	=	Kilograms-force per square centimetre (kgf/cm^2; kg/cm^2)	X	14.223	=	Pounds-force per square inch (psi; lbf/in^2; lb/in^2)
Pounds-force per square inch (psi; lbf/in^2; lb/in^2)	X	0.068	=	Atmospheres (atm)	X	14.696	=	Pounds-force per square inch (psi; lbf/in^2; lb/in^2)
Pounds-force per square inch (psi; lbf/in^2; lb/in^2)	X	0.069	=	Bars	X	14.5	=	Pounds-force per square inch (psi; lbf/in^2; lb/in^2)
Pounds-force per square inch (psi; lbf/in^2; lb/in^2)	X	6.895	=	Kilopascals (kPa)	X	0.145	=	Pounds-force per square inch (psi; lbf/in^2; lb/in^2)
Kilopascals (kPa)	X	0.01	=	Kilograms-force per square centimetre (kgf/cm^2; kg/cm^2)	X	98.1	=	Kilopascals (kPa)
Millibar (mbar)	X	100	=	Pascals (Pa)	X	0.01	=	Millibar (mbar)
Millibar (mbar)	X	0.0145	=	Pounds-force per square inch (psi; lbf/in^2; lb/in^2)	X	68.947	=	Millibar (mbar)
Millibar (mbar)	X	0.75	=	Millimetres of mercury (mmHg)	X	1.333	=	Millibar (mbar)
Millibar (mbar)	X	0.401	=	Inches of water (inH_2O)	X	2.491	=	Millibar (mbar)
Millimetres of mercury (mmHg)	X	0.535	=	Inches of water (inH_2O)	X	1.868	=	Millimetres of mercury (mmHg)
Inches of water (inH_2O)	X	0.036	=	Pounds-force per square inch (psi; lbf/in^2; lb/in^2)	X	27.68	=	Inches of water (inH_2O)

Torque (moment of force)

Unit		Factor		Result		Factor		Result
Pounds-force inches (lbf in; lb in)	X	1.152	=	Kilograms-force centimetre (kgf cm; kg cm)	X	0.868	=	Pounds-force inches (lbf in; lb in)
Pounds-force inches (lbf in; lb in)	X	0.113	=	Newton metres (Nm)	X	8.85	=	Pounds-force inches (lbf in; lb in)
Pounds-force inches (lbf in; lb in)	X	0.083	=	Pounds-force feet (lbf ft; lb ft)	X	12	=	Pounds-force inches (lbf in; lb in)
Pounds-force feet (lbf ft; lb ft)	X	0.138	=	Kilograms-force metres (kgf m; kg m)	X	7.233	=	Pounds-force feet (lbf ft; lb ft)
Pounds-force feet (lbf ft; lb ft)	X	1.356	=	Newton metres (Nm)	X	0.738	=	Pounds-force feet (lbf ft; lb ft)
Newton metres (Nm)	X	0.102	=	Kilograms-force metres (kgf m; kg m)	X	9.804	=	Newton metres (Nm)

Power

Unit		Factor		Result		Factor		Result
Horsepower (hp)	X	745.7	=	Watts (W)	X	0.0013	=	Horsepower (hp)

Velocity (speed)

Unit		Factor		Result		Factor		Result
Miles per hour (miles/hr; mph)	X	1.609	=	Kilometres per hour (km/hr; kph)	X	0.621	=	Miles per hour (miles/hr; mph)

*Fuel consumption**

Unit		Factor		Result		Factor		Result
Miles per gallon, Imperial (mpg)	X	0.354	=	Kilometres per litre (km/l)	X	2.825	=	Miles per gallon, Imperial (mpg)
Miles per gallon, US (mpg)	X	0.425	=	Kilometres per litre (km/l)	X	2.352	=	Miles per gallon, US (mpg)

Temperature

Degrees Fahrenheit = (°C x 1.8) + 32

Degrees Celsius (Degrees Centigrade; °C) = (°F - 32) x 0.56

**It is common practice to convert from miles per gallon (mpg) to litres/100 kilometres (l/100km), where mpg (Imperial) x l/100 km = 282 and mpg (US) x l/100 km = 235*

Safety first!

Regardless of how enthusiastic you may be about getting on with the job at hand, take the time to ensure that your safety is not jeopardized. A moment's lack of attention can result in an accident, as can failure to observe certain simple safety precautions. The possibility of an accident will always exist, and the following points should not be considered a comprehensive list of all dangers. Rather, they are intended to make you aware of the risks and to encourage a safety conscious approach to all work you carry out on your vehicle.

Essential DOs and DON'Ts

DON'T rely on a jack when working under the vehicle. Always use approved jackstands to support the weight of the vehicle and place them under the recommended lift or support points.
DON'T attempt to loosen extremely tight fasteners (i.e. wheel lug nuts) while the vehicle is on a jack — it may fall.
DON'T start the engine without first making sure that the transmission is in Neutral (or Park where applicable) and the parking brake is set.
DON'T remove the radiator cap from a hot cooling system — let it cool or cover it with a cloth and release the pressure gradually.
DON'T attempt to drain the engine oil until you are sure it has cooled to the point that it will not burn you.
DON'T touch any part of the engine or exhaust system until it has cooled sufficiently to avoid burns.
DON'T siphon toxic liquids such as fuel, antifreeze and brake fluid by mouth, or allow them to remain on your skin.
DON'T inhale brake lining dust — it is potentially hazardous (see *Asbestos* below)
DON'T allow spilled oil or grease to remain on the floor — wipe it up before someone slips on it.
DON'T use loose fitting wrenches or other tools which may slip and cause injury.
DON'T push on wrenches when loosening or tightening nuts or bolts. Always try to pull the wrench toward you. If the situation calls for pushing the wrench away, push with an open hand to avoid scraped knuckles if the wrench should slip.
DON'T attempt to lift a heavy component alone — get someone to help you.
DON'T rush or take unsafe shortcuts to finish a job.
DON'T allow children or animals in or around the vehicle while you are working on it.
DO wear eye protection when using power tools such as a drill, sander, bench grinder, etc. and when working under a vehicle.
DO keep loose clothing and long hair well out of the way of moving parts.
DO make sure that any hoist used has a safe working load rating adequate for the job.
DO get someone to check on you periodically when working alone on a vehicle.
DO carry out work in a logical sequence and make sure that everything is correctly assembled and tightened.
DO keep chemicals and fluids tightly capped and out of the reach of children and pets.
DO remember that your vehicle's safety affects that of yourself and others. If in doubt on any point, get professional advice.

Asbestos

Certain friction, insulating, sealing, and other products — such as brake linings, brake bands, clutch linings, torque converters, gaskets, etc. — contain asbestos. *Extreme care must be taken to avoid inhalation of dust from such products since it is hazardous to health.* If in doubt, assume that they *do* contain asbestos.

Fire

Remember at all times that fuel is highly flammable. Never smoke or have any kind of open flame around when working on a vehicle. But the risk does not end there. A spark caused by an electrical short circuit, by two metal surfaces contacting each other, or even by static electricity built up in your body under certain conditions, can ignite fuel vapors, which in a confined space are highly explosive. Do not, under any circumstances, use gasoline for cleaning parts. Use an approved safety solvent.

Always disconnect the battery ground (–) cable *at the battery* before working on any part of the fuel system or electrical system. Never risk spilling fuel on a hot engine or exhaust component.

It is strongly recommended that a fire extinguisher suitable for use on fuel and electrical fires be kept handy in the garage or workshop at all times. Never try to extinguish a fuel or electrical fire with water.

Torch (flashlight in the US)

Any reference to a "torch" appearing in this manual should always be taken to mean a hand-held, battery-operated electric light or flashlight. It DOES NOT mean a welding or propane torch or blowtorch.

Fumes

Certain fumes are highly toxic and can quickly cause unconsciousness and even death if inhaled to any extent. Fuel vapor falls into this category, as do the vapors from some cleaning solvents. Any draining or pouring of such volatile fluids should be done in a well ventilated area.

When using cleaning fluids and solvents, read the instructions on the container carefully. Never use materials from unmarked containers.

Never run the engine in an enclosed space, such as a garage. Exhaust fumes contain carbon monoxide, which is extremely poisonous. If you need to run the engine, always do so in the open air, or at least have the rear of the vehicle outside the work area.

If you are fortunate enough to have the use of an inspection pit, never drain or pour fuel and never run the engine while the vehicle is over the pit. The fumes, being heavier than air, will concentrate in the pit with possibly lethal results.

The battery

Never create a spark or allow a bare light bulb near a battery. They normally give off a certain amount of hydrogen gas, which is highly explosive.

Always disconnect the battery ground (–) cable *at the battery* before working on the fuel or electrical systems.

If possible, loosen the filler caps or cover when charging the battery from an external source (this does not apply to sealed or maintenance-free batteries). Do not charge at an excessive rate or the battery may burst.

Take care when adding water to a non maintenance-free battery and when carrying a battery. The electrolyte, even when diluted, is very corrosive and should not be allowed to contact clothing or skin.

Always wear eye protection when cleaning the battery to prevent the caustic deposits from entering your eyes.

Mains electricity (household current in the US)

When using an electric power tool, inspection light, etc., which operates on household current, always make sure that the tool is correctly connected to its plug and that, where necessary, it is properly grounded. Do not use such items in damp conditions and, again, do not create a spark or apply excessive heat in the vicinity of fuel or fuel vapor.

Secondary ignition system voltage

A severe electric shock can result from touching certain parts of the ignition system (such as the spark plug wires) when the engine is running or being cranked, particularly if components are damp or the insulation is defective. In the case of an electronic ignition system, the secondary system voltage is much higher and could prove fatal.

Index

HAYNES AUTOMOTIVE MANUALS

NOTE: New manuals are added to this list on a periodic basis. If you do not see a listing for your vehicle, consult your local Haynes dealer for the latest product information.

ALFA-ROMEO
531 Alfa Romeo Sedan & Coupe '73 thru '80

AMC
Jeep CJ – *see JEEP (412)*
694 Mid-size models, Concord, Hornet, Gremlin & Spirit '70 thru '83
934 (Renault) Alliance & Encore '83 thru '87

AUDI
162 100 '69 thru '77
615 4000 '80 thru '87
428 5000 '77 thru '83
1117 5000 '84 thru '88
207 Fox '73 thru '79

AUSTIN
049 Healey 100/6 & 3000 Roadster '56 thru '68
Healey Sprite – *see MG Midget Roadster (265)*

BLMC
260 1100, 1300 & Austin America '62 thru '74
527 Mini '59 thru '69
*646 Mini '69 thru '88

BMW
276 320i all 4 cyl models '75 thru '83
632 528i & 530i '75 thru '80
240 1500 thru 2002 exceptTurbo '59 thru '77
348 2500, 2800, 3.0 & Bavaria '69 thru '76

BUICK
Century (front wheel drive) – *see GENERAL MOTORS A-Cars (829)*
*1627 Buick, Oldsmobile & Pontiac Full-size (Front wheel drive) '85 thru '90 Buick Electra, LeSabre and Park Avenue; Oldsmobile Delta 88 Royale, Ninety Eight and Regency; Pontiac Bonneville
*1551 Buick Oldsmobile & Pontiac Full-size (Rear wheel drive) Buick Electra '70 thru '84, Estate '70 thru '90, LeSabre '70 thru '79 Oldsmobile Custom Cruiser '70 thru '90, Delta 88 '70 thru '85, Ninety-eight '70 thru '84 Pontiac Bonneville '70 thru '86, Catalina '70 thru '81, Grandville '70 thru '75, Parisienne '84 thu '86
627 Mid-size all rear-drive Regal & Century models with V6, V8 and Turbo '74 thru '87
Skyhawk – *see GENERAL MOTORS J-Cars (766)*
552 Skylark all X-car models '80 thru '85

CADILLAC
Cimarron – *see GENERAL MOTORS J-Cars (766)*

CAPRI
296 2000 MK I Coupe '71 thru '75
283 2300 MK II Coupe '74 thru '78
205 2600 & 2800 V6 Coupe '71 thru '75
375 2800 Mk II V6 Coupe '75 thru '78
Mercury in-line engines – *see FORD Mustang (654)*
Mercury V6 & V8 engines – *see FORD Mustang (558)*

CHEVROLET
*1477 Astro & GMC Safari Mini-vans '85 thru '90
554 Camaro V8 '70 thru '81
*866 Camaro '82 thru '89
Cavalier – *see GENERAL MOTORS J-Cars (766)*
Celebrity – *see GENERAL MOTORS A-Cars (829)*
625 Chevelle, Malibu & El Camino all V6 & V8 models '69 thru '87
449 Chevette & Pontiac T1000 '76 thru '87
550 Citation '80 thru '85
*1628 Corsica/Beretta '87 thru '90
274 Corvette all V8 models '68 thru '82
*1336 Corvette '84 thru '89
704 Full-size Sedans Caprice, Impala, Biscayne, Bel Air & Wagons, all V6 & V8 models '69 thru '90
319 Luv Pick-up all 2WD & 4WD '72 thru '82
626 Monte Carlo all V6, V8 & Turbo '70 thru '88
241 Nova all V8 models '69 thru '79
*1642 Nova & Geo Prizm front wheel drive '85 thru '90
*420 Pick-ups '67 thru '87 – Chevrolet & GMC, all V8 & in-line 6 cyl 2WD & 4WD '67 thru '87
*1664 Pick-ups '88 thru '90 – Chevlolet & GMC all full-size (C and K) models, '88 thru '90
*831 S-10 & GMC S-15 Pick-ups '82 thru '90
*345 Vans – Chevrolet & GMC, V8 & in-line 6 cyl models '68 thru '89
208 Vega except Cosworth '70 thru '77

CHRYSLER
*1337 Chrysler & Plymouth Mid-size front wheel drive '82 thru '88
K-Cars – *see DODGE Aries (723)*
Laser – *see DODGE Daytona (1140)*

DATSUN
402 200SX '77 thru '79
647 200SX '80 thru '83
228 B-210 '73 thru '78
525 210 '78 thru '82
206 240Z, 260Z & 280Z Coupe & 2+2 '70 thru '78
563 280ZX Coupe & 2+2 '79 thru '83
300ZX – *see NISSAN (1137)*
679 310 '78 thru '82
123 510 & PL521 Pick-up '68 thru '73
430 510 '78 thru '81
372 610 '72 thru '76
277 620 Series Pick-up '73 thru '79
235 710 '73 thru '77
720 Series Pick-up – *see NISSAN Pick-ups (771)*
376 810/Maxima all gasoline models '77 thru '84
124 1200 '70 thru '73
368 F10 '76 thru '79
Pulsar – *see NISSAN (876)*
Sentra – *see NISSAN (982)*
Stanza – *see NISSAN (981)*

DODGE
*723 Aries & Plymouth Reliant '81 thru '88
*1231 Caravan & Plymouth Voyager Mini-Vans '84 thru '89
699 Challenger & Plymouth Saporro '78 thru '83
236 Colt '71 thru '77
419 Colt (rear wheel drive) '77 thru '80
610 Colt & Plymouth Champ (front wheel drive) '78 thru '87
*556 D50 & Plymouth Arrow Pick-ups '79 thru '88
234 Dart & Plymouth Valiant all 6 cyl models '67 thru '76
*1140 Daytona & Chrysler Laser '84 thru '88
*545 Omni & Plymouth Horizon '78 thru '89
*912 Pick-ups all full-size models '74 thru '90
*349 Vans – Dodge & Plymouth V8 & 6 cyl models '71 thru '89

FIAT
080 124 Sedan & Wagon all ohv & dohc models '66 thru '75
094 124 Sport Coupe & Spider '68 thru '78
087 128 '72 thru '79
310 131 & Brava '75 thru '81
038 850 Sedan, Coupe & Spider '64 thru '74
479 Strada '79 thru '82
273 X1/9 '74 thru '80

FORD
*1476 Aerostar Mini-vans '86 thru '88
788 Bronco and Pick-ups '73 thru '79
*880 Bronco and Pick-ups '80 thru '90
014 Cortina MK II except Lotus '66 thru '70
295 Cortina MK III 1600 & 2000 ohc '70 thru '76
268 Courier Pick-up '72 thru '82
789 Escort & Mercury Lynx all models '81 thru '90
560 Fairmont & Mercury Zephyr all in-line & V8 models '78 thru '83
334 Fiesta '77 thru '80
754 Ford & Mercury Full-size, Ford LTD & Mercury Marquis ('75 thru '82); Ford Custom 500, Country Squire, Crown Victoria & Mercury Colony Park ('75 thru '87); Ford LTD Crown Victoria & Mercury Gran Marquis ('83 thru '87)
359 Granada & Mercury Monarch all in-line, 6 cyl & V8 models '75 thru '80
773 Ford & Mercury Mid-size, Ford Thunderbird & Mercury Cougar ('75 thru '82); Ford LTD & Mercury Marquis ('83 thru '86); Ford Torino, Gran Torino, Elite, Ranchero pick-up, LTD II, Mercury Montego, Comet, XR-7 & Lincoln Versailles ('75 thru '86)
*654 Mustang & Mercury Capri all in-line models & Turbo '79 thru '90
*558 Mustang & Mercury Capri all V6 & V8 models '79 thru '89
357 Mustang V8 '64-1/2 thru '73
231 Mustang II all 4 cyl, V6 & V8 models '74 thru '78
204 Pinto '70 thru '74
649 Pinto & Mercury Bobcat '75 thru '80
*1026 Ranger & Bronco II gasoline models '83 thru '89
*1421 Taurus & Mercury Sable '86 thru '90
*1418 Tempo & Mercury Topaz all gasoline models '84 thru '89
1338 Thunderbird & Mercury Cougar/XR7 '83 thru '88
*344 Vans all V8 Econoline models '69 thru '90

GENERAL MOTORS
*829 A-Cars – Chevrolet Celebrity, Buick Century, Pontiac 6000 & Oldsmobile Cutlass Ciera '82 thru '89
*766 J-Cars – Chevrolet Cavalier, Pontiac J-2000, Oldsmobile Firenza, Buick Skyhawk & Cadillac Cimarron '82 thru '89
*1420 N-Cars – Pontiac Grand Am, Buick Somerset and Oldsmobile Calais '85 thru '87; Buick Skylark '86 thru '87

GEO
Tracker – *see SUZUKI Samurai (1626)*
Prizm – *see CHEVROLET Nova (1642)*

GMC
Safari – *see CHEVROLET ASTRO (1477)*
Vans & Pick-ups – *see CHEVROLET (420, 831, 345, 1664)*

HONDA
138 360, 600 & Z Coupe '67 thru '75
351 Accord CVCC '76 thru '83
*1221 Accord '84 thru '89
160 Civic 1200 '73 thru '79
633 Civic 1300 & 1500 CVCC '80 thru '83
297 Civic 1500 CVCC '75 thru '79
*1227 Civic except 16-valve CRX & 4 WD Wagon '84 thru '86
*601 Prelude CVCC '79 thru '89

HYUNDAI
*1552 Excel '86 thru '89

ISUZU
*1641 Trooper & Pick-up all gasolime models '81 thru '89

JAGUAR
098 MK I & II, 240 & 340 Sedans '55 thru '69
*242 XJ6 all 6 cyl models '68 thru '86
*478 XJ12 & XJS all 12 cyl models '72 thru '85
140 XK-E 3.8 & 4.2 all 6 cyl models '61 thru '72

JEEP
*1553 Cherokee, Comanche & Wagoneer Limited '84 thru '89
412 CJ '49 thru '86

LADA
*413 1200, 1300. 1500 & 1600 all models including Riva '74 thru '86

LANCIA
533 Lancia Beta Sedan, Coupe & HPE '76 thru '80

LAND ROVER
314 Series II, IIA, & III all 4 cyl gasoline models '58 thru '86
529 Diesel '58 thru '80

MAZDA
648 626 Sedan & Coupe (rear wheel drive) '79 thru '82
*1082 626 & MX-6 (front wheel drive) '83 thru '90
*267 B1600, B1800 & B2000 Pick-ups '72 thru '90
370 GLC Hatchback (rear wheel drive) '77 thru '83
757 GLC (front wheel drive) '81 thru '86
109 RX2 '71 thru '75
096 RX3 '72 thru '76
460 RX-7 '79 thru '85
*1419 RX-7 '86 thru '89

MERCEDES-BENZ
*1643 190 Series all 4-cyl. gasoline '84 thru '88
346 230, 250 & 280 Sedan, Coupe & Roadster all 6 cyl sohc models '68 thru '72
983 280 123 Series all gasoline models '77 thru '81
698 350 & 450 Sedan, Coupe & Roadster '71 thru '80
697 Diesel 123 Series 200D, 220D, 240D, 240TD, 300D, 300CD, 300TD, 4- & 5-cyl incl. Turbo '76 thru '85

MERCURY
See FORD Listing

MG
475 MGA '56 thru '62
111 MGB Roadster & GT Coupe '62 thru '80
265 MG Midget & Austin Healey Sprite Roadster '58 thru '80

MITSUBISHI
Pick-up – *see Dodge D-50 (556)*

MORRIS
074 (Austin) Marina 1.8 '71 thru '80
024 Minor 1000 sedan & wagon '56 thru '71

NISSAN
*1137 300ZX all Turbo & non-Turbo '84 thru '86
*1341 Maxima '85 thru '89
*771 Pick-ups/Pathfinder gas models '80 thru '88
*876 Pulsar '83 thru '86
*982 Sentra '82 thru '90
*981 Stanza '82 thru '90

OLDSMOBILE
Custom Cruiser – *see BUICK Full-size (1551)*
658 Cutlass all standard gasoline V6 & V8 models '74 thru '88
Cutlass Ciera – *see GENERAL MOTORS A-Cars (829)*
Firenza – *see GENERAL MOTORS J-Cars (766)*
Ninety-eight – *see BUICK Full-size (1551)*
Omega – *see PONTIAC Phoenix & Omega (551)*

OPEL
157 (Buick) Manta Coupe 1900 '70 thru '74

PEUGEOT
161 504 all gasoline models '68 thru '79
663 504 all diesel models '74 thru '83

PLYMOUTH
425 Arrow '76 thru '80
For all other PLYMOUTH titles, see DODGE listing.

PONTIAC
T1000 – *see CHEVROLET Chevette (449)*
J-2000 – *see GENERAL MOTORS J-Cars (766)*
6000 – *see GENERAL MOTORS A-Cars (829)*
1232 Fiero '84 thru '88
555 Firebird all V8 models except Turbo '70 thru '81
*867 Firebird '82 thru '89
Full-size Rear Wheel Drive – *see Buick, Oldsmobile, Pontiac Full-size (1551)*
551 Phoenix & Oldsmobile Omega all X-car models '80 thru '84

PORSCHE
*264 911 all Coupe & Targa models except Turbo '65 thru '87
239 914 all 4 cyl models '69 thru '76
397 924 including Turbo '76 thru '82
*1027 944 including Turbo '83 thru '89

RENAULT
141 5 Le Car '76 thru '83
079 8 & 10 with 58.4 cu in engines '62 thru '72
097 12 Saloon & Estate 1289 cc engines '70 thru '80
768 15 & 17 '73 thru '79
081 16 89.7 cu in & 95.5 cu in engines '65 thru '72
598 18i & Sportwagon '81 thru '86
Alliance & Encore – *see AMC (934)*
984 Fuego '82 thru '85

ROVER
085 3500 & 3500S Sedan 215 cu in engines '68 thru '76
*365 3500 SDI V8 '76 thru '85

SAAB
198 95 & 96 V4 '66 thru '75
247 99 including Turbo '69 thru '80
*980 900 including Turbo '79 thru '88

SUBARU
237 1100, 1300, 1400 & 1600 '71 thru '79
*681 1600 & 1800 2WD & 4Wd '80 thru '88

SUZUKI
1626 Samurai/Sidekick and Geo Tracker '86 thru '89

TOYOTA
*1023 Camry '83 thru '90
150 Carina Sedan '71 thru '74
229 Celica ST, GT & liftback '71 thru '77
437 Celica '78 thru '81
*935 Celica except front-wheel drive and Supra '82 thru '85
680 Celica Supra '79 thru '81
1139 Celica Supra in-line 6-cylinder '82 thru '86
201 Corolla 1100, 1200 & 1600 '67 thru '74
361 Corolla '75 thru '79
961 Corolla (rear wheel drive) '80 thru '87
*1025 Corolla (front wheel drive) '84 thru '88
*636 Corolla Tercel '80 thru '82
230 Corona & MK II all 4 cyl sohc models '69 thru '74
360 Corona '74 thru '82
*532 Cressida '78 thru '82
313 Land Cruiser '68 thru '82
200 MK II all 6 cyl models '72 thru '76
*1339 MR2 '85 thru '87
304 Pick-up '69 thru '78
*656 Pick-up '79 thru '90
787 Starlet '81 thru '84

TRIUMPH
112 GT6 & Vitesse '62 thru '74
113 Spitfire '62 thru '81
028 TR2, 3, 3A, & 4A Roadsters '52 thru '67
031 TR250 & 6 Roadsters '67 thru '76
322 TR7 '75 thru '81

VW
091 411 & 412 all 103 cu in models '68 thru '73
036 Bug 1200 '54 thru '66
039 Bug 1300 & 1500 '65 thru '70
159 Bug 1600 all basic, sport & super (curved windshield) models '70 thru '74
110 Bug 1600 Super (flat windshield) '70 thru '72
238 Dasher all gasoline models '74 thru '81
*884 Rabbit, Jetta, Scirocco, & Pick-up all gasoline models '74 thru '89 & Convertible '80 thru '89
451 Rabbit, Jetta & Pick-up all diesel models '77 thru '84
082 Transporter 1600 '68 thru '79
226 Transporter 1700, 1800 & 2000 all models '72 thru '79
084 Type 3 1500 & 1600 '63 thru '73
1029 Vanagon all air-cooled models '80 thru '83

VOLVO
203 120, 130 Series & 1800 Sports '61 thru '73
129 140 Series '66 thru '74
244 164 '68 thru '75
*270 240 Series '74 thru '90
400 260 Series '75 thru '82
*1550 740 & 760 Series '82 thru '88

SPECIAL MANUALS
1479 Automotive Body Repair & Painting Manual
1054 Automotive Electrical Manual
1480 Automotive Heating & Air Conditioning Manual
482 Fuel Injection Manual
299 SU Carburetors thru '88
393 Weber Carburetors thru '79
300 Zenith/Stromberg CD Carburetors thru '76

See your dealer for other available titles

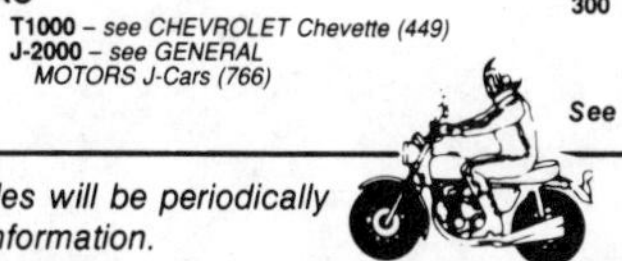

** Listings shown with an asterisk (*) indicate model coverage as of this printing. These titles will be periodically updated to include later model years — consult your Haynes dealer for more information.*

Over 100 Haynes motorcycle manuals also available

6-1-90

Haynes Publications Inc., P.O. Box 978, Newbury Park, CA 91320 • (818) 889-5400 • (805) 498-6703

Printed by
J H Haynes & Co Ltd
Sparkford Nr Yeovil
Somerset BA22 7JJ England